INTERNATIONAL LIST OF GEOGRAPHICAL SERIALS

Third Edition, 1980

INTERNATIONAL LIST OF GEOGRAPHICAL SERIALS

LISTE INTERNATIONALE DES PÉRIODIQUES GÉOGRAPHIQUES

INTERNATIONALE LISTE GEOGRAPHISCHER ZEITSCHRIFTEN UND SERIEN

LISTA INTERNACIONAL DE REVISTAS GEOGRÁFICAS

ELENCO INTERNAZIONALE DI PERIODICI GEOGRAFICI

МЕЖДУНАРОДНЫЙ СПИСОК ГЕОГРАФИЧЕСКИХ ПЕРИОДИЧЕСКИХ И ПРОДОЛЖАЮЩИХСЯ ИЗДАНИЙ

國際地理學期刊彙目

地理学雑誌国際総目録

القائمة الدولية للدوريات الجغرافية

INTERNATIONAL LIST OF GEOGRAPHICAL SERIALS THIRD EDITION, 1980

Compiled by
CHAUNCY D. HARRIS
University of Chicago
AND
JEROME D. FELLMANN
University of Illinois

Revised, Expanded, and Updated
A Comprehensive Retrospective Inventory of
3,445 Geographical Serials from 107 Countries
in 55 Languages with Locations in Union Lists

THE UNIVERSITY OF CHICAGO
DEPARTMENT OF GEOGRAPHY
RESEARCH PAPER NO. 193

1980

Published 1980 by the Department of Geography

The University of Chicago

Research Papers are available from

Department of Geography
The University of Chicago
5828 University Avenue
Chicago, Illinois 60637, U.S.A.

Price: $8.00 ($6.00 by series subscription)

Library of Congress Cataloging in Publication Data

Harris, Chauncy Dennison, 1914-
International list of geographical serials.

(Research paper - The University of Chicago, Department of Geography ; no. 193)
Bibliography: p. 13-25
Includes index.
1. Geography--Periodicals--Bibliography--Union lists. I. Fellmann, Jerome Donald, 1926- joint author. II. Title. III. Series: Chicago. University. Dept. of Geography. Research paper ; no. 193.
H31.C514 no. 193 [Z6003] [G115] 910s [016.91'05]
ISBN 0-89065-100-0 80-16392

The predecessor of this publication was HARRIS, CHAUNCY D., and FELLMANN, JEROME D., with the Assistance of Jack A. Licate. International List of Geographical Serials, 2nd ed. Chicago: University of Chicago, Department of Geography, Research Paper No. 138. 1971. xxvii plus 267 p. 2,415 entries from 90 countries.

Financial assistance from the Center for International Studies of the University of Chicago in the preparation and publication of this study is gratefully acknowledged.

The compilers wish to thank Miss VIOLET MILICEVIC who patiently and painstakingly typed and retyped the manuscript and suffered through numerous emendations, that required skill not only in typing but also in page composition for photographic reproduction. She deserves our particular gratitude.

TABLE OF CONTENTS

LISTED COUNTRIES USING NON-ROMAN SCRIPTS

COLLABORATORS IN THE PREPARATION OF THIS LIST

Collaborator	Area
AKHTAR, Dr. Rais, Center for the Study of Regional Development, Jawaharlal Nehru University, New Mehrauli Road, New Delhi 110067, India	INDIA
AMIRAN, Prof. David H. K., Department of Geography, The Hebrew University of Jerusalem, Jerusalem, Israel	ISRAEL
BEAVON, Prof. Keith S. O., Department of Geography and Environmental Studies, University of the Witwatersrand, 1 Jan Smuts Avenue, Johannesburg 2001, South Africa	SOUTH AFRICA
BECKER, Prof. Bertha K., Rua Santa Clara, 192/1001 (Copacabana) 20,000 Rio de Janeiro, RJ, Brazil	BRAZIL
BERNARD, Deryck. Head, Department of Geography, University of Guyana, Box 841, Georgetown, Guyana	GUYANA
BJELOVITIĆ, Prof. Milos, Geografsko Društvo Bosne i Hercegovine, Prirodno Matematički Fakultet, Putnika 43, Sarajevo V, Yugoslavia	YUGOSLAVIA
BRIEND, Mlle. Anne-Marie, Chef de Services, Laboratoire d'Information et de Documentation en Géographie "Intergéo," 191, rue Saint-Jacques, 75005 Paris, France (Algeria, Benin, Ivory Coast, Laos, Lebanon, Madagascar, Morocco, Senegal, Tunisia, Zaïre).	FRANCE Francophone Countries
CÁC, Nguyên Tho', Dôc lâp-Tu' do-Hanh phuc, Cuc do dac va Ban Do Nha Nuoc, Vietnam	VIETNAM
CENTRO Nacional de Documentação e Informação de Moçambique, Caixa Postal 4116, Maputo, Mozambique	MOZAMBIQUE
CHENG, James, K. M., Assistant Curator, Far Eastern Library, University of Chicago, 1100 East 57th Street, Chicago, Illinois 60637, USA	CHINA
CHENG, Prof. Kenneth T. C., 17-A, Lane 31, Yung Kang Street, Taipei, Taiwan 106, Republic of China	CHINA (Taiwan)
CONACHER, Arthur, Head, Department of Geography, The University of Western Australia, Nedlands, Western Australia 6009, Australia	AUSTRALIA (Western)
COOPER, Malcolm, Lecturer in Geography, Department of Geography, The University of New England, Armidale, N.S.W. 2351, Australia	AUSTRALIA (Armidale)
DAUS, Dr. Federico A., Presidente, Sociedad Argentina de Estudios Geográficos, Av. Santa Fe 1145, 1059 Buenos Aires, Argentina	ARGENTINA

DAVIES, Prof. D. H., National Chairman, Geographical Association of Rhodesia, Department of Geography, University of Zimbabwe-Rhodesia, P.O. Box M. P. 167, Mount Pleasant, Salisbury, Zimbabwe-Rhodesia	ZIMBABWE-RHODESIA
DIFRIERI, Prof. Horacio A., Director, Departamento de Geograía, Universidad de Buenos Aires, 25 de Mayo 217, 3' piso, 1003 Buenos Aires, Argentina	ARGENTINA
DRAZNIOWSKY, Dr. Roman, Curator, American Geographical Society Collection of the University of Wisconsin-Milwaukee Library, P. O. Box 399, Milwaukee, Wisconsin 53201, USA	UNITED STATES (AGS Publications)
DUGDALE, G. S., Librarian, Royal Geographical Society Kensington Gore, London SW7 2AR, England	UNITED KINGDOM
DUMONT, Prof. Maurice E., Sint-Denijslaan 76, B-9000 Gent, Belgium	BELGIUM
FLORES CASTELLI, Col. Octavio, Jefe Secretaría Geográfica Instituto Geográfico Militar, Nueva Santa Isabel No. 1640, Santiago, Chile	CHILE (IGM)
FOGELBERG, Dr. Paul, Department of Geography, University of Helsinki, Hallituskatu 11-13, SF-00100 Helsinki 10, Finland	FINLAND
FOUCHARD, E., Laboratoire d'Information et de Documentation en Géographie "Intergéo," 191, rue Saint-Jacques, 75005 Paris, France	FRANCE Francophone Countries
FRASER, Dr. J. Keith, Executive Secretary, Canadian Committee for Geography, Environment Canada, Ottawa K1A 0H3, Canada	CANADA
GRITZNER, Dr. Charles F., Executive Director, National Council for Geographic Education, University of Houston, Houston, Texas 77004, USA	UNITED STATES (NCGE Publications)
GUTIÉRREZ DE MACGREGOR, Dr. Teresa, Instituto de Geografía, Universidad Nacional Autónoma de México, México 20, D. F., Mexico	MEXICO
GANJI, Prof. Mohammed H., 10 Koocheh Shahrdar, Ave. Shah-Reza, Tehran 16, Iran	IRAN
GRANT, Prof. Charles J., Department of Geography and Geology, University of Hong Kong, Hong Kong	HONG KONG
HADISUMARNO, Surastopo, Faculty of Geography Gadjah Mada University, Bulaksumur, Yogyakarta, Indonesia	INDONESIA
HARMS, Miss I. J. C., Librarian, Geografisch Instituut, Rijksuniversiteit Utrecht, Heidelberglaan 2, Postbus 80.115, 3584 CS-Utrecht, The Netherlands	NETHERLANDS (Utrecht)
ter HART, Dr. H. W., Economisch Geografisch Institut, Universiteit van Amsterdam, Burg Teilegenhuis, kamer 2329, Jodenbreestraat 23, Amsterdam, The Netherlands	NETHERLANDS
HAUGHTON, Prof. Joseph P., Department of Geography Trinity College, University of Dublin, Dublin 2, Ireland	IRELAND

HOCK, Prof. Khoo Soo, Department of Geography, University of Malaya, Kuala Lumpur 22-11, Malaysia	MALAYSIA
JACOBSEN, Prof. N. Kingo, Department of Geography, University of Copenhagen, Haraldsgade 68, DK-2100 Copenhagen Ø, Denmark	DENMARK
JOHNSON, David S., Assistant Editor, Zambia Geographical Association, P. O. Box RW 287, Lusaka, Zambia	ZAMBIA
al-KHALAF, Dr. Jassim M., 39 Higaz Street, Mohandissen, Doki, Cairo, Egypt	ARABIC COUNTRIES
KÖKSAL, Prof. Dr. Aydoğan, Coğrafya Araştirmalari Enstitüsü, Dil ve Tarih-Coğrafya Fakültesi, Ankara, Turkey	TURKEY (Ankara)
KUO, Young, Chief Librarian, Institute of Geography, Academia Sinica, Peking, People's Republic of China	CHINA (People's Republic)
LAOUINA, A., Secrétaire général, Comite National de Géographie du Maroc, Institut Scientifique, Avenue Moulay-Cherif, B. P. 1014, Rabat, Morocco	MOROCCO
LAŠKA, Vaclav, Bibliographer for Slavic Studies, University of Chicago Library, 1100 East 57th Street, Chicago, Illinois 60637, USA	SLAVIC COUNTRIES
LEE, Prof. Chan, Department of Geography, College of Social Sciences, Seoul National University, Seoul 151, Korea	KOREA
LEIMGRUBER, Dr. Walter, Geographisches Institut, Universität Basel, Bernoullianum, Klingelbergstrasse 16, CH-4056 Basel, Switzerland	SWITZERLAND
LICATE, Dr. Jack A., Area Information Research Center, 1123 Wayburn Avenue, Detroit, Michigan 48230, USA	LATIN AMERICA
LOGHKOMOEVA, Nina M., Director, Library, Geographical Society of the USSR, Per Grivtsova, 10, 190000 Leningrad-tsentr, USSR	USSR (Geographical Society)
LÜDEMANN, Dr. Heinz, Direktor, Institut für Geographie und Geoökologie, Akademie der Wissenschaften der DDR, Georgi-Dimitroff-Platz 1, 701 Leipzig, German Democratic Republic (DDR)	GERMANY (Democratic Republic)
MABOGUNJE, Prof. Akin L., Department of Geography, University of Ibadan, Ibadan, Nigeria	NIGERIA
MAHALI, Dr. S., Director, The Syrian Geographical Society, Damascus, Syria	SYRIA
MARSHALL, Brian, Geography Librarian, The University of Auckland, Private Bag, Auckland, New Zealand	NEW ZEALAND
MARTINEZ RODRIGUEZ, Prof. Dr. Ignacio, Secretario, Instituto Nacional de Investigaciones Geográficas de la Universidad de la Republica, Avda. Cataluña 3180 Montevideo, Uruguay	URUGUAY
MAZÚR, Prof. Emil, Geografický Ústav Slovenskej Akadémie Vied, Ul. Obrancov mieru 49, 886 25 Bratislava, Czechoslovakia	CZECHOSLOVAKIA

MEYNEN, Prof. Dr. Emil, Langenbergweg 82, D-5300 Bonn-Bad Godesberg 1, Federal Republic of Germany	GERMANY (Federal Republic)
MICHIE, W. C., Vice-Chairman, Geographical Association, The Ministry of Education, P. O. Box 8024, Causeway, Salisbury, Zimbabwe-Rhodesia	ZIMBABWE-RHODESIA
MIGLIORINI, Prof. Elio, Via Vitelleschi, 26, 00193 Roma, Italy	ITALY
MIKLAVC, Janja, Librarian, PZE za Geografijo, Filozofska Fakulteta, Aškerčeva 12, 61000 Ljubljana, Yugoslavia	YUGOSLAVIA
MULCHANSINGH, Dr. Vernon C., Chairman, Department of Geography, University of the West Indies, Mona, Kingston 7, Jamaica	JAMAICA
ONGWENY, Dr. George S., Secretary, Geographical Society of Kenya, Department of Geography, University of Nairobi, P. O. Box 30197, Nairboi, Kenya	KENYA
OOI, Prof. Jin Bee, Department of Geography, University of Singapore, Bukit Timah Road, Singapore 10	SINGAPORE
PATTERSON, Maureen L. P., Bibliographic Specialist for South Asian Materials, University of Chicago Library, 1100 East 57th Street, Chicago, Illinois 60637, USA	SOUTH ASIA
PEDRAZA Ortega, Omar, Jefe Sección Información Geográfica, Instituto Geografico "Agustin Codazzi," Carrera 30 no. 48-51, Apartado Aéro 6721, Bogotá, D. E. Colombia	COLOMBIA
ROBINSON, Prof. Kenneth W., Department of Geography, University of Newcastle, Newcastle, N.S.W. 2308, Australia	AUSTRALIA (Newcastle)
ROSEIRA, Maria João Queiroz, Faculdade de Ciências Sociais e Humanas, Universidada Nova de Lisboa, Avenida Berna, Lisboa-1, Portugal	PORTUGAL
ROSELL, Dominador Z., Editor-in-Chief, Philippine Geographical Journal, P. O. Box 2116, Manila, Philippines	PHILIPPINES
RUDBERG, Prof. Sten, Naturgeografiska Institutionen, Göteborgs Universitet, Dicksongatan 4, 412 56 Göteborg, Sweden	SWEDEN
SALINAS, Mrs. Lea, Bibliographical Services, Instituto Panamericano de Geografia e Historia, Ex-Arzobispado 29, México 18, D. F., Mexico	LATIN AMERICA
SEADLE, Michael, Assistant to Bibliographic Specialist for South Asian Materials, University of Chicago Library, 1100 East 57th Street, Chicago, Illinois 60637, USA	SOUTH ASIA

SIMONFAI, Mrs. Judit, Chief Librarian, Földrajztudományi Kutató Intézet, Magyar Tudományos Akadémia, Népköztársaság útja 62, Budapest VI, Hungary — HUNGARY

SINGH, Dr. Rana P. B., No. 19 Aiyar Hostel, Banaras Hindu University, Varanasi 221005, India — INDIA

TOLEDO PALOMO, Ricardo, Primer Secretario, Sociedad de Geografía e Historia de Guatemala, 3a Avenida 8-35, zona 1, Ciudad de Guatemala, Guatemala — GUATEMALA

TURCHÁNYI, Mrs. Eszter, Librarian, Földrajztudományi Kutató Intézet, Magyar Tudományos Akadémia, Népköztársaság utja 62, Budapest VI, Hungary — HUNGARY

TUSZYŃSKA-RĘKAWEK, Mgr. Halina, Director of the Library, Instytut Geografii i Przestrzennego Zagospodarowania PAN Krakowskie Przedmieście 30, 00-927 Warszawa, Poland — POLAND

TYSON, Prof. Peter D., Department of Geography and Environmental Studies, University of the Witwatersrand, 1 Jan Smuts Avenue, Johannesburg 2001, South Africa — SOUTH AFRICA

VÁZQUEZ MAURE, Prof. Francisco, Real Sociedad Geográfica, Valverde, 22, Madrid-13, Spain — SPAIN

VIELHABER, Mag. Christian, Institut für Geographie der Universität Wien, Universitätsstrasse 7, A-1010 Wien, Austria — AUSTRIA

VIVÓ ESCOTO, Dr. Jorge A., Centro de Investigaciones Geográficas, Facultad de Filosofia y Letras, México, D. F., Mexico — MEXICO

VRIŠER, Prof. Dr. Igor, PZE za Geografijo, Filozofska Fakulteta, Aškerčeva 12, 61000 Ljubljana, Yugoslavia — YUGOSLAVIA

YONEKURA, Nobuyuki, Research Associate, Department of Geography, Faculty of Science, University of Tokyo, Hongo, Tokyo, 113, Japan — JAPAN

ZAR, Kathleen, Bibliographer, Geography, University of Chicago Library, 1100 East 57th Street, Chicago, Illinois 60637, USA — GENERAL

ACKNOWLEDGEMENTS

First of all the compilers wish to express profound gratitude to the above-named collaborators who have greatly increased the accuracy and coverage of this list by their generous and knowledgeable help.

The compilers also wish to thank others who aided in providing information, expertise, services, access, or facilities.

In the University of Chicago Library the following were helpful: Kathleen Zar, Bibliographer for Geography and Head of the Map Collection, James Silk, Assistant, and Scott Sprinzen, Reference Librarian, for geography and related fields; Patricia Swanson, Head of Reference Services, and Janet L. Fox, Inter-Library Loan; Maureen L. P. Patterson, Bibliographer, and Michael Seadle, Assistant, South Asia Reference Center; Luc Kwanten, Curator, and James K. M. Cheng, Assistant Curator, Far Eastern Collection; Bruce D. Craig, Bibliographer, and Paul Sprachman, Assistant Librarian, Middle Eastern Collection; Vaclav Laška, Bibliographer, and Halyna Pankiw, Assistant, Slavic Collection, and Pearl Kahan, Head, Order Section, Serials Department.

At the American Geographical Society Collection of the University of Wisconsin-Milwaukee Library Roman Drazniowsky, Curator, and Howard A. Deller, Susan Ewart Peschel, and Kathleen Pett provided access and assistance.

At the Library of Congress, John Wolter, Chief of the Geography and Map Division, kindly made arrangements for access to serial records, to the stack collection of current periodicals, and to the official catalogue for records of monographic series.

John D. Stephens of Virginia Polytechnic Institute and State University and Bruce Young of Wilfrid Laurier University provided information on a number of serials.

Over the years five individuals have been repeatedly helpful with bibliographical information: G. S. Dugdale, Librarian of the Royal Geographical Society in London; Anne-Marie Briend, Chef de Services, Laboratoire d'Information et de Documentation en Géographie, Paris; Emil Meynen, retired Director of the Institut für Landeskunde of the Bundesanstalt für Landeskunde und Raumforschung, Bonn; Elio Migliorini, the distinguished Italian geographer and bibliographer, Rome; and Nina M. Loshkomoeva, Librarian of the Geographical Society of the USSR in Leningrad. Nobuyuki Yonekura, Research Associate in the Department of Geography, Faculty of Science, University of Tokyo, kindly checked Japanese titles, and James K. M. Cheng, Assistant Curator of the Far Eastern Library of the University of Chicago confirmed titles in Chinese, Japanese, or Korean.

Special thanks are due to individuals who provided copy in non-Roman scripts for titles in various languages: James K. M. Cheng, C. H. Fieg and Yoko Kuki for Chinese, Japanese, and Korean; Igor Kleopov, Russian; Halina Pankiw, Ukrainian; Kostas Kazazis, Greek; Paul Sprachman and Faiez Mossad, Arabic and Persian; Channa Cohen, Hebrew; Michael Seadle, South Asian languages, Mani Reynolds, Thai; and Donald N. Levine, Amharic.

INTRODUCTION

It has been nine years since the appearance of the second edition of the International List of Geographical Serials in 1971. In the intervening years, a large number of new geographical serials have made their appearance. The birth rate of new geographical serials has been accelerating, from an average of about 45 a year in the 1950s, to 70 a year in the 1960s, to more than 100 a year in the 1970s, as recorded in successive editions of this list. Just over 1,000 titles are new to this third edition. Thus, of the 3,445 titles, almost one-third are new. Nearly all entries from the previous edition have been revised. A few entries have been dropped.

The inventory of geographical serials for each country is of particular value to geographers interested in that specific country or area. Since most geographical serials report most fully on the regions near at hand, serials published in any country are likely to be an especially valuable source of regional geographic information on that country. Since the geographer is interested in specific areas, he is much more likely to be concerned with foreign serials than are scholars in other disciplines.

The fundamental purpose of this list is to provide a comprehensive inventory of all known geographical serials, both those currently being published and those no longer active (closed).

This list aims to include only serials which are primarily geographic in content. In many cases it has been difficult to decide which serials should be considered "geographical." In general we have been lenient in including older serials, often geographical in name but general in content, from the nineteenth century, before the rise of a specialized academic discipline of geography. With respect to current serials we have applied more rigorous standards. As a result some entries included in earlier editions have been excluded from the present edition. Many of the decisions to delete were made as a result of an examination of the serials in libraries with major geographical collections or on the basis of advice from our collaborators in each country.

COMPILATION OF THIS INTERNATIONAL LIST OF GEOGRAPHICAL SERIALS

Summary

The compilation of this International List of Geographical Serials involved essentially three tasks: (1) identifying titles suitable for inclusion; (2) determining data for a proper bibliographical description of each title; and (3) locating sources of information on library holdings of each serial. The sources utilized often contributed to more than one of the above tasks.

The first task involved securing information of serials published in all countries and in scores of languages from diverse sources and selecting titles to be included by some definition of the boundaries of the term geographical.

The initial basic stock of titles was built up from an examination of retrospective inventories of geographical serials, including previous editions of this work. But any such list begins immediately to become out of date.

It thus has been important to pay particular attention to means of discovering each new periodical as it appears. This has been accomplished by examination of recent lists of geographical serials currently received or held by leading libraries, by lists of current geographical serials, either general or specialized by region or systematic fields, and by lists of serials cited in current geographical bibliographies. Particularly valuable have been lists specifically of _new_ geographical serials.

Bibliographical details were secured from each of the above types of sources; from union lists of periodical holdings of libraries; from directories of current serials; from published records of library holdings of serials; from personal examination of the serials themselves; from information supplied by an international group of collaborators.

In aggregate thousands of issues of the listed serials were examined in libraries to confirm titles, publishers, place of publication, beginning dates, frequency, title changes, the number and date of last issue for closed serials, languages of text or supplementary tables of contents or abstracts, cumulative indexes, and addresses for current serials. Such physical examination of the serials has taken place mainly in five libraries: The University of Chicago Library in Chicago, the American Geographical Society Collection of the University of Wisconsin-Milwaukee Library in Milwaukee (formerly the Library of the American Geographical Society in New York), the Library of Congress in Washington, D. C., the Library of the Royal Geographical Society in London, and La Bibliothèque du Centre Interuniversitaire de l'Institut de Géographie in Paris. We are deeply grateful to the librarians in each of these libraries for their courtesy, help, and for permission to examine manuscript records of serial holdings and shelf lists and to have access directly to stacks.

Collaborators from all over the world helped in the compilation of this _International List of Geographical Serials_. Country lists of serials listed in the second edition together with lists of possible additions were submitted to geographers, bibliographers, and librarians for comments, suggestions, corrections, additions, and deletions. These knowledgeable collaborators eliminated false or misinterpreted titles and added valuable new serials, which had not yet found their way into the bibliographical sources utilized. They also corrected many errors. After the final entries for this edition were typed xerographic copies were again sent to collaborators for verification and further corrections. They are listed at the beginning of this volume.

For three countries collaborators for this list made independent original questionnaire surveys of departments of geography, geographical societies, and other geographical institutions or publishers for up-to-date information of current geographical serials: J. Keith Fraser in Canada, specially for this edition, A. M. Briend and M. Pineau in France, and Emil Meynen for Germany, Austria, and Switzerland.

Certain major union lists of serials were searched for each title in this list for location of information on library holdings [See the section below "Union Lists of Serials against which Entries have been Checked," pp. 28-31].

Discussion of Individual Sources

The retrospective lists of geographical serials that provided the basic stock of titles for this edition of International List of Geographical Serials were, of course, previous editions of this list, issued in 1949, 1950, 1960, and 1971 [1-4] (Numbers in brackets refer to entries in the List of Principal Sources at the end of this section). These in turn leaned heavily on a wide variety of previous retrospective lists of geographical serials by Elizabeth T. Platt [5], and by John K. Wright and Elizabeth T. Platt [6], former librarians of the American Geographical Society, Hugh Robert Mill [7] of the Royal Geographical Society, the International Catalogue of Scientific Literature: List of Journals [8], Inventaire des Périodiques scientifiques des bibliothèques de Paris, by Alfred Lacroix and Léon Bultingaire for the Académie des Sciences, Paris [9], and lists by Paul Dinse [10], Georg Kollm [11], Helene Müller [12], Rolf Diedrich Schmidt and Charlotte Streumann [13], and Zofia Kaczorowska [14].

Serials received by geographical libraries or geographical serials received by more general libraries provided many additional entries. Among the lists utilized were those of the deposit library of the Association of American Geographers at the University of Cincinnati Library [15], the Library of the Royal Geographical Society [16], of the Gesellschaft für Erdkunde zu Berlin, compiled by Hans Leonhardy and Wolfgang Scharfe [17], the University Library in Frankfurt-am-Main [18], the Geographical Institute in Copenhagen [19], Sociedad Geográfica de Colombia [20], the Institute of Geology and Geography of the Academy of the Socialist Republic of Romania [21], and the libraries of the Academy of Sciences of the USSR [22].

Recent lists that have been particularly valuable have been the extensive Katalog der Zeitschriften und Serien für das Fachgebiet Geographie...of the Niedersächische Staats- und Universitätsbibliothek, Göttingen [23] and Serial Publications in Geography held by the Virginia Polytechnic Institute and State University, by John D. Stephens [24]. The recent lists of publications received on exchange by Belgian Society for Geographic Study [25] and the Real Sociedad Geográfica in Spain [26] were also useful.

Lists of current geographical serials, prepared at various dates, were also reviewed. Some are universal in covering all types of geographical periodicals in seven libraries of France prepared by A. M. Briend [27] or lists in Geographisches Taschenbuch 1953 [28] and 1958/59-1962/63 (29]. Others are devoted to a particular country, as the United States by John Fraser Hart [30], the United Kingdom by Margaret A. Bass [31] and by John Connell and K. Hoggart [32], France by A. M. Briend and M. Pineau [33], Daniel Faucher [34], and Myriem Foncin [35], Germany, Austria, and Switzerland by Emil Meynen [36-37], Poland [38], the Soviet Union by D. J. Welsh [39], or Japan by Robert B. Hall and Toshio Noh [40], or the Tokyo Library Bureau [41]. The related field of cartography is covered by lists by K. A. Salishchev [42-43], Donald A. Wise [44], and Karl-Heinz Meine [45]. Elio Migliorini has published a list of journals for teachers of geography in Italy [46]. Bruce Young has compiled an extensive bibliography of a particular category of geographical serials--discussion papers, occasional papers, and monographs of departments of geography--with listing of individual titles [47]; this is particularly valuable since some of these fall into the categories of fugitive materials not easily identified or obtained.

Lists of serials analyzed by major comprehensive geographical bibliographies were reviewed and independent investigations made of actual citation patterns: Bibliographie Géographique Internationale [48-49], Geo Abstracts [50], Current Geographical Publications [51-54], Documentatio Geographica [55], Dokumentation zur Raumentwicklung [56], Literaturinformation: Territorialforschung, Territorialplanung [57], Referativnyi Zhurnal: Geografiia[58], and Bibliographia Cartographica [59]. The most-cited geographical serials in such bibliographies have been separately analyzed in a sister publication to this list [60].

Several publications regularly note the appearance of new geographical serials. New Serial Titles: Classed Subject Arrangement [61] is especially valuable in being relatively comprehensive. A 21-year subject guide to New Serial Titles for 1950-1970 [62] was particularly useful in covering a longer period and in making a more discriminating classification of geographical periodicals, than in the uncumulated monthly issues in the classed subject arrangement.

Specialized lists of new geographical periodicals in Current Geographical Publications [63] in New Geographical Literature and Maps [64] and later in the Geographical Journal [65] were of high value in providing information from the two most comprehensive libraries of geographical societies. The three best continuing lists which provide information on new geographical serials as they commence publication are currently contained in New Serials Titles: Classed Subject Arrangement, Current Geographical Publications, and the Geographical Journal. A series of annual reviews of new geographical periodicals by Wilma B. Fairchild, Mary Alice Lamberty, Lynn S. Mullins and others appeared in the Geographical Review, 1952-1974 [66].

Union lists not only record the holdings of libraries of specified serials, they also provide bibliographical information supplied by many co-operating libraries. Entries in this International List of Geographical Serials have been checked against certain major union lists: Union List of Serials in Libraries of the United States and Canada [67], if the serial began publication by December 31, 1949; New Serial Titles: A Union List of Serials Commencing Publication after December 31, 1949 [68-70], if the serial began publication later; British Union-Catalogue of Periodicals [71-75]; Catalogue Collectif des Périodiques du Début du XVIIe Siècle à 1939 [76], if the serial had begun publication by 1939; Gesamtverzeichnis Deutschsprachiger Zeitschriften und Serien [77], if the serial is in the German language; Gesamtverzeichnis der Zeitschriften und Serien... Neue und geänderte Titel seit 1971...[78], if the serial began publication or changed title after December 31, 1970; and Retrospektivnyi Svodnyi Ukazatel' Inostrannykh Periodicheskikh i Prodolzhaiushchikhsia Izdanii imeiushchikhsia v krupneishikh Bibliotekakh SSSR 1750-1965 [79], selectively for certain socialist countries, members of the Council for Mutual Economic Assistance, or areas using the Cyrillic alphabet, for which this Soviet Union list of serials provides useful additional information. Two lists of international congresses have been consulted occasionally: International Congresses and Conferences [80] and Gesamtverzeichnis der Kongress-Schriften [81]. For some government publications the List of Serial Publications of Foreign Governments 1815-1931, edited by Winifred Gregory [82] has been helpful.

Among general guides to current serials Ulrich's International Periodicals Directory [83], Irregular Serials and Annuals: An International Directory [84]. and Ulrich's Quarterly [85] have been the most useful both for the listing of current geographical periodicals and serials and for current information on addresses, frequency of publication, and changes. Other more specialized and sometimes somewhat dated lists utilized were Verzeichnis Ausgewählter Wissenschaftlicher Zeitschriften des Auslandes by the Deutsche Forschungsgemeinschaft [86], Guide to Current Latin American Periodicals [87] by Irene Zimmerman, a similar list by the Pan-American Union [88], Brazilian Serial Documents, by Mary Lombardi [89], and Guide to Current British Journals by David P. Woodworth [90].

Published library records of holdings or acquisitions of regional serials were helpful for checking the larger areas of Asia: Library of Congress Accessions List India. Quinquennial Serials Cumulation [91], Library of Congress Accessions Lists for Southeast Asia [92-94], Cornell University Libraries Southeast Asia Catalog. Serials [95], Chinese Periodicals in the Library of Congress [96], and the Far Eastern Serials in the University of Chicago Library [97].

Two general catalogs of serial holdings of individual libraries were helpful in checking: the 1979 Union Catalog of Serials Currently Received in the Libraries of the University of Wisconsin, Madison [98], and Serial Holdings in the Linda Hall Library, 1979 Annual [99].

In view of the very large number of Soviet geographical serials and the difficulties of access to them on the part of many, special efforts have been made to make this section on the Soviet Union as complete as possible.

Three western union lists of Soviet serials were checked: Half a Century of Soviet Serials 1917-1968 by Rudolf Smits [100], Gesamtverzeichnis russischer und sowjetischer Periodika und Serienwerke by Peter Bruhn and Werner Philipp [101], and Catalog Collectif des Périodiques...Périodiques Slaves en Caractères Cyrilliques [102], for holdings in libraries of the United States, Federal Republic of Germany, and France.

Successive inventories of serials published in Russia or the Soviet Union were checked for forms of title and for issues actually published and their dates. These inventories covered the dates 1703-1900 [103], 1901-1916 [104], 1917-1949 [105], and 1950-1973 [106-114]. These were helpful especially for irregular serials. Lists of journals for 1966-1975 [115-116] provided a check but little new information. A special list of Ukrainian serials for 1918-1950 [117] was of value in getting forms correct for the period covered.

The Geographical Society of the USSR and its numerous branches have published more geographical serials than any other body. Fortunately special lists are available of these publications for the first century, 1845-1945 by L. S. Berg [118], for the period 1917-1966 by I. B. Kostrits and D. M. Pinkhenson [119], and successive lists of publications of the Geographical Society and its many branches received by the Library of the Society by its librarian, N. M. Loshkomoeva [120-121].

One special feature of this edition has been the systematic checking of Soviet bibliographical sources for fuller information, particularly for the irregular serials or for geography subseries of general university series. For some of these, the numbers and dates of confirmed publications have been indicated; the provision of such information for irregular Soviet serials does not indicate that the serial is closed unless explicitly so stated.

In spite of all this checking various entries are still incomplete. Two types of publications have been especially difficult. Whether or not irregular serials are still being published sometimes is not ascertainable since the publishing institution itself does not always know whether another number will be issued. Thus the major bibliographical sources typically leave open many such serials, of which no numbers have appeared for a long time and for some of which no further numbers are likely to appear. Semi-publications such as discussion papers are also difficult since they are not present in many libraries; often it is not clear whether a discussion paper or a working paper is considered only a draft, which in revised form is to be published in a regular serial, or whether it is simply a less expensive form of publication in a small edition.

LIST OF PRINCIPAL SOURCES

Earlier Retrospective Inventories of Geographical Serials

1 Harris, Chauncy D., and Fellmann, Jerome D., A Comprehensive Checklist of Serials of Geographic Value: A Chicago-centered union list indicating the complete holdings of the University of Chicago Library (with call numbers) and the holdings of other libraries insofar as necessary to obtain a complete set of each listed serial, Part I. Geographical Serials Proper. Chicago: Department of Geography, University of Chicago, March, 1949. xvii plus 100 p. Mimeographed. 1,024 titles.

2 Harris, Chauncy D., and Fellmann, Jerome D., with the Active Collaboration of Burton W. Adkinson, Director of the Reference Department, Library of Congress, Washington, D. C., Nordis Felland, Librarian, American Geographical Society of New York, N. Y., G. R. Crone, Librarian, Royal Geographical Society, London, England, and M. Foncin, Conservateur, Département des Cartes et Plans, Bibliothèque Nationale, Paris, France. A Union List of Geographical Serials, 2nd ed., Chicago, Illinois: University of Chicago, Department of Geography, Research Paper No. 10, June 1950. xix plus 124 p.
1,197 entries. Index. Records the complete holdings in geographical serials of four libraries, the Library of Congress, the Library of the American Geographical Society of New York, the University of Chicago Library, and the Library of the Royal Geographical Society, and the holdings of French geographical serials by three libraries in Paris, Bibliothèque Nationale, Bibliothèque de la Société de Géographie, and Bibliothèque de l'Institut de Géographie de l'Université de Paris. Holdings of other libraries are given only insofar as necessary to fill gaps in the holdings of the above-named libraries.

3 Harris, Chauncy D., and Fellmann, Jerome D. International List of Geographical Serials. Chicago, Illinois: University of Chicago, Department of Geography, Research Paper No. 63, June, 1960. lix plus 189 p. 1,651 entries. Index.

4 Harris, Chauncy D., and Fellmann, Jerome D. With the Assistance of Jack A. Licate. International List of Geographical Serials. 2nd ed. Chicago, Illinois: University of Chicago, Department of Geography, Research Paper No. 138. 1971. xxvi plus 267 p. 2,415 entries from 90 countries.

5 Platt, Elizabeth T. Typed list of geographical serials giving about 1,100 entries in alphabetical order. Based on a number of sources: 1) the periodical collection of the American Geographical Society's library; 2) the Union List of Serials...and a number of earlier check lists of periodicals; 3) various published catalogues of European libraries; 4) lists such as those contained in the Geographen Kalender for 1912; and 5) regional bibliographies." After Miss Platt's death in 1943, this manuscript list was updated through additions by Miss Nordis Felland in the 1940s. Many titles in related topical fields or general regional periodicals are excluded from the present list.

6 Wright, John K., and Platt, Elizabeth T. Aids to Geographical Research, second edition, "American Geographical Society Research Series No. 22," New York: Columbia University Press, 1947. 331 p. Reprinted: Westport, Connecticut: Greenwood Press, 1971. About 350 periodicals listed in this volume have been incorporated into the present list.

7 Mill, Hugh Robert. Catalogue of the Library of the Royal Geographical Society, London: John Murray, 1895, appendix III "List of Periodicals including Proceedings and Transactions of Academies and Scientific Societies," compiled by James Murie, p. 773-833. This source lists about 1,200 items, of which about 200 have been included in the present list.

8 International Catalogue of Scientific Literature. J. Geography. v1-14 (1903-1919). London: Harrison and Sons, 1903-1919. "Lists of journals with abbreviated titles," at end of each volume. Covers periodicals with citations 1901-September 1916.

9 Académie des Science, Paris. Inventaire des Périodiques scientifiques des Bibliothèques de Paris, dressé sous la direction de Alfred Lacroix, par M. Léon Bultingaire. Paris: Masson et Cie., 1924. Under "Géographie" in the index, pp. 1087-1088, about 450 periodicals are listed, of which 296 are included in this list.

10 Dinse, Paul. Katalog der Bibliothek der Gesellschaft für Erdkunde zu Berlin. Berlin: Mittler, 1903. "Zeitschriften und Periodische Veröffentlichen" pp. 775-795.

11 Kollm, Georg. "Geographische Zeitschriften," Geographisches Jahrbuch, v.32 (1909), p. 419-429, 175 periodicals.

12 Müller, Helene, "Verzeichnis der periodischen Schriften der Bücherei der Gesellschaft für Erdkunde zu Leipzig," Wissenschaftliche Veröffentlichungen der Gesellschaft für Erdkunde zu Leipzig, v. 11 (1938), 163 p. 754 periodicals many of which are nongeographic in nature.

13 Schmidt, Rolf Diedrich, and Streumann, Charlotte. Verzeichnis der Geographischen Zeitschriften, Periodischen Veröffentlichungen und Schriftenreihen Deutschlands und der in den Letzteren Erschienenen Arbeiten. Bearbeitet im Institut für Landeskunde. Direktor: Prof. Dr. E. Meynen. Bad Godesberg: Selbstverlag Bundesanstalt für Landeskunde und Raumforschung, 1964. 303 p. Berichte zur Deutschen Landeskunde. Sonderheft 7. 393 entries. Detailed information on editors, publishers, dates, and contents with indexes.

14 Kaczorowska, Zofia. Spis Zagranicznych Czasopism i Wydawnictw Seryjnych z Zakresu Nauk o Ziemi, Znajdujących się w Bibliotekach Polskich. [The List of Foreign Periodicals and Serial Publications concerning Sciences of the Earth, Being in Possession of Polish Libraries] (Polska Akademia Nauk. Instytut Geografii). Warszawa: Panstwowe Wydawnictwo Naukowe, 1957. 376 p.

15 "Current Holdings of the AAG Depository Library at the University of Cincinnati and the University of Chicago." Professional Geographer, v18, no6 (November, 1966), p. 369-376.

16 Royal Geographical Society. Current Geographical Periodicals: a Handlist and Subject Index of Current Periodicals in the Library of the R.G.S. London, 1961. 19 p. (R.G.S. Library series no. 5). 689 items, of which 313 are included in this list.

17 Leonhardy, Hans, and Scharfe, Wolfgang, "Periodica: Verzeichnis der Zeitschriften und Periodischen Veröffentlichungen in der Bibliothek der Gesellschaft für Erdkunde zu Berlin (Stand 1.10.1965), "Die Erde: Zeitschrift der Gesellschaft für Erdkunde zu Berlin, v97, no1 (1966), p. i-xxxvi. 859 entries, of which 265 are included in this list.

18 "Zeitschriftenverzeichnis der Städtischen und Universitäts-Bibliotheken, Frankfurt am Main. Laufende gehaltene Zeitschriften der Stadt- und Universitäts-Bibliothek, Senckenbergischen Bibliothek und Medizinischen Hauptbibliothek. Stand vom 1.1.1956." 20. Geographie, p. 166-186. Mimeographed.

19 "Tidsskrifter og Årbøger. Fortegnelse over Løbende Serier, som Indigår til Geografisk Instituts Bibliotek," B. Henning Hansen. Geografisk Tidsskrift, v66, no1 (June 1967), p. 90-118.

20 "Catálogo de Publicaciones Periódicas." Boletín Bibliográfico de la Sociedad Geográfica de Colombia (Academia de Ciencias Geográficas), no. 9 (November, 1968), p. 1-37. 225 titles of which 178 are included in this list.

21 "Geographical Periodicals in the Library of the Institute of Geology and Geography of the Academy of the Socialist Republic of Romania," Revue Roumaine de Géologie, Géophysique et Géographie: Série de Géographie, v11, no2 (1967, no2), p. 177-185. 274 serials for the period 1959-1966, of which 199 are included in this list.

22 Svodnyi Ukazatel' Inostrannykh Periodicheskikh Izdanii poluchaemykh Bibliotekoi Akademii Nauk SSSR. Leningrad: Izdatel'stvo otdel Biblioteki Akademii Nauk SSSR, 1969. 426 p.

Geography, entries 2411-2593, p. 94-100, with 182 titles. Also earth sciences as a whole, entries 1778-1792, p. 70-71; geodesy and cartography, entries 1793-1845, p. 71-73; geophysics (seismology, meteorology and climatology, and oceanography), entries 1846-2075, p. 73-81; geology, entries 2074-2410, p. 81-94; geomorphology, entries 2411-2593, p. 100-101; hydrology, entries 2607-2649, p. 101-103; glaciology and permafrost studies, entries 2650-2654, p. 103; and Polar countries, entries 2655-2669, p. 103.

Recent Lists of Geographical Serials Currently Received or Held by Individual Libraries

23 Niedersächische Staats- und Universitätsbibliothek, Göttingen. Katalog der Zeitschriften und Serien für das Fachgebiet Geographie und einige fachverwandte Disziplinen. Stand: 1. April 1979. Göttingen: Niedersächische Staats- und Universitätsbibliothek Göttingen, 1979. 493 p. Especially Teil I. Geographie und Reisen und Entdeckungen, p. 11-194.

24 Stephens, John D. Serial Publications in Geography. Blacksburg, Virginia: Department of Geography, Virginia Polytechnic Institute and State University, 1978, 88 p. List of 292 geographical and 23 regional science serials from 45 countries in the Library of Virginia Polytechnic Institute and State University in alphabetical order, with geographical index.

The 5th edition, dated December, 1979, 121 pages, lists 354 geographical serials and 27 regional science serials from 46 countries.

25 "Échanges avec Notre Bulletin. In ruil ontvangen Tijdschriften," Société Belge d'Etudes Géographiques. Bulletin; Belgische Vereniging voor Aardrijkskundige Studies. Tijdschrift. v48 (1979), no1, p. 13-14.

26 "Publicaciones periódicas que se reciben actualmente en la R.S.G.," Boletin de la Real Sociedad Geográfica," tomo 112, parte 2 (1976), p. 622-626.

Lists of Current Geographical Serials

General

27 Briend, A. M. et al. "Répertoire Collectif des Périodiques Vivants d'Intérêt Géographique prossédés en France par 7 Bibliothèques. Intergéo, v8 no4, whole no36 (1974), p. 303-368.

758 current geographical serials based on serials cited in the 1974 Bibliographie Géographique Internationale and available in the geographical collections of the Bibliothèque du Centre de Géographie des Universités de Paris I, IV, VII, Bibliothèque de la Société de Géographie, and Département des Cartes et Plans de la Bibliothéque Nationale. Holdings recorded for seven libraries: (1) Bibliothèque du Centre de Géographie des Universités de Paris I, IV, VII [now Bibliothèque du Centre Interuniversitaire de l'Institut de Géographie], 191, rue Saint-Jacques, 75005 Paris; (2) Bibliothèque Nationale, Departement des Périodiques; (3) Bibliothèque Nationale, Département des Cartes et Plans; (4) Bibliothèque de la Société dę Géographie, 58 rue de Richelieu, 75084 Paris, Cedex 2; (5) Centre d'Etudes de Géographie Tropicale du C.N.R.S. Service de Documentation, Domaine Universitaire de Bordeaux, 33405 Talence; (6) Bibliothèque du Centre de Documentation du C.N.R.S., 26, rue Boyer, 75971 Paris; and (7) Service de Documentation et de Cartographie Géographiques [now Laboratoire d'Information et de Documentation en Geographie], C.N.R.S., Section de Documentation, 191, rue Saint-Jacques, 75005 Paris.

28 "Führende geographische Zeitschriften der Erde," Geographisches Taschenbuch, 1953, pp. 213-222.

29 Geographisches Taschenbuch, 1958/59-1962/63.
(1) "Geographische Zeitschriften: Europa," by Charlotte Streumann, 1958-1959, p. 269-278. 106 entries of which 100 are included in this list.
(2) "Führende Geographische Zeitschriften: Afrika, Asien, Australien und Ozeanien, Sowjet-Union," 1960-1961, p. 152-163. 115 entries of which 106 are included in this list. Especially valauble is the list of 56 periodicals in the Soviet Union prepared by Dora Fischer.
(3) "Führende Geographische Zeitschriften: Amerika," 1962-1963, p. 21-26. 66 entries, all included in this list.

Individual Countries

30 Hart, John Fraser, Selected Geographical Serials Published in the United States. (International Geographical Union. United States National Committee). Urbana, Illinois: Department of Geography, University of Illinois, 1972. 17 p.

60 geographical serials published in the United States current in 1972, with bibliographical details.

31 Bass, Margaret A., "Periodical Literature Survey in the Fields of Geography, Geomorphology and Geology. Interim report I. Great Britain and Northern Ireland. January 1968." Professor K. M. Clayton, School of Environmental Sciences, University of East Anglia, Norwich NR4 7TJ, England. iii+22 pp. Mimeographed.

32 Area. "Geography published underground: 1. The output from British Universities," by John Connell, v6 (1974), no2, p. 121-127; "2. The output from British polytechnics and cognate disciplines," by K. Hoggart, v6 (1974), no2, p. 128-129; "Geographical papers from British departments," by John Connell, v6 (1974), no3, p. 195-199; and "Geographical papers from British departments," v7 (1975), no3, p. 165-166.

33 Briend, A. M., and Pineau, M. "Liste des Périodiques Géographiques Français," Intergéo bulletin, no47 (1977, 11e année 3e trimestre), p. 12-36.
94 current French geographical periodicals and other serials based on questionnaire to all French centers of geographic work, checked against other sources of information, and completed by examination of copies of each listed publication.

34 Faucher, Daniel. "L'Activité Géographique Française: Les Revues," L'Information Géographique: La Géographie Française au Milieu du XXe Siècle. G. Chabot, R. Clozier, and J. Beaujeu-Garnier, eds. Paris, 1957. p. 299-304.

35 France. Bibliothèque Nationale, Paris. Cabinet de la Géographie. "Periodiques Géographiques Vivants édités en France et dans les Colonies Française." 1944, 17 p. Mimeographed. Compiled by M. Foncin. 37 current French geographical periodicals and 78 titles in related fields.

36 Meynen, Emil. "Geographische Zeitschriften und Jahrbücher sowie Schriftenreihen der Hochschulwissenschaft in der Bundesrepublik Deutschland, DDR, Österreich und der Schweiz," Geographisches Taschenbuch und Jahrweiser für Landeskunde 1977-1978, p. 20-26.

37 Meynen, Emil. "Geographische Zeitschriften und Jahrbücher sowie Schriftenreihen in der Bundesrepublik Deutschland, Berlin (West), DDR, Österreich und der Schweiz auf grund einer Umfrage zusammengestellit von E. Meynen," Geographisches Taschenbuch und Jahrweiser für Landeskunde, 1979/1980, p. 48-61.

38 "Wykaz czasopism i innych wydawnictw ciągłych wykorzystanych w bibliografii geografii polskiej za lata 1971-1972 oraz ich skróty: Wydawnictwo geograficzne," Bibliografia Geografii Polskiej 1971-1972 Warszawa: Polska Akademia Nauk. Instytut Geografii i Przestrzennego Zagospodarowania, 1974 p. 15-16.

39 Welsh, D. J. "Hand-List of the Russian Serials in the Library of the Royal Geographical Society." London: Royal Geographical Society, 1956. Library Series No. 4. 15 p. Mimeographed.

40 Hall, Robert B. and Noh, Toshio. Japanese Geography: A Guide to Japanese Reference and Research Materials. University of Michigan. Center for Japanese Studies. Bibliographical Series No. 6. Revised edition, Ann Arbor: University of Michigan Press, 1970. Chapter 5 "Periodicals," p. 12-18. (1st ed. 1956). 45 periodicals of interest to geographers.

41 Directory of Japanese Learned Periodicals. 1957. Compiled by National Diet Library. Fourth Part. Generalities, Humanities, and Social Sciences. 2nd edition, Tokyo: Tokyo Library Bureau, 1958. Section 91: Geography. Chirigaku, 9-53/64.

Cartography

42 Salishchev, K. A. "Periodicheskie i prodolzhaiushchiesia izdaniia po kartografii," Itogi nauki i tekhniki: kartografiia, tom 8 (1978), p. 189-218, and Canadian Cartographer, v16 no2 (D 1979), p. 109-132.

43 Salishchev, K. A. "Die Kartographischen Zeitschriften der Erde," Petermanns Geographische Mitteilungen, v110 (1966), no2, p. 147-159.

44 Wise, Donald A. "Cartographic Sources and Procurement Problems. Appendix A. Selected Serials Containing Lists and/or Reviews of Current Maps and Atlases." Special libraries association, Geography and map division. Bulletin, no112 (June 1978), p. 19-22.
53 serials listed.

45 Meine, Karl-Heinz, "Zeitschriften und Schriftenreihe für Kartographie sowie geodätische und geographische Zeitschriften mit kartographischen Beiträgen, Stand 1967/68," Geographisches Taschenbuch 1966/1969, p. 26-32. Section on cartographical periodicals and serials listed 32 entries.

Other

46 Migliorini, Elio. "Riviste de geografia destinate agli insegnanti," Geografia nelle scuole, v22 (1977), p. 150-154.

47 Young, Bruce. Bibliography and review of geography department discussion papers, occasional papers, and monographs. Waterloo, Ontario, Canada: Department of geography, Wilfrid Laurier university. vol. 1, 1974, v.2, 1976, v.3, 1978. Volume 3 includes previous entries and lists individual publications in geographical series published by 83 universities in Australia, New Zealand, the United Kingdom, Canada, the United States, and some other countries, in English (with some in French in Canada).

Useful both in including many series not listed in other sources and in providing titles of individual papers in such series. New edition in preparation in 1980.

Lists of Serials Analyzed

or Cited in Current Geographical Bibliographies

48 Bibliographie Géographique Internationale. International Geographical Bibliography. v.82 (1977). Tables annuelles. Liste des périodiques dépouillés au cours de l'année 1977. List of periodicals analyzed during the year 1977." p. 1-15.
v.83 (1978). Tables annuelles. "Liste des périodiques dépouillés au cours de l'année," p. 1-15.
"Liste des périodiques dépouillés dans ce fascicule," v.84 (1979), no.1, p.95-99; no.2, p. 99-103; no.3, p. 99-102; no. 4, p. 139-143.

49 [Grivot, Françoise]. Bibliographie Géographique Internationale: Répertoire des Principaux Périodiques d'Intérét Géographique Cités dans la Bibliographie Géographique Internationale. Paris: Éditions du Centre National de la Recherche Scientifique, 1966. 74 p.

968 serials both geographical and in related fields, of which 243 are incorporated in this list.

50 Geo Abstracts. Annual index volumes. List of journals from which abstracts have been published each year.
1966, p. i-xvii. 798 serials.
1967, p. 7-27. 1,313 serials.
1968, p. 7-28. 1,339 serials.
1969, p. 7-23. 1,048 serials.
1970, p. 7-23. 1,043 serials.
1971, p. 7-31, about 1,500 serials.
1972, v. 1, A, B, p. 7-22. About 900 serials.
v. 2, parts C, D, and F, p. 7-17. About 650 serials.
1973, v. 1, p. 9-20. About 700 serials.
v. 2, p. 7-17. About 700 serials.
1974, v. 1, p. 7-18. About 680 serials.
v. 2, p. 7-16. About 550 serials.
1975, v. 1, p. 791-807. About 920 serials.
v. 2, p. 443-464. About 1,170 serials.
1976, v. 1, p. 1-14. About 840 serials.
v. 2, p. 473-490. About 1,040 serials.
1977, v. 1, p. 677-692. About 950 serials.
v. 2, p. 551-569. About 1,130 serials.
1978, v. 1, p. 725-743. About 1,260 serials.

5-year cumulative indexes, 1966-1970.
Series A. Geomorphology, p. 7-26. About 1,150 serials.
Series B. Biogeography and climatology, p. 7-24. About 1,050 serials.
Series C. Economic geography, p. 1-20. About 1,200 serials.
Series D. Social geography and cartography, p. 7-21. About 900 serials.

51 Gerould, Albert C., and Warman, Henry J., "Most Cited Periodicals in Geography," The Professional Geographer, vol. 6, no. 2 (March 1954), p. 6-12.

52 Strumm, Robert H., "Most Cited Serials in Geography as Represented by Current Geographical Publications," The Professional Geographer, vol. 18, no. 4 (July 1966), p. 248-251.

53 Aiyepeku, Wilson O., "The Periodical Literature of Geography," LIBRA: International Library Review and IFLA Communications, vol. 22, no. 3 (1972), p. 169-182.

55 Documentatio Geographica. "Abkürzungsverzeichnis der im Jahresband... der DOCUMENTATIO GEOGRAPHICA ausgewerteten Zeitschriften und Serien." Jahresband 1966. Teil I. Titelband, p. xli-xlix. 1967... p. xxiii-xxxii. 1968... p. xxi-xxvii. 1969... p. xxi-xxvii. 1970... p. xxi-xxviii. 1971/1972... p. xxiii-xxxv. 1973... p. xi-xxv. Closed 1973.

56 Dokumentation zur Raumentwicklung. Vierteljahreshefte zur Literaturdokumentation aus Raumforschung, Raumordnung, Regionforschung, Landeskunde und Sozialgeographie. A current and annotated bibliography of regional science, regional planning and social geography. Herausgegeben von/Edited by Bundesforschungsanstalt für Landeskunde und Raumordnung. Jahrgang 1974/1975. Teil 1: Titelband. Bonn-Bad Godesberg: Bundesforschungsanstalt für Landeskunde und Raumordnung, 1976. "Abkürzungsverzeichnis der im Jahresband 1974/1975 der Dokumentation zur Raumentwicklung ausgewerteten Zeitschriften und Series," p. vi-xx.
1976. Jahrgang 1976. Teil I. Titelband (1977), p. ix-xx.
1977. Jahrgang 1977. Teil I. Titelband (1978), p. ix-xx.
1978. Jahrgang 1978. Teil I. Titelband (1979), p. ix-xx. Closed 1978.

57 Literaturinformation: Territorialforschung, Territorialplanung. Verzeichnis der ausgewerteten Zeitschriften. Stand 1.12.1977. Herausgegeben und bearbeitet vom Institut für Geographie und Geoökologie der Akademie der Wissenschaften der DDR in Zusammenarbeit mit der Forschungsleitstelle für Territorialplanung der Staatlichen Plankommission. Leipzig: Institut für Geographie und Geoökologie, 1978. 20 p.

58 Referativnyi Zhurnal: Geografiia. "Osnovnye periodicheskie i prodolzhaiushchiesia izdaniia, referiruemye v vypuskakh svodnogo toma "Geografiia," Sostavil' M. P. Saburova. Title varies slightly. 1961, no. 1, 4 p. insert; 1967, no. 1, p. 1-5; 1968, no. 1, p. 2-6; 1969, no. 1, p. 5-11; 1970, no. 1, p. 5-10; 1971, no. 1, p. 3-9; 1972, no. 1, p. 6-12; 1973, no. 1, p. 6-12; 1977, no. 1, p. 5-14; 1978, no. 1, p. 7-15; and 1979, no. 7, p. 3-11.

59 "Verzeichnis der ausgewerteten Zeitschriften," Bibliographia Cartographica, vol. 1 (1974), p. 140-158; v.2 (1975), p. 161-173; v.3 (1976), p. 177-185; v.4 (1977), p. 175-183; v.5 (1978), p. 180-189.

60 Harris, Chauncy D., "Geographical Serials Most Cited in Geographical Bibliographies," in his Annotated World List of Selected Current Geographical Serials, 4th edition. Chicago, Illinois: University of Chicago, Department of Geography, Research Paper No. 194, 1980.
Geographical serials most cited in Current Geographical Publications, Geo Abstracts, Bibliographie Géographique Internationale, Referativnyi Zhurnal: Geografiia, Literaturinformation: Territorialforschung, Territorialplanung, and Dokumentation zur Raumentwicklung.

Lists of New Geographical Serials

61 New Serial Titles: Classed Subject Arrangement (U.S. Library of Congress. Prepared under the Sponsorship of the Joint Committee on the Union List of Serials). January 1955-December 1979. Monthly (does not cumulate). Selected titles from classes 551.4 and 910-919. Classes 910-919 contain without discrimination very disparate material from serious scholarly serials in geography to commercial directories and tourist publications.

62 New Serial Titles 1950-1970 Subject Guide. New York: R. R. Bowker Company, 1975. 2 vols. 3,692 p.
Geography, Dewey class 910-912, vol. 2, p. 3315-3331. Geography of Specific Areas, Dewey class 914-919, vol. 2, p. 3347-3378. Arranged by countries.

63 Current Geographical Publications (American Geographical Society of New York). "Aids to Geographical Research," class 85, section "New Periodicals." Monthly except July and August. January 1948-December 1977; also January-September 1978, nos 1-7 and January-September 1979, nos 1-7. From vol. 41, no. 3 (March 1978) issue published by the American Geographical Society Collection, The University of Wisconsin-Milwaukee Library, Milwaukee, Wisconsin.

64 "Geographical Periodicals Added to the Library," New Geographical Literature and Maps (Royal Geographical Society). n.s. vol. 3, no 25-vol. 6, no 46 (June 1963-December 1973) 2 numbers a year.
vol. 3, no 25 (June 1963), p. iii; 26 (December 1963), p. i; 27 (June 1964), p. iii; 28 (December 1964), p. iii; 29 (June 1965), p. iii; 30 (December 1965), p. i; vol. 4, no 31 (June 1966), p. i; 32 (December 1966), p. i; 33 (June 1967), p. 295; 34 (December 1967), p. 400; 35 (June 1968), p. 508; 36 (December 1968), p. 622; 37 (June 1969), p. 728; 38 (December 1969), p. 846; 39 (June 1970), p. 955; 40 (December 1970), p. 1060; vol. 5, no 41 (June 1971), p. 80; 42 (December 1971), p. 188; 43 (June 1972), p. 314; 44 (December 1972), p. 424; vol. 6, no 45 (June 1973), p. 84; 46 (December 1973), p. 196.
For 1974 and later years see Geographical Journal [61].

65 "Geographical periodicals. New periodicals added to the Society's library," 1973-1978. Geographical journal,
v. 139, part 2 (June 1973), p. 374
v. 140, part 2 (June 1974), p. 328-329
v. 141, part 1 (March 1975), p. 138-139
v. 141, part 2 (July 1975), p. 307
v. 141, part 3 (November 1975), p. 495
v. 142, part 2 (July 1976), p. 352-353
v. 142, part 3 (November 1976), p. 544
v. 143, part 1 (March 1977), p. 130
v. 143, part 2 (July 1977), p. 337
v. 143, part 3 (November 1977), p. 487
v. 144, part 1 (March 1978), p. 153
v. 144, part 2 (July 1978), p. 364
v. 145, part 2 (July 1979), p. 352

66 "New geographical periodicals" (Title varies), Geographical Review. By Wilma B. Fairchild, Mary Alice Lamberty, Lynn S. Mullins, and others. v. 42 (1952), p. 150-151; v. 43 (1953), p. 124-126; v. 44 (1954), p. 155-157; v. 45 (1955), p. 125-126; v. 46 (1956), p. 273-274; v. 48 (1958), p. 122-124; v. 49 (1959), p. 128-130; v. 50 (1960), p. 125-127; v. 51 (1961), p. 311-314; v. 52 (1962), p. 302-304; v. 53 (1963), p. 317-319; v. 54 (1964), p. 278-281; v. 55 (1965), p. 290-292; v. 56 (1966), p. 449-451, 594-596; v. 57 (1967), p. 434-437, 569-572; v. 59 (1969), p. 147-151, 290-294; v. 60 (1970), p. 129-132, 271-274; v. 61 (1971), p. 451-454, 608-611; v. 62 (1972), p. 128-133, v. 63 (1973), p. 569-577, v. 64 (1974), p. 145-150, 289-296.

Union Lists of Periodical Holdings of Libraries

67 Union List of Serials in Libraries of the United States and Canada. Edited by Edna Brown Titus. 3rd ed. New York: H. W. Wilson, 1965. 5 vols. 4,649 p.

68 New Serial Titles: A Union List of Serials Commencing Publication after December 31, 1949. 1950-1970 Cumulation. New York: R. R. Bowker, 1973. 4 vols. 6,713 p.

69 New Serial Titles. 1971-1975 Cumulation. Washington, D. C.: Library of Congress, 1976. 2 vols. 2,462 p.

70 New Serial Titles, 1976-1977 Cumulation, and later monthly issues or quarterly cumulations through December 1979. Washington, D.C.: Library of Congress.

71 British Union-Catalogue of Periodicals: A Record of the Periodicals of the World from the Seventeenth Century to the Present Day in British Libraries. Edited by James D. Stewart with Muriel E. Hammond and Erwin Saenger. London: Butterworth Scientific Publications, 1955-1958. 4 vols.

72 British Union-Catalogue of Periodicals. Supplement to 1960. London: Butterworths, 1962. 991 p.

73 British Union-Catalogue of Periodicals incorporating World List of Scientific Periodicals. New Periodical Titles 1960-1968. Edited by Kenneth I. Porter and C. J. Koster. London: Butterworths, 1970. 603 plus 128 p.

74 British Union-Catalogue of Periodicals incorporating World List of Scientific Periodicals. New Periodical Titles 1969-1973. Edited by J. Gascoigne. London: Butterworths, 1976. 374 plus 114 p.

75 British Union-Catalogue of Periodicals incorporating World List of Scientific Periodicals. New Periodical Titles. Annual volumes 1974, 1975, 1976, and 1977 and quarterly issues 1978 nos 1-4 and 1979 nos 1-2.

76 Bibliothèque Nationale. Département des Périodiques. Catalogue Collectif des Périodiques du Début du XVIIe Siècle à 1939 conservés dans les Bibliothèques de Paris et dans les Bibliothèques universitaires des Départements. Paris: Bibliothèque Nationale, 1967-1977. 4 vols.

77 Gesamtverzeichnis Deutschsprachiger Zeitschriften und Serien in Bibliotheken der Bundesrepublik Deutschland einschliesslich Berlin (West). Title vor 1971 mit Besitznachweisen Stand: April 1978. Bearbeitet und herausgegeben von der Staatsbibliothek Prussischer Kulturbesitz, Abteilung Gesamtkataloge und Dokumentation. [2nd ed.] München: Verlag Dokumentation Saur KG, 1978. 2 vols. 2,512 p.

78 Gesamtverzeichnis der Zeitschriften und Serien in Bibliotheken der Bundesrepublik Deutschland einschliesslich Berlin (West). Neue und geänderte Titel seit 1971 mit Besitznachweisen Stand April 1978. Bearbeitet und herausgegeben Staatsbibliothek Preussischer Kulturbesitz, Abteilung Gesamtkataloge und Dokumentation. [8th ed.] München: Verlag Dokumentation Saur KG, 1978. 1,381 p.

79 Retrospektivnyi Svodnyi Ukazatel' Inostrannykh Periodicheskikh i Prodolzhaiushchikhsia Izdanii, imeiushchikhsia v krupneishikh Bibliotekakh SSSR 1750-1965. Estestvennye Nauki, Tekhnika, Meditsina, Sel'skoe Khoziaistvo. Moskva: "Kniga," 1974-1979. 6 vols.

80 International Congresses and Conferences: A Union List of their Publications Available in Libraries of the United States and Canada. Edited by Winifred Gregory. New York: H. W. Wilson, 1938. 229 p.

81 Gesamtverzeichnis der Kongress-Schriften in Bibliotheken der Bundesrepublik Deutschland einschliesslich Berlin (West). Schriften von und zu Kongressen, Konferenzen, Kolloquien, Symposien, Tagungen, Versammlungen und dergleichen vor 1971 mit Besitznachweisen Stand 1976. Bearbeitet und herausgegeben von der Staatsbibliothek Preussischer Kulturbesitz. Abteilung Gesamtkataloge und Dokumentation. München: Verlag Dokumentation Saur KG. Hauptband. 1976. 563 p. Registerband, 1976. 344 p. Supplement für Kongresse 1977, 1978.

82 List of Serial Publications of Foreign Governments 1815-1931. Edited by Winifred Gregory. New York: H. W. Wilson, 1932. 720 p.

Directories of Current Serials

83 Ulrich's International Periodicals Directory: a classified guide to current periodicals, foreign and domestic. New York and London: R. R. Bowker, 1932- . Biennial. "Geography," 1st ed. 1932, p. 130-133; 2nd ed., 1935, p. 163-166; 3rd, 1938, p. 193-196; Inter-American edition, 1943, p. 133-135; 5th ed., 1947, p. 160-162; 6th ed., 1951, p. 219-222; 7th ed., 1953, p. 270-273; 8th ed., 1956, p. 270-275; 9th ed., 1959, p. 297-303; 10th ed., 1963, p. 246-251; 11th ed., vol. 1, 1965, p. 173-174 and vol. 2, 1966, p. 674-679; 12th ed., vol. 2, 1968, p. 786-791; 13th ed., 1969, vol. 1, p. 602-608; 14th ed., 1971, vol. 1, p. 741-748; 15th ed., 1973, p. 881-891; 16th ed., 1975, p. 654-662; 17th ed., 1977, p. 720-727; and 18th ed., 1979, p. 738-746.

84 Irregular Serials and Annuals: An International Directory. New York and London; R. R. Bowker, 1967- . Biennial. "Geography," 1st. ed., 1967, p. 239-244; 2nd ed., 1972, p. 2347-2356; 3rd ed., 1974, p. 313-321; 4th ed., 1976, p. 346-354; 5th ed., 1978, p. 399-407; 6th ed., 1980, p. 412-421.

85 Ulrich's Quarterly: A Supplement to Ulrich's International Periodicals Directory and Irregular Serials and Annuals. vl- (1977-). Quarterly. vol. 1 (1977), p. 37-38, 189-190, 390, 595; vol. 2 (1978), p. 25, 119, 240, 357; vol. 3 (1979), no. 1, p. 27; no. 2, p. 128; no. 3, p. 255; no. 4.

86 Deutsche Forschungsgemeinschaft. Verzeichnis Ausgewählter Wissenschaftlicher Zeitschriften des Auslandes. VAZ. Wiesbaden: Franz Steiner Verlag, 1957. Section 14. Geographie, pp. 516-529.

87 Zimmerman, Irene. A Guide to Current Latin American Periodicals: Humanities and Social Sciences. Gainesville, Florida: Kallman Publishing Company, 1961. "Geography," p. 246-248, list 24 geographical serials.

88 Pan-American Union. Directorio de Publicaciones Periódicas en el Campo de las Ciencias Sociales. Primera parte: América Latina. Directorios. III. Washington: Pan-American Union, 1955. 83 p.

89 Lombardi, Mary. Brazilian serial documents: a selective and annotated guide. Bloomington, Indiana: Indiana University Press, 1974. 445 p., especially p. 263-265.

90 Woodworth, David P., ed. Guide to Current British Journals 2nd ed. London: The Library Association, 1973. "Geography," volume 1, p. 327-328.

Library Holdings of Regional Serials

91 U. S. Library of Congress. Accessions List India. Quinquennial Serials Cumulation 1976. Volume 3. Subject Arrangement of Titles Listed in Vols. 1-2. New Delhi: Library of Congress Office, 1978. "Geography," p. 303-304.

92 U. S. Library of Congress. Accessions List: Indonesia, Malaysia, Singapore and Brunei. Cumulative List of Indonesian Serials 1964--December 1973. Jakarta, Indonesia, The Library of Congress Office, April 1974. 316 p.

93 ______. Accessions List Southeast Asia. Cumulative List of Indonesian Serials 1974-1976.

94 ______. ______. 1977 Supplement to Cumulative List of Indonesian Serials.

95 Cornell University Libraries. Southeast Asia Catalog. Serials. Asia General, Burma, Cambodia, Indonesia, Laos, Malaysia-Singapore-Brunei, Philippines, Portuguese Timor, Thailand, Vietnam. Boston, Massachusetts: G. K. Hall, 1976, vol. 6, p. 643-742, and vol. 7, p. 1-740.

96 U. S. Library of Congress. Chinese Periodicals in the Library of Congress. Compiled by Han Chu Huang, Chinese and Korean Section, Orientalia Division. Washington, D. C.: Library of Congress, 1978. 521 p.
More than 6,400 titles published 1868-1975. In alphabetical order with titles romanized according to the modified Wade-Giles system and with characters in traditional Chinese form, other bibliographical data, and holdings of the Library of Congress. The Library of Congress is said to have the largest collection of Chinese serial publications outside China.

97 University of Chicago Library. Far Eastern Library. Far Eastern Serials. "Reference List No. 2." Chicago: Far Eastern Library, University of Chicago Library, 1977. 370 p.
2,400 serial titles in Chinese, and 1,300 in Japanese, and 50 in Korean held by the Far Eastern Library. Also 1,250 in Western languages. In alphabetical order of romanized forms separately for Chinese, Japanese, Korean, and Western languages.

General Catalogs of Serial Holdings of Individual Libraries

98 1979 Union Catalog of Serials currently received in the libraries of the University of Wisconsin-Madison. Madison, Wisconsin: 6th ed. Madison, Wisconsin, 1979. 4 vols. 1,699 p. Current through May 4, 1979.

99 Linda Hall Library Science and Technology. Serial Holdings in the Linda Hall Library. 1979 Annual. Kansas City, Missouri: Linda Hall Library, 1979. 622 p.
List of all serials in the Library's records through April 30, 1979. Particularly strong on scientific and technical serials from the Soviet Union.

Serials of the Soviet Union

Union Lists

100 Smits, Rudolf. Half a Century of Soviet Serials 1917-1968: A Bibliography and Union List of Serials Published in the USSR. Washington, D.C.: Library of Congress, 1968. 2 vols. 1,667 pp. 29,761 entries and 28,000 cross references.

101 Bruhn, Peter. Gesamtverzeichnis russischer und sowjetischer Periodika und Serienwerke in Bibliotheken der Bundesrepublik Deutschland und West-Berlin. Union list of Russian and Soviet periodicals and serial publications in libraries of the Federal Republic of Germany and West Berlin. Herausgegeben von Werner Philipp. Wiesbaden: in Kommission bei O. Harrassowitz. 3 vols. (19 nos), 1960-1973. Register, 1976. (Berlin. Freie Universität. Osteuropa-Institut. Bibliographische Mitteilungen, Heft 3). A union list of about 10,000 main entries, 1702-1956, in alphabetical order, with holdings in libraries. Vol. 3, no. 19 (1973) consists of addenda. Vol. 4 (1976) is index.

102 Bibliothèque Nationale. Département des Périodiques. Catalogue Collectif des Périodiques conservés dans les Bibliothèque de Paris et dans les Bibliothèques Universitaires de France. Periodiques Slaves en Caractères Cyrilliques. Etat des Collections en 1950. Paris: Bibliothèque Nationale, 1956. 2 vols. 873 p.
Supplément 1951-1960. 1963. 495 p.
Addenda et errata. État général des collections en 1960. 1965. 223 p.
Holdings by 46 libraries of Paris and of French universities of periodicals in Russian, Ukrainian, Belorussian, Bulgarian, and Serbian.

Official Inventories of Published Periodicals and Serials

103 Lisovskii, N. M. Russkaia Periodicheskaia Pechat' 1703-1900gg. (Bibliografiia i Graficheskiia Tablitsy). Petrograd: Tip. G. A. Shumakheria i B. D. Brukera, 1915. 267 p.

104 Gosudarstvennaia Publichnaia Biblioteka imeni M. E. Saltykova-Shchedrina. Bibliografiia Periodicheskikh Izdanii Rossii 1901-1916. Pod obshchei redaktsiei V. M. Barashenkova, O. D. Golubevoi, N. Ia. Morachevskogo. Leningrad; 1958-1961. 4 vols.

105 Vsesoiuznaia Knizhnaia Palata. Periodicheskaia Pechat' SSSR. 1917-1949. Bibliograficheskii Ukazatel': Zhurnaly, Trudy i Biulleteni po Estestvennym Naukam i Matematike. Moskva: Izdatel'stvo Vsesoiuznoi Knizhnoi Palaty, 1956. Geografiia, p. 49-56. Entries 447-527.

106 Vsesoiuznaia Knizhnaia Palata. Letopis' Periodicheskikh Izdanii SSSR. 1950-1954 gg. Bibliograficheskii Ukazatel'. Moskva: Izdatel'stvo Vsesoiuznoi Knizhnoi Palaty, 1955. p. 20-28.

107 ______. ______. 1955-1960 gg. Chast' I. Zhurnaly, Trudy, Biulleteni. 1963. p. 68-82 and passim.

108 ______. ______. 1961-1965 gg. Chast' I. Zhurnaly, Trudy, Biulleteni. Kniga 1. Opisanie Izdanii. Moskva: "Kniga," 1973. p. 57-74 and passim.

109 ______. ______. Trudy, Uchenye Zapiski, Sborniki i Drugie Prodolzhaiushchiesia Izdaniia 1966, 1967. Moskva: "Kniga," 1971- . p. 19-27 and passim.

110 ______. ______. ______. 1968. Moskva: "Kniga," 1974. p. 17-23 and passim.

111 ______. ______. ______. 1969, 1970. Moskva: "Kniga," 1975. p. 28-38 and passim.

112 ______. Letopis' Periodicheskikh i Prodolzhaiushchikhsia Izdanii. Sborniki. 1971. Moskva: "Kniga," 1976. p. 21-28 and passim.

113 ______. ______. ______. 1972. Moskva: "Kniga," 1977. p. 21-28 and passim.

114 ______. ______. ______. 1973. Moskva: "Kniga," 1978. p. 24-29 and passim.

115 Vsesoiuznaia Knizhnaia Palata. Letopis Periodicheskikh Izdanii SSSR. 1966-1970 gg. Chast' I. Zhurnaly. Moskva: "Kniga," 1972. p. 23-24, 72, 73, 75, and 77.

116 ______. ______. 1971-1975. Chast' I. Zhurnaly. Moskva: "Kniga," 1977. p. 27-28, 105-109, and 112.

117 Knyzhkova Palata Ukrainskoi RSR. Periodychni Vydannia URSR 1918-1950. Zhurnaly. Bibliohrafichnyi dovidnyk. Kharkiv: Vydavnytstvo Knyzhkovoi Palaty URSR, 1956. 461 p.

Serial Publications of the Geographical Society of the USSR

118 Berg, L. S. Vasesoiuznoe Geograficheskoe Obshchestvo za Sto Let (Moskva: Izdatel'stvo Akademii Nauk SSSR, 1946). "Periodicheskie izdaniia obshchestva," pp. 249-250.

119 "Seriinye i Periodicheskie Izdaniia Geograficheskogo Obshchestva SSSR za 1917-1966 gg.," in Geograficheskoe Obshchestvo Soiuza SSR 1917-1969, by I. B. Kostrits and D. M. Pinkhenson. Moskva: Izdatel'stvo "Mysl'," 1968, pp. 246-251. Lists the 71 principal periodical and serial publications of the Geographical Society of the USSR and its sections and branches, 1917-1966.

120 "Perechen' Izdanii Geograficheskogo Obshchestva Soiuza SSR Postupivshikh v Biblioteku Obshchestva za 1961 god...1967g." Geograficheskoe Obshchestvo SSSR. Izvestiia. v. 94, no 5 (1962), p. 451-458; v. 95, no 6 (1963), p. 566-578; v. 96, no 6 (1964), p. 541-552; 1964, v. 97, no 4 (1965), p. 392-399; 1965, v. 98, no 2 (1966), p. 190-203; 1966, v. 99, no 4 (1967), p. 338-368; and 1967, v. 101, no 2 (1969), p. 174-190.

121 Loshkomoeva, N. M. "Perechen' Izdanii Geograficheskogo Obshchestva Soiuza SSR, postupivshikh v Biblioteku Obshchestva za 1968-1970gg.," Geograficheskoe obshchestvo SSSR. Izvestiia, v. 104, no 2 (1972), p. 144-156; 1971, v. 105, no 1 (1973), p. 81-94; 1972, v. 106, no 6 (1974), p. 529-544; 1973-1975, v. 109, no 5 (1977), p. 459-469; 1976, v. 110, no 1 (1978), p. 83-90; and 1977, v. 110, no 5 (1978), p. 469-477; 1978, v. 111, no 4 (1979), p. 369-382.

STUDIES OF THE CORPUS OF GEOGRAPHICAL SERIALS

For a detailed analysis of the characteristics of geographical serials published up to 1970 based on serials with entries in International List of Geographical Serials, 2nd edition, 1971, one can consult the thorough study by Myrna W. P. Chak, "Geographical Serial Literature: Size, Growth, and Characteristics."

Brief over-all reviews of the corpus of geographical serials have been made by Harris and Fellmann in 1950, 1960, and 1970, and of supplementary use of international languages in serials by Harris.

Chak, Myrna W. P. "Geographical Serial Literature: Size, Growth, and Characteristics," M. S. Thesis, Loughborough University of Technology, Loughborough, England, 1978. 213 p.
Detailed analysis of 2,497 geographical serials which had been published up to 1970 and of 1189 considered to be current as of that date. Based on International List of Geographical Serials, 2nd edition, but updated in certain respects. Processed through a computer for studies of growth rates, mortality, longevity, frequency of publication, type of publication, type of issuing body, geographical distribution, basic and supplementary languages utilized, abstracts, and secondary serial literature. Extensive tables. Numerous figures with graphs of growth rates.

Harris, Chauncy D., and Fellmann, Jerome D. "Geographical Serials," Geographical Review, Vol. 41, No. 4 (October, 1950), p. 649-656.

Harris, Chauncy D., and Fellmann, Jerome D. "Current Geographical Serials, 1960," Geographical Review, Vol. 51, No. 2 (April, 1961), p. 284-289.

Harris, Chauncy D., and Fellmann, Jerome D. "Current Geographical Serials, 1970," Geographical Review, Vol. 65, No. 1 (January, 1973), p. 99-105.

Harris, Chauncy D. "English, French, German, and Russian as Supplementary Languages in Geographical Serials," Geographical Review, Vol. 49, No. 3 (July, 1959), p. 387-405.

LOCATION OF LIBRARY HOLDINGS THROUGH UNION LISTS OF SERIALS

No individual or library can expect to collect more than a fraction of all geographical serial publications. The number of serials is too great, their sources and languages of publication too diverse, the period of publication of some too long, and the number of existing sets of some too small for any library to have a complete collection of all.

Even the very greatest of the libraries have large gaps in their collections of the geographical serials in this list. Such gap occur in the best libraries of geographical societies, such as the American Geographical Society Collection in the University of Wisconsin-Milwaukee Library in Milwaukee, Wisconsin, formerly in New York, in the United States; the Library of the Royal Geographical Society in London, England; of the Société de Géographie in Paris, France; the Società Geografica Italiana in Rome, Italy; the Gesellschaft für Erdkunde zu Berlin in Berlin, Germany; or the Geograficheskoe Obshchestvo S.S.S.R. in Leningrad, USSR. Such gaps also characterize the greatest national libraries such as the Library of Congress in Washington, the British Library (formerly British Museum) in London, Bibliothèque Nationale in Paris, or the Lenin State Library in Moscow. Many geographical serials are absent from even the strongest of the university libraries such as those of Harvard, Yale, Columbia, Cornell, Michigan, Illinois, Chicago, Louisiana State, California (Berkeley) or California (Los Angeles) in the United States; London, Cambridge, or Oxford in England; Paris or Strasbourg in France; Göttingen in the Federal Republic of Germany; or Moscow in the USSR. Specialized geographical libraries, such as those of the Bundesforschungsanstalt für Landeskunde und Raumordnung in Bonn-Bad Godesberg, Federal Republic of Germany, or the Institut für Geographie und Geoökologie in Leipzig, German Democratic Republic, tend to concentrate on applied geography and planning within these countries. Municipal libraries, such as the New York Public, are also important but not comprehensive resources.

Evidence of the incompleteness of the holdings of any single library comes from an analysis of reported collections of libraries in the United States. In an investigation of library holdings of geographical serials as reported in New Serials Titles, 1950-1970 Cumulation and New Serials Titles, 1971-1975 Cumulation, of 1,876 serials listed in this International List of Geographical Serials that commenced publication in the years 1950-1975, 642, or 34 per cent, were not reported as being held by any American library, 354, or 19 per cent, were held by only one library (involving some 44 different libraries), 294, or 16 per cent, were held by two, three, or four libraries (involving 86 different libraries), and 586, or 31 per cent were held by five or more libraries.[1] Even the Library of Congress, which held far more reported geographical serials than any other library, listed holdings of only 48 per cent of these serials. None of four university libraries with the largest reported holdings, Illinois, Harvard, Chicago, and California, listed holdings of more than 26 per cent of these serials.

A detailed study by J. Cicchini examines the holdings by six major libraries in Paris of the core group of major geographical serials (defined as the ones analyzed by six principal current geographical bibliographies).[2] This study reveals clearly that no single library in France has more than a fraction of even current major geographical periodicals and other serials. Of the 866 titles selected, only 478 or 55 per cent, were represented in the holdings of the library with the most comprehensive collection, La Bibliothèque du Centre Interuniversitaire de l'Institut de Géographie, 191 rue Saint-Jacques; 385, or 44 per cent, were held at least in part by La Bibliothèque du Centre National de la Recherche Scientifique (Centre de Documentation Scientifique et Technique); 233, or 25 per cent, were held in part by both of these libraries; and 640 or 75 per cent were found to be held in part in one or the other of these two very strong libraries. Somewhat astonishing is the rather low percentage held by the combination of the great national library, La Bibliothèque Nationale, and the large Bibliothèque de la Société de Géographie, Paris, housed at the Bibliothèque Nationale: only 211 or 24 per cent. When it comes to complete holdings of geographical serials, records are even more fractional. Of the 866 major current geographical serials, La Bibliothèque du Centre Interuniversitaire de l'Institut de Géographie has complete runs of only 184 or 21 per cent; La Bibliothèque du Centre National de la Recherche

[1]Thanks are due to M. Justin Wilkinson for making these tabulations.

[2]Cicchini, J. "Enquête sur les Périodiques specialisés en Géographie," Intergéo Bulletin, no. 52 (1978), p. 25-38.

Scientifique, only 116 or 14 per cent; and La Bibliothèque Nationale in combination with La Bibliothèque de la Société de Géographie, only 135 or 13 per cent.

The holdings of geographical serials by individual libraries are not recorded in this list. However, to assist individuals wishing information about which libraries have holdings of given serials, and also sources of confirmed bibliographical information, each entry in this list has been checked against certain union lists. Since each such list differs from others in the form of titles used in the main entries, location by volume, page, and column (or in some cases by entry number) has been recorded in this list.

It has been possible to include references for all periods and all areas only to union lists covering the United States, Canada, and the United Kingdom. The French union list covers serials only through 1939 and thus is not helpful for the last 40 years. The German union lists utilized cover either German-language serials for all periods or for all serials only for those beginning publication after 1970. The union list of the Soviet Union covers only foreign, i. e. non-Soviet, serials up to 1965. The American, British, and French union lists were checked for all serials in this list and locations in them noted under each entry. German-language serials were checked against the appropriate German union list but locations were not noted, since this list printed from computer records is still in the process of rapid evolution and each edition rapidly replaces the previous one, making page notations quickly out of date. The Soviet union list of foreign periodicals in libraries of the USSR was checked only for the serials from certain countries: Poland, German Democratic Republic, Czechoslovakia, Hungary, Romania, Bulgaria, Yugoslavia, the Mongolian Peoples Republic, North Korea, and Vietnam. This union list is particularly valuable for serials using the Cyrillic alphabet. For serials of the Soviet Union itself, holdings of Russian and Soviet serials in the libraries of the United States, France, and the Federal Republic of Germany are reported by three union lists; all Soviet serials have been checked against them and locations of each entry in these union lists noted. The entries in Chinese were checked against the list of holdings of the Library of Congress.

Union List of Serials against which Entries have been Checked

Identifying Abbreviation	Name of Union List
U	Union list of serials in libraries of the United States and Canada. Third Edition. Edited by Edna Brown Titus. Under the Sponsorship of the Joint Committee on the Union List of Serials with the Co-operation of the Library of Congress. Funded by a Grant from the Council on Library Resources, Inc. 3rd ed. New York: H. W. Wilson, 1965. 5 vols., 4,649 pages in 3 colums. (1st ed., 1927; 2nd ed., 1943). Contains 156,449 titles of serials that commenced publication prior to 1950, and 70,538 cross references based on the holdings of 956 libraries in the United States and Canada.

N70 New serial titles: a union list of serials commencing publication after December 31, 1949. 1950-1970 cumulative. Washington, D. C.: Library of Congress, and New York and London: R. R. Bowker Company, 1973. 4 vols., 6,713 pages in 3 columns.
About 220,000 titles of serials commencing publication in the 21-year period 1950-1970, with holdings in approximately 500 libraries of the United States and 140 libraries in Canada.

N75 ______. 1971-1975 Cumulation. Prepared under the Sponsorship of the Joint Committee on the Union List of Serials. Supplement to the Union List of Serials, Third Edition. Washington, D. C.: The Library of Congress, 1976. 2 vols., 2,462 pages in 3 columns.
About 70,000 titles.

N76, N77 N78, N79 ______. Monthly issues or quarterly, annual, and multiyear cumulations covering the years indicated. At time of publication there was a cumulation covering 1976-1977 and quarterly cumulations for 1978-1979. These will ultimately be cumulated into a single volume for 1976-1979, and later larger cumulations.

B British union-catalogue of periodicals: a record of the periodicals of the world from the seventeenth century to the present day, in British libraries. Edited for the Council of the British Union-Catalogue of Periodicals by James D. Stewart, with Muriel E. Hammond, and Erwin Saenger. London: Butterworth Scientific Publications, 1955-1958. Reprinted. Hamden, Conn.: Archon Books, 1968. 4 vols. Vol. I, A-C, 691 p.; Vol. II, D-K, 677 p.; Vol. III, L-R, 767 p.; Vol. IV, S-Z, 630 p. 2 columns.
About 140,000 titles that commenced publication up to about 1950, held by 441 libraries in the United Kingdom. Listed under the earliest title or name of issuing agency.

BS ______. Supplement to 1960. Edited for the Council of the British Union-Catalogue of Periodicals by James D. Stewart with Muriel E. Hammond and Erwin Saenger. London: Butterworths; Hamden, Conn.: Archon Books, 1962. 991 p.
About 50,000 titles, most of which began publication 1950-1960.

B68 British union-catalogue of periodicals incorporating World list of scientific periodicals. New periodical titles 1960-1968. Edited for the National Central Library by Kenneth I. Porter and C. J. Koster. London: Butterworths, 1970. 603 p. plus Index of Sponsoring Bodies, 128 p.
About 10,000 titles of serials beginning publication 1960-1967. This cumulative volume updates and re-collates material issued in annual volumes of New periodical titles 1964 to 1968, covering the period 1960-1967.
Radical shift from previous editions of British union-catalogue of periodicals in form of entry for publications of societies and institutions with nondistinctive titles from listing under organization to listing by cover title. Contains separate index of sponsoring bodies.

B73 ______. New periodical titles 1969-1973. Edited for the British Library by J. Gascoigne. London: Butterworths, 1976. 374 p. plus Index of Sponsoring Bodies, 114 p.
About 10,000 titles of serials beginning publication 1969-1973.

B74, B75 B76, B77 ______. ______. Annual cumulations or quarterly issues, 1974- . London: Butterworths, 1964- . Quarterly with annual cumulations. Holdings by British libraries of serials which began publication after 1959. Annual volumes replaced by cumulations for 1960-1968 and 1969-1973 listed above. Annual volumes for 1974, 1975, 1976, and 1977, and later quarterly issues.

F Bibliothèque Nationale. Département des Périodiques. Catalogue collectif des périodiques du début du XVIIe siècle à 1939 conservés dans les bibliothèques de Paris et dans les bibliothèques universitaires des départements. Paris: Bibliothèque Nationale, 1967-1977. Tome I. A-B. 1977. 1,055 p. Tome II. C-I. 1973. 945 p. Tome III. J-Q. 1969. 1,123 p. Tome IV. R-Z. 1967. 1,063 p.
About 75,000 periodicals published 1700-1939 in 71 leading French libraries. Especially full coverage of French serials. Excludes periodicals in Cyrillic. Valuable for bibliographical descriptions.

GZS Gesamtverzeichnis der Zeitschriften und Serien in Bibliotheken der Bundesrepublik Deutschland einschliesslich Berlin (West). Neue und geänderte Titel seit 1971 mit Besitznachweisen Stand April 1978. Union List of Serials in Libraries of the Federal Republic of Germany including Berlin (West). New and changed titles after 1970 with holdings locations as of April 1978. Bearbeitet und herausgegeben Staatsbibliothek Preussischer Kulturbesitz, Abteilung Gesamtkataloge und Dokumentation. München: Verlag Dokumentation Saur KG; New York, London, and Paris: K. G. Saur, 8th ed., 1978. 1,351 p. (1st ed. 1973. 125 p.).
New computerized world inventory of serials that began publication in 1971 or later held by libraries in the Federal Republic of Germany including West Berlin. About 25,000 new titles in the period 1971-1977. Future editions to be microfiche.

R Retrospektivnyi svodnyi ukazatel' inostrannykh periodicheskikh i prodolzhaiushchikhsia izdanii, imeiushchikhsia v krupneishikh bibliotekakh SSSR 1750-1965. Estestvennye nauki, Tekhnika, Meditsina, Sel'skoe khoziaistvo. Union list of foreign serials in the largest libraries of the USSR, 1750-1965. Science, Technology, Medicine, Agriculture. Gosudarstvennaia Biblioteka SSSR imeni V. I. Lenina, Biblioteka po Estestvennym Naukam Akademii Nauk SSSR, Vsesoiuznaia Gosudarstvennaia Biblioteka Inostrannoi Literatury, and Gosudarstvennaia Publichnaia Nauchno-Tekhnicheskaia Biblioteka SSSR. Moskva: "Kniga," 6 vols. 1974-1979. vol. 1. Russian alphabet. Latin alphabet A, 1974, 496 p. vol. 2 B-C, 1975, 452 p. vol. 3 D-I, 1976, 468 p. vol. 4 J-M, 1977, 480 p. vol. 5 N-R, 1978, 537 p. vol. 6 S-Z. Dopolneniia. Supplement; 1979, 599 p.
About 30,000 periodicals and other serials from outside the Soviet Union published up to 1965 held by 112 libraries in the USSR.

ULC International congresses and conferences: a union list of their publications available in libraries of the United States and Canada. Edited by Winifred Gregory. New York: H. W. Wilson, 1938. 229 p.

ULG List of serial publications of foreign governments 1815-1931. Edited by Winifred Gregory. New York: H. W. Wilson, 1932. 720 p. Reprinted. Millwood, N. Y.: Kraus Reprint, 1973. 720 p.
30,000 titles in 75 libraries.

G Gesamtverzeichnis deutschsprachiger Zeitschriften und Serien in Bibliotheken der Bundesrepublik Deutschland einschliesslich Berlin (West). Titel vor 1971 mit Besitznachweisen. Stand: April 1978. Union List of German Language Serials in Libraries of the Federal Republic of Germany including Berlin (West). Titles before 1971 with holdings locations as of April 1978. Bearbeitet und herausgegeben von der Staatsbibliothek Preussischer Kulturbesitz, Abteilung Gesamtkataloge und Dokunentation. München: Dokumentation Saur; New York, London, and Paris: K. G. Saur. 2nd ed., 1978. 2 vols. 2,512 p. (1st ed. 1976. 1,617 p.).
About 47,000 titles of German-language serials published before 1971, with holdings in libraries of the Federal Republic of Germany including West Berlin.

D — Schmidt, Rolf Diedrich, and Streumann, Charlotte. Verzeichnis der Geographischen Zeitschriften, Periodischen Veröffentlichungen und Schriftenreihe Deutschlands und der in den Letzteren Erschienen Arbeiten. Bearbeitet im Institut für Landeskunde. Direktor: Prof. Dr. E. Meynen. Bad Godesberg: Selbstverlag Bundesanstalt für Landeskunde und Raumforschung, 1964. 303 pp. Berichte zur Deutschen Landeskunde, Sonderheft 7.
393 entries. Detailed information on editors, publishers, authors, dates, and contents with indexes.
Not actually a union list but entries for German-language serials in it are recorded since they provide detailed and definitive information on both periodicals and monographic series.

S — Half a Century of Soviet Serials, 1917-1968. A Bibliography and Union List of Serials Published in the USSR. Compiled by Rudolph Smits. Washington, D. C.: Library of Congress, 1968. 2 vols. 1,661 p.
29,761 numbered entries with some information on holdings of libraries in the United States and Canada.

FC — Catalogue Collectif des Périodiques Conservés dans les Bibliothèques de Paris et dans les Bibliothèques Universitaires de France. Périodiques Slaves en Caractères Cyrilliques. Etat des Collections en 1950. Bibliothèque Nationale. Départment des Périodiques. Paris: Bibliothèque Nationale, 1956. 2 vols. 873 p. Tome I. A-H. Tome II. O-IA.
About 7,000 titles and cross references of serials held by 46 French libraries (20 in Paris, 26 in provinces).

FC-S — ______. ______. Supplément 1951-1960. 1963. 495 p.
About 4,000 additional titles and cross references.
______. ______. Addenda et Errata. État général des Collections en 1960. 1965. 222 p.
Mostly corrections or additional notes.

GR — Bruhn, Peter. Gesamtverzeichnis russischer und sowjetischer Periodika und Serienwerke in Bibliotheken der Bundesrepublik und West-Berlin. Union list of Russian and Soviet periodicals in libraries of the Federal Republic of Germany and West-Berlin. Herausgegeben von Werner Philipp. Berlin. Freie Universität. Osteuropa-Institut. Bibliographische Mitteilungen, Heft 3. Wiesbaden: Otto Harrassowitz, 1960-1973. 3 vols. 19 nos. 1,843 p.; no.20 (1976). Register.
About 10,000 titles, 1702-1956, with holdings in the libraries of the Federal Republic of Germany and West Berlin.

C — Chinese Periodicals in the Library of Congress. Compiled by Han Chu Huang, Chinese and Korean Section, Orientalia Division. Library of Congress. Washington, D. C.: Library of Congress, 1978. 521 p.
More than 6,400 titles of Chinese periodicals published 1868-1975 in the collection of the Library of Congress, said to have the largest holdings of Chinese serial publications outside China.

DP — Bibliography and review of geography department discussion papers, occasional papers, and monographs. Compiled by Bruce Young. Waterloo, Ontario: Wilfrid Laurier University. Department of Geography. 1- 1974- . Vol. 3, 1978. 113 p. Vol. 4 expected in April 1980.
Not a union list but bibliographically useful in listing 83 university series in geography and in providing titles of individual papers in each of the listed series.

Other Major Union Lists of Serials by Countries

The above union lists indicate locations of library holdings only in the United States and Canada, the United Kingdom, France, the Federal Republic of Germany, and the Soviet Union.

Individuals may utilize other sources for information on location of holdings of specified geographical serials: computerized data bases, records of individual libraries, and other national union lists, particularly for other countries. Among the more extensive of such other union lists, not utilized in the preparation of this list, are the following:

Argentina

Sociedad argentina de bibliotecarios de instituciones sociales, científicas, artísticas y técnicas. Catálogo colectivo de publicaciones periódicas existentes en bibliotecas científicas y técnicas argentinas. 2nd ed. Dirigido por Ernesto G. Gietz. Buenos Aires: Consejo nacional de investigaciones científicas y técnicas, 1962. 1,726 p.
25,129 titles of 142 libraries in Argentina.

Australia

National Library of Australia, Canberra. Serials in Australian libraries: social sciences and humanities: a union list. 2nd rev. ed. Canberra: National Library, 1968-1974. (1st ed., 1963-1967. 3v.).
34,000 titles in 200 Australian libraries.

Australia. Commonwealth Scientific and Industrial Research Organization. Union catalogue of the scientific and technical periodicals in the libraries of Australia, edited by Ernest R. Pitt. 2nd ed. Melbourne: 1951. 735 p. (1st ed. 1930).
About 25,000 titles with holdings for 248 libraries.

______. ______. Supplement. New titles, January 1946-December 1952, edited by Adelaide L. Kent. Melbourne, 1954. 111 p.
1,926 new titles with holdings for 193 libraries.

______. Scientific serials in Australian libraries, edited by Adelaide L. Kent. Supplementing and designed to supersede Union catalogue of scientific and technical periodicals in the libraries of Australia. Melbourne, 1958- . 1 vol. loose-leaf.
Holdings for more than 300 libraries.

Austria

Nationalbibliothek, Wien. Zentralkatalog neuerer ausländischer Zeitschriften und Serien in österreichischen Bibliotheken (ZAZ). Im Auftrag des Bundesministeriums für Unterricht herausgegeben von der Österreichischen Nationalbibliothek. Wien, 1962-63. 6 vols.
Foreign serials received by about 450 libraries in Austria. Subject index.

______. Neue ausländische Periodica in österreichischen Bibliotheken. Wien. 1- (1963-). Semiannual.

Belgium

Bibliothèque Royale de Belgique. Catalogue collectif belge et luxembourgeois des périodiques étrangers en cours de publication, rédigé sous la direction de A. Cockx. Belgische en Luxemburgse centrale catalogus van lopende buitenlandse tijdschriften, samengesteld onder de leiding van A. Cockx. Bruxelles: Culture et Civilisation, 1965. 2 v., 1982 p.
46,195 title entries for foreign periodicals held in about 450 libraries of Belgium or Luxembourg.

Bulgaria

Narodna Biblioteka, Sofiya. Chuzhdi periodichni izdaniia v po-golemite nauchni biblioteki v Bulgariia, 1900-1958; svoden katalog. Sustavili: Liliana Albanska, Snezhina Tosheva, Lora Daskalova-Ribarska. Pod red. na Khristo Trenkov. Sofiya. 1966. 348 p.
Foreign periodicals in the major scientific libraries in Bulgaria, 1900-1958.

Canada

National Library of Canada, Ottawa. Periodicals in the social sciences and humanities currently received by Canadian Libraries. Inventaire des périodiques de sciences sociales et d'humanités que possèdent les bibliothèques canadiennes. Ottawa: Queen's Printer. 1968. 2 vols.
About 12,000 titles held by 179 libraries.

National Research Council, Canada. Library.
Union list of scientific serials in Canadian libraries. 2nd ed. Ottawa, 1967. 2 vols. (NRC 9520). (1st ed., 1957).

Chile

Centro Nacional de Información y Documentación. Catálogo colectivo nacional de publicaciones periódicas. Santiago, Chile, 1968. 1 vol. various pagings.

People's Republic of China

China. Chiao yü pu. Kao teng chiao yü chiao hsüeh. 47 so kao teng hsüeh hsiao t'u shu kuan kuan ta'ang wai wen ch'i k'an lien ho mu lu. Peking: Kao teng chiao yü ch'u pan she 1958. 645 p.
Foreign language periodicals in 47 college and university libraries of China. 958 titles in cyrillic and 11,086 titles in roman alphabet. Holdings of 47 libraries.

Costa Rica

Costa Rica. Universidad, San Pedro. Servicios de Biblioteca, Documentación y Información. Cátalogo colectivo de publicaciones periódicas existentes en Costa Rica. Hilda Pacheco Gurdián, Coordinadora.
Ciudad Universitaria Rodrigo Vacio, San José: La Universidad, 1976. 479 p.

Czechoslovakia

Bečka, Josef, and Foch, Václav. Soupis cizozemských periodik v knihovnách Ceskoslovenské republiky. Praha: Ministerstvo školství a národní osvěty, 1929. 2 vols. 1,609 p.
Foreign serials in 650 libraries of Czechoslovakia. Titles in alphabetical order but with a subject listing also.

Praha. Knihovna vysokých škol technických.
Soupis cizozemských periodik technických a příbuzných v knihovnách Československé republiky: základní soupis z let 1928-1953, a doplatky k Soupisu Bečkovu-Fochovu. Redigoval Josef Lomský. Praha: Nakl. Ceskoslovenské akademie věd, 1955-1958. 3 vols.
List of foreign technical and related serials in the libraries of Czecho-slovakia. Basic list for the years 1928-1953 and supplement to the 1929 list of Bečka anf Foch. 11,835 serials held by 700 libraries.

Praha. Státni teckhnická knihovna.
Souborný seznam odborných zahraničních časopisů (z devizových oblastí) 1960. Zpracoval kolektiv za redakce A. Vejsové a Zd. Jirousové. Praha, 1960. 624 p.
Union list of specialized periodicals from foreign areas requiring foreign exchange.

Bratislava. Univerzita. Knižnica. Zahraničné periodiká v ČSSR. 1969- . Bratislava.

Denmark

Periodicakatalog. Fortegnelse over løbende udenlandske periodica ved danske videnskabelige og faglige biblioteker. Foreløbig udgave, pr. 1 januar 1953. Red. af Torkil Olsen. København: Dansk bibliografisk kontor, 1955. 281 p.

______. Supplement, omfattende de af Det Kongelige Bibliotek, Universitetsbiblioteket og Statsbiblioteket i Århus i tiden fra i. january 1953 til 31. december 1958 nyerhvervede, løbende tidsskrifter. Red. af T. W. Langer. København: Dansk bibliografisk kontor, 1959. 104 p.

East Africa

Periodicals in East African libraries: a union list. 8th ed. Morgantown: West Virginia University Library, 1974. 815 p.

Egypt

Egypt. al-Markas al-Qawmī lil-Buḥūth.
Union catalogue of scientific periodicals in Egypt up to end of 1949 by Fouad I National Research Council. Cairo: Government Press, 1951. 379 p. 3,094 titles and holdings in 77 libraries.

Finland

Yhteisluettelo Suomen tieteellisten kirjastojen ulkomaisesta kirjallisuudesta, B: aikakauslehdet ja sarjat. Samkatalog över utländsk litteratur i Finlands vetenskapliga bibliotek, B: tidskrifter och serier. Union catalogue of foreign literature in the research libraries of Finland, B: periodicals and serials. Helsinki: Helsingin Yliopiston Kirjasto.

France

Bibliothèque Nationale, Paris. Département des Périodiques. Inventaire des périodiques étrangers reçus en France par les Bibliothèques et les Organismes de Documentation. 4th ed. Paris: Bibliothèque Nationale, 1969. 1,207 p. (1st ed. 1955 [1956], 21,000 titles; 2nd ed. 1957-1958 [1959], 25,000 titles; 3rd ed. 1960-1961 [1962], 30,000 titles).
About 43,000 foreign current serials received in 1968 by 2,300 libraries. Provides title, subtitles, place of publication, and location of holdings but not dates of publication.

Germany

Gesamt-Zeitschriften-Verzeichnis. Berlin: Königliche Bibliothek, 1914. 355 p. 17,190 serial titles located in about 350 libraries.

Gesamtverzeichnis der ausländischen Zeitschriften (GAZ) 1914-1924. Berlin: Preussische Staatsbibliothek, 1929. 784 p.
14,573 serial titles 1914-1924 held by about 1,100 libraries.

Federal Republic of Germany

Gesamtverzeichnis ausländischer Zeitschriften und Serien 1939-1958 (GAZS). Bearbeitet und herausgegeben von der Staatsbibliothek der Stiftung Preussischer Kulturbesitz. Wiesbaden: Harrassowitz, 1959-1968. v.1-5, nos. 1-43, 5,152 p. (Nos. 1-10 by Westdeutsche Bibliothek). Nachträge. 1-30 (1966-1978).
Union list of about 45,000 foreign periodicals 1939-1958 in about 120 libraries of the Federal Republic of Germany.

German Democratic Republic

Deutsche Staatsbibliothek, Berlin. Auskunftsbüro der deutschen Bibliotheken. Gesamtverzeichnis ausländischer Zeitschriften 1939-1959 (GAZ). Berlin, 1961- . v1 A-C. 1964. v2 D-L. 1965. v3 M-Q. 1968. v4 R-S. 1975. v5 T-Z in manuscript.
About 25,000 titles held by 200 libraries of the German Democratic Republic.

Deutsche Staatsbibliothek, Berlin. Neueste ausländische Zeitschriften (NAZ). Berlin, 1969. 3 vols. 1,601 p. (Zeitschriften-Bestandsverzeichnisse, 13). Serials beginning 1960 or later, or with new titles, or first acquired after 1959.

Zentralkatolog: der DDR: Zeitschriften und Serien des Auslandes. Bearbeitet und herausgegeben vom Institut für Leihverkehr und Zentralkataloge bei der Deutschen Staatsbibliothek. Jahrgang 1971- . Berlin: Deutsche Staatsbibliothek. 1971. Issued in 5 vols.

Hungary

Országos Széchényi Könyvtár. Kurrens külföldi folyóiratok a magyar könyvtárakban; lelöhelyjegyzék. 1962. Budapest, 1962. 366 p.
Current foreign periodicals in Hungarian libraries. About 12,000 titles based on subscription orders by Hungarian libraries.

Indonesia

Indonesia. Biro Perpustakaan. Check list of serials in Indonesian libraries. Katalogus induk sementara madjalah pada perpustakaan di Indonesia. Djakarta, Biro Perpustakaan, Dep. P. D. dan K., 1962. 3 vols.
More than 5,000 foreign serials (vols. 1-2) and Indonesian serials (vol. 3) with holdings in more than 100 libraries.

Irish Republic

Union list of current periodicals and serials in Irish libraries. 5th ed. Compiled by L. Duignan. Dublin: Irish Association for Documentation and Information Services, 1974. 2 vols. (1st ed., 1929).
About 15,000 titles in 74 libraries.

Israel

Union list of serials in Israel libraries: natural, applied and social sciences. Prepared under the sponsorship of the Standing Committee of the National and University Libraries.
4th ed. Jerusalem: Jewish National and University Library, 1975.
2 vols. 1,747 p. Added title page: Reshimat kitve-'et be-sifriyot Yisrael (1st ed., 1955, 123 p. 2nd ed., 1964, 538 p.).
More than 30,000 serial titles in the biological, physical, and social sciences and technology.

Japan

Gakujutsu zasshi sōgō mokuroku.
Jimbun kagaku ōbun hen. 1958-nen han. Mombushō Daigaku Gakujutsu Kyoku hen. Tōkyō: Nippon Gakujutsu Shinkōkai, 1958. 462 p.
Union list of periodicals in humanities in western languages. 10,200 titles and holdings of 300 libraries.

______. ______. Sappurimento-han 1. 1962-nen han...1962. 232 p.
1,600 new titles.

______. Jimbun kagaku Wabun hen. 1959-nen han. Mombushō Daigaku Gakujutsu Kyoku kanshū...1959. 560 p.
Union list of periodicals in humanities in Japanese.
12,900 titles in 400 libraries.

______.Shizen kagaku ōbun hen. 1957-nen han. Mombushō Daigaku Gakujutsu Kyoku hen...1957. 858 p.
Union list of periodicals in natural sciences in western languages.
14,000 titles and holdings of 250 libraries.

______. Shizen kagaku Wabun hen. 1959-nen han. Mombushō Daigaku Gakujutsu Kyoku kanshū...1959. 635 p.
Union list of periodicals in natural sciences in Japanese.
13,000 titles and holdings of 400 libraries.

Mexico

Mexico. Consejo Nacional de Ciencia y Tecnología. Departamento de Sistemas de Información. Catalogo colectivo de publicaciones periódicas existentes en bibliotecas de la República Mexicana, compilada por el Departamento de sistemas de Información, CONACYT. 2nd ed. México: Consejo Nacional de Ciencia y Tecnología, 1976. 861 p. (Serie Directorios y catálogos, 2).

Catálogo colectivo de publicaciones periódicas existentes en Bibliotecas de la República Mexicana, por Pablo Velásquez Gallardo and Ramon Nadurille. México: Instituto Nacional de Investigaciones Agrícolas, 1968. 2 vols. Serial holdings of 130 libraries in Mexico.

The Netherlands

Koninklijke Bibliotheek, The Hague. Centrale catalogus van periodieken en seriewerken in Nederlandse bibliotheken (CCP). 2nd ed. 's Gravenhage: De Bibliotheek, 1978. 7 vols.
(1st ed., 1971-1973. 2 vols.).

New Zealand

Union list of serials in New Zealand libraries. 3rd ed. Wellington: National Library of New Zealand, 1969-1970. 6 vols.
(1st ed., 1953; 2nd ed., 1964-1968).
About 40,000 titles and holdings of nearly 2,000 libraries.

Norway

Oslo. Universitet. Bibliotek. Utenlandske Periodika i Norge: samfunnsvitenskap, historie, geografi, rettsvitenskap. Universitetsbiblioteket i Oslo. 2nd ed. Oslo: Biblioteket, 1976. 650 p. (Norsk Samkatalog). (1st ed. 1973).

Philippines

Quezon. University of the Philippines. Inter-Departmental Reference Service. Union List of Serials of Government Agency Libraries of the Philippines. Compiled by Maxima M. Ferrer and others. Revised and enlarged edition. Manila, 1960. 911 p. (1st ed., 1955).
Nearly 8,000 titles of foreign and domestic periodicals and holdings of 79 libraries. Includes classed list of serials.

Poland

Kaczorowska, Zofia. Spis zagranicznych czasopism i wydawnictw seryjnych z zakresu nauk o ziemi, znajdujacych się w bibliotekach polskich. Warszawa: Państwowe Wydawn. Naukowe, 1957. 376 p.
Foreign serials in the earth sciences with holdings for 134 libraries.

Biblioteka Narodowa, Warszawa. Zakład Katalogów Centralnych. Centralny katalog bieżących czasopism zagranicznych w bibliotekach polskich w roku 1957. Warszawa, 1961. 530 p.
Central catalog of current foreign periodicals in Polish libraries for the year 1957. 7,678 periodicals in 360 libraries.

______. ______. w roku 1959. 1962. 806 p.
14,317 periodicals in 635 libraries. Subject index.

______. ______. ______. Nowe tytuły za lata 1960-1961. 1963. 318 p.

Portugal

Centro de Documentação Cientifica, Lisboa. Publicações periódicas estrangeiras, inventariadas nas bibliotecas portuguesas. Lisboa: Instituto para a alta Cultura, 1948-1959. 6 vols.
Serial holdings of 267 Portuguese libraries. 21,690 serials are listed in the six volumes. Subject index. Geographical serials are listed in volume 5 (1959).

Romania

Academia Republicii Populare [later Socialiste] Romîne. Biblioteca. Repertoriul general, al periodicelor ştiinţifice şi tehnice străine aflate în principalele biblioteci din R.P.R. Bucureşti. 1957- . 6 vols.
I. Periodice ştiinţice generale
II. Matematica şi ştiinţele naturii
III. Medicină..1957. 358 p.
IV. Tehnică
V. Chimie. 1963
VI. Agricultura--Zootehnie--Medicină veterinară.
Holdings of more than 300 libraries. Geographical and subject indexes.

Catalogul colectiv al periodicelor străine intrate în principalele bibliotheci din R.P.R. in anul 1957- . Bucureşti, 1958- . Annual. (Issued jointly by Biblioteca Academiei R.P.R. and Biblioteca Centrală de Stat).

South Africa

Periodicals in South African libraries. Tydskrifte in Suid-Afrikaanse biblioteke. 2nd ed. Pretoria: South African Council for Scientific and Industrial Research and Human Sciences Research Council. 1972-1973. Issued in parts. (1st ed., 1961, loose leaf).
Supersedes Catalogue of Union periodicals, 1943-1953, 2 vols and 2 supplements. A 3rd ed., 1974, in microfiche.

South Asia

Ranganathan, Shiyali Remamrita, and others. Union catalogue of learned periodical publications in South Asia. Published with the assistance of UNESCO. Delhi: Indian Library Association; London: G. Blunt, 1953. (Indian Library Association. English series, 7).
Volume 1, Physical and biological sciences. Only one published. Periodicals held by 249 libraries in Indonesia, Malaya, Thailand, Burma, Sri Lanka, and India.

Spain

Spain. Dirección General de Archivos y Bibliothecas. Catálogo colectivo de publicaciones periódicas en bibliotecas españolas. Madrid: Ministerio de Educación y Ciencia, Dirección General de Archivos y Bibliotecas, 1971- . vl- .
Union list of serials in Spanish libraries as of 1969. In alphabetical order in each volume. Titles in classed arrangement at end of each volume.

Sri Lanka

Ceylon Association for the Advancement of Science.
List of the scientific periodicals in the libraries of Ceylon.
Compiled by M. U. S. Sultanbawa. 1953. 143 p.
Entries in classed order with holdings of 35 libraries.

Switzerland

Vereinigung Schweizerischer Bibliothekare. Verzeichnis ausländischer Zeitschriften und Serien in Schweizerischen Bibliotheken. Répertoire des Périodiques étrangers reçus par les bibliothèques suisses. Red: Schweizerische Landesbibliothek. 5th ed. Bern: Vereinigung Schweizerischer Bibliothekare, 1973. 768 p. (1st ed., 1904; 2nd ed., 1912, 3rd ed., 1925; 4th ed., 1955).

Sudan

Khartum. University. Library.
Sudanese union catalogue of periodicals. Khartoum, 1961. 116 p.
Holdings of 19 libraries.

Sweden

Accessionskatalog över utländsk litteratur i svenska forskningsbibliotek, AKP: nytillkomna tidskrifter och serier. Union catalogue of foreign literature in Swedish research libraries, AKP: newly acquired periodicals. 1954- (Stockholm: Kunggliga Biblioteket, annual).
Union list of accessions by more than 100 Swedish libraries of foreign periodicals and serials from 1954 on.

United Kingdom

World list of scientific periodicals published in the years 1900-1960. 4th ed. Edited by Peter Brown and George Burder Stratton. London: Buttersworths, 1963-1965. 3 vols. 1824 p.
59,961 numbered entries.

Uruguay

Montevideo. Universidad. Catálogo colectivo de publicaciones periódicas existentes en la bibliotecas universitarias del Uruguay. Montevideo: La Universidad, 1970. 2 vols. 606 p.

Yugoslavia

Katalog strane periodike u bibliotekama Jugoslavijie od 1919-1968. Urednik Slobodan Komadinić; urednik Ljiljana Milovanović. Beograd: Jugoslovenski bibliografski institut, 1972- .

Zimbabwe

Phillips, Sally. Periodicals in Rhodesian libraries. Salisbury: University College of Rhodesia, 1968. 384 l.

Union Lists of Serials from Specified Areas

Periodicals from the Soviet Union

Union list of Russian scientific and technical periodicals available in European libraries. Liste des périodiques scientifiques et techniques russe existant dans les bibliothèques européennes. Editors: L. J. van der Wolk and S. Zandstra. Amsterdam: Netherlands University Press, 1963- . Supplement 1965- .
vol. 1 (1963). 463 p. The Netherlands by W. C. Smit and S. Zandra. Holdings for 135 libraries in the Netherlands.

France. Direction des bibliothèques de France.
Liste des périodiques soviétiques reçus en France par les bibliothèques et les organismes de documentation en 1960. Paris: Imprimerie nationale, 1961. 83 p.
1,102 Russian-language periodicals received in 1960 by 275 libraries in France.

U. S. Library of Congress. Processing Department.
Russian periodicals in the Helsinki University Library: a checklist. Washington, D. C.: Library of Congress, 1959. 120 p.
Holdings of Russian periodicals in the Helsinki University Library and also for Harvard College Library, the New York Public Library, and the Library of Congress in the United States.

Periodicals from Africa

Travis, Carole, and Alman, Miriam. Periodicals from Africa: a Bibliography and Union List of Periodicals published in Africa. Boston: G. K. Hall, 1977. 619 p. (Standing Conference on Library Materials on Africa).
Holdings by 60 university, national, government, and private libraries in the United Kingdom of periodicals published in Africa. Bibliographic information also on other titles. Arranged alphabetically by country of publication, then by title. Title index.

Serials in Arabic

Union catalogue of Arabic serials and newspapers in British libraries. Paul Auchterlonie and Yasin Safadi, editors. London: Mansell, 1977. 146 p.
Holdings of 29 libraries reported in alphabetical order in romanized format, 1,011 entries.

Periodicals from Southeast Asia

Nunn, Godfrey R. Southeast Asian periodicals: an international union list. (London: Mansell, 1977). 456 p.
26,000 periodicals published since the beginning of the nineteenth century, arranged by country: Burma; Cambodia; Indonesia; Laos, Malaysia, Singapore,

and Brunei; Philippines; Thailand; Timor; Vietnam. Within countries arranged by title. Holdings shown in libraries in these countries and also in selected libraries in the United States, Canada, Great Britain, France, India, The Netherlands, Portugal, Spain, and Australia.

Periodicals from China

Hervouet, Y. Catalogue des périodiques chinois dans les bibliothèques d'Europe. Preparé avec la collaboration de J. Lust et R. Pelissier. Paris: Mouton, 1958. 102 p. (Le Monde d'Outre-Mer, passé et présent. 4. série: Bibliographies, 2).
Entries for 600 periodicals with holdings of 35 libraries (11 in the United Kingdom, 7 in Germany, 6 in France, 3 each in Belgium, Italy, and Sweden, and one each in Denmark and the Netherlands).

Chinese periodicals in British libraries. Introduction by E. D. Grinstead. London: British Museum, 1965. 102 p. Handlist no. 2.
Holdings of libraries in the United Kingdom.

Shih Bernadette P. N., and Snyder, Richard L. International union list of Communist Chinese serials: scientific, technical and medical, with selected social science titles. Cambridge, Massachusetts: Massachusetts Institute of Technology. Libraries. 1963. 148.
874 titles in 28 libraries (18 in the United States, 3 each in Canada, United Kingdom, and Japan, and one in Hong Kong). 499 titles in science, technology, and medicine; 102 titles in social sciences; and 273 unidentified titles without reports of holdings.

The Problem of Diverse Forms of Entries in Union Lists

The user of only a single union list quickly adapts to the form of notation and practices of that list. If one utilizes a variety of such lists, however, he must pay close attention to differences among them. A serial may be listed under the title exactly as given on the title page or under the issuing institution, if the title is nondistinctive, i.e. either as Annals of the Association of American Geographers, or as Association of American Geographers. Annals. A serial which has changed titles may be listed only under the original title or issuing agency, only under the current or latest title, or separately under each successive title or institution. Titles in forms of writing other than the Latin alphabet may be listed in the original form or may be transliterated or romanized, according to various systems.

An example of alternate forms of entry is provided by entry 2313 in this International List of Geographical Serials

AKADEMIIA nauk S.S.S.R. Moskva. 1725 as Peterburgskaia akademiia nauk; 1915-1925 Rossiiskaia akademiia nauk; 1925- Akademiia nauk S.S.S.R. Institut geografii.
Trudy. 1-81 (1931-1962). Closed 1962. (U-1:116b; B-1:25a; B68:568a; S620; FC-1:16, 18; GR-2:1245).
Supersedes the academy's Komissiia po izucheniiu estestvennykh proizvoditel'nykh sil S.S.S.R. Geograficheskii otdel [2316].
v1-12 (1931-1934) as Geomorfologicheskii institut. Trudy;
v14-24 (1935-1937) as Institut fizicheskoi geografii. Trudy.
1-11 (1931-1934) published in Leningrad.
Includes subseries: ...

Union List of Serials in Libraries of the United States and Canada, third edition, volume 1, page 116, column b, lists this entry in the same general form, but because of the date of publication cannot list closing date or number nor specify the subseries.

Akademiia nauk SSSR. Institut geografii
Trudy. 1,, 1931-

Finding this title in British Union-Catalogue of Periodicals: A Record of the Periodicals of the World from the Seventeenth Century to the Present Day, in British Libraries takes some detective work. BUCP lists institutions under original name but one may not know the founding name of academies now well known and frequently referred to by contemporary names. Starting with the current name one finds in volume one on page 72 the following cross-reference:

> Академия наук (СССР), И. [Akademiya nauk (SSSR), I.]= Academia scientiarum i petropolitana.
>
> This heading occurs in the same volume on page 17
>
> Academia scientiarum i petropolitana (Academia scientiarum Unionis Rerum publicarum Sovieticarum Socialisticarum: И. Академия наук; Российская академия наук; Академия наук СССР [I. Akademiya nauk; Rossiiskaya akademiya nauk; Akademiya nauk SSSR]...

One then proceeds down the long list of institutes under the Academy of Sciences to page 22, where another cross reference is found:

> Институт географии [Institut geografii]=Геоморфологический институт [Geomorfologichesky institut].
>
> Following this lead one finds another cross reference:
>
> Геоморфологический институт.[Geomorfologichesky institut].Труды [Trudui]= Труды Географического отдела [Trudui Geograficheskogo otdela]. Комиссия по изучению естественных etc. [Komissiya po izucheniyu estestvennykh etc.].
>
> The heading for this commission is found on page 24:
>
> Комиссия по изучению естественных производительних сил России [Komissiya po izucheniyu estestvennuikh sil Rossii].

Under this heading one proceeds down to the title of the series on page 25 under the Geographical section of this commission:

> _____. Географический отдел [Geografichesky otdel]. Труды [Trudui]. 1-2. 1928-30. then b Труды [Trudui]. (Геоморфологический институт [Geomorfologichesky institut]). 1-13. 1931-33. then Труды [Trudui]. (Институт физической географии [Institut fizicheskoi geografii]; Институт географии [Institut geografii].) 14- . 1935- .

Leaving out the material in brackets with transliterations but filling in for dashes of headings on previous pages and omitting complex name changes the basic entry is thus:

Academia scientiarum i petropolitana. Комиссия по изучению естественных производительных сил России. Географический отдел. Труды.

Thus this serial is listed under a former Latin name of the Academy of Sciences of the USSR, not in use during the period when this serial was published, and under the heading of a geographical section of a commission, which was the publisher not of this serial but of a predecessor.

This same serial is also listed in British Union-Catalogue of Periodicals incorporating World List of Scientific Periodicals. New Periodical Titles 1960-1968 which changed radically the form of entry to the title appearing on the cover and which provided information on the closing date on page 568, column a, with the following entry:

TRUDY INSTITUTA GEOGRAFII, AKADEMIJA NAUK SSSR.
AKADEMIJA NAUK SSSR: INSTITUT GEOGRAFII. MOSCOW (CEASED WITH) 81, 1962.//

In Half a Century of Soviet Serials, 1917-1968. A Bibliography and Union List of Serials Published in the USSR, entry 620 on page 28 reads:

Akademiia nauk SSSR. Institut geografii.
TRUDY
Moscow. 1931- Vols. 1-11 (1931-34) issued by the Academy's Geomorfologicheskii institut· vols. 14-24 (1935-37) by its Institut fizicheskoi geografii.

In the Catalogue Collectif des Périodiques Conservés dans les Bibliothèques de Paris et dans les Bibliothèques Universitaires de France. Périodiques Slaves en Caractères Cyrilliques. État des Collections en 1950, on pages 16 and 18 appear the following entries:

Академия наук СССР. Труды геоморфологического института.
1931-1934 (1-11).

devenu Академия наук СССР. Труды института физической географии. 1935-1937.

devenu Академия наук СССР. Труды института географии.

In Peter Bruhn, Gesamtverzeichnis russischer und sowjetischer Periodika und Serienwerke in Bibliotheken der Bundesrepublik und West-Berlin, separate entries appear for each title change, or rather change in the name of the issuing institution. The entries appear on pages 1210, 1242, and 1245 of volume 2:

TRUDY GEOMORFOLOGIČESKOGO INSTITUTA. Akademija nauk SSSR. Leningrad.
1. 1931 - 11. 1934.

TRUDY INSTITUTA FIZIČESKOJ GEOGRAFII. Akademija nauk SSSR. Moskva.
Leningrad. 14. 1935 - 24. 1937.

TRUDY INSTITUTA GEOGRAFII. -- (TRANSACTIONS OF THE INSTITUTE OF GEOGRAPHY.)
Akademija nauk SSSR. --(Academy of Sciences of the USSR.) Moskva.
25. 1937 - 67. 1956 ff.

Subseries are also listed.

CHAUNCY D. HARRIS

EXPLANATION OF ENTRIES

The entries are numbered to facilitate reference and cross-references. The entry numbers in the present edition are new and do not correspond to numbers in the previous editions.

General Organization

This list is arranged alphabetically by countries according to the English version of the individual country name. The national form of the country names, as listed in The Statesman's Year-Book or the Encyclopaedia Britannica has been added, but is not used in alphabetization. Since new geographical serials are appearing constantly, some blank space has been left following each country's entries so that new titles can be added by the user.

Within each country entries are listed in alphabetical order of titles or of issuing agencies, transliterated into the Latin alphabet or romanized, if necessary. The system of transliteration or romanization follows that typically used in American libraries as recorded in the Union List of Serials or in New Serial Titles.

For titles in Chinese or Japanese characters or non-Latin alphabets, such as Russian, Serbian, Bulgarian, or Arabic, an introductory section after the name of the country records the original form of the titles, but the full bibliographical data are given only in the following main body of transliterated entries, which also determines the order and the numbering. Unfortunately, in some cases the publications themselves were not accessible for examination. Some original titles therefore have been inferred from transliterated forms and the compilers fear that a number of errors may have crept into the rendition of these titles. Particularly for Chinese and Japanese publications, the original characters provide the only positive identification.

Form of Title

The form of titles is generally that used in the Union List of Serials or New Serial Titles.

(a) Publications of a society are listed under the society name unless the title clearly is independent of the society's name.

AMERICAN geographical society of New York. New York, New York.
Bulletin. (not BULLETIN of the American geographical society).

GEOGRAFICHESKOE obshchestvo S.S.S.R. Leningrad.
Izvestiia. (not IZVESTIIA Vsesoiuznogo geograficheskogo obshchestva, as it appears on the title page).

ÖSTERREICHISCHE geographische Gesellschaft. Wien.
Mitteilungen. (not MITTEILUNGEN der Österreichischen geographischen Gesellschaft).

ASSOCIATION de géographes français. Paris.
Bulletin.

GEOGRAPHICAL review. (American geographical society of New York).

ERDE. (Gesellschaft für Erdkunde zu Berlin. Zeitschrift). Berlin.

GEOGRAPHICAL journal. (Royal geographical society, London). London.

ANNALES de géographie. Bulletin de la Société de géographie. Paris.

(b) Titles are given in the original language or, if necessary in Latin transliteration or romanization, but not in translation.

GEOGRAFICHESKOE obshchestvo, S.S.S.R. Leningrad.
Izvestiia. (not BULLETIN of the Geographical society of the USSR).

CHIRIGAKU hyoron (not GEOGRAPHICAL review of Japan).

PRZEGLĄD geograficzny (not POLISH geographical review).

(c) University publications are listed under the city, state, province, or country in which the university is located, if the university name is identical with the place name.

CHICAGO. University. Department of geography. Chicago, Illinois.
Research papers.

MOSKVA. Universitet.
Vestnik. Seriia 5. Geografiia.

BERLIN. Freie Universität. Geographisches Institut. Anthropogeographie.
Abhandlungen.

HULL. University. Department of geography. Hull.
Occasional papers in geography.

An attempt has been made to render all city and provinical names according to local usage as recorded in Leon E. Seltzer (ed.), The Columbia Lippincott Gazetteer of the World (New York: Columbia University Press, 1952) or in various, more recent, sources: Wien not Vienna, Moskva not Moscow.

Order of Data in Entries

Data on each serial are given in the following order.

(1) Number in this list; if the serial is thought to be actively current as of January 1, 1980, its entry number is enclosed by this symbol: (____

(2) Title (if distinctive; if title is not distinctive it is given under 5, below).

(3) Issuing agency (in parentheses if title is given above).

(4) Placc of publication.

(5) Title (if not distinctive and hence listed under the issuing agency in 3, above). Title begins a new line.

(6) Volumes or numbers published.

(7) Dates of publication (in parentheses if volume data are given).

(8) Frequency of publication (for current serials only).

(9) References to the entry's location is selected union catalogues. (see abbreviations on pages 28-30).

(10) Miscellaneous bibliographical information, such as names of ancestors or descendants or changes in name. These data begin a new line.

(11) Cumulative indexes, if any. These data begin a new line.

(12) Language or languages used in articles or in abstracts or supplementary tables of contents (current serials), if not apparent from title or from language of the country. These data begin a new line.

(13) Address of publication (current serials only), if known and useful. The address begins a new line and is enclosed by brackets.

Abbreviations and Signs
(in order of usual occurrence in an entry)

66 reference number of a serial in this list.

(714 reference number of a serial confirmed as current.

*(14 This title is listed in Chauncy D. Harris, Annotated World List of Selected Current Geographical Serials, Fourth edition 1980. Chicago: University of Chicago, Department of Geography, Research Paper No. 194, 1980.

no number

ns new series

nsv new series, volume number

s series

v volume

pt part

__. and continuation, from that volume number or date to the present and still in progress.

- from and including the preceding number to and including the number which follows.

Ja, F, Mr, Ap, My, Je, Jl, Ag, S, O, N, and D. The months January. . . December.

U, N70, N75, B, BS, B68, B73, F, GZS, R, ULC, ULG, G, D, S, FC, GR, and DP. These abbreviations, in parentheses, indicate union catalogues where information is given on library holdings of the serial. These symbols are identified on pages 28-30.

BL. British Library. Lending Division. Current serials received. Used rarely for current serials not in any union list, as, for example, from Vietnam.

[ ] Employed primarily to set off addresses of current serials from other bibliographic data.

Sample Entries

*(1201 ERDE. (Gesellschaft für Erdkunde zu Berlin. Zeitschrift). Berlin.
1-8 (1949-1956); 88- (1957-) Quarterly. (U-2:1469b; B-4:598b; D:30; G).
Supersedes Gesellschaft für Erdkunde zu Berlin. Zeitschrift [1011] and continues its volume numbering from v88 (1957).
English abstracts at beginning of each article from 1957 no1.
[Arno-Holz Strasse 14, D-1000 Berlin 41, BRD.]

The reference number of this entry, 1201, is semi-enclosed to indicate that the serial is current. Since the serial title is clearly distinct from the issuing agency, the title assumes first position followed by the name of the issuing agency in parentheses and--in this instance--the relation of the serial to the issuing agency. The serial has been published continuously since 1949 (although a change occurred in the numbering system of the title in 1957) and at present is being issued four times a year. Bibliographical details and/or information about library holdings may be found in the Union list of serials, volume 2, page 1469, second column; in the British union-catalogue of periodicals, volume IV, page 598, second column; and as item 30 in Verzeichnis der geographischen Zeitschriften, periodischen Veröffentlichungen und Schriftenreihe Deutschlands und der in den letzteren erschienen Arbeiten, and in Gesamtverzeichnis deutschsprachigen Zeitschriften und Serien in Bibliotheken der Bundesrepublik Deutschland einschliesslich Berlin (West). This serial supersedes and now continues the numbering of a previous publication of the issuing agency; its predecessor's reference number is 1011. English is used as a supplementary language. The editorial and publication address of the Society is Arno-Holz-Strasse 14 in West Berlin, Germany. The serial is listed and annotated in Annotated World List of Selected Current Geographical Serials.

(1847 UTRECHT. Rijksuniversiteit. Geografisch instituut. Utrecht.
Publicaties. sA. Sociale geografie. Human geography. v1- (1955-)
Irregular monographs and reprints. (N70-4:5364a; BS:18a).
Title varies.
In Dutch or English with some abstracts in English or French.
[Heidelberglaan 2, 3584 CS Utrecht, The Netherlands.]

The title of this current serial is subordinate to the issuing agency. The agency designation, therefore, assumes first position in the entry, which is listed by city, university, and geographical institute. This serial constitutes series A of the Geografisch instituut's Publicaties. Since Dutch and English are used interchangeably in the publication (as indicated by the last line of the entry), both Dutch and English titles are given. Issued continuously since 1955, the series is composed of irregularly appearing monographs and reprints. Bibliographic data and information on some American and Canadian library holdings are given in the 1950-1970 cumulation, volume 4, page 5364, first column, of New serial titles.

Information on holdings of British libraries is given in British Union-Catalogue of Periodicals, Supplement to 1960, page 18, first column. The current address of the Institute is Heidelberglaan 2, 3584 CS Utrecht, The Netherlands.

183 LIÈGE. Université. Société d'histoire et de géographie. Liège.
Bulletin. 1885-1887, 1890.

This entry is a title unlisted in any of the union lists of serials, cited in this List. The entry is under the issuing agency listed by city, university, society, and then the title of the publication. Information in the hands of the compilers indicates that this serial was published in 1885-1887, and 1890. Since the serial must be presumed closed, no further data on supplementary languages or address are given.

*(680 ANNALES de géographie. Bulletin de la Société de géographie. Paris.
1- (1891-) 6 nos a year. (U-1:374c; B-1:150b; F-1:203b-204a).
v1-24 (1891/92-1913/14) included v1-24 of Bibliographie géographique internationale [696], then called in turn Bibliographie de l'année... Bibliographie annuelle, and Bibliographie géographique annuelle.
Indexes: 1891-1901, 1902-1911, 1912-1921, 1922-1931, 1932-1951, 1952-1961, 1962-1971.
English abstracts of articles from v76 no413 (Ja/F 1967).
[Librairie Armand Colin,103 Boulevard St.-Michel, 75240 Paris, Cedex 05, France.]

The reference number of this entry, 680, is semi-enclosed indicating that the periodical is current. The serial, although independently published is also the Bulletin of the Société de Géographie, Paris. It has been published continuously and regularly since 1891 and is currently published six times a year. Bibliographical details and library holdings can be found in Union list of serials, British union-catalogue of periodicals, and in Catalogue collectif des périodiques du début du XVIIe Siècle à 1939 conservés dans les bibliothèques de Paris et dans les bibliothèques des départements, at volume and page numbers indicated. From 1891/92 to 1913/14 Annales de géographie included volumes 1-24 of the predecessors of the Bibliographie géographique internationale (listed as entry 696), which during these years had three slightly different names. Seven cumulative indexes by decades have been published for the years from 1891 to 1971. Although the periodical is in French it has included abstracts in English from the beginning of 1967. The current address of publication, given in brackets, is Librairie Armand Colin, 103 Boulevard St.-Michel, 75240 Paris, Cedex 05, France.

*(1253 GEOLIT. Rezensionen: Geographie, Raumwissenschaften. (Westermann). Braunschweig, 1977- 4 a year. (N79; G; GZS). 2 nos in 1977. [Westermann Verlag, Postfach 3320, D-3300 Braunschweig, BRD.]

This new current geographical serial is listed in one of the monthly issues of New serial titles in 1979, in Gesamtverzeichnis deutschsprachiger Zeitschriften und Serien..., and in Gesamtverzeichnis der Zeitschriften und Serien..Neue und geänderte Titel seit 1971...Although it currently appears four times a year, only two numbers were published in the first year, 1977. It is published in Braunschweig by Westermann Verlag.

*(311 BŬLGARSKO geografsko druzhestvo, Sofiya. Izvestiia. 1-10 (1933-1942); nsv 1- (11-) (1953-) Irregular. (U-1:830b; B-1:445a; BS:157b; FC-1:301; R-1:402). In Bulgarian. 1-10 (1933-1942) with French or German abstracts and titles in tables of contents; nsvl- (1953-) with English, French, or German abstracts and titles in tables of contents.

This serial is current and is listed under the society, Bŭlgarsko geografsko druzhestvo, since the title, Izvestiia, is not distinctive. Volumes 1 to 10 were published during the years 1933-1942. A new series with number of the previous series in parentheses began with volume 1 (11) in 1953 and appears at irregular intervals. It is listed in Union list of serials, British union-catalog of periodicals and its Supplement to 1960, in Catalogue collectif des périodiques... Périodiques slaves en caractère cyrilliques, and in Retrospektivnyi svodnyi ukazatel' inostrannykh periodicheskikh i prodolzhaiushchikhsia Izdanii... in volumes and at pages indicated. It is in Bulgarian but abstracts and titles in tables of contents are provided in supplementary international languages, French or German, 1933-1942, and English, French, or German since 1953.

*(2657 MOSKVA (Moscow). Universitet. Moskva. Vestnik. Seriia 5. Geografiia. v15- (1960-) 6 nos a year. (N70-3:3911c; BS:573b; B68:580a; S12395; FC-S:81; GR-3:1517). Supersedes in part university's Vestnik. Seriia biologii, pochvovedeniia, geologii, geografii [2656] and continues its numbering. Supplementary table of contents in English from v15 (1960) and English abstracts from v23 (1968).

This current serial is listed under issuing institution, Moscow University, since the title, Vestnik, is not distinctive. The Vestnik of the University is published in many series; the separate series for geography, series 5, began in 1960 and continued the numbering of a previous series. It appears six times a year and is listed in New serial titles...1950-1970, British union-catalogue of periodicals...Supplement to 1960, and New periodical titles 1960-1968, Half a century of Soviet serials, 1917-1968, Catalogue collectif des périodiques...Périodiques slaves en caractères cyrilliques, and Gesamtverzeichnis russischer und sowjetischer periodika und Serienwerke... A supplementary table of contents in English has been provided since the beginning of the geography series in 1960. Supplementary abstracts in English were added in 1968.

*(1685 CHIRIGAKU hyoron. Geographical review of Japan. (Nippon chiri gakkai. Association of Japanese geographers). Tokyo. 1- (1925-) 12 a year. (U-2:1027a). Indexes: 1-10 (1925-1934); 11-20 (1935-1944); 21-30 (1947-1957); 31-40 (1958-1967); 41-50 (1968-1977). In Japanese with English table of contents from v1 (1925); occasional English abstract from v8 (1932) and regular abstracts from v23 (1950). [Nippon Chiri-Gakkai, Building of Japan Academic Societies Center, 2-4-16, Yayoi, Bunkyo-ku, Tokyo 113, Japan.]

This current serial is listed under its distinctive title, Chirigaku hyoron, but the supplementary English title, Geographical review of Japan, also helps to identify it. It is published by the Nippon Chiri Gakkai, the Association of Japanese Geographers, and has been published since 1925.

It is now published monthly. It is listed in Union list of serials. Five cumulative indexes have been published, each covering a decade. Although basically in Japanese, supplementary English has been used in the table of contents from its beginning in 1925, and in abstracts irregularly from 1932 and regularly from 1950. The current publishing address is in the Building of Japan Academic Societies Center in Tokyo.

INTERNATIONAL

1 ACTA americana. (Inter-American society of anthropology and geography. Review). Washington, D.C. 1-6 (1943-1948). Closed 1948. (U-1:50a). In English, Spanish, French, and Portuguese.

(2 BOLETÍN aéreo. (Instituto panamericano de geografía e historia; Pan American institute of geography and history). México, D. F. 1- (1955-). (N70-1:807b). 6 nos. a year.
In Spanish.
[Instituto Panamericano de Geografía e Historia, Ex-Arzobispado 29, México 18, D.F. Mexico.]

3 CONGRÈS des géographes et des ethnographes slaves.
Comptes rendus. (B-4:93a).
1 Praha (1924); 2 Kraków (1927); 3 Beograd (1930); 4 Sofiya (1936).
In Belorussian, Croatian, Czech, English, French, German, Polish, Russian, Serbian, Slovak, Slovenian, and Ukrainian.

4 CONGRÈS international de géographie commerciale.
Comptes rendus. 1-2 (1878-1879). (ULC:26c; B-1:633a; F-2:221a).
1 Paris (1878) 1v; 2 Bruxelles (1879) 1v.

5 CONGRÈS international de géographie économique et commerciale. Paris.
Procès-verbaux. 1 (1900).

6 CONGRÈS international de géographie historique.
Comptes rendus. 1 (1930). (ULC:27a; B-1:633a).
1 Bruxelles (1930) 3 v. (1931-1935) v1 as Compte rendu; v2 as Mémoires.

7 CONGRESO geográfico hispano-portugués-americano.
Actas. 1 Madrid (1893). (ULC:49b).

8 CONGRESO internacional de historia y geografía de América.
Annaes, announcements, comunicación, etc. 1922, 1926, 1936. (ULC:50a).
1 Rio de Janeiro (1922); 2 Asunción (1926); 3 Buenos Aires (1936).

(9 INSTITUTO panamericano de geografía e historia. (Pan-American institute of geography and history).
Proceedings, actas, etc. 1- (1932-) Irregular.
[IPGH, Ex-Arzobispado 29, México 18, D.F. Mexico.]

(10 ________.
Publicación. 1- (1930-) Irregular. (U-4:3250a: B-2:508a).
no 370 (1978) lists publications 1-370 and gives a history.
Text in English, Spanish, Portuguese, or French.

11 ________. Centro de entrenamiento para la evaluación de recursos naturales. México.
Los estudios sobre recursos naturales en las Américas. 1-9 (1953-1956).
In Spanish.

12 ________. Comisión de geografía (Pan-American institute of geography and history. Commission on geography). Rio de Janeiro.
Publicação. Publicación. Publication. Irregular. Monographs.
(N70-3:4520a).

*(13 INTERNATIONAL geographical congress. (Congrés international de géographie). Place of publication varies.
Comptes rendus...proceedings, atti, Verhandlungen, or Abstracts of papers. 1- (1872-). (ULC:134c; N70-2:2915b; B-1:636b-637a; F2:213a).
1 Anvers (1871) 2v; 2 Paris (1875) 2v; 3 Venezia (1881) 2v; 4 Paris (1889) 2v; 5 Bern (1891) 2v; 6 London (1895) 1v; 7 Berlin (1899) 2v; 8 Washington (1904) 1v; 9 Genève (1908) 3v; 10 Roma (1913) 1v; 11 Cairo (1925) 5v in 3; 12 Cambridge (1928) 1v; 13 Paris (1931) 3v; 14 Warszawa (1934) 3v; 15 Amsterdam (1938) 14v; 16 Lisboa (1949) 4v; 17 Washington (1952) 1v; 18 Rio de Janeiro (1956) 4v; 19 Stockholm (1960) 1v; 20 London (1964) 1v; 21 New Delhi (1968) 4v; 22 Montreal (1972) 2v; (papers), 1v (proceedings); 6v (Studies in Canadian geography); 23 Moscow (1976) 12v.
For numerous miscellaneous guides, bibliographies, circulars, announcements, etc., see ULC:134c-136b. Index: 1-23 (1872-1976).
Papers in English or French, and in some congresses also German, Italian, Portuguese, Spanish, or Dutch depending on country in which held.

*(14 INTERNATIONAL geographical union. (Union géographique internationale). Place of publication varies, New York, Zürich, Chicago, London.
IGU bulletin. Bulletin de l'UGI. v1 no1- (Ja 1950-). (N70-2:2915c; B-2:435b; B-1:435b).
v1-7 (1950-1956) in English and French duplicate texts: 8- (1957-) in either English or French. v1-19 (1950-1968) as IGU newsletter. Bulletin de nouvelles de l'UGI.
[Distributed in each country by the national committee for the IGU of that country.]

15 ______. Reports of commissions.
Most reports are single publications rather than numbered serials. N70-2:2915c-2916a lists some of them. Publications of commissions are listed in Union géographique internationale. International geographical union. La géographie à travers un siècle de congres internationaux. Geography through a century of international congresses. 1972. "Publication des commissions," p. 239-247, and IGU Bulletin, v. 27, 1976, no. 1 "Publications of IGU commissions and working groups 1969-1975," p. 125-135.
These reports were published by each commission independently and the commissions had many different locations and were of limited durations. These publications properly fall into the category of fugitive materials.

*(16 REVISTA cartográfica. (Instituto panamericano de geografía e historia. Comisión de cartografía). México. 1- (1952-). 2 nos a year.
1-23 (1952-1973) published in Buenos Aires; 24- (1973-) in México. (N70-3:5001c; BS:732a).
In English, Portuguese, or Spanish.
[IPGH, Ex-Arzobispado 29, México 18, D.F., Mexico.]

*(17 REVISTA geográfica. (Instituto panamericano de geografía e historia. Comisión de geografía). México. 1- (1941-). (U-4:3613a: B-3:699a; BS:733b).
v1-4 (no1-12) (1941-1944) published in México; v5-8 (no13/24); no 25-75 (1949-1973) published in Rio de Janeiro; no 76- (1973-) published in México.
In English, Spanish, Portuguese, or French with summaries in a second language. Indexes: 1-63 (1941-1965), 64-77 (1966-1972).
[IPG, Ex-Arzobispado 29, México 18, D.F., Mexico.]

*(18 WORLD cartography. (United Nations. Department of social affairs). New York. 1- (1951-) Irregular. (N70-4:6380c; B-4:567a; BS:968b).
In English.
Issued also in French: Cartographie mondiale. (Nations unies. Département des questions sociales). (N70-1:1151c).

AFGHÁNISTÁN (Pushtu or Dari [Persian])

Dawlat-i Jumhūrī-i Afghānistān

دولت جمهوری افغانستان

20 جغرافیا (دافغانستان جغرافیې موسسې خپرونه)

19 BULLETIN for teachers of geography. (Kabul. University. Institute of geography). Kabul. 1- (1962-). (B68:115a).
In Persian.

20 GEOGRAPHICAL review of Afghanistan. (Kabul. University. Institute of geography). Kabul. v1-12 no4 (1962-Mr 1973). (N70-2:2281c; B58:225b).
In Persian or Pushtu with supplementary English title, table of contents, and some articles.
[Institute of Geography, Faculty of Letters and Humanities, University of Kabul, Kabul, Afghanistan.]

ALGERIA (Arabic or French)

al-Jumhūrīyah al-Jazā'irīyah ad-Dīmuqrātīyah ash-Sha'bīyah
République Algérienne Démocratique et Populaire

الجمهورية الجزائرية الديموقراطية الشعبية

21 AFRICA. (Bulletin de la société de géographie d'Alger). Alger. v1 no1-4. (1880). Closed 1880. (U-1:80b; B-1:56a; F-1:90a).
Superseded by Société de géographie d'Alger et de l'Afrique du Nord. Bulletin [26].

*(22 ANNALES algériennes de géographie. (Alger. Université. Institut de géographie). Alger. 1- (1966). (N70 1:355b).
In French with abstracts in Arabic and English.
[3 rue du Professeur Vincent, Alger, Algeria.]

(23 CAHIERS de l'aménagement de l'espace. (Centre national d'études et de recherche pour l'aménagement du territoire, Alger). Hydro-Alger. 1- (1978-). Quarterly.
[Office des Publications universitaires, 29, rue Abou Nouas, Alger, Algeria.]

(24 CAHIERS de la recherche. (Constantine. Centre universitaire de recherche d'études et de réalisations). Constantine. 1- (1978-). Quarterly.
[Centre Universitaire de Recherche d'Études et de Réalisations, Route de Ain-El-Bey, Constantine, Algeria.]

25 SOCIÉTÉ archéologique, historique, et géographique du département de Constantine. Constantine?
Bulletin mensuel. 1-8 (no1-66) (1926-1933). Closed 1933. (U-5:3931c; F-1:809b).

26 SOCIÉTÉ de géographie d'Alger et de l'Afrique du Nord. Alger.
Bulletin. 1-50 (no1-174) (1896-1945). Closed 1945. (U-5:3945a; B-4:135a; F-1:834b-835a). Name varies.
1896-1899 as Société de géographie d'Alger. Bulletin; 1900-1940 as Société de géographie d'Alger et de l'Afrique du Nord. Bulletin. no165-174 (1941-1945) as Bulletin provisoire. Supersedes society's Africa [21].
Indexes: 1-10 (1896-1905) in 11; 11-27 (1906-1922) in 27; 28-37 (1923-1932) in 38.

27 SOCIÉTÉ de géographie de Constantine. Constantine.
Bulletin. 1-3 (1883-1885). Closed 1885. (U-5:3945b; B-4:135b).

(28 SOCIÉTÉ de géographie et d'archéologie [de la province d'Oran]. Oran.
Bulletin. 1- (1878-). (U-5:3945c; B-1:455a; F-1:835b; F-1:1021a).
1-5 (1878-1885) as Bulletin; 6-80 (1886-1957/1960) as Bulletin trimestriel de géographie et d'archéologie; 1967, no1- as Bulletin.
Not published 1885, 1965.
Indexes: 1-17 (1878-1897); 18-27 (1898-1907); 28-48 (1908-1927), 1928-1956.
In French or Arabic.
[7, Boulevard de Tripoli, Oran, Algeria.]

29 SOCIÉTÉ historique et géographique de la région de Sétif. Toulouse, Sétif.
Bulletin. 1 (1935). Closed 1935. (F-1:864a).

ANGOLA (Portuguese)

30 SOCIEDADE propagadora de conhecimentos geográphico-africanos. Luanda.
Boletim. v1 no1-2 (1881). Closed 1881. (B-4:109b).
In Portuguese.

ARGENTINA (Spanish)

República Argentina

31 ACADEMIA argentina de geografía, Buenos Aires. Buenos Aires.
Anales. 1-6 (1957-1962) Annual. (N70-1:35b; BS:5a).

32 ANUARIO geográfico argentino. (Comité nacional de geografia). Buenos Aires. 1941, 1942 supplement only. Closed. (U-1:414a).

33 ARGENTINA. Instituto geográfico militar. Buenos Aires.
Anuario. 1-8 (1912-1932); 9-16 (1943-1958/1962); 17 (1963-1975). ULG:7c; B-2:506a; F-1:344a).

34 ATLAS. (Argentina. Instituto geográfico militar. Órgano oficial).
Buenos Aires. 1-3 (1954-1956). Closed 1956. (N70-1:535b; BS:93b).

35 BOLETÍN de estudios geográficos. (Mendoza. Universidad nacional de Cuyo.
Instituto de geografía). Mendoza. v1-19, no1-74/77 (1948-1972). Quarterly.
(U-1:728b; N70-3:3726a; BS:131b).
Suspended 1952-1955.

*(36 BUENOS AIRES. Universidad nacional. Instituto de geografía "Romualdo
Ardissone."
Cuadernos de geografía. 1- (1976-).

37 BUENOS AIRES. Universidad nacional. Instituto de geografía "Romualdo
Ardissone." (1964-1967 as Centro de estudios geográficos). Buenos Aires.
Publicaciones.
sA. Memorias originales y documentos. 1-28 (1917-1969). (U-1:822a;
N68-2-2414c).
1-13 (1917-1930) as Instituto de investigaciones geográficas.
Publicaciones.
sB. Documentos cartográficos, planimétricos e iconográficos.
v2-4 (1949-1952). (U-1:822a).
1 never published.
sC. Métodos de la geografía. 1-4 (1955-1960). (N70-1:921c).
[25 de Mayo 217, Buenos Aires, Argentina.]

38 COLECCIÓN Nadir. (Sociedad geográfica americana editorial y cultural).
Buenos Aires. (U-2:1089b; N70-1:1352b). Closed.
sA. América. 1- (1946-).
sB. Miscelánea. 1- (1946-).
sD. no1-8 (1946).

39 CORRIENTES. (Instituto histórico y geográfico. Publicación).
(N70-1:1551b).

*(40 GAEA. (Sociedad argentina de estudios geográficos "Gaea." Anales).
Buenos Aires. 1- (1922-) Irregular. (U-2:1661b; B2:248b;
F-2:677b; F-2:757a).
[Santa Fe 1145, Buenos Aires, Argentina.]

*(41 ______. Serie especial. 1- (1973-). (N77).

42 GEOGRAFÍA de la República Argentina. (Sociedad argentina de estudios
geográficos "Gaea"). Buenos Aires. 1- (1947-). (U-5:3902c).

(43 GEOGRÁFICA: revista del Instituto de geografía (Resistencia. Universidad
nacional de nordeste. Instituto de geografía). Resistencia.
1- (1972-). (N75-1:904c).
[Instituto de Geografía, Las Heras 727, Resistencia, Chaco, Argentina.]

44 INSTITUTO geográfico argentino, Buenos Aires. Buenos Aires.
Boletín. v1-25 no8 (1879/81-1911); s2 no1-12 (My/Ag 1926-1930).
Closed 1930. (U-3:2025c; B-2:506a; F-1:669a).
Index: 1-16 (1879/81-1895) in v17, no4/6; 17-22 (1896-n.d.) in v23.

45 MAGAZINE geográfico argentino. Argentine geographic magazine. Buenos Aires.
v1-2 no8 (1936-Ja 1938). Closed 1938. (U-3:2511a).

46 MENDOZA. Universidad nacional de Cuyo.
Instituto de historia y disciplinas auxiliares. Sección de estudios
geográficos. Mendoza.
Serie especial. 1 (1952). All published. (N70-3:3726a).

47 REUNIÓN de trabajos y communicaciones de ciencias naturales y geografía del litoral argentino.
[Trabajos]. 1- (1961-). (N70-3:4992c).

48 REVISTA geográfica americana. (Sociedad geográfica americana, Buenos Aires). Buenos Aires. 1-252 (1933-1956). Closed 1956. (U-4:3613a; B-3:699b; F-4:140b).

49 ROSARIO. Universidad nacional del Litoral. Facultad de ciencias, ingeniería y arquitectura. Instituto de fisiografía y geología.
Publicaciones. 1- (1937-) Annual. (U-4:3708c; B-4:428b; F-3:1080b).
Summaries in English and French.
[Universidad Nacional de Rosario, Avenida Pellegrini 250, Rosario, Argentina.]

50 ______. ______. Sección de historia y geografía. Rosario.
Publicaciones. 1-14 (1922-1928). Closed 1928? (U-4:3708b).

*(51 SOCIEDAD argentina de estudios geográficos "Gaea." Buenos Aires.
Boletín. 1- (1934-) Irregular. (U-5:3902c; BS:795a).
Suspended Ap 1935-Jl 1943.
[Santa Fe 1145, Buenos Aires, Argentina.]

52 ______.
Semana de geografía. Actas. (BS:795a).

53 ______. Filial Rosario. Rosario. Boletín. 6 (1972).

54 SOCIEDAD de historia y geografía de Cuyo. Mendoza.
Revista. 1-2 (1946). (U-5:3906c; B-4:106a).

55 SOCIEDAD geográfica argentina, Buenos Aires. Buenos Aires.
Revista. 1-7 (1881-1890). Closed 1890. (U-5:3909b; F-4:130a).

56 TUCUMÁN. Universidad. Departamento de geografía. Tucumán.
Publicaciones. Serie monográfica. 1-19 (1942-1971) Irregular. (U-5:4274a; N70-4:5873b; B-4:356b).
Earlier by Tucumán. Universidad. Instituto de estudios geográficos.
Subseries of Tucumán. Universidad. Publicaciones (not in this list).
Title varies: Monografías.

57 ______. ______. ______.
Serie de geografía matemática y física. 1-3 (1946-1948). Closed. (U-5:4274a; B-4:356b).

58 ______. ______. ______.
Serie didáctica. 1-9 (1948-1956) Closed. (U-5:4274a).

59 ______. ______. Instituto de estudios geográficos. Tucumán.
Publicaciones especiales. 1-3 (1950-1952). (N70-4:5873c).

60 UMWELT des Auslandsdeutschen in Südamerika. Buenos Aires.
1: Bücherei zur Landeskunde Südamerikas. 1-22 (1933-1942). Closed. U-5:4294c).
In German.

61 ZEITSCHRIFT für argentinische Volks- und Landeskunde. (Deutscher Lehrerverein). Buenos Aires. 1-3 (1911-1914). Closed 1914. (U-5:4596a).
1-2 as Zeitschrift für argentinische Volkskunde.
In German.

AUSTRALIA (English)

The Commonwealth of Australia

62 ARMIDALE, New South Wales. University of New England. Department of geography. Armidale.
Monograph series. 1-2 (1963-1966). Closed 1966. (N70-1:466a).
Superseded by New England monographs in geography, 3- (1975-) [90].

62a ______. ______. ______.
Occasional papers in geography. 1-3 (1971-1973) Irregular.

63 ______. ______. ______.
Research series in applied geography. 1-43 (1965-1976) Irregular.
Closed 1976. (N70-3:4984c; DP).
Superseded by Studies in applied geographical research [65].

64 ______. ______. ______.
Research series in physical geography. 1-? (1968- ?). Closed. (N75-1:179a).
Superseded by Research series in applied geography [63].
[Department of Geography, The University of New England, Armdale, N.S.W. 2351, Australia].

(65 ______. ______. ______.
Studies in applied geographical research. 1- (1977-) Irregular.
Not numbered. (N77).
Supersedes Research series in applied geography [63].
[Department of Geography, University of New England, Armidale, N.S.W. 2351, Australia.]

(65a ______. ______. ______.
Geoview. 1- (1976-). 2 nos a year. Geography student journal.
[Department of Geography, University of New England, Armidale, N.S.W. 2351, Australia.]

66 AUSTRALIAN and New Zealand association for the advancement of science. Sydney.
Reports [of the meetings. These include presidential addresses of Section P. Geography; list of papers presented; abstracts up to v25 (1946)]. 1-30 (1888-1954). Closed 1954. Meeting every 1-2 years. (U-1:563c; B-1:259a).
Through 1926 as Australasian association for the advancement of science. Geography was in Section E to v18 (1926). No meetings 1914-1920, 1933-1934, 1936, 1940-1947. Continued as special issue of the Australian journal of science (not in this list).
Index: 1-16, 1888-1923.

*(67 AUSTRALIAN geographer. (Geographical society of New South Wales, Sydney. Journal). Sydney. 1- (1928-). 2 nos. a year. (U-1:565a; B-1:264a; BS:99a; F-1:514b).
Abstracts. Cumulative index every 3 years.
[Dr. Philip Tilley, Editor, Department of Geography, University of Sydney, N.S.W. 2006, Australia.]

68 AUSTRALIAN geographical record. (Institute of Australian geographers). (N70-1:572a). Canberra. nol-5 (1959-1964). Closed 1964.
Superseded by Australian geographical studies [70].

69 AUSTRALIAN geographical society. Melbourne.
Reports. Irregular. (N70-1:572a).

*(70 AUSTRALIAN geographical studies. (Institute of Australian geographers. Journal). Melbourne. 1- (1963-) Semiannual. (N70-1:572a; B68:63b).
Supersedes Australian geographical record [68].
Abstracts.
[Mr. B. O'Rourke, Business Manager, Australian Geographical Studies, Department of geography, University of Sydney, Sydney, N.S.W. 2006, Australia.]

71 AUSTRALIAN map curators circle. Canberra.
Newsletter. 1977- . (N78).

(72 AUSTRALIAN national university, Canberra. Research school of Pacific studies. Department of biogeography and geomorphology. Canberra.
Publications. BG 1- (1969-). Mimeographed. (N70-1:578b).
[Australian National University Press, P.O. Box 4, Canberra, A. C. T. 2600, Australia.]

73 ______. ______. Department of geography. Canberra.
Publications. G 1-6 (1965-1968) Irregular. Closed 1968. (N70-1:578b; DP). With division of the Department of geography, superseded by two separate series: Department of biogeography and geomorphology. Publications [72]; and Department of human geography. Publications [74].

*(74 ______. ______. Department of human geography. Canberra.
Publications. HG 1- (1969-). (N70-1:578b; DP).
[Australian National University Press, P.O. Box 4, Canberra, A. C. T. 2600, Australia.]

75 ______. School of general studies. Department of geography. Canberra.
Occasional papers. 1- (1964-) Irregular. (N70-1:578b).

76 BULLETIN incorporating G(eography) T(eachers) A(ssociation) Newsletter. (Geographical society of New South Wales, Sydney). Sydney. 1- (1969-) 8 issues a year.

77 CAPRICORNIA. (Queensland. University, Brisbane. Geographical society. Journal). Brisbane.

*(78 CARTOGRAPHY. (Australian institute of cartographers). Canberra.
1- (D 1954-) 2 nos. a year, sometimes combined; 4 nos. per v. (N70-1:1151c; BS:184b).
[The Australian Institute of Cartographers, Box 1292, Canberra City, A. C. T. 2601, Australia.]

(79 GEOGRAPHER. Roleystone, Western Australia. 1- (1969-) 6 nos. a year.
[Carlson Marsh and associates, Urch Road, Roleystone, W.A. 6111 Australia.]

(80 GEOGRAPHICAL education. (Australian geography teachers' association). Newtown. 1- (Je 1969-) Annual. (N75-1:906b; B73:128a).
[Mrs. B. E. Riley, Business Manager, Geographical Education, Sydney Teachers College, P.O. Box 63, Camperdown, N. S. W. 2050, Australia.]

81 GEOGRAPHICAL society of New South Wales. New England branch. (University of New England. Department of geography). Armidale.
Occasional papers in geography. 1-3 (1971-1973) Irregular. (N75-1-906c; DP).
[Department of Geography, University of New England, Armidale, N. S. W. 2351 Australia.]

82 GEOGRAPHY bulletin. (Geographical society of New South Wales; Geography teachers' association of New South Wales). Sydney. 1- (1969-) Quarterly. (N77).
[Geography Teachers Association of N.S.W., 35 Clarence Street, Sydney N.S.W. 2000, Australia.]

83 GEOGRAPHY teacher. Parkville. 1961- 2 nos. a year.
[Geography Teachers' Association of Victoria, Geography Department, State College of Victoria, Rusden, Blackburn Road, Clayton North, Victoria 3168, Australia.]

84 GEOGRAPHY teachers' association of Queensland. Brisbane.
Journal. 7- (1972-). (N75-1:907a).

(85 GEOWEST. (Western Australia. University, Nedlands. Department of geography. Occasional papers). Perth, Western Australia. 1- (1975-) Irregular. (DP). [Department of Geography, University of Western Australia, Nedlands. W.A. 6009, Australia.]

(86 GLOBE. (Australian map curators circle). Melbourne. 1- (1974-) Annual (N77). [Australian Map Curators Circle, P.O. Box E133, Canberra, A.C.T., Australia.]

*(87 JAMES COOK university of North Queensland. Department of geography. Townsville, Queensland.
Monograph series. 1- (1970-) Irregular. (N75-1:1218c; B73:219b; DP).
Includes separately numbered occasional papers. 1- (1974-) Irregular.
[Department of geography, James Cook university of North Queensland, Townsville 4811, Australia].

*(88 MONASH university, Melbourne. Department of geography. Melbourne.
Monash publications in geography. 1- (1972-). (N75-2:1475a; DP).
[Department of Geography, Monash University, Melbourne, Victoria 3168, Australia.]

89 NETWORK geography. (Geography teachers' association of New South Wales). Sydney. 4-7 (1975-1976). Mimeographed.

90 NEW ENGLAND monographs in geography. (Armidale, New South Wales. University of New England. Department of geography). Armidale, New South Wales. 3-? (1975- ?). Closed 1977 (N78).
Supersedes and continues numbering of Armidale, New South Wales. University of New England. Department of geography. Monograph series [62].
Superseded by Studies in applied geographical research [65].

(91 NEW SOUTH WALES. University, Kensington. School of applied geography.
Occasional papers. 1- (1972-). (DP).
[School of Geography, University of New South Wales, P.O. Box 1, Kensington, N. S. W. 2033, Australia.]

92 NEWCASTLE, New South Wales. University. Department of Geography. Newcastle.
Publications in geography. 1-? (1971- ?). Closed 1975. (B73:276b).
Superseded by Research papers in geography [93].

(93 ______. ______. ______.
Research papers in geography. 1- (1975-) Irregular.
[Department of Geography, University of Newcastle, Newcastle, N.S.W. 2308, Australia.]

94 OONDOONA. (Sydney university geographical society). Sydney. 1- (1959-) Irregular.
[Sydney University Geographical Society, University of Sydney, Sydney, N. S. W. 2006, Australia.]

95 QUEENSLAND. University, Brisbane. Department of geography. Brisbane.
Papers. 1-2 (1960-1961) Irregular. (N70-3:4883a; B68:575a; B73:250b).
[University of Queensland Press, St. Lucia, Queensland, Australia.]

*(96 QUEENSLAND geographical journal. (Royal geographical society of Australasia. Queensland branch). Brisbane. 1- (1885/86-) Irregular. (U-4:3506c; B-2:269a; F-3:1107b; F-3:1037b). Not published 1976-1977.
1-14 (1885/86-1898/99) as Proceedings and transactions, Queensland branch of Royal geographical society of Australasia. v15-41 also as ns.
[Royal Geographical Society of Australia, Queensland Branch, 117 Anne Street, Brisbane, Queensland 4000, Australia.]

97 ROYAL geographical society of Australasia. New South Wales Branch. Sydney. Journal. v1-6 no6 (1883-1898). Closed 1898. (U-4:3721a; B-2:269a; F-3:209a; F-3:1031b). 1-2 (1883-1884) as Geographical society of Australasia. New South Wales and Victorian branches. Proceedings; 3-4 (1885-1886) in one volume, Transactions and proceedings. v6 no1 repeated in numbering. v5 (1891/92 and 6 (1895/97) marked New series.

98 ______. Queensland branch. Brisbane.
Transactions. 1 (1924). (U-4:3721a; B-2:269a).
Consists of v1 of Reports of the Great Barrier Reef Committee.

*(99 ______. South Australian branch. Adelaide.
Proceedings. 1- (1885-) Annual. (U-4:3721a; B-2:269a; F-3:1038a).
[RGS of Australasia, S. A. Branch, c/o State Library of South Australia, North Terrace, Adelaide, South Australia 5000, Australia.]

100 SYDNEY. University. Sydney.
Publications in geography. 1-4 (1938-1939). (U-5:4137a; B-4:273b).

101 ______. ______.
University reprints. s6: Geology and geography. 1-4 (1924-1938). (U-5:4137b; B-4:273b).

102 ______. ______. Department of geography; and Geographical society of New South Wales.
Research paper in geography. 1-16 (1961-1971). Closed 1971. (N70-4:5652c).
Title varies: Early numbers as Research paper.

103 TAMINGA. (Geography teachers' association of South Australia). [Adelaide]. 1- (1960-) Irregular. (N70-4:5690b).
[Geography Teachers' Association of South Australia, c/o 163a Greenhill Road, Parkside, S.A. 5063, Australia.]

104 VICTORIAN geographical journal. (Royal geographical society of Australasia. Victorian branch). Melbourne. 1-34 (1883-1918). Closed 1918. (U-5:4390b; B-2:269a; F-4:905a; F-3:1031b; F-4:707b). v1-2 (1883-1884) as Geographical society of Australasia. New South Wales and Victorian branches. Proceedings; v3/4-19 (1885-1901) as Royal geographical society of Australasia. Victorian branch. Melbourne. Transactions and proceedings; v20-34 (1902-1918) as Victorian geographical journal. Merged into Victorian historical magazine (not in this list).
Index: 1-34 (1883-1918) in 34.

105 WESTERN Australia. University, Nedlands. Department of geography. Nedlands, Western Australia.
Theses and dissertations presented and currently being undertaken in geography. 1976- . (78).

106 ______. University, Perth. Economics department. Geographical laboratory.
Research report. 1-46 (1949-1963) Irregular (mimeographed). (U-5:4483a; B-4:531a).

(107 WESTERN geographer. (Geographical Association of Western Australia. Journal). Perth. 1- (1973-) 2 nos a year from 1979.
[Department of Geography, University of Western Australia, Nedlands, W.A. 6009, Australia.]

AUSTRIA (German)

Republik Österreich

(108 ABHANDLUNGEN zur Humangeographie. (Wien. Universität. Geographisches Institut). Wien. 1- (1972-) Irregular. (N75-1:12c)
[Geographisches Institut der Universität Wien, Universitätsstrasse 7, A-1010 Wien, Austria.]

(109 ALPENKUNDLICHE Studien. (Innsbruck. Universität. Geographisches Institut...). v1- (1968-) Irregular. (G).
Subseries of Innsbruck. Universität. Veröffentlichungen. 1 as 1 of whole series, 2 (2), 3 (7), 4 (8), 5 (9), 6 (13), 7 (18), 8 (26).

110 ARCHIV für Geographie, Historie, Staats-und Kriegskunst. Wien. 1-13 (1810-1822). (U-1:443c; B-1:191b; F-1:376b; G).
Succeeded by Archiv für Geschichte, Statistik, Literatur und Kunst, 14-19 (1823-1828) which continues its numbering (not in this list).

111 ARCHIV für Geographie und Statistik, ihre Hülfswissenschaften und Litteratur. Wien. v1-3 (1801-1804). Closed 1804. (G).

112 ARCHIV für Welt-, Erd- und Staatenkunde und ihre Hilfswissenschaften und Litteratur. Wien. 1-4 (1811-1813). Closed 1813. (G).

113 BEITRÄGE zur alpinen Karstforschung. (Bundesministerium für Land- und Forstwirtschaft. Speläologisches Institut). Wien. no1 - (1956-) Irregular. (N70-1:667b): No 2 issued in 1953.
Reprint series.
[Hofburg, Bettlerstiege, Wien I, Austria.]

114 CARINTHIA II. Naturwissenschaftliche Beiträge zur Heimatkunde Kärntens. Mitteilungen des Naturwissenschaftlichen Vereines für Kärnten. Klagenfurt. 81- (1891-) Annual. (U-2:928b; B-1:506a; F-2:31b; G).

Subtitle earlier: Mitteilungen des Naturhistorischen Landesmuseums für Kärnten.
Supersedes Carinthia. Zeitschrift für Vaterlandskunde, Belehrung und Unterhaltung 1-80 (1811-1890) (not in this list).
Indexes: 1-100 (1811-1910), 101-125 (1911-1935).
[Naturwissenschaftlicher Verein für Kärnten, Museumsgasse 2, A-9020 Klagenfurt, Austria.]

115 ______.
Sonderhefts. Klagenfurt. no1- (1930-) Irregular. (G).

116 DEUTSCHE Rundschau für Geographie. Leipzig: Wien. 1-37 (1878-1915). Closed 1915. (U-2:1315b; B-2-37a; F-2:365b; G).
1-32 (1878-1910) as Deutsche Rundschau für Geographie und Statistik).

117 DONAURAUM. (Wien. Forschungsinstitut für Fragen des Donauraumes. Zeitschrift). Salzburg. 1- (1956-) Quarterly. (N70-2:1778c; BS:266a; G).
Supersedes institute's Mitteilungsblatt (not in this list).
Includes a separate English-language synopsis.
[Schottergasse 10, A1010 Wien, Austria.]

(118 GW-UNTERRICHT: Eine Zeitschrift für Lehrer, die Geographie und Wirtschaftskunde unterrichten. (Zentralsparkasse der Gemeinde Wien). Wien. 1- (1978-). 3 nos a year.
[Gegergasse 1, A-1030 Wien, Austria.]

119 GEOGRAPHISCHE Informationen. (Freytag, Berndt und Artaria). Wien. 1- (D 1958-). (N70-2:2282a).

*(120 GEOGRAPHISCHER Jahresbericht aus Österreich. Wien. 1- (1894-) Biennial. (U-2:1696c; B-2:270b; F-2:729b).
4-13 include Wien. Universität. Verein der Geographen. Berichte 29-50 (1902-1924) [157]. Title varies slightly.
[Geographisches Institut der Universität Wien, Universitätsstrasse 7, A-1010 Wien, Austria.]

(121 GEOGRAPHISCHES Jahrbuch Burgenland. (Vereinigung Burgenländischer Geographen). Neusiedl am See. 1- (1977-) Annual. (N78).
[Hauptplatz 50, Neusiedl am See, Austria.]

122 GESELLSCHAFT für Salzburger Landeskunde. Salzburg.
Mitteilungen. 1- (1861-) Annual. (U-2:1718b; B-2:291a; F-3:568b; G).
Indexes: each 10th v has index to preceding 10. 1-100 (1860-1960).
[Universitätsplatz 1, Salzburg, Austria.]

123 ______.
______. Beiheft. 1- (1946-). (U-2:1718b: G).

124 ______.
______. Ergänzungsband. 1- (1960-). (N70-2:2338b; G).

(125 GLOBUSFREUND. Publikation. (Coronelli Weltbund der Globusfreunde; Coronelli world league of friends of the globe). Wien. 1- (1952-) 3 nos a year. (N70-2:2360c; N70-1:1545b; BS:354a).
In English, German, or French.
[Coronelli Weltbund der Globusfreunde, Erdbergstr. 32, A-1030 Wien, Austria.]

(126 GRAZ. Universität. Geographisches Institut. Graz.
Arbeiten. 1- (1951-) Irregular. (N70-2:2388b).
[Geographisches Institut der Universität, Universitätsplatz 2, A-8010 Graz, Austria.]

127 ______. ______. ______.
Veröffentlichungen. 1-3 (1925-1930). Closed 1930. (U-2:1759b).

(128 INNSBRUCKER geographische Studien (Innsbruck. Universität. Geographisches Institut). Innsbruck. 1- (1973-) Irregular. (G).
[Geographisches Institut der Universität, Innrain 52, A-6020 Innsbruck, Austria.]

129 JAHRBUCH für Landeskunde von Niederösterreich. (Verein für Landeskunde von Niederösterreich). Wien. nsl- (1902-). (U-3:2141c; B-1:362a; F-3:18b; G).
Succeeds society's Blätter (1865-1901) [150].
Index: 1902-1927, 1928-1938.

130 KARTOGRAPHISCHE Mitteilungen. Wien. v1-2 no2 (1930-1932). Closed 1932. (U-3:2273b).

131 KARTOGRAPHISCHE und schulgeographische Zeitschrift. Wien. 1-10 (1912-1922). Closed 1922. (U-3:2273b; F-3:236a; G).

132 LINZ. Oberösterreichischer Musealverein. Linz.
Jahrbuch. 1- (1835-) Annual. (U-4:3127b; B-4:447a).
Name of society varies: 1835-1839 as Verein des vaterländischen Museums für Österreich ob der Enns und das Herzogthum Salzburg; 1839-1920 as Museum Francisco-Carolinum; 1937-1944 as Verein für Landeskunde und Heimapflege im Gau Oberdonau. Title of publication varies: 4-7, 10-80 include monographic material called Beiträge zur Landeskunde von Österreich ob der Enns (varies); 1835-1893/94 as Bericht; 1894/95-1922/23 as Jahresbericht.

133 MITTEILUNGEN für Erdkunde; Heimatkundliche Fachzeitschrift fur Geographie, Geologie, etc. Linz. v1-13 (1932-1949). Closed 1949. (U-3:2707a). Suspended 1940-1945.

134 MONATSBLATT für Landeskunde von Niederösterreich und Wien. (Verein für Landeskunde von Niederösterreich). Wien. 1-26 (1902-1927). Closed 1927. (U-5:4337a: B-4:459a; F-3:600b; G).
Supersedes Verein für Landeskunde von Niederösterreich. Blätter 1-2, [150].
Superseded by Unsere Heimat [149].
Index: 1902-1927.

(135 NATURWISSENSCHAFTLICHER Verein für Steiermark. Graz.
Mitteilungen. 1- (1863-) Annual. (U-4:2937c; B-3:327a).
Index: 1-20 (1863-1883) as supplement to 20; 21-40 (1884-1903).

136 OBERÖSTERREICHISCHE Landesbaudirektion. Linz.
Schriftenreihe. 1- (1946-) Irregular.

137 ÖSTERREICHISCHE geographische Gesellschaft. Wien.
Abhandlungen. 1-12 (1899-1922); 13- (1938-) Irregular. (U-4:3141a; B-2:270a; F-1:13b; G).
Name of issuing agency until 1959: Geographische Gesellschaft in Wien; earlier as Kaiserlich-königliche geographische Gesellschaft in Wien.
Index: v1-6 (1899-1907).
[Österreichische Geographische Gesellschaft, Karl-Schweighofer Gasse 3, A-1071 Wien, Austria.]

*(138 ______.
Mitteilungen. 1- (1857-) 3 nos. a year. (U-4:3141b; B-2:270a; F-3:570b; G).
Index: 1857-1907; 1908-1959.
Summaries in English, French, and German v111-115 (1969-1973).
Supplementary table of contents and abstracts in English and French at end of each number. v118- (1976-).
[Österreichische Geographische Gesellschaft, Karl Schweighofergasse 3, A-1070 Wien, Austria.]

138a ______. Zweigverein Innsbruck. Innsbruck
Jahresbericht. 1973- . (N79).

(139 ÖSTERREICHISCHE Gesellschaft für Raumforschung und Raumplanung. (ÖGRR). Wien.
Berichte zur Raumforschung und Raumplanung. 1- (1957-) Quarterly. (N70-3:4352a).
Title varies: v1-8 (1957-1964) as Österreichische Gesellschaft zur Förderung von Landesforschung und Landesplanung. Berichte zur Landesforschung und Landesplanung. Supersedes society's Berichte [141].
In German with English summary at end of each no.
[Springer Verlag, Wien, New York.]

(140 ______.
Schriftenreihe. 1- (1955-) Irregular. (N70-3:4352a).

141 ÖSTERREICHISCHE Gesellschaft zur Förderung von Landesforschung und Landesplanung. Klagenfurt.
Berichte. no1-13 (1954-1956). Closed 1956. (N70-3:4352a).
Continued as Österreichische Gesellschaft für Raumforschung und Raumplanung. Berichte [139].

(142 ÖSTERREICHISCHER Alpenverein. Innsbruck.
Alpenvereinsjahrbuch. 1- (1865-) Annual. (U-4:3144c; U-2:1322a; B-2:40b; F-3:14b; F-4:1002b; G). v1-73 (1870-1942) as Deutscher und österreichischer Alpenverein. Zeitschrift, alternating amoung München, Wien, Berlin, Graz; v83-88 (1958-1963) as Deutscher Alpenverein. Jahrbuch. München. v95- (1970-) as Alpenvereinsjahrbuch.
Indexes: 1-8 in 8; 1-17 (1869-1886).

(143 ______.
Wissenschaftliche Alpenvereinshefte. Innsbruck. 1- (1897-) Irregular. (U-5:4527b; B-2:40b; F-4:977b).
1-4 as Wissenschaftliche Ergänzungshefte zur Zeitschrift des deutschen und österreichischen Alpenvereins; 5-11 as Wissenschaftliche veröffentlichungen des deutschen und österreichischen Alpenvereins; 12- (1952-) under present title.

(144 ÖSTERREICHISCHES Institut für Raumplanung. Wien.
Mitteilungen. 1- (1959-). Monthly. (N70-2:2807b; G).
Name of issuing agency until 1961: Wien. Institut für Raumplanung.
[Österreichisches Institut für Raumplanung, Franz-Josefs-Kai 27, A-1011 Wien, Austria.]

145 ______.
Tätigkeitsbericht. 1957- Annual. (N70-2:2807b; G).

(146 ______.
Veröffentlichungen. 1- (1957-) Irregular. (N70-2:2807b).

(147 SALZBURG. Universität. Geographisches Institut. Salzburg.
Arbeiten. 1- (1970-) Irregular. (N77, GZS).
[Geographisches Institut der Universität Salzburg, Akademiestr. 20, A-5020, Salzburg, Austria.]

(148 TIROLER Wirtschaftsstudien. Schriftenreihe der Jubiläumsstiftung der Kamer der gewerblichen Wirtschaft für Tirol. Innsbruck. 1- (1955-) Irregular. (N70-4:5794a; G).

(149 UNSERE Heimat. (Verein für Landeskunde von Niederösterreich and Wien). Wien. 1- (1928-) Monthly. (U-5:4337a; B-4:436b; B-4:459a; F-4:846b; G).

Supersedes Monatsblatt für Landeskunde von Niederösterreich [134].
Name of issuing body varies; subtitle varies. 1944-1945 issued as Verein für Landeskunde von Niederösterreich.
Mitteilungen an die Mitglieder.
Index: 1-13 (1928-1940).
[Herrengasse 13, A-1010 Wien, Austria.]

150 VEREIN für Landeskunde von Niederösterreich. Wien.
Blätter. 1-2 (1865-1866); nsv1-35 (1867-1901). Closed 1901. (U-5:4370b; B-1:362a; F-1:646a; G). 1-2 (1865-1866) as Blätter für Landeskunde von Niederösterreich. Merged into Jahrbuch für Landeskunde von Niederösterreich [129].
Index: 1-nsv14 (1865-1880); nsv15-19 (1881-1885).

151 VIERTELJAHRSHEFTE für den geographischen Unterricht. Wien. 1-2 (1902-1903). Closed 1903. (U-5:4397b; F-4:921b).

(152 WETTER und Leben. (Österreichische Gesellschaft für Meteorologie. Bioklimatische Sektion). Wien. 1- (1948-) Bimonthly. (U-5:4500a; B-4:537a).

153 WIEN (Vienna). Hochschule für Welthandel, now Wirtschaftsuniversität Wien. Institut für Raumforschung. Wien.
Arbeiten. 1- (1956-) Irregular.

154 ______. Militärgeographisches Institut in Wien. Wien.
Mitteilungen. 1-34 (1881-1919). Closed 1919. (U-5:4395a; B-1:267b; F-3:577b).
Index: 1-25 (1881-1905) in 25.

155 ______. Universität. Geographisches Institut. Wien.
Geographische Abhandlungen. v1-8 (1891-1905). Closed 1905. (U-5:4396a; G).
Also issued as Arbeiten.

(156 ______. ______. ______. Seminarbetrieb. Wien.
Beiträge. 1- (1972-) Irregular. (GZS).
[Geographisches Institut der Universität Wien, Universitätsstrasse 7, A-1010 Wien, Austria.]

157 ______. ______. Verein der Geographen. Wien.
Berichte. 1-29 (1875-1903). Closed 1903. (U-5:4396c; B-4:475b).
1875-1893 as society's Jahresbericht. 12-29 (1885/86-1903) the reports contained two sections: A. Geschäftlicher Theil; B. Wissenschaftlicher Theil. 29-50 (1902-1924) in Geographischer Jahresbericht aus Österreich, v4-13 [120].

*(158 WIENER geographische Schriften. (Wien. Hochschule für Welthandel, now Wirtschaftsuniversität Wien. Geographisches Institut). Wien. 1- (1957-) Irregular. (N70-4:6343c; BS:964a).
English summaries and titles in table of contents.
[Verlag Ferdinand Hirt, Wien, Austria.]

159 WIENER geographische Studien. Wien. 1-23 (1933-1953). Closed 1953. (U-5:4505b; B-4:543b; G).

(160 WIENER geographische Studienbehelfe. (Wien. Universität. Geographisches Institut). Wien. 1- (1975-) Irregular. (G).
[Geographisches Institut der Universität Wien, Universitätsstrasse 7, A-1010 Wien, Austria.]

(161 WIENER Quellenhefte zur Ostkunde. Reihe Landeskunde. (Arbeitsgemeinschaft Ost). Wien. 1958- Quarterly. (N70-4:6344a).
[Österreichisches Ost- und Südost-Europa Institut, Josefplatz 6, A-1010 Wien, Austria.]

(162 WIRTSCHAFTSGEOGRAPHISCHE Schriften. (Österreichische Gesellschaft für Wirtschaftsraumforschung). Wien. 1- (1977-).
[Verlag Ferd. Hirt, KG, Widerhofergasse 8, A-1090 Wien, Austria.]

(163 ZEITSCHRIFT für Gletscherkunde und Glazialgeologie. Innsbruck. 1- (1949-). 2 nos. a year. (U-4:4605c; B-4:604b).
[Institut für Geographie, Universität Innsbruck, Innrain 52, A-6020 Innsbruck, Austria.]

164 ZEITSCHRIFT für Schulgeographie. Wien. 1-32 (1880-1911). Closed 1911. (U-5:4617c; F-4:1032b).
Merged into Geographischer Anzeiger [1001].

বাংলাদেশ BANGLADESH (Bengali; English)

People's Republic of Bangladesh

Gana Prajatantri Bangladesh

166 ভূগোল সাময়িকী 170 উপকূল

165 BANGLADESH national geographical association. Dacca.
Journal. 1-2 (1973-1974) Semiannual. (N75-1:252b).
In English.
[c/o Department of Geography, Jahangirnagar University, Savar, Dacca, Bangladesh.]

166 BHŪGOLA sāmayikī. (Bāṃlādeśa jātīya bhūgola samiti). Dacca.
1 (1974). English title: Bengali journal of the Bangladesh national geographical association.
In Bengali.
[Bangladesh National Geographical Association, c/o Department of Geography, Dacca College, Dhanmondi, Dacca 2, Bangladesh.]

167 DELTA; annual bulletin of the Dacca university geographical association.
Dacca.

168 EAST Pakistan geographical society. (From 1972 as Bangladesh geographical society). Dacca.
Monographs. 1-3 (1961-1967) Irregular. (N70-2:1823b; B68-366b).
In English.
[Department of Geography, University of Dacca, Dacca, Bangladesh.]

*(169 ORIENTAL geographer. (Bangladesh geographical society). Dacca.
1- (1957-) Semiannual. (N70-3:4449c; BS:649b). v1-14 (1957-1970) issued by the East Pakiston geographical society.
Index: v1-4 in v4.
In English.
[Bangladesh Geographical Society, Department of Geography, University of Dacca, Dacca, Bangladesh.]

170 UPAKŪLA. Upokul. Geographical magazine. (Dhākā biśvabidyālaẏa bhūgāla samiti. Dacca university geography association). Dacca. 1- (1972-) Irregular. (N78).
In Bengali.
[Dacca University Geographical Association, Ramna, Dacca 2, Bangladesh.]

BELGIUM (French or Dutch)

Royaume de Belgique--Koninkrijk België

*(171 AARDRIJKSKUNDE. (Vereniging leraars aardrijkskunde). De Pinte.
nsv1- (1977-) Quarterly.
Succeeds in part Géographie - Aardrijkskunde [181].
In Dutch. Summaries in English
[W. Vlassenbroeck. Latemstraat 5, 9720 De Pinte, Belgium.]

*(172 ABSTRACTS of Belgian geology and physical geography. (Gent. University. Geological institute). Gent. 1- (1967-) Annual. (N70-1:131b).
In English.
[Geological Institute, University, Gent, Belgium.]

*(173 ACTA geographica lovaniensia. (Leuven. Katholieke universiteit. Geografisch institut. Louvain-la-Neuve. Université catholique. Institut de géographie). 1- (1961-) Irregular. (N70-1:58b; B-68:7a).
In Dutch or French typically with summaries in other language and in English.
[Geografisch Instituut, K.U.L. Celestijnenlaan, 300, B3030 Heverlee, Belgium, or Institut de Geographie, U.C.L., Place Louis Pasteur, 3, B-1348 Louvain-la-Neuve, Belgium.]

(174 BIBLIOGRAPHIE géographique de la Belgique. Aardrijkskundige bibliografie van België. (Belgium. Commission belge de bibliographie). Bruxelles.
By M. E. Dumont and L. De Smet. 4th supplement by M. E. Dumont, L. De Smet, and W. Vlassenbroeck.
no14-17 (1954-1956) of Bibliographia belgica (not in this list).
By Seminarie voor menselijke aardrijkskunde Rijksuniversitet, Gent.
1st supplement, 1960 (Bibliographia belgica no. 48); 2nd supplement, 1965 (Bibliographia belgica no. 82); 3rd supplement, 1970. 4th supplement, 1978; By Seminarie voor menselijke en ekonomische geografie, Rijksuniversiteit, Gent. (N70-1:726b; B-1:331a; BS:121b).
[Publicaties ME Geografie, Blandijnberg 2, Gent, Belgium.]

175 BRUXELLES. Université libre. Institut géographique. Bruxelles.
[Publications]. v1-10 (1899-1905). Closed 1905. (U-1:809a; B-4:432b).

176 CERCLE des géographes liégeois. (Liège. Université). Liège.
Bulletin. 1-11 (1928-1939). Closed 1939. (U-2:966c; B-1:526a; F-1:910b).
Superseded by Société géographique de Liège. Bulletin [190].

177 ______.
______. Supplément. no1-13 (1932-1940). Closed 1940. (U-2:966c; F-1:910b).

178 COMITÉ spécial du Katanga. Bruxelles.
Publications. sA. Géographie, géologie et mines. 1- (1953-).
(N70-1:1418b; BS:219b).

179 GENT (Ghent). Rijksuniversiteit. Seminarie voor menselijke en ekonomische geografie. Gent.
Publikaties. 1- (1968-) Irregular.
In Dutch.
[Publicaties ME Geografie, Blandijnberg 2, Gent, Belgium.]

*(180 GÉOGRAPHIE. Écologie. Environnement. Organisation de l'espace. G.E.O. (Fédération des professeurs de géographie). Liège. 1- (1977-) 2 nos a year (N79). Succeeds in part Géographie-Aardrijkskunde [181].
In French. Abstracts in English and French.
[E. Mérenne, rue Côte d'Or, 190. 4200 Liège, Belgium.]

181 GÉOGRAPHIE - Aardrijkskunde. (Fédération belge des professeurs de géographie A. S. B. L. Belgische federatie van leraars in de aardrijkskunde V. Z. W.). Bruxelles. v1-22, no1-87; no88-111 (1948-1976). Closed 1976. (U-2:1695c; N70-2:2282a; N75-1:11c).
In Flemish or French.
Succeeded by two separate series: Aardrijkskunde [171] in Dutch and Géographie (G.E.O.) [180], in French.

182 LIÈGE. Université. Séminaire de géographie. Liège.
Travaux. 1-145 (1905-1963) Irregular. (U-3:2415a; B1:14a).
no21-145 (1929-1962) carry also title Cercle des géographes liègeois; 1-125 with double numbering. Reprints individually numbered. Superseded by Travaux géographiques de Liège [196].

183 ______. ______. Société d'histoire et de géographie. Liège.
Bulletin. 1885-1887, 1890.

184 ______. ______. ______.
Exposé des travaux de la société. 1883/84-1884/85.

185 LOUVAIN. Université catholique. Institut géographique Paul Michotte (earlier: Séminaire de géographie). Louvain.
sB. Bibliothèque. 1 (1955) Reprint. (N70-1:765b).

186 MOUVEMENT géographique. (Journal populaire des sciences géographiques. Organe des intérêt belges au Congo). Bruxelles. 1-35 (1884-1922).
Closed 1922. (U-3:2778a; B-3:270a; F-3:649ab).

*(187 REVUE belge de géographie. (Société royale belge de géographie). Bruxelles.
1- (1876-) 3 per annum. (U-5:3970c; B-4:126a).
1876-1882 as Société belge de géographie. Bulletin; 1883-1961 as Société royale belge de géographie. Bulletin; present title from v86 (1962). Includes Compte-rendu, paged separately [192].
Index: v1-25 (1876-1901); v26-38 (1902-1914).
In French with occasional English abstracts from v88 (1964).
[87, avenue Adolphe Buyl, 1050 Bruxelles 5, Belgium.]

(187a REVUE internationale d'écologie et de géographie tropicales. GEO-ECO-TROP.
Tervuren. 1- (1977-) 2 per annum.

*(188 SOCIÉTÉ belge d'études géographiques. Belgische vereeniging voor aardrijkskundige studies. Bruxelles.
Bulletin. Tijdschrift. 1- (My 1931-) 2 nos. a year. (U-5:3932b; B-4:126a; BS:801b; F-1:810b).
Indexes: 1931-1940, 1941-1950, 1951-1960, 1961-1970.
In French or Dutch with summaries in other language or English.
[Institut géographique de l'Université, 2 Blandijnberg, Gent, Belgium.]

189 ______.
Mémoires. 1- (1935-) Irregular monographs. (U-5:3932b).
In French, Flemish, or English.

*(190 SOCIÉTÉ géographique de Liège. Liège.
Bulletin. 1- (1965-) Annual. (N70-4:5384c; B73:60a).
Supersedes Cercle des géographes liégeois. Bulletin [176].
In French. Abstracts in English and French from v13 (1977).
[Secrétariat de la Société Géographique de Liège, Séminaire de Géographie de l'Université, place du Vingt-Août 7, 4000 Liège, Belgium.]

191 SOCIÉTÉ royale belge de géographie. Bruxelles. Bruxelles.
Année coloniale. 1- (1950-). (N70-4:5386c).

192 ______.
Compte-rendu. 1-82 (1877-1958). (U-5:3971a).
Issued with separate paging with Société royale belge de géographie. Bulletin [187].
From v83 (1959) included as integral part of Société royale belge de géographie. Bulletin, later Revue belge de géographie [187].

193 ______. Section de géographie commerciale.
Renseignements commerciaux. 1922-1924. Closed 1924. (U-5:3971a; B-4:126a; F-4:75a).
Supplement to the society's Bulletin [187].

*(194 SOCIÉTÉ royale de géographie d'Anvers. Koninklijk aardrijkskundig genootschap, Antwerpen. Antwerpen.
Bulletin. Tijdschrift. v1-60 (1876-1940); v61- (1946-) Annual. (U-5:3971c; B-4:135a; F-1:835a).
Suspended between May 1940 and 1946.
Index: 1-50 (1876-1926).
In Dutch, French, or occasionally English.
[Prins Albertlei 3, bus 6, 2600 Berchem, Belgium.]

195 ______.
Mémoires. 1-4 (1879-1895). Closed 1895. (U-5:3972a; B-4:135a; F-3:455b).

(196 TRAVAUX géographiques de Liège. (Société géographique de Liège; and Liège. Université. Séminaire de géographie). Liège. 146- (1963-) Annual or semiannual. (U-3:2415a; N70-4:6608b). Collections of reprints.
Supersedes and continues numbering of Liège. Université. Séminaire de géographie. Travaux [182].
Most articles have abstracts in English and French from v166 (1978).
[Séminaire de Géographie de l'Université, place du Vingt-Août 7, 4000 Liège, Belgium.]

BENIN (French)

République Populaire du Benin

Formerly Dahomey

197 ASSOCIATION dahoméenne de géographie. Cotonou.
Bulletin de liaison. 1- (1974-) Quarterly. (N76).
[Association Dahoméenne de Géographie, B. P. 526, Cotonou, Benin.]

(198 BENIN. Université, Lomé. Section sciences. Lomé.
Annales. 1- (1975-).
Supersedes Annales de l'école des sciences (not in this list).

BOLIVIA (Spanish)

República de Bolivia

199 BIBLIOTECA boliviana de geografía e historia. La Paz.
1-5 (1887-1894). Closed 1894. (U-1:675b).

200 BOLIVIA. Dirección general de estadística y estudios geográficos.
Anuario nacional estadístico y geográfico de Bolivia. 1917 (1918).

201 ______. ______.
Anuario geográfico y estadístico de la República de Bolivia. 1919 (1920).

202 SOCIEDAD de estudios geográficos e históricos, Santa Cruz. Santa Cruz.
Boletín. v1- (1904-) Irregular. (U-5:3906b).
v1-3 no1-7 (1904-1906) and v30 (no26-28) (1945-1947) known to have been published. v1-2 no3 (1904-1905) as Sociedad geográfica e histórica, Santa Cruz.

203 SOCIEDAD de geografía e historia, "Cochabamba." Cochabamba.
Boletín. 1-6 Irregular. (U-5:3906c).
v3-5 carry dates 1930-1941/42.

204 SOCIEDAD geográfica de La Paz. La Paz.
Boletín. 1-34 (no1-61/62) (1898-1931); 52-54 (no63-70) (1941-1949).
(U-5:3909b; B-4:106a; F-1:664b).
8 (1910), 35-51 (1932-1940) never published.

205 SOCIEDAD geográfica e histórica "Sucre." Sucre.
Boletín. 1-34 (no1-61/62) (1898-1931); no63- (1941-) Irregular.
(U-5:3909c; N70-4:6676b; N75-2:1988c; B-4:106a).
1-445 as Sociedad geográfica de Sucre. Present title from 446 (1962).

206 SOCIEDAD geográfica y de historia. Tarija.
Revista. v1-3 (no1-5) (1944-1946). (U-5:3909c).

207 SUR, Boletín de la Sociedad geográfica y de historia "Potosí." Potosí.
1-2 (no1-5) (Ag 1913-Mr 1914); s2 no1 (D 1943); (U-5:4120b).
1-2 (no1-5 (1913-1914) as Sociedad geográfica Potosí. Boletín.

BRAZIL (Portuguese)

República Federativa do Brasil

208 ANUÁRIO geográfico do Brasil. (Brazil. Instituto brasileiro de geografia e estatística. Conselho nacional de geografia). Rio de Janeiro.
1-3 (1953-1961). Closed 1961. (N70-1:393b).

209 ASSOCIAÇÃO dos geógrafos brasileiros. São Paulo.
Anais. Annual. (U-1:514c; B-1:232a; BS:87a).
v1 (1945-1946) was published 1949. Supersedes the association's Boletim [211].

210 ______.
Avulso. no1- (1961-) Irregular. (N70-1:502b).
[Caixa Postal 8105, São Paulo, S. P., Brazil.]

211 ______.
Boletim. 1-5 (1941-1944). Closed 1944. (U-1:514c; B-1:232a).
no1 (1941) issued as integral part of Revista brasileira de geografia, v3 no2 (Ja/Mr 1941). Superseded by Associação dos geógrafos brasileiros. Anais [209].

212 ATLAS de relações internacionais (Serviço-grafico de Fundação IBGE).
Lucas. 1- (Ja/Mr 1967-). (N75-1:205c).
no1- issued also as Caderno especial da Revista brasileira de geografia, v29- . [279].

213 BAHIA (state). Directoria de serviços geográficos, geológicos e meteorológicos. Salvador.
Boletim. v1- (1930/1931-).

214 ______. Universidade. Laboratório de geomorfologia e estudos regionais. Salvador.
Publicações. 1- (1958-). (N70-1:613a).

215 ______. ______. ______.
Trabalhos. (N70-1:613a).

216 BIBLIOGRAFIA geográfico-estatística brasileira. (Brazil. Instituto brasileiro de geografia e estatística). Rio de Janeiro. 1- (1936/50-). (N70-1:722a).

217 BOLETIM baiano de geografia. (Associação dos geógrafos brasileiros. Seção regional do Rio de Janeiro. Núcleo de Salvador). Salvador. v1-9/11 no1-15/17 (1960-Jl 1970). (N70-1:806a).

218 BOLETIM carioca de geografia. (Associação dos geógrafos brasileiros. Secção regional do Rio de Janeiro). Rio de Janeiro. 1- (1948-). (U-1:725b; BS:131a).
v1-2 (1948-1949) as the section's Boletim.
[Av. Beira-Mar 436-10° and., Rio de Janeiro, Brazil.]

219 BOLETIM de geografia teorética. (Associação de geografia teorética [AGETEO]). Rio Claro, SP. 1- (1971-) 2 per annum. (N76).
[Associação de Geografia Teorética, Caixa postal 178, 13500 Rio Claro, S.P., Brazil.]

220 BOLETIM geográfico. (Brazil. Secretaria de planejamento da presidência da república. Fundação Instituto brasileiro de geografia e estatística. IBGE. Departamento de documentação e divulgação geográfica e cartografica). Rio de Janeiro. 1-258/259 (1943-1978). Closed 1978. (U-1:764b; B-1:367a; BS:138b; N75-1:311c).
no1-105 (1943-1951) monthly; no106-243 (1952-1974) bimonthly; no244-258/259 (1975-1978) quarterly.
v1 no1-3 as Boletim do conselho nacional de geografia: v1-8 with 12 nos per volume; v9 (1951) with 9 nos. no1-198 (1943-1967) issued by Instituto brasileiro de geografia e estatística. Conselho nacional de geografia; no199-214 (1967-1970) by Fundação IBGE. Instituto brasileiro de geografia; no215-233 (1970-1973) by Ministério de planejamento e coordenação geral. Fundação IBGE. Instituto brasileiro de geografia; no234-239 (1973-1974) by Ministério de planejamento e coordenação geral; no240- (1974-) by Secretaria de planejamento de presidencia da república. IBGE.
Indexes: nos1-33 in no34; nos34-45 in no46; 1943-1949 in no83 (F 1950).

221 BOLETIM geográfico de Minas Gerais. (Minas Gerais. Departamento geográfico). Belo Horizonte. 1-? (1958-1969). Closed 1969. (N70-1:806c).

222 BOLETIM geográfico do estado do Rio Grande do Sul. (Diretorio regional de geografia do Rio Grande do Sul). Pôrto Alegre. 1- (1955-) Irregular. (N70-1:807a).
Title varies: Boletim geográfico do Rio Grande do Sul.

223 BOLETIM mineiro de geografia. (Associação dos geógrafos brasileiros. Seção regional de Minas Gerais). Belo Horizonte. 1-? (1957-1969). Closed 1969. (N70-1:807a).
Summaries in English and French.

224 BOLETIM paranaense de geociências. (Paraná. Universidade. Setor de tecnologia. Instituto de geociências). Curitiba. 1-33 (1960-1975) Irregular. (N70-1:807a).
1-20 (1960-1966) as Boletim paranaense de geografia. (Associação dos geógrafos Brasileiros. Secção regional do Paraná; and Universidade do Paraná. Conselho de pesquisas).
In Portuguese with many English summaries.
[Instituto de geociências, Universidade do Paraná, Caixa postal 756, Curitiba, Paraná, Brazil.]

*(225 BOLETIM paulista de geografia. (Associação dos geógrafos brasileiros. Seção regional de São Paulo). São Paulo. 1- (1949-). 3 nos a year. (U-1:726b; B-1:367a; BS:131b).
Supplementary English titles in table of contents and abstracts in English and Portuguese from no52 (O 1976).
Some English summaries from no26 (1957).
[Departamento de Geografia da Universidade de São Paulo, Caixa Postal 8105, 01000 São Paulo, SP, Brazil.]

226 BRAZIL. Exército. Diretoria do serviço geográfico.
Anuário. 1- (1948-) Annual.

227 ______. Instituto brasileiro de geografia e estatística.
Conselho nacional de geografia. Rio de Janeiro.
Biblioteca geográfica brasileira.
sA. Livros. 1-26 (1945-1971).
sB. Folhetos. 1-18 (1943-1958).
sC. Manuais. 2 (1948).

228 ______. ______. ______. São Paulo (state). Diretório regional. São Paulo.
Boletim. 1-3 (1942-1943). Closed 1943.

229 ______. ______. ______. Rio de Janeiro.
Curso de ferias para aperfeiçoamento de professores de geografia do ensino medio. 1966- (N70-1:846c; N70-1:849b).
1966 by Brazil. Conselho nacional de geografia.

230 ______.
Ministerio das relações exteriores. Mapoteca. Rio de Janeiro.
Bibliografia cartográfica. Serie A. 1- (1960-). (N70-1:851b; B68:74a).

231 CADERNOS de geografia. (Recife. Universidade católica de Pernambuco). Recife. v1 no1-3 (1967-1971). (N70-1:985b).

232 CONGRESSO brasileiro de geografia.
Anais. 1-10 (1909-1944).

233 CONGRESSO sul-riograndense de história e geografia. Rio Grande do Sul.
Anais. 1 (1936); 2 (1937) 3v; 3 (1940) 4v; 4 (1946) 2v. (U-2:1172b).

234 CURITIBA. Instituto histórico, geográfico, e etnográfico paranaense. Curitiba.
Boletim. v1-8 (?-1963) Irregular. (N70-1:1605b).
Superseded by Institute's Revista [235].

235 ______. ______.
Revista.
Succeeds Institute's Boletim [234].
[Rua Quinze de Novembro 1050, Curitiba, Paraná, Brazil.]

236 ______. Universidade do Paraná. Departamento do geografia. Curitiba.
Geografia do Brasil. 1 (1962). Only one issued. Closed 1962.
(N70-1:1606a).

237 CURSO de informações geográficas(para professores de geografia do ensino médio). (Brazil. Instituto brasileiro de geografia. Departamento de documentação, divulgação geográfica e cartográfica). Rio de Janeiro.
1961- (1962-). (N70-1:1618b).

238 FUNDAÇÃO I B G E. Rio de Janeiro.
Noticias. Quarterly. (N77).
[Avenida Augusto Severo 8, Rio de Janeiro, Brazil.]

239 GEGHP. (Gabinete de estudinhos de geografia e história da Paraíba).
João Pessoa. v3 (1937/1942?) Irregular.

*(240 GEOGRAFIA. (Associação de geografia teorética). Rio Claro. 1- (1976-)
Irregular.
[Associação de geografia teorética, Caixa postal 178, 13500 Rio Claro SP, Brazil.]

241 GEOGRAFIA. (Associação dos geógrafos brasileiros). São Paulo. vl-2 no4.
(1935-1936). Closed 1936. (U-2:1692a; F-2:727a).

242 GEOGRAFIA. (Associação dos geógrafos brasileiros). São Paulo. 1-
(Ap 1976-). (N77).

243 GEOGRÁFICA. (Sociedade geográfica brasileira). São Paulo. 1- (1951-).
(N70-2:2278a).
[Rua 24 de Maio 104, 3º andar, São Paulo, S.P., Brazil.]

244 INSTITUTO archeológico e geográphico alagoano, Maceió. Maceió.
Revista. 1-15 (1872-1931). Closed 1931. (U-3:2020c; F-4:137b).
None published 1888-1900; 1917-1923.

245 INSTITUTO arqeológico, histórico e geográfico pernambucano, Recife. Recife.
Revista. 1-42 (1863-1949). Closed 1949? (U-3:2020c; B-2:504a;
F-4:137b). nol-51 (1863-1901) lack volume numbers. Suspended My 1870-D 1882; F 1883-AP 1886. 1863-1919 as Instituto archeológico e geographico pernambucano.
Index: nol-100 (1863-1918) in 100.

246 INSTITUTO geográfico e histórico da Bahia. Salvador.
Revista. nol-79 (1894-1955) Irregular. (U-3:2025c; B-2:506a; F-4:147b).
1-7 as Revista trimensal. nol-41 also as vl-10; 1-46, año 1-27.
Index: vl-68 (1894-1942).

247 INSTITUTO geográfico e histórico dos Amazonas, Manaus. Manaus.
Revista. 1-7 (1917-1948) Irregular. (U-3:2025c).
Suspended 1918-1931, 1939-1947.

248 INSTITUTO histórico e geográfico brasileiro, Rio de Janeiro. Rio de Janeiro.
Boletim. 1930. Closed 1930. (B-2:506b).

(249 ______.
Revista. 1-109 (1839-1930); 164- (1939-) 4v a year. (U-3:2026b;
B-3:704b; F-4:147a).
1-13 (1839-1850) as Revista trimensal de história e geographia.
8-13 (1846-1850) as s2 vl-6; 14-19 (1851-1856) as s3 vl-6. v85-109 also as v139-163.
Indexes: 1-67 (1839-1906) in v68; 1-90 (1839-1921).

250 ______.
Revista. Tomo especial. 1- (1956-) Irregular. (N70-2:2838b).
Some issues unnumbered.

251 ______. Commissão central de bibliographia brasileira.
Boletim. 1 (1895). Closed 1895. (U-3:2026b; B-2:506b).
Also issued in French.

252 INSTITUTO histórico e geográfico de Alagoas. Maceió.
Revista. 16- (1932-). (N75-1:1130c).
Continues Instituto histórico de Alagoas. Revista (not in this list).

253 INSTITUTO histórico e geográfico de Goiás. Goiâna.
Revista. 1- (1972-) Irregular.

254 INSTITUTO histórico e geográfico de Juiz de Fora. Juiz de Fora, MG.
Revista. v1- (1965-). (N70-2:2838b; N75-1:1130c).
[Caixa postal 438, Juiz de Fora, Brazil.]

255 INSTITUTO histórico e geográfico de Minas Gerais. Belo Horizonte.
Revista. 1-3 (1943-1947); 4- (1957-) Irregular. (U-3:2026b; BS:428b).
[Rua dos Carijos 244, 9º andar, Caixa Postal, 122, Belo Horizonte, Minas Gerais, Brazil.]

256 INSTITUTO histórico e geográfico de Santa Catarina. Florianópolis.
Revista. v1-4 (1902-1915); 1943-1944. Closed 1944? (U-3:2026b).

257 INSTITUTO histórico e geográfico de Santa Maria.
Revista. 1- (1962-) Annual. (N70-2:2838b).

258 INSTITUTO histórico e geográfico de Santos.
Revista. 1- (1959-). (N70-2:2838b).

259 INSTITUTO histórico e geográfico de São Paulo. São Paulo.
Revista. 1- (1894/95) Irregular, about once a year. (U-3:2026b).

260 INSTITUTO histórico e geográfico de Sergipe. Aracajú.
Revista. 1-9 (no1-14) (1913-1929); no15- (1939-) Irregular. (U-3:2026b).
Suspended 1920-Je 1925; 1927-1930. 1-9 also as año 1-14.

261 INSTITUTO histórico e geográfico do Espírito Santo. Vitória.
Revista. v1- (1917-) Irregular (annual).
[Instituto histórico e geográfico do Estado do Espírito Santo, Rua D. Fernando 236, Vitória, Espírito Santo, Brazil.]

262 INSTITUTO histórico e geográfico do Pará. Belém.
Revista. v1-12 (1917-1951) Irregular. (U-3:2026b; F-4:138a).
Index: v1-6 (1917-1931) in v7.

263 INSTITUTO histórico e geográfico do Rio Grande do Norte. Natal.
Revista. 1-48/49 (1902-1951/52) Irregular. (U-3:2026b).

264 INSTITUTO histórico e geográfico do Rio Grande do Sul. Pôrto Alegre.
Revista. v1-29 (no1-120) (1921-1950); 31- no121- (1975-). (U-3:2026b; B-2:506b).
Not published 1951-1974.

265 INSTITUTO histórico e geográfico paraibano. João Pessoa.
Revista. 1- (1909-) Irregular, about every 5 years.
(U-3:2026b; B-2:506b).
Suspended 1912-22.

266 INSTITUTO histórico, geográfico e etnográfico do Pará. Belém.
Revista. v1 (no1-3) (1900). Closed 1900. (U-3:2026b).

267 INSTITUTO histórico, geográfico e genealógico de Sorocaba. Sorocaba, Minas Gerais.
Revista. v1- (1956-). (N70-2:2838b).
[Rua Rui Barbosa 78, Sorocaba, Brazil.]

268 MANÁUS. Instituto geográfico e histórico dos Amazonas. Manáus.
Boletim. 1- (1967?-). (N70-3:3627c).

269 MINAS GERAIS (state). Commissão geográfica e geológica.
Boletim. 1-2 (1894-1934). (ULG:72a; B-3:213a).

270 ______. Departamento geográfico. Belo Horizonte.
Boletim.

*271 NOTÍCIA geomorfológica. (Campinas. Universidade católica. Departamento de geografia). Campinas.
1- (1958-). (N70-1:1054a).
Abstracts in English.
[Notícia Geomorfológica, Universidade Católica de Campinas, Rua Marechal Deodoro, 1099, Caixa Postal 317, 13100, Campinas, São Paulo, Brazil.]

272 PANORAMA. (Santa Cruz do Rio Pardo. Instituto de educação Leônidas do Amaral Vieira. Centro de estudos geográficos "Moraes Rego").
v1 (no1-6) (Ag 1954-Ap 1955). (N70-3:4531b).
[Panorama, R. Prudente de Moraes, 1133, Santa Cruz do Rio Pardo, E.F.S., Estado de São Paulo, Brazil.]

273 PARANÁ (state). Departamento de geografia, terras e colonização. Curitiba.
Boletim. 1- (Jl 1953-). (N70-3:450c).

274 ______. ______. Divisão de geografia. Curitiba.
Boletim. 1-2 (1964-1967). Mimeographed. (N70-3:4540c).

275 ______. Universidade federal. Instituto de geologia. Curitiba.
Geografia física. 1-8 (1959-1967). Closed 1967? (N70-1:4541c).
no1 issued by university's Faculdade de filosofia, ciências e letras. Curso de geografia e história.
Name of university varies: Curitiba. Universade; Universidade do Paraná.
Some English summaries

276 PRESIDENTE PRUDENTE. Universidade estadual Paulista. Faculdade de filosofia, ciências e letras. Departamento de geografia. Presidente prudente.
Boletim. 1- (1969-) Annual. (N75-2:1770b).
[Departamento de Geografia, Universidad Estadual Paulista, Rua Roberto Simonsen 305, Caixa Postal 957, Presidente Prudente, SP, Brazil.]

277 RECIFE. Universidade. Departamento de geografia. Recife.
Serie VI. Geografia. 1- (S 1961-). (N70-3:4937b).
Subseries of Faculdade de filosofia de Pernambuco. Cadernos (not in this list).

278 REVISTA brasileira de cartografia. (Sociedade brasileira de cartografia). Rio de Janeiro.
1- (1970-). (N75-2:1868).

*(279 REVISTA brasileira de geografia. (Brazil. Secretaria de planejamento de presidência da república. Fundação instituto brasileiro de geografia e estatística). Rio de Janeiro. 1- (1939-) Quarterly. (U-4:3591c; B-3:695b; F-4:119b).
vl nol-v29 no2 (1939-1967) issued by Instituto brasileiro de geografia e estatística. Conselho nacional de geografia; v29 no3-v32 nol (1967-1970) Fundação IBGE. Instituto brasileiro de geografia; v32 no2-v35 nol (1970-1973) by Ministério de planejamento e coordenação geral. Fundação IBGE. Instituto brasileiro de geografia; v35 no2-v36 nol (1973-1974) by Ministério de planejamento e coordenação geral. IBGE; v36 no2- (1974-) by Secretaria de planejamento da presidência da república. IBGE.
v3 no2 (Ja/Mr 1941) contains Associação dos geógrafos brasileiros. Boletim nol. Contains separately paged Atlas de relações internacionais. (Caderno especial da revista brasileira de geografia). 1- (1967-) Quarterly.
Index: 1-10 (1939-1948).
Summaries in English and French.
[Av. Beira Mar 436, 11° andar, ZC-06 Rio de Janeiro RJ, Brazil.]

280 REVISTA de geografia e história. (Maranhão [state]. Diretorio regional de geografia). 1-2 (1946-1947). (U-4:3602b).

281 REVISTA geográfica universal. (Bloch). Rio de Janeiro. 1- (Ag 1974-) Monthly. (N77).
[Bloch Editores, Rua Frei Caneca, 511, Rio de Janeiro, Brazil].

282 RIO DE JANEIRO (city). Universidade do Brasil. Faculdade nacional de filosofia. Cadeira de geografia do Brasil. Rio de Janeiro. Publicação avulsa. 1-2 (1950). Closed 1950. (N70-3:5067b).

283 ______. ______. ______. ______.
Serie trabalhos discentes. 1-3 (1961). Closed 1961. (N70-3:5067a).

284 ______. ______. Centre de pesquisas de geografia do Brasil. Rio de Janeiro. Série bibliográfica. Publicação.
s1. Cartografia. Bibliografia cartográfica brasileira. 1-5 (1951-1955). Closed 1955. (N70-3:5067a).
s2. Geografia. Bibliografia geográfica brasileira. 1-5 (1951-1960). Closed 1960. (N70-3:5067a; N70-1:722a).

(285 ______. Universidade federal do Rio de Janeiro. Instituto de geociências. Rio de Janeiro.
Anuário. 1- (1977-) Annual.

286 ______. ______. ______.
Cadernos de mestrado. Programa de pós-gradução em geografia. CCMN. 1- (1975-) Irregular.

287 RIO DE JANEIRO (state). Departamento geográfico. Rio de Janeiro. Anuário geográfico do estado do Rio de Janeiro. 1-21 (1948-1968) Annual. Closed 1968.
Earlier issued by: Brazil. Instituto brasileiro de geografia e estatística. Conselho nacional de geografia.

288 SANTA CATARINA (state). Departamento estadual de geografia e cartografia. Florianópolis.
D.E.G.C. Boletim geográfico. v1-3 (nol-6) (1947-1949). Closed 1949. (U-2:1252a).

289 ______. ______.
Publicação.
Série 1. 1- (1952-) Irregular monographs. (N70-4:5177b).
Série 2. v1 (1955). (N70-4:5177b).

290 SÃO PAULO (city). Universidade. Faculdade de filosofia, ciências e letras. São Paulo.
Boletim. Geografia. 1-7 (1944-1951. Closed. (U-5:3780c; B-4:26b).
Irregular monographs.

291 ______. ______. Instituto (formerly Departamento) de geografia. São Paulo.
Publicações. no1 (1946). (U-5:3780c).

(292 ______. ______. ______.
Publicações. (N70-4:5185a). In separately numbered series:
Aerofotografia. 1- (1966-) Irregular. (24 [1976]). (N70-4:5185a).
Atlas de geografia. 1- (1975-).
Biblio-geo. 1- (1977-) Irregular. (2 [1978]).
Biogeografia. 1- (1969-) Irregular. (12 [1977]). (N70-1:779b; B73:129a).
Cadernos de ciência da terra. 1- (1969-) Irregular. (63 [1977]).
Cartografia. 1- (1972-) Irregular. (4 [1977]).
Climatologia. 1- (1971-) Irregular. (7 [1977]). (N75-2:1921b).
Geografia econômica. 1- (1966-) Irregular. (12 [1973]).
Geografia das indústrias. 1- (1969-) Irregular. (6 [1978]).
Geografia e planejamento. 1- (1969-) Irregular. (30 [1977]). N75-2:1921b).
Geografia urbana. 1- (1969-) Irregular. (19 [1977]). (N70-2:2278a).
Geomorfologia. 1- (1966-) Irregular. (54 [1977]). (N70-4:5185a; B73:129a).
Métodos em questão. 1- (1971-) Irregular. (16 [1977]). (N75-2:1921b).
Orientação. 1969?- . (N75-2:1672c).
Paleoclimas. 1- (1971-) Irregular. (3 [1977]). (N75-2:1921b).
Sedimentologia e pedologia. 1- (1971-) Irregular. (12 [1978]). (N75-2:1921b).
Serie teses e monografias. 1- (1969-) Irregular.
[Caixa postal 11.154. São Paulo (Capital), S.P., Brazil.]

293 ______. ______. Museu paulista. São Paulo.
Coleção. Serie de geografia. 1 (1975). Only one issued.
Supersedes in part Museu Paulista coleção (not in this list).

294 SÃO PAULO (State). Instituto geográfico e geológico. São Paulo.
Boletim. no1-21 (1889-1906); no22-31 (1930-1945); no32-33 (1952-1953). (ULG:74b; B-4:26b; F-1:653b; F-2:155b).
no1-21 (1889-1906) and 22 (1930) as Brazil. São Paulo. Commissão geográphica e geológica. Boletim. Not published 1907-1929.

295 ______. ______.
Notas previas. 1-6 (1938-1965) Irregular. Monographs. (N70-4:5186c).
[Caixa Postal 8770, São Paulo (capital), S.P., Brazil.]

296 ______. ______.
O.I.G.G. Revista do Instituto geográfico e geológico. São Paulo.
1- (1943-). (U-5:3779c; B-4:26a).
Scientific articles preceded by English summaries.
[Rua Antonio de Godoy, 122 8° andar, São Paulo, Brazil.]

297 ______. ______.
Relatório. (N77).
[Secretaria da Agricultura, Coordenadoria da Pesquisa de Recursos Naturais, São Paulo, Brazil.]

298 SOCIEDADE brasileira de cartografia. Rio de Janeiro.
Boletim. 1- ? (1950-1969). (N70-4:5376b).
Superseded in 1970 by society's Revista [300].

299 ______.
Relatorio anual do presidente. Annual. (N70-4:5376b).

300 ______.
Revista. 1- (1970-) 3-4 a year. (N70-4:5376b; B68:481a).
Succeeds Society's Boletim [298].
[Rua Mexico 41, Sala 706, Rio de Janeiro, Brazil.]

(301 SOCIEDADE brasileira de geografia. Rio de Janeiro.
Boletim. 1- (Jl/Ag 1950-) Irregular. (N70-4:5376b).
Suspended Jl 1951-Mr 1963. Resumed with v3 no1 (Ap/S 1963).
1885-1945 as Sociedade de geografia do Rio de Janeiro.

302 ______.
Geographia do Brasil. 1-10 (1922). Closed 1922. (U-5:3913c; B-2:268b).
By the society under its earlier name, Sociedade de geografia do Rio de Janeiro.

303 ______.
Revista. 1- (1885-) Annual. Publication suspended at various times; resumed v55- (1958-). (U-5:3913c; N70-4:6676b; F-4:121b-122a).
1- (1885) as Boletim; 1-52 (1885-1945) as Sociedade de geografia do Rio de Janeiro. Revista.
Index: v1-9 (1885-1893) in 23/24 (1909/1911).
[Praça da República 54, 1° andar, Rio de Janeiro, Brazil.]

304 SOCIEDADE cearense de geografia e história, Fortaleza. Fortaleza.
Revista. 1-7 (1935-1941). (U-5:3914a).

305 SOCIEDADE de geographia de Lisboa. Seção no Brasil. Rio de Janeiro.
Revista. 1-3 (Ap 1881-Ja/F 1885); s2 no1-4 (S1885-Ja/Mr 1886).
Closed 1886. (U-5:3914b; B-4:108b).

306 UNIVERSIDADE federal de Pernambuco. Departamento de geografia. Recife.
s6. Geografia. 1- (S 1961-) Bimonthly. (N70-3:4937c; N70-4:6094b).
Subseries of: Faculdade de filosofia de Pernambuco. Cadernos (not in this list). 1961-1966 as: Recife. Universidade.

BULGARIA

Narodna Republika Bŭlgariya (Bulgarian)
НАРОДНА РЕПУБЛИКА БЪЛГАРИЯ

*(307 РЕФЕРАТИВНЫЙ бюлетень Болгарской научной литеруры: геология и география.

308 БЪЛГАРИЯ. Министерство на войната. Географски институт. Годишник.

309 БЪЛГАРСКА академия на науките. Географски институт. София. Известия.

310 БЪЛГАРСКО географско дружество. София. Географска библиотека.

*(311 _____. Известия.

312 ГЕОГРАФИЯ. Научно-популярно списание. (Българско географско дружество). София.

313 ГЕОГРАФСКИ преглед. Научно-популярно списание. (Българско географско дружество). София.

314 ИСТОРИЯ и география. София.

315 ОБУЧЕНИЕТО по география: методическо списание. София.

*(316 ПРОБЛЕМИ на географията. (Българска академия на науките). София.

317 СБОРНИК от статии по картография. (България. Главно управление по геодезия и картография). София.

*(318 СОФИЯ. Университет. Геолого-географски факултет. София. Годишник. Книга 2. География.

319 СТРАНИ и народи. Популярно списание по география. (Българско географско дружество). София.

*(307 ABSTRACTS of Bulgarian scientific literature: geology and geography. (Bulgarian academy of sciences. Centre for scientific information of natural, mathematical and social sciences). Sofiya. v1- (1957/58-). 2 nos a year. (N70-1:32a; BS:724a; R-1:78a).
Issued in 2 editions: English and Russian. Russian title: Referativnyi biulleten' bolgarskoi nauchnoi literatury: geologiia i geografiia.

308 BULGARIA. Ministertvo na voinata. Geografski institut.
Godīshnīk. 1922-1932. Closed 1932.
In Bulgarian with some French abstracts.

309 BŬLGARSKA akademiia na naukite, Sofiya. Geografski institut. Sofiya.
Izvestiia. 1-16 (1951-1974). Irregular. (N70-1:931c; R-1:40b).
In Bulgarian with summaries in French and Russian.
[Ul Ivan Vazov no. 13, Sofiya, Bulgaria.]

310 BŬLGARSKO geografsko druzhestvo, Sofiya. Sofiya.
Geografska biblioteka. no1-5 (1920-1935). Closed 1935. (U-1:830b).

*(311 ______.
Izvestiia. 1-10 (1933-1942); nsv1- (11-) (1953-). Irregular. (U-1:830b; B-1:445a; BS:157b; FC-1:301; R-1:402).
In Bulgarian. 1-10 (1933-1942) with French or German abstracts and titles in tables of contents: nsv1- (1953-) with English, French, or German abstracts and titles in tables of contents.

312 GEOGRAFIIA. Nauchno-populiarno spisanie. (Bŭlgarsko geografsko druzhestvo). Sofiya. 1- (1950-) 10 nos. a year. (N70-2:2279b; BS:341b; R-1:25b-26a).
Supersedes Geografski pregled [313].

313 GEOGRAFSKI pregled. Nauchno-populiarno spisanie. (Bŭlgarsko geografsko druzhestvo). Sofiya. 1-4 (1946-1950). Closed 1950. (BS:341b; R-1:26a).
Superseded by Geografiia [312].

314 ISTORIIA i geografiia. Sofiya. 1- (1958-) Bimonthly. (N70-2:3030b; R-1:52b).
[D. N. "Naroda Prosveta," Ul. V. Drumev 37, Sofiya 5, Bulgaria.]

315 OBUCHENIETO po geografiia; metodichesko spisanie. 2- (1969-). Sofiya. (N70-3:4340a).

*(316 PROBLEMI na geografiiata (Bulgarska akademiia na naukite). Sofiya. 1- (1974-). (N76).
English title: Problems of geography. Supplementary English table of contents and abstracts.
[Izd-vo na Bulgarskata Akademiia na Naukite, Ul. "G. Penkovski," 3, Sofiya, Bulgaria.]

317 SBORNIK ot statii po kartografiia. (Bulgaria. Glavno upravlenie po geodeziia i kartografiia). Sofiya. 1959- Annual. (N70-4:5205a; R-1:80b).

*(318 SOFIYA (Sofia). Universitet. Geologo-geografski fakultet. Sofiya.
Godīshnīk. Annuaire. Kniga 2. Geografiia. v44- (1948-). (N70-4:5405c; N70-4:5405a; BS:812a).
To 1948 name of issuing agency was Istoriko-filologicheski fakultet; v44-56 (1948-1963) by Biologo-geologo-geografski fakultet, as Kniga 3. Geografiia.
In Bulgarian with supplementary titles in table of contents and abstracts in English, French, or German.

319 STRANI i narodii. Populiarno spisanie po geografiia. (Bŭlgarsko geografsko druzhestvo). Sofiya. 1-3 (1947-1949).

CAMEROON (French or English)

République Unie du Cameroun

(320 UNIVERSITÉ de Yaoundé. Département de géographie. Yaoundé. Cahiers. 1- (1977-).

CANADA (English or French)

(321 ALBERTA. University, Edmonton. Department of geography. Edmonton. Studies in geography. In 3 series.
Bibliography. 1- (1971-) Irregular. (N75-2:2062c: N76; DP).
[Department of Geography, University of Alberta, Edmonton, Alta., T6G 2H4, Canada.]

(322 ______. ______. ______.
Monograph. 1- (1974-) Irregular.

(323 ______. ______. ______.
Occasional paper. 1- (1970-) Irregular.

*(324 ALBERTAN geographer. (Publication of the graduate students, Department of geography, University of Alberta). Edmonton. 1- (1964/65-) Annual. (N70-1:209c; B68:27b).
Index: v1-5 (1964/65-1969) in v6 (1970); v8-13 (1972-1977) in v14 (1978), p. 119-121.
[Department of Geography, University of Alberta, Edmonton, Alberta T6G 2H4, Canada.]

325 ARCTIC bibliography (Arctic institute of North America). Washington, D.C., Montreal.
1-16 (1953-1975) Irregular. Closed 1975. (N70-1:432b).
v1-11 (1953-1963), prepared for and in cooperation with the U.S. Department of defense.
v12-16 (1965-1975), prepared with support of government agencies of the United States and Canada.

326 ARROW. (Geography club). Montreal. 1- (1970-). (N75-1:180b).
Supersedes: Montreal-Quebec arrow (not in this list).
[Gallery Publications, Ltd., 1165 Green Avenue, Montreal, Quebec H3Z 2A2, Canada.]

(327 ASSOCIATION des géographes du Québec. Québec.
Bulletins. 1- (1962-) Annual. (N70-1:506b).
10-12 (1966-1969) as Association des géographes de l'Amérique française.
[L'Association des géographes du Québec, Case postale 189, Saint-Redempteur, Cte Levis, P.Q., GOS 3BO Canada.]

*(328 ASSOCIATION of Canadian map libraries. Association des cartothèques canadiennes. Ottawa.
Bulletin 1- (1968-) 3 times a year. (N75-1:198c-199a).
1-15 (1968-1974) as Association's Newsletter.
[c/o National Map Collection, Public Archives of Canada, Ottawa, Ontario K1A 0N3, Canada.]

(329 ______. Proceedings of the annual conference. Comptes rendus des conférences annuelles. 1- (1967-) Annual. (N75-1:199a).
1 (1967) as National conference on Canadian map libraries. Proceedings. Comptes rendus.

(330 ASSOCIATION québecoise pour l'étude du Quaternaire. Sherbrooke, Québec.
Bulletin AQQUA. 1- (1975-) 4 nos. a year.
In French.
[Département de Géographie, Université de Sherbrooke, Sherbrooke, Québec J1K 2R1, Canada.]

*(331 B. C. GEOGRAPHICAL series. (Tantalus research, ltd.). Vancouver. 1- (1965-) Irregular. (N70-1:602b).
Supersedes Canadian association of geographers. British Columbia division. Occasional papers [354].
[Tantalus Research Ltd., P.O. Box 34248, 2405 Pine Street, Vancouver, B.C., V6J 4NB, Canada.]

332 B I G. (Québec. Université Laval. Institut de géographie). Québec.
1- (D 1959-). (N70-1:603c).
Initials stand for Bulletin de l'institut de géographie.

*(333 CAHIERS de géographie du Québec. (Québec. Université Laval. Institut de géographie). ns1- (O 1956-) 3 nos. a year. (N70-1:989bc; BS:170b).
v1-21, no1-54 (1956-1977) as Cahiers de géographie de Québec.
Supersedes the institute's two series: Cahiers de géographie [406]; and Notes de géographie [407].
In French. Abstracts in French and English.
Index 1-20 (1956-1966) as no21 (1966).
[Les Presses de l'Université Laval, Boîte postale 2447, Québec, P.Q. G1K 7R4, Canada.]

334 CANADA. Board on geographical names. Ottawa.
Memoir. 1- (1953-). (N70-1:1056b).

(335 ______. Canadian permanent committee on geographical names. Ottawa.
Gazeteer of Canada. 1952- Irregular. (N70-1:1087c).
By province and territory with irregular revision.
Annual cumulative supplements, 1973- . (N70-1:1087c; B68:222b).
Special supplements: 1-2 (1964-1967). (N70-1:1087c; B68:222b).
v1-9 issued by Board on geographical names.
v10- issued by geographical branch.
[Surveys and Mapping Branch, Department of Energy, Mines and Resources, Ottawa, Ontario K1A 0E4, Canada.]

(336 ______. ______.
Toponymy study series. 1973- Irregular.
[Surveys and Mapping Branch, Department of Energy, Mines and Resources, Ottawa, Ontario K1A 0E4, Canada.]

337 ______. ______.
CANOMA. 1- (1975-) Semiannual.
Index: v1-4 (1975-1978) in v4 no2.
[Surveys and Mapping Branch, Department of Energy, Mines and Resources, Ottawa, Ontario K1A 0E4, Canada.]

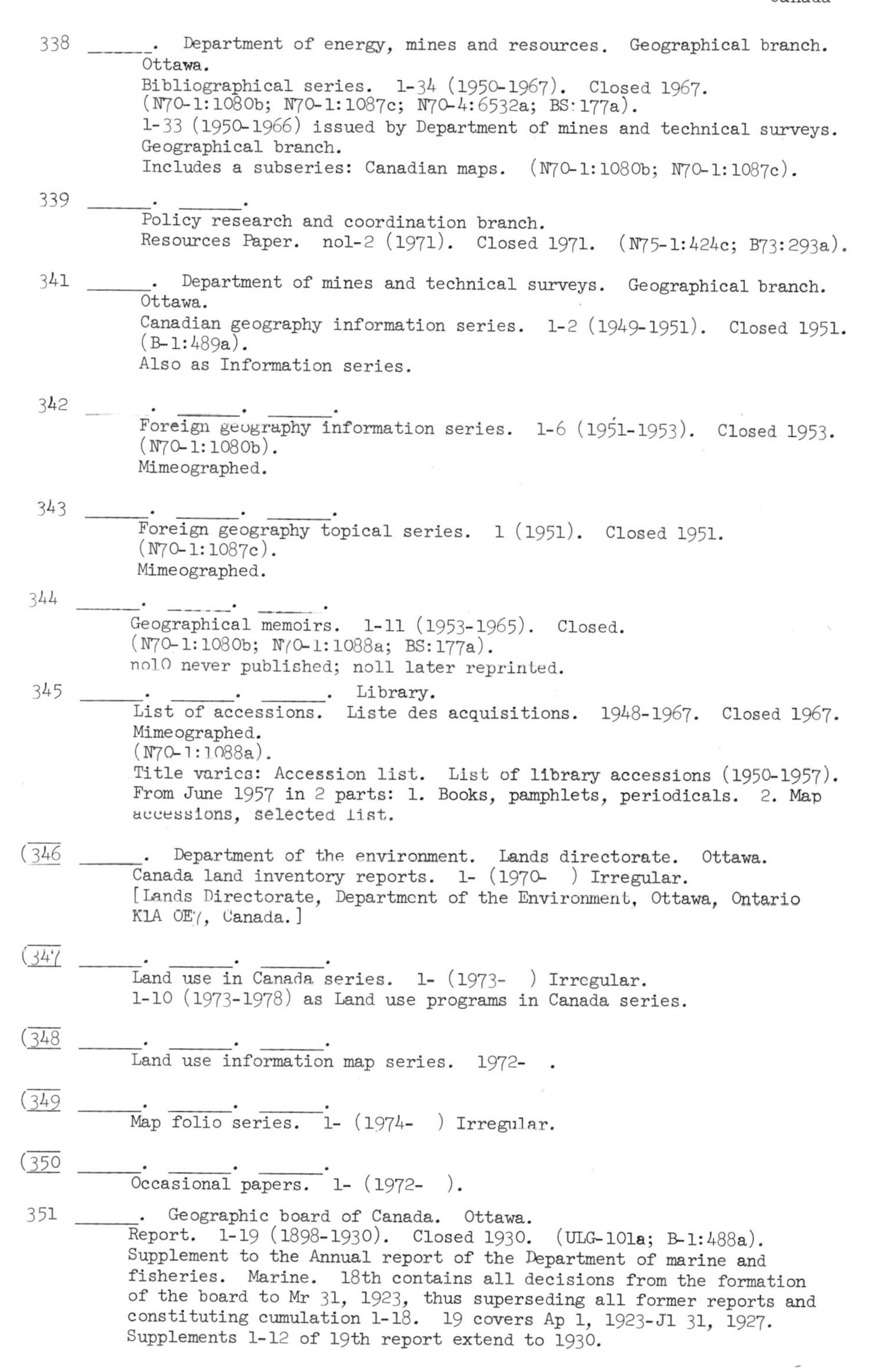

338 ______. Department of energy, mines and resources. Geographical branch. Ottawa.
Bibliographical series. 1-34 (1950-1967). Closed 1967. (N70-1:1080b; N70-1:1087c; N70-4:6532a; BS:177a).
1-33 (1950-1966) issued by Department of mines and technical surveys. Geographical branch.
Includes a subseries: Canadian maps. (N70-1:1080b; N70-1:1087c).

339 ______. ______.
Policy research and coordination branch.
Resources Paper. no1-2 (1971). Closed 1971. (N75-1:424c; B73:293a).

341 ______. Department of mines and technical surveys. Geographical branch. Ottawa.
Canadian geography information series. 1-2 (1949-1951). Closed 1951. (B-1:489a).
Also as Information series.

342 ______. ______. ______.
Foreign geography information series. 1-6 (1951-1953). Closed 1953. (N70-1:1080b).
Mimeographed.

343 ______. ______. ______.
Foreign geography topical series. 1 (1951). Closed 1951. (N70-1:1087c).
Mimeographed.

344 ______. ______. ______.
Geographical memoirs. 1-11 (1953-1965). Closed. (N70-1:1080b; N70-1:1088a; BS:177a).
no10 never published; no11 later reprinted.

345 ______. ______. ______. Library.
List of accessions. Liste des acquisitions. 1948-1967. Closed 1967.
Mimeographed.
(N70-1:1088a).
Title varies: Accession list. List of library accessions (1950-1957).
From June 1957 in 2 parts: 1. Books, pamphlets, periodicals. 2. Map accessions, selected list.

(346 ______. Department of the environment. Lands directorate. Ottawa.
Canada land inventory reports. 1- (1970-) Irregular.
[Lands Directorate, Department of the Environment, Ottawa, Ontario K1A 0E7, Canada.]

(347 ______. ______. ______.
Land use in Canada series. 1- (1973-) Irregular.
1-10 (1973-1978) as Land use programs in Canada series.

(348 ______. ______. ______.
Land use information map series. 1972- .

(349 ______. ______. ______.
Map folio series. 1- (1974-) Irregular.

(350 ______. ______. ______.
Occasional papers. 1- (1972-).

351 ______. Geographic board of Canada. Ottawa.
Report. 1-19 (1898-1930). Closed 1930. (ULG-101a; B-1:488a).
Supplement to the Annual report of the Department of marine and fisheries. Marine. 18th contains all decisions from the formation of the board to Mr 31, 1923, thus superseding all former reports and constituting cumulation 1-18. 19 covers Ap 1, 1923-Jl 31, 1927.
Supplements 1-12 of 19th report extend to 1930.

(351a ______. National commission for cartography. Chronicle. Chronique. Ottawa.
1- (1976-) Quarterly. (GZS).
In English or French.
[Department of Geography, Carleton University, Ottawa K1S 5B6, Canada.]

(352 CANADIAN association of geographers. Association canadienne des géographes.
Directory. Annuaire. 1964- Irregular. (N75-1:453b; N79).
Also as Membership list. Liste des membres.
1965-1976 in Canadian association of geographers. Newsletter [353].
Present title from 1978.

353 ______.
Newsletter. Nouvelles. 1965-1976. Closed 1976. Montréal.
(N70-1:1104c).
Succeeded by Association's Directory [352].
[Canadian Association of Geographers, Burnside Hall, McGill University,
805 Sherbrooke Street West, Montreal H3A 2K6, Canada.]

354 ______. British Columbia division. Vancouver.
Occasional papers in geography. no1-7 (1960-1965). Closed 1965.
(N70-1:1104c; B68:397a).
Called also Occasional papers. Superseded by B. C. geographical series
[331].

355 ______. Education committee.
Bulletin. 1- (1956-) Irregular. (N70-1:1104c; BS:179a).
Mimeographed. Title varies: Education bulletin; Bulletin for Canadian
high school teachers.
[Canadian Association of Geographers, Education Committee, Burnside
Hall, McGill University, 805 Sherbrooke Street West, Montréal, Québec
H3A 2K6, Canada.]

(356 ______. Prairie division.
Newsletter. 1- (1977-) Semiannual.
v2 no2 (1978) is a collection on conference papers.
[Department of Geography, University of Winnipeg, 515 Portage Avenue,
Winnipeg, Manitoba R3B 2E9, Canada.]

357 ______. Western division. Vancouver.
Papers from British Columbia meetings. (N70-1:1104c).
Earlier listed as British Columbia division.

*(358 CANADIAN cartographer. Toronto. 1- (1964-) Semiannual. (N70-1:1106b;
B73:64a).
1-4 (1964-1967) as the Cartographer. (Ontario institute of chartered
cartographers). 14- (1977-). Official journal of the Canadian
cartographic association.
Index and abstracts: v1-10 (1964-1973) in Cartographica, monograph
no12 (1974) [363]; v11-15 (1974-1978) in Cartographica 24 (1979).
Abstracts in English, French, German, and Spanish.
[Bernard V. Gutsell, Department of Geography, York University,
4700 Keele Street, Downsview, Ontario M3J 1P3, Canada].

(359 CANADIAN committee for geography. Comité canadien de géographie
(International geographical union. Union géographique internationale).
Ottawa.
Newsletter. Bulletin de nouvelles. 1- (S 1972-) 2-3 a year
(N75-1:463b; N76).
Earlier as Canadian national committee for geography. Newsletter.
In English and French.
Issued in co-operation with the Canadian association of geographers.
[Canadian Committee for Geography, Environment Canada, Ottawa, K1A OH3,
Canada.]

*(360 CANADIAN geographer. Géographe canadien. (Canadian association of geographers. Association canadienne des géographes). Toronto.
nol-17 (1950-1960) Irregular; v5- (1961-) Quarterly. (N70-1:1111bc; BS:179b).
Index: 1951-1967.
In English or French; abstracts at back of each number in English and French.
[The Secretary, Canadian Association of Geographers, Burnside Hall, McGill University, 805 Sherbrooke Street West, Montréal, Québec H3A 2K6, Canada.]

*(361 CANADIAN geographic (Royal Canadian geographical society).
Ottawa. 1- (1930-) Bimonthly. (U-2:908a; N78; B-1:495b-496a; F-2:23a).
1930-June 1978 as Canadian geographical journal. Monthly until 1976.
Index: 1-59 (1930-1959); 1-99 (1930-1979) in preparation.
[Royal Canadian Geographical Society, 488 Wilbrod Street, Ottawa K1N 6M8, Canada.]

362 CARLETON university, Ottawa. Department of geography. Ottawa.
Laboratory technique reports. LTR-1- (1972-). (DP).
[Department of Geography, Carleton University, Colonel By Drive, Ottawa, Ontario K1S 5B6, Canada.]

*(363 CARTOGRAPHICA. (York university, Toronto. Department of geography. B. V. Gutsell). 1- (1971-). (N75-1:477a). Index: 11-22 (1974-1978).
[Bernard V. Gutsell, Department of Geography, York University, 4700 Keele Street, Downsview, Ontario M3J 1P3, Canada.]

364 CÉTOF. Bulletin de liaison des professeurs de géographie de la région de Québec. 1- (1974-). Mimeographed.

365 COMITÉ international des historiens et géographes de langue française. Montréal.
Bulletin de liaison. 1- (1971-).

366 CONTACT: Journal of urban and environmental affairs. (Waterloo, Ontario. University. Division of environmental studies). 3- (1971-). (N75-1:599a).

367 DIDACTIQUE: géographie (Société des professeurs de géographie du Québec). Montréal. 1- (1972-) 3 nos a year. (N78).
[Laboratoire de didactique-géographie, Faculté des science de l'education, Université de Montréal, C.P. 6128, Montréal 101, P.Q., Canada.]

368 DIRECTORY of Canadian geography. Répentoire de la géographie canadienne. (Canadian committee for geography. Environment Canada). 1- (1972-) Irregular. (N76).
[Canadian Committee for Geography, Environment Canada, Ottawa K1A 0H3, Canada.]

369 ÉTUDE toponymique. (Québec [Province]. Commission de géographie). ns1- (1968-). (N70-2:1968b).

370 GEOGRAPHICAL association of Montreal. Montreal.
Bulletin. no1 (1943). Only one issued. (U-2:1694a).

371 GEOGRAPHICAL bulletin. (Canada. Department of mines and technical surveys. Geographical branch). Ottawa. no1-22 (1951-1964) without v nos; v7-9 (1965-1967). Closed 1967. (N70-2:2281ab; B-1:489a; BS:177a).
v8 no3-v9 no4 (1966-1967) issued by Department of energy, mines and resources.

Indexes: 1-4 (1951-1953), 5-8 (1954-1956); 9-12 (1956-1959), 13-16 (1959-1961), 17-20 (1962-1963), 21-22 (1964); v7 (1965), v8 (1966), v9 (1967).
In English or French.

*(372 GEOGRAPHICAL paper. Étude géographique. (Canada. Department of the environment. Lands directorate). Ottawa. 1- (1950-) Irregular. (N70-1:1080b; N70-1:1087c-1088a; N70-4:6532a; N76; BS:177a).
nol-10 (1950-1957) as Miscellaneous series. 1-39 (1950-1955) issued by Department of mines and technical surveys. Geographical branch; 40-49 (1967-1971) issued by the Department of energy, mines and resources, first by the Geographical branch, later by the Policy planning branch or Policy research and coordination branch; 50- (1972-) issued by the Department of the environment. Lands directorate.
Includes subseries: Gulf of St. Lawrence ice survey (N70-1:1088a).
[Land Directorate, Department of the Environment, Ottawa K1A OH3, Canada.]

*(373 GÉOGRAPHIE physique et quaternaire. (Montréal. Université. Département de géographie). Montréal. 1- (1947-) 4 per annum. (U-4:3628b; N78; B-3:709a; B68:483a).
Supersedes the Bulletin des sociétés de géographie de Québec et de Montréal [418]. 1-17 (1947-1963) as Revue canadienne de géographie. (Société de géographie de Montréal [and] Montréal. Université. Institut de géographie); 18-30 (1964-1976) as Revue de géographie de Montréal.
Index: 1947-1963 [v1-17] in v18, nol (1964), p. 125-129.
Mainly in French but some articles in English. Abstracts in English, French, and a third language (which varies) precede each major article.
[Les Presses de l'Université de Montréal, B.P. 6128, "Succ. A," Montréal, Québec H3C 3J7, Canada.]

374 GEOLAB. Bulletin de la société des professeurs de géographie du Québec. Québec. 1- (1966-).

(375 GEOSCOPE. (Ottawa. University. Geographers' association). Ottawa. 1- (1970-) Semiannual. (N75-1:916a; B73:129b).
Index: 1-9 (1970-1978) in v10 nol (Ap 1979), p99-109.
In English or French. Abstracts in French and English.
[Department of geography and regional planning, University of Ottawa, Ottawa, Ontario K1N 6N5, Canada.]

(376 GEOSCOPE: a journal for the geography teacher. (Québec Provinical association of geography teachers). Montréal. 1- (1966-) 3 nos a year.
v1-6 (1966-1973) contained nos1-16 continuously numbered. From v7 (1974) numbers are internal to each volume.
Preceded by the Association of geography teachers of Montreal. Newsletter (Ja 1965-Ja 1966), 4 nos (not in this list).
[Provincial Association of Geography Teachers (Québec), P.O. Box 32, Station Notre Dame de Grâce, Montréal, P.Q., H4A 3P4, Canada.]

(377 GUELPH, Ontario. University. Centre for resources development. Guelph. Occasional papers. 1- (1967-) Irregular.

(378 ______. ______. Department of geography. Guelph.
Geomorphology series. 1- (1969-) Biennial. (DP).
Part of this series also listed as Geographical publication in N75-1:906c and B73:128a but that should not be considered a separate series.
[Department of Geography, University of Guelph, Guelph, Ontario N1G 2W1, Canada.]

(379 ______. ______. ______.
Rural geography series. 1- (1975-). (DP).
[Department of Geography, University of Guelph, Guelph, Ontario N1G 2W1, Canada.]

380 INDEX de revues de géographie de langue française. (Comité international des historiens et géographes de langue française). Chicoutimi, Québec 1-v3 no4 (1973/74-1976) Quarterly, with annual cumulation in no 4. Closed 1976. (N75-1:1061a).
Preliminary no. 0, Ap 1973.

(381 LAURENTIAN university, Sudbury. Department of geography. Sudbury, Ontario. 1- (1978-) Irregular.
In English or French.
[Department of Geography, Laurentian University, Sudbury, Ontario P3E 2C6, Canada.]

382 LIST of theses and dissertations on Canadian geography. Liste des thèses et dissertations sur la géographie du Canada. Ottawa. 1972 (N76).
Supplement 1972-1975.
[Lands Directorate, Department of the Environment, Ottawa, Ontario K1A 0E7, Canada.]

*(383 McGILL university, Montréal. Department of geography. Montréal.
Climatological bulletin. 1- (1967-) 2 nos a year. (N70-3:3694b; B68:138a).
In English or French.
[Department of Geography (Climatology), McGill University, 805 Sherbrooke Street West, Montréal, P.Q. H3A 2K6, Canada.]

*(384 ______. ______.
Climatological research series. 1- (1966-) Irregular. Processed. (N70-3:3694b; B68:138a: DP).
[Department of Geography (Climatology), McGill University, 805 Sherbrooke Street West, Montréal, P.Q. H3A 2K6, Canada.]

385 ______. ______.
Publications in tropical geography. Savanna research series. 1- (1964-). (N70-3:3694b).
[Department of Geography, McGill University, 805 Sherbrooke Street West, Montréal, P.Q. H3A 2K6, Canada.]

386 ______. Subarctic research laboratory.
McGill subarctic research papers. 1- (1956-) Irregular. (N70-3:3695a).
Title varies. Miscellaneous papers.
Publication. Earlier published by the Department of geography (N70-3:3694b) later by the Centre for northern studies and research.
[Centre for Northern Studies and Research, McGill University, 1020 Pine Avenue West, Montréal, P.Q. H3A 1A2, Canada.]

387 McMASTER university, Hamilton, Ontario. Department of geography.
Discussion papers no 1- (1974-). (N77).
[Department of Geography, McMaster University, Hamilton, Ontario, L8S 4K1, Canada.]

(388 MANITOBA geographical studies (Manitoba. University. Department of Geography). Winnipeg. 1- (1973-). (N75-2:1399b; DP).
[Department of Geography, University of Manitoba, Winnipeg, Manitoba, R3T 2N2, Canada.]

389 MANITOBA geography teacher. (Manitoba teachers' society). Winnipeg. 1- (fall 1968-). (N77).
[191 Harcourt Street, Winnipeg 12, Manitoba, Canada.]

(390 MONOGRAPH. (Ontario association for geographic and environmental education). Place of publication varies. 1960- Quarterly. (N70-3:3873c).
Earlier issued by Ontario geography teachers association.
[c/o Doug McLeod, Ministry of Education, 759 Hyde Park Road, London, Ontario, Canada.]

(391 MONTRÉAL. Université. Département de géographie. Montréal.
Notes et documents. 1- (1977-) Irregular.
In French.
[Département de Géographie, Université de Montréal, Case Postale 6128, Succursale "A," Montréal, P.Q. H3C 3J7, Canada.]

392 ______. ______. Service de biogéographie. Montréal.
Bulletin. 1-26 (D 1945-O 1961). (U-3:2752b; BS:570b).

(393 MUSK-OX. (Saskatchewan. University. Institute of northern studies, Musk-ox circle). Saskatoon. 1- (1967-) Semiannual. (N70-3:3951c-3952a; B73:222b).
[Director, Institute of Northern Studies, University of Saskatchewan, Saskatoon, Saskatchewan, S7N OWO, Canada.]

*(394 NATURAL hazard research working paper. (Toronto; Boulder, Colorado. 1- (1968-) Irregular. Mimeographed. (N75-2:1542a; DP).
[Gilbert F. White, Institute of Behavioral Science, University of Colorado, Boulder, Colorado 80302 USA.]

(395 NEWFOUNDLAND. Memorial university, St. John's. Department of geography. St. John's.
Research notes. 1- (1976-) Irregular.
[Department of Geography, Memorial University of Newfoundland, St. John's, Newfoundland, A1B 3X9, Canada.]

(396 ONTARIO association for geographic and environmental education. London. (Place of publication varies).
Monograph. 1- (1960-) Quarterly. (N77).
Continues Ontario geography teachers association. Monograph [390].
[Ontario association for geographic and environmental education, c/o Doug McLeod, Ministry of Education, 759 Hyde Park Road, London, Ontario, Canada.]

*(397 ONTARIO geography. London, Ontario. (University of Western Ontario, Department of geography). London. no1- (1967-) 2 nos a year. (N70-3:4396ab; B68:401a). Index: 1-12 (1967-1978) in 12 (1978), p. 102-120.
[Department of Geography, The University of Western Ontario, London, Ontario N6A 5C2, Canada.]

(398 OTTAWA. University. Department of geography and regional planning.
Occasional paper. Travaux. 1- (1971-). (N75-2:1679b: B73:240b; DP). no1 issued by Department of geography.
In English or French.
[Département de Géographie et d'Aménagement Régional, 78. est rue Laurier, Ottawa, Ontario, K1N 6N5, Canada.]

399 ______. ______. ______.
Research notes. Notes de recherches. 1- (1972-). (N75-2:1679b; DP).
In English or French.

400 QUÉBEC (city). Université Laval. Centre d'études nordiques. Québec.
Collection nordicana. no1-(1964-) Irregular. (N70-3:4876a; B68-565a).
1-29 (1964-1970) as Travaux divers.
In English or French, with abstract in other language.

401 ______. ______. ______.
Travaux et documents. 1- (1963-) Irregular. (N70-3:4876a; B68:565a).
In French with English abstracts.

402 ______. ______. Centre de recherches en aménagement et en développement.
Documents de travail.
In French.

(403 ______. ______. Département de géographie. Québec.
Notes et documents de recherche. 1- (1974-) Irregular.
In French or English.
[Département de géographie (publications), Université Laval, Ste-Foy, P.Q. G1K 7P4, Canada.]

404 ______. ______. ______.
Travaux. 1- (1970-) Irregular. (N75-2:1812c).
1 (1970) issued by Institut de géographie.
In French.
(Les Presses de l'Université de Laval, La Librairie, B.P. 2447, Québec, P.Q. G1K 7R4, Canada.]

405 ______. ______. Groupe d'étude de choronymie et de terminologie géographique (GECET). Québec.
Publications [Choronoma]. 1- (1966-) Irregular. (N70-2:2423c).
In French.
[Les Presses de l'Université Laval, La Librairie, B.P. 2447, Québec, P.Q. G1K 7R4, Canada.]

406 ______. ______. Institut de géographie. Québec.
Cahiers de géographie. 1-7 (1952-1956). Closed 1956. (N70-3:4876b; BS:170b).
Formerly as Institut d'historie et de géographie. Superseded by Cahiers de géographie de Québec [333].
In French.

407 ______. ______. ______.
Notes de géographie. 1-8 (1952-1956). Closed 1956. (N70-3:4876b; BS:632b).
Formerly as Institut d'historie et de géographie. Superseded by Cahiers de géographie du Québec [333].
In French.

408 QUÉBÉC (province). Ministère de l'industrie et du commerce. Service de géographie. Montréal.
Publication. 1-3 (1955-?). Closed with no3. (N70-3:4879b).
In French

(409 REGINA geographical studies (Regina. University. Department of geography). Regina.
1- (1977-) Irregular.
[Department of Geography. University of Regina, Regina, Saskatchewan S4S 0A2, Canada.]

(410 SAINT MARY'S university. Department of Geography. Halifax, Nova Scotia.
Atlantic region geographical studies. 1- (1977-) Irregular. (DP).
[Department of Geography, Saint Mary's University, Halifax, N.S. B3H 3C3, Canada.]

(411 ______. ______.
Occasional papers in geography. 1- (1977-) Irregular. (DP).
[Department of Geography, Saint Mary's University, Halifax, N.S. B3H 3C3, Canada.]

(412 SHERBROOKE, Québec. Université. Association des diplômes en géographie de l'Université de Sherbrooke. Sherbrooke.
Bulletin d'information de l'ADGUS. 1- (1975-) 3 or 4 nos a year.
In French.
[Association des diplômes en géographie de l'Université de Sherbrooke, Sherbrooke, C.P. 394, Québec J1H 5J7, Canada.]

*(413 ______. ______. Département de géographie. Sherbrooke.
Bulletin de recherche. 1- (1972-) 6 nos a year. (DP).
In French.
[Département de géographie, Université de Sherbrooke, Sherbrooke, Québec J1K 2R1, Canada.]

414 SHIELD. (Toronto. University. Graduate geographers' association). Toronto. 1-7 (1950-1956). Closed 1956. (N70-4:5308c).

(415 SIMON FRASER university. Department of geography. Burnaby.
Discussion papers. 1- (1978-) Irregular.
[Department of Geography, Simon Fraser University, Burnaby, B.C. V5A 1S6, Canada.]

416 SIR George Williams university. Department of geography. Montréal.
Publication. 1 (1972). Only one issued. Closed 1972.

417 SOCIÉTÉ de géographie de Québec. (Geographical society of Quebec). Québec.
Bulletin (Transactions). 1-28 no1/2 (1880-1934). Closed 1934.
(U-5:3945c; B-4:135b; F-1:836a).
1886/89 in one unnumbered volume. Suspended 1898-1907. nsv3-28 (1908-1934); Succeeded 1942-44 by Sociétés de géographie de Québec et de Montréal. Bulletin [418].
Index: 1880-1934.
In French.

418 SOCIÉTÉS de géographie de Québec et de Montréal. (Quebec and Montreal geographical societies). Québec.
Bulletin. nsv1-3 no5/6 (Ja 1942-Je 1944). Closed 1944. (U-5:3975c; B-4:135b). Superseded ultimately by Géographie physique et quaternaire [373], and by Cahiers de géographie du Québec [333].
In French.

419 SYMPOSIUM on cartography. Symposium on cartography (proceedings).
Ottawa. 1- (1962-). (N70-4:5655c-5656a; B68:122a).
Proceedings for 1962-1964 have also title: Canadian cartography.
1962-1964 issued by Canadian institute of surveying; 1967- issued by Ontario institute of chartered cartographers.

(420 TORONTO. University. Department of geography. Toronto.
Discussion paper series. 1- (1969-) Irregular. (N76; B73-396a; DP).
[Department of Geography, University of Toronto, Toronto, Ontario M5S 1A1, Canada.]

(421 ______. ______. ______.
Research publication. 1- (1968-) Irregular (N70-4:5821b; B73:292a; DP).
[Department of Geography, University of Toronto, Toronto, Ontario M5S 1A1, Canada.]

(422 TRAVAUX géographiques du Saguenay. (Université du Québec à Chicoutimi. Module de géographie.) Chicoutimi. 1- (1976-) Annual.
In French.
[Module de Géographie, Université du Québec à Chicoutimi, Chicoutimi, Québec G7H 2B1, Canada.]

(423 TRENT student geographer. (Trent university geographical society).
Peterborough 1- (1972-) Annual. (N78).
[Department of Geography, Trent University, Peterborough, Ontario K9J 7B8, Canada.]

424 TRENT university. Department of geography. Peterborough.
Occasional papers. 1- (1972-). (DP).
[Department of Geography, Trent University, Peterborough, Ontario K9J 7B8, Canada.]

425 VICTORIA, British Columbia. University. Department of geography. Victoria.
Occasional paper in geography. (N77).

*(426 WATERLOO, Ontario. University. Department of geography. Waterloo.
Department of geography publication series. no1- (1971-). (N77; DP).
[Department of Geography, Faculty of Environmental Studies, University of Waterloo, Waterloo, Ontario, Canada N2L 3GI.]

(427 ______. ______. ______.
Occasional papers. 1- (1971-).
[Department of Geography, University of Waterloo, Waterloo, Ontario N2L 3G1, Canada.]

428 ______. ______. ______. Division of environmental studies. Waterloo.
Occasional papers. 1- (1972-). (DP).
[Faculty of Environmental Studies, University of Waterloo, Waterloo, Ontario N2L 3G1, Canada.]

(429 ______. ______. School of urban and regional planning.
Working paper series. 1- (1977-) Irregular.
[University of Waterloo, Waterloo, Ontario N2L 3G1, Canada.]

*(430 WESTERN geographical series. (Victoria, B.C. University. Department of geography). Victoria. 1- (1970-)Irregular. (N75-2:2283c; B73:364a; DP).
[Department of Geography, University of Victoria, Box 1700, Victoria, B.C. V8W 2Y2, Canada.]

(431 WESTERN Ontario, University of. Department of geography.
Geographical papers). London. 1- (1964-) Irregular.
[University Bookstore, University of Western Ontario, London, Ontario N6A 3K7, Canada.]

(432 WILFRID LAURIER university. Department of geography. Waterloo.
Bibliography and review of geography department discussion papers, occasional papers and monographs. 1- (1974-) Biennial.
v1 (1974) had titles: Geography: discussion papers, occasional papers, monographs (on cover), Bibliography of geography department discussion papers, occasional papers and monographs (inside at head of text).
[Department of Geography, Wilfrid Laurier University, Waterloo, Ontario N2L 3C5, Canada.]

433 ______. ______.
Geographical inter-university resource management seminars (GIRMS).
1- (1970/71-) Annual. (DP).
[Department of Geography, Wilfrid Laurier University, Waterloo, Ontario N2L 3C5, Canada.]

(434 ______. ______.
Special publication. no1- (1978-) Irregular.
[Department of Geography, Wilfrid Laurier University, Waterloo, Ontario N2L 3C5, Canada.]

(435 YORK university, Toronto. Department of geography. Downsview.
Discussion paper. 1- (1972-). (N77; DP).
[Department of Geography, York University, 4700 Keele Street, Downsview, Ontario M3J 1P3, Canada.]

(436 ______. Atkinson college. Department of geography.
Geographical monographs. Toronto. 1- (1973-). (N75-1:906b; DP).
[Department of Geography, Atkinson College, York University, 4700 Keele Street, Downsview, Ontario, Canada M3J 2R7.]

CHILE (Spanish)

República de Chile

437 ASOCIACIÓN de geógrafos de Chile. Santiago de Chile.
Boletín. 1- (1967-) 2 per annum.

438 CHILE. Inspección de geografía y minas. Santiago de Chile.
Boletín. 1-11 (no1-46) (1905-1915). Closed 1915. (ULG:136b; F-1:661b).

439 ______. Instituto geográfico militar. Santiago de Chile.
Anuario. 1- (1891-1932). About every 5 years.

(440 ______. ______.
Boletín informativo. 1- (1960-) Monthly, later quarterly.
(N70-1:1251c).
[Instituto Geográfico Militar, Nueva Santa Isabel No. 1640, Santiago, Chile.]

441 ______. Universidad, Santiago. Instituto de geografía. Santiage de Chile.
Publicación. 1-4 (1946-1954). (H-2:1017b).
In Spanish with occasional English and German summaries.

442 ______. ______. ______. Sección geomorfología aplicada.
Publicación. 1-5 (1960-1965). Closed 1965. (N70-1:1255a).

443 CUADERNOS geográficos del sur. (Concepción. Universidad. Departamento de Geografía). Concepción. 1- (1971-). (N78; B73-3:87b).
No1 (1971) issued by university's Instituto central de geografía.
[Departamento de Geografía, Casilla 1257, Concepción, Chile.]

444 GEOCHILE. (Sociedad geográfica de Chile). Santiago de Chile.
v1 (no1-2) (1951-1952). All issued? (N70-2:2275c).

* (445 INFORMACIONES geográficas. (Chile. Universidad, Santiago. Instituto de geografía). Santiago de Chile. 1- (1951-) Annual. (N70-2:2781b; BS:412a).

446 LINARES: Revista trimestrial de historia y geografía de la provincia de Linares. (Sociedad linarense de historia y geografía). Linares.
v1-15 (no1-60) (1933-1947). Closed 1947? (U-3:2425c; F-3:321b).

447 MEMORIAL-técnico del ejército de Chile. (Chile. Instituto geográfico militar). Santiago de Chile. v1-16 (no1-61) (1932-1948). Closed 1948. (U-3:2601b; B-3:178a).
Superseded by Revista geográfica de Chile [449].

(448 REVISTA chilena de historia y geografía. (Sociedad chilena de historia y geografía). Santiago de Chile. 1- (1911-) Irregular. (U-4:3592c; B-3:696a; F-4:120a).
Title varies slightly. Suspended 1924-1925.
Index: v1-50 in 52.
[Sociedad Chilena de Historia y Geografía, Casilla postal, 1386, Santiago de Chile, Chile.]

449 REVISTA geográfica de Chile "Terra australis." (Chile. Instituto geográfico militar. Comité nacional de geografía, geodesía y geofísica [and] Sección nacional del Instituto panamericano de geografía e historia). Santiago de Chile. 1-19 (1948-1961) Irregular. (U-4:3613a; B-3:178a).
Supersedes Memorial-técnico del ejército de Chile [447].

*(450 REVISTA geográfica de Valparaíso. (Universidad católica de Valparaíso. Departamento de geografía). 1- (1967-). (N70-3:5017a; B73:296a).

451 VALPARAÍSO (city). Universidad católica. Departamento de geografía. Valparaíso.
Notas geográficas. 7 (1976) Irregular. Monographs.

CHINA

(Chinese)

The entries for China are divided into three groups:

(a) China through 1949
(b) People's Republic of China
(c) China: Taiwan

CHINA through 1949
中國（到一九四九年止）

452 地理集刊
453 海洋集刊
454 地理專刊
455 方志
456 福建省地理學會年報
457 史地雜誌
460 國立中央大學理科研究所地理學部專刊
462 國立清華大學理科報告 地質·地理·氣象
463 史學與地學
464 史地學報
465 史地叢刊
466 中山大學地理系報告
468 地學集刊
469 地學集刊
470 地學雜誌
471 地理
472 地理季刊
473 地理集刊
474 地理教學
475 地理學季刊
476 禹貢半月刊

PEOPLE'S REPUBLIC OF CHINA
中华人民共和国

(477 氣象学报
477a 中国地理学会年会論文選集：自然地理学
478 海洋与湖沼
479 国内报刊有关地理資料索引
480 地理
(481 地理集刊
482 地理教学参考資料
(483 地理知识
(484 地理学报
485 地理学資料
486 地理譯报
487 测绘学报
488 测绘通报
489 中华地理志丛刊

CHINA: TAIWAN
台灣

(490 氣象學報

(491 中國地理學會會刊

492 中國文化學院地學研究所研究報告

493 中國文化學院地理學系學訊

494 敦明產業地理研究所論文摘要

495 敦明產業地理研究所中國區域地理叢刊

496 敦明產業地理研究所研究報告

496a 敦明產業地理研究所畜牧地理叢刊

496b 敦明產業地理研究所作物地理叢刊

497 台北市文獻

498 海洋彙刊

499 江蘇文獻

500 華岡地學

(501 台灣師範大學地理研究所地理研究報告

(502 台灣大學理學院地理學系研究報告

(503 台灣文獻

(504 地學彙刊

(505 地理教育

506 地理學研究

507 地理與產業

CHINA (Chinese)

through 1949

452 CHUNG-kuo ti-li yen-chiu-so. (China institute of geography). Pei-p'ei, Szechwan.
Bulletin. sA 1-2 (1943-1944); sB 1-2 (1942).

453 ______.
Hai-yang shu-k'an. [Oceanographic papers]. v1 no1 (1942).

454 ______.
Ti-li chuan-k'an. [Geographical memoirs]. Chungking? 1-2 (1942-1944); ns no1-4 (1946).
Supplementary English titles.

455 FANG-chih. [Geographical review]. (Nanking. National central university. Department of geography). Nanking. v1-9 no3/4 (Jl 1928-Jl 1936). Closed 1936. (U-2:1525c; C:140a).
v1-5 no2 as Ti-li tsa-chih; v5 no3-v8 no12 as Fang-chih yüeh-k'an.

456 FU-chien shêng ti li hsüeh hui. (Geographical society of Fukien province). Foochow.
Nien pao. [Yearbook]. no5-6 (1945-1946). Others?

457 HANGCHOW. Kuo li chê-chiang tā hsüeh. (National Chekiang university. Hangchow).
Shih ti tsa chih. [History and geography]. v1-2 no2 (My 1937-Mr 1942). (C:391b).
Supplementary English table of contents.

458 HARBIN. Klub estestvoznaniia i geografii KhSML.
Izvestiia. 1 (1941), 2 (1944)? No more issued? (U-3:1796b).
In Russian.

459 KLUB estestvoznaniia i geografii, Harbin. Harbin.
Ezhegodnik. 1 (1933). Closed 1933. (U-3:2299b).
English title: Club of natural science and geography of the YMCA.
In Russian.

460 NANKING. Kuo li chung yang ta hsüeh. (National central university. Nanking).
Ti li hsüeh pu chuan k'an. [Bulletin, Department of geography].
sA. 1-7 (1942-1945). Closed 1945.
no7 with English subtitle and abstract.
sB. 1-8 (1943-1945). (U-4:2818c).
no2 has English subtitle.

461 OBSHCHESTVO izucheniia Man'chzhurskogo kraia, Harbin. Harbin.
Trudy. Geologiia i fizicheskaia geografiia. no1 (1928).
Closed 1928. (U-4:3130c; B-3:433a).
In Russian.

462 PEKING (Peiping). National Tsing hua university.
Science reports. sC. Geographical, geological and meteorological sciences.
v1 no1-4 (1936-1948). Closed 1948? (U-4:3286a; B-4:355a; F-4:452b).
v1 no1 as sC: Geological and geographical sciences.
Suspended 1936-1947.
In English.

463 SHIH-hsüeh yü ti-hsüeh. [Journal of history and geography]. (Chung-kuo shih-ti hsüeh-hui). (Historical and geographical society of China). Shanghai. no1-4 (1926-1929). Closed 1929. (U-5:3862b; F-2:78a; C:385b).

464 SHIH-ti hsüeh-pao. (Tung-nan ta hsüeh shih ti yen chiu hui; Nanking. National central university. Historical and geographical society). Nanking. v1-4 no1 (1921-26). Closed 1926. (U-5:3862b; C:391b).

465 SHIH-ti ts'ung-k'an. [Historical geographical magazine]. (Peking. National normal university. Historical and geographical society). Peking. v1-2 no2/3 (1920-1923). Closed 1923. (U-4:3285c; C:391b).

466 SUN Yat-sen university, Kwangchow. (Kuo-1: Chung-shan ta-hsüeh). Geographical department.
Report. v1 no1-2 (1930-1931). Closed 1931.
In English.

467 ______. Geographisches institut.
Mittcilungen. v1 no1-2 (1930-1931). Closed 1931. (U-5:4113c; B-4:251a).
In German.

468 TI hsüeh chi k'an. Geo-quarterly. (Ya-hsin geo-institute. National Tsing hua university. Geographical society series). Sinhwa; Wuchang. v1-6 no1 (1943-1948). Closed 1948. (U-5:4213b; C:444a).
Supplementary English table of contents.

469 ______.
Special publication. 1-9 (1943?-1947). Closed 1947?
Supplementary English title.

470 TI-hsüeh tsa-chih. [Geographical magazine; Geographical journal]. (Chung-kuo ti-hsüeh-hui). Peking. 1-24 (no1-181) (1910-1937). Closed 1937. (U-5:4213b; B-4:318b; F-4:683b; C:444a).
Supplementary English table of contents and abstracts.

471 TI-li. [Geography]. (Chung-kuo ti-li yen-chiu-so. China institute of geography). Nanking. (Wartime as Pei-p'ei, Szechwan). v1-6 no1 (1941-1948). Closed 1948. (U-5:4213b; B-4:320a; C:444b).
Supplementary English table of contents (except v1 no1).

472 TI li chi k'an. [Geographical magazine]. Chungking. 1-4 (1942-1946).

473 TI-li chi-k'an. Kuo-li chung-shan ta hsüeh. Li-hsüeh-yuan. Ti-li hsüeh hsi. (Sun Yat-sen university. Geographical department. Bulletin). Kwangchow. 1 (1931); ns no1-11 (1937-1943). Closed 1943. (U-5:4213b; C:445a).
Supplementary table of contents and abstracts in English, French, or German.

474 TI-li chiao hsüeh. [Teaching of geography]. (National normal university, Peking. Department of geography). Peking. v1-2 no4 (1937-1947). (C:445a).

475 TI-li hsüeh chi-k'an. [Quarterly journal of geography]. (Kuo-li chung-shan ta-hsüen. Ti-li hsüeh-hsi). (Sun Yat-sen university. Geographical department). Kwangchow. v1 no1-4 (1933-1934). Closed 1934. (U-5:4213b; B-4:320a; F-4:683b; C:445a).
Supplementary English table of contents.

476 YÜ kung pan-yüeh-k'an. [Chinese historical geography]. (Yü kung hsüeh-hui). Peking. 1-7 (no1-82) (1934-1937). Closed 1937. (U-5:4585c; F-4:996b; C:512b).
1-2 as Evolution of Chinese geography.

CHINA (Chinese)

PEOPLE'S REPUBLIC OF CHINA

Chung-Hua Jen-Min Kung-Ho Kuo

(477 CH'I hsiang hsüeh pao. Acta meteorologica sinica. (Meteorological society of China). Peking. 1-36 no2 (1925-1966); 37- (S1979-). Quarterly. (U-2:996b; BS:23b; C:27b). Not published 1966-1978.
1-15 (1925-1941) as Ch'i hsiang tsa chih; 1-10 (1925-1935) with English title Meteorological society bulletin. Annual; 11-14 (1935-1940) as Meteorological magazine; 15- (1941-), Acta meteorological sinica. English abstracts.

477a CHUNG-KUO ti li hsüeh hui nien hui lun wen hsüan chi: tzu jan ti li hsüeh. (Chung-kuo ti li hsüeh hui tzu jan ti li chuan yeh wei yüan hui). Peking. 1963. (C:117a).

478 HAI yang yü hu chao. (Chung-kuo hai yang hu chao hsüeh hui). Peking. 1- (1957-) Quarterly. (N70-2:2460c; C:155a).
Title also in Latin: Oceanologia et limologia sinica.

479 KUO nei pao k'an yu kuan ti li tzu liao so yin (K'o hsüeh ch'u pan she). Peking.
1953-1959 Annual. Closed 1959. (N70-2:3373a; N75-1:1318a; C:299b).

480 TI li. [Dili.] (Chung-kuo ti li hsüeh hui). Peking. 1961-1966 no1. Closed 1966. (N70-4:5784b; B68:172b; C:444b).
English title: Geography. (Association of Chinese Geographers).
See Ti-li chih-shih [483].

(481 TI-li chi-k'an. (Chung-kuo. K'o hsüeh yuan. Ti li yen chiu so. Peking. 1- (1957-) Irregular. (N70-4:5784b; C:444b).
Issued by Academia sinica. Institute of geography.

482 TI li chiao hsüeh ts'an kao tzu liao. Shanghai. Ja 1959-My 1960 Semimonthly. Closed 1960.

(483 TI-li chih-shih. (Chung-kuo k'o hsüeh yuan. Ti li yen chiu so). [Geographical society of China and Academia sinica. Institute of geography]. Peking. Ja 1950-Jn 1966; S 1972- Monthly. (N70-4:5784b; C:445a). 1961-Ja 1966 as Ti li. Publication suspended 1966-1972.

*(484 TI-li hsüeh-pao. (Acta geographica sinica). (Academia sinica. Institute of geography). Peking. 1-32 no2 (1934-1966); 33- (S 1978-) Quarterly. Suspended 1966-1978. (U-5:4213b; B-4:320a; C:445a).
v1-17 (1934-1950) as Journal of the geographical society of China. (Ti-li hsüeh-pao [Chung-kuo ti-li hsüeh-hui]). Chungking; Nanking. v1-15 (1934-1948) with supplementary English table of contents and abstracts; v16-17 (1950) with English table of contents; 1956-1966 with Russian or English abstracts with titles listed in table of contents; 33- (1978-), supplementary English titles in table of contents and abstracts in English.

485 TI-li hsüeh tzu liao. [Memoirs of geography]. (Chung-kuo k'o hsüeh yuan. Ti-li yen chiu so, Nanking). Peking. no1-7 (1958-1960). Closed 1960. (N70-4:5784b; C:445b).
Issued by Academia sinica. Institute of geography.
Supplementary English table of contents.

486 TI-li i pao. (K'o hsüeh ch'u pan she). [Geographical society of China]. Peking. 1956- Quarterly. (N70-4:5784b; C:445b).
[Translations into Chinese of articles in other languages].

487 TS'Ê hui hsüeh pao. Acta geodetica et cartographica sinica.
(K'o hsüeh ch'u pan she). Peking. v1-v4 no1 (N 1957-Ja 1960);
v5-v9 no1/2 (1962-My 1966) Quarterly. (N70-4:5868c; BS23a; C:460a).

488 TS'Ê hui t'ung pao. (K'o hsüeh ch'u pan she). Peking. v1-10 no6 (1955-Jn 1966).
Bimonthly. (N70-4:5868c; C:460a).
[Cha'o-yang-men 117, Peking.]

489 CHUNG-hua ti li chih cong-k'an. (Chung-hua ti li chih pien chi pu).
Peking. (N70-4:5878a).
English title: Materials on the physical geography of the Tung-pei
region.

CHINA: TAIWAN (Chinese)

Chung-hua Min-kuo

(490 CH'I hsiang hsüeh pao. [Meteorological bulletin]. (Taiwan. Ch'i hsiang
so). T'ai-pei. 1- (1955-) Quarterly. (N70-1:1236a; C:27b).
English abstracts.
[Central Weather Bureau, 64, Kung Yuen Road, T'ai-pei, Taiwan 100,
China.]

(491 CHUNG-KUO ti-li hsüeh hui hui k'an. [Geographical society of China.
Bulletin]. T'ai-pei, 1- (O 1970-) Annual.
In Chinese or English.
[Geographical Society of China, c/o Department of Geography, The
National Taiwan Normal University, Ho-ping Tung Road, Taipei, Taiwan
107, China.]

492 CHUNG-KUO wen hua hsüeh yüan ti hsüeh yen chiu so yen chiu pao kao.
[College of Chinese Culture. Institute of Geography. Science reports].
1- (1963-).
English abstracts.
[Hwa Kang, Yangmingshan, Taiwan 113, China.]

493 CHUNG-KUO wen hua hsüeh yüan ti-li hsüeh hsi hsüeh-hsün. [Geographical
information]. 1- (1974) Annual.
[Department of Geography, College of Chinese Culture, Hwa Kang,
Yangming-shan, Taiwan 113, China.]

494 FU-MING ch'an yeh ti li yen chiu so. [Fu-ming geographical institute of
economic development]. T'ai-pei.
Occasional abstracts.1-9 (1960-1964). Closed 1964. (N70-2:2229a).
Formerly Fu-ming institute of agricultural geography).

495 ______.
China regional geography series. 1-15 (1954-1964). Closed 1964.
(N70-2:2229a).

496 ______.
Research report. 1-124 (1947-1964). Closed 1964. Irregular, about 10 a year.
(N70-2:2229a; N70-4:5681a).
Includes a subseries: Urban and rural survey in Taiwan.
In Chinese or English.

496a ______. Animal geography series. 1-7 (1948-1955). Closed 1955.

496b ______. Crop geography series. 1-10 (1947-1954). Closed 1954.

497 REGIONAL geographical studies on T'ai-pei. T'ai-pei.
1- (1952-) Quarterly.
[Commission on the Studies of Regional Geography of Taipei, 3rd floor,
No. 85, Ming Chuang Road, Taipei, Taiwan, China.]

498 HAI yang hui k'an [Journal of the institute of oceanography]. T'ai-pei.
1- (1965-). 2 nos. a year.
In Chinese or English.
[The Institute of Oceanography, College of Chinese Culture, Hwa Kan, Yan-Ming-Shan, Taiwan 113, China.]

499 REGIONAL geographical review of Kiangsu. T'ai-pei.
1- (Ja 1963-) Quarterly.
[Association on the Study of Regional Geography in Kiangsu, No. 415, Tingchow Road, Taipei, Taiwan, China.]

500 HUA-KANG ti hsüeh. [Journal of earth sciences]. T'ai-pei.
no1- (Mr 1968-) Irregular.
Supplementary English titles.
[Department of geography, College of Chinese Culture. Hwa K'ang, Yang-Ming-Shan, Taiwan 113, China.]

*(501 T'AI-WAN shih fan ta hsüeh ti li yen chiu so. Ti li yen chiu pao kao.
T'aipei. 1- (Ja 1975-) Annual. (N76; C:431b).
English title: Geographical research. (National Taiwan normal university. Institute of geography).
English abstracts.
[Institute of geography, National Taiwan University, Roosevelt Road, Taipei, Taiwan 107, China.]

*(502 T'AI-WAN ta hsüeh, T'ai-pei. Li hsüeh yuan. Ti li hsüeh hsi.
T'ai-wan ta hsüeh li hsüeh yuan ti li hsüeh hsi yen chiu pao kao.
1- (Ag 1962-). (N70-4:5683c; C:432b).
Supplementary English title, table of contents, and summaries.
Title also in English: Science reports.
Alternative listings: (1) T'ai-pei (city). National Taiwan university. Department of geography and meteorology. Science reports. (2) T'ai-wan ta hsüeh, T'ai-pei. Li hsüeh yuan. Department of geography and meteorology. Science reports. (3) T'ai-wan ta hsüeh, T'ai-pei. Li hsüeh yuan. Ti li hsüeh hsi. Science reports. (4) National Taiwan university. Department of geography and meteorology, T'ai-pei. Science reports.
[Department of Geography, National Taiwan University, Roosevelt Road, T'ai-pei, Taiwan 107, China.]

(503 T'AI-WAN wen hsien. Report of historico-geographical studies of Taiwan.
[Taiwan. Historical research commission of Taiwan province].
T'ai-pei. v1 no1- (1949-) Quarterly. Issues sometimes combined.
(N60-2:1978a; C:434b). Title varies.
Supplementary English subtitle and table of contents.

(504 TI hsüeh hui k'an. [Annals of the association of Chinese geographers].Tai-pei.
no1- (1969-) Annual.
[The China Academy, Hwa Kang, Yang-Ming-Shan, Taiwan 113, China.]

(505 TI-LI chiao yü. [Geographical education]. T'ai-pei.
1- (1967-) Irregular.
In Chinese or English.
[Department of Geography, National Taiwan Normal University, Hoping Tung Road, Taipei, Taiwan 107, China.]

506 TI-li hsüeh yen chiu. (T'ai-wan sheng li shih fan ta hsüeh, T'ai-pei. Ti-li hsüeh hsi). [Taiwan normal university. Department of geography]. T'ai-pei. no1- (Je 1966-) Annual. (N70-4:5784b).
English title: Geographical studies.
Supplementary English title, table of contents, and abstracts.
[National Taiwan Normal University, Ho-ping Tung Road, T'ai-pei, Taiwan, China.]

507 TI-li yü ch'an yeh. (Fu-ming ch'an yeh ti li yen chiu so; Fu-ming geographical institute of economic development). T'ai-pei. 1-8 (1956-1963). Closed 1963. (N70-4:5784c; C:445b).
English title: Geography and industries. Formerly as Fu-ming institute of agricultural geography.
In Chinese, English, or Japanese.

COLOMBIA (Spanish)

República de Colombia

508 ASSOCIACIÓN colombiana de geógrafos (ACOGE). Tunja. (Tunja. Universidad pedagógica y tecnológica de Colombia). Noticias bimestrales. no4 (Diciembre 1975).
[Apartado aereo no. 053119, Bogotá 2, Colombia.]

509 COLOMBIA. Instituto geográfico "Agustín Codazzi." Bogotá. Informe de labores. 1970/73- . (N76).

510 ______. ______.
Memoria sobre los trabajos ejecutados. 1969- . (N75-1:547b).

511 ______. ______.
Publicación. 1- (1959-) Irregular. (N70-1:1389a).

512 ______. ______.
Publicación especial. (N70-1:1389a; BS:428).

*(513 COLOMBIA geográfica. (Colombia. Instituto geográfico "Agustín Codazzi." Revista). Bogotá. 1- (1970-). (N75-1:548b).
[Carrera 30 no. 48-51, Apartado Aéreo 6721, Bogotá, D.E., Colombia.]

514 CORREO geográfico. (Asociación colombiana de geógrafos ACOGE). Tunja. 1- (1968-). (N70-1:1551a).

515 CUADERNOS de geografía de Colombia. (Sociedad geográfica de Colombia. Colombia. Academia de ciencias geográficas). Bogotá. 1- (1955-) Irregular. (N70-1:1591a).
Supplement to the society's Boletín [521]. Title varies. Cuadernos de geografía colombiana.

516 EXTENSIÓN cultural. Serie textos. (Colombia. Universidad, Bogotá. Departamento de geografía). (N70-2:2020a).

517 REVISTA coogeográfico. Bogotá. 1975- . (N76).

518 REVISTA geográfica. (Universidad del Atlántico. Instituto de investigación etnológica). Barranquilla. 1 (1952). Closed 1952. (N70-3:5016c).

519 REVISTA geográfica de Colombia. (Colombia. Instituto geográfico militar). Bogotá. no1-9 (1936-1939). Closed 1939. (U-4:3613a; B-3:699b; F-4:140b).

520 SOCIEDAD cartográfica de Colombia. (Colombia. Instituto geográfico "Agustín Codazzi"). Bogotá.
Boletín. 1- (1964-) Irregular.

*(521 SOCIEDAD geográfica de Colombia. (Colombia. Academia de ciencias geográficas). Bogotá.
Boletín. Ap 1907; 1- (1934-) 4 nos a year. (U-5:3909b; B-4:106a; F-1:664b). My 1944-Ag 1947 not published. v1-6 no2/3, v7 no2, Je 1934-Je/D 1939, S 1942 called s2.
Author index: nos1-100 (1903-1973).
[Observatorio Astronómico Nacional, Carrera 8º No.8-00. Apartado No. 25-84, Bogotá.]

522 ______.
Boletín bibliográfico. 1- (1963-) Irregular. (N70-4:5374c).
1-8 (1963-1968) originally as mimeographed lists but printed as a single volume in 1968.
[Observatorio Astronómico Nacional, Carrera 8º No. 8-00, Apartado No. 25-84, Bogotá.]

523 ______.
Revista. v1 no1 (N 1921); v1 no1 (N 1924). Only two nos issued. (U-5:3909b).

COSTA RICA (Spanish)

República de Costa Rica

524 ACADEMIA de geografía e historia de Costa Rica. San José.
Anales. 1959/62- . (N70-1:36c).
Supersedes Academia costarricense de la historia. Revista (Not in this list).

525 COSTA RICA. Instituto físico-geográfico nacional. San José.
Anales. 1-9 (1888-1896). Closed 1896. (U-2:1213c; B-1:661a; F-1:179ab).
1 as Museo nacional de Costa Rica, San José. Anales; 2-4 as Instituto físico-geográfico y museo nacional de Costa Rica, San José. 8 never published.

526 ______. ______.
Boletín. 1-3 (no1-36) (1901-1904). Closed 1904. (U-2:1214a; B-2:506a; F-1:669a).

527 ______. ______.
Circular. 1-2 (1901). Closed 1901. (ULG:157c).

528 ______. ______.
Informe sobre los trabajos prácticos. 1890/91-1897/98. Closed 1898. (ULG:157c).

* (529 ______. Instituto geográfico nacional. San José.
Informe semestral. 1954- . Twice a year. (N70-1:1556a; BS:428b).
1954-1957 as Informe cuadrienal; O 1956-D 1957 as Informe trimestral.
Also as Instituto geográfico de Costa Rica.
Indexes: 1954-1972, 1973-1977.

530 ______. ______. San José.
Investigaciones geográficas. 1-2 (1962-1964). (N70-1:1556a).

531 ______. ______.
Publicación. 1- (1953-) Irregular. (N70-1:1556a).

532 ______. ______.
Serie geográfica. 1- (1964-). (N70-1:556a).

533 ______. Universidad, San Pedro.
Serie historia y geografía. 1- (1955-) Irregular. (N70-1:557c).
Title varies: no1-3: Editorial universitaria. Sección historia.

*(534 REVISTA geográfica de América Central. (Universidad nacional. Escuela de geografía). Heredia. 1- (1974-) 2 nos. a year. (N77).
Title varies slightly: Revista de geografía de América Central.
Abstracts in English and French.

535 SAN JOSÉ. Museo nacional de Costa Rica. San José.
Serie geográfica. v1 no1 (1938). Closed 1938? (U-5:3773a; B-1:661a).

536 SOCIEDAD de geografía e historia de Costa Rica, San José.
Boletín. no5-6 (1946-1947). (U-1:17a; U-5:3906c).
Name changed about 1956 to Academia costarrincense de la historia.

537 ______.
Publicación. no1-5 (1941-1946). (U-1:17a).

CUBA (Spanish)

República de Cuba

538 ACTUALIDADES de la geografía. (La Habana. Universidad. Centro de información científica y técnica). La Habana. 1- (1970-). (N75-1:27c; N76).
1 (1970) as Actualidades técnico-científicas: geografía.

539 BOLETÍN nacional de historia, geografía y ciencias naturales.
La Habana. v1-4 no6/8 (1912-1927). Closed 1927. (U-1:731b).
3-4 also as s2 v3-4.

*(540 La HABANA. Universidad. Centro de información científica y técnica.
La Habana.
Ciencias. serie 7. Geografía. 1- (1972-) Irregular.

541 La HABANA. Universidad. Escuela de geografía. La Habana.
Boletín. v1 no1-3 (1964). All issued? Mimeographed.

542 SOCIEDAD geográfica de Cuba. La Habana.
Memorias. v1 (1923). Closed 1923? (U-5:3909b; F-3:496b).

543 ______.
Revista. 1-29 (1928-1957), 30 (1960). Closed 1960. Quarterly, but issues often combined. (U-4:3613a; B-4:106a; F-4:130a; R-5:455a).
v30 as Revista geográfica.
Suspended 1958-59.

CURAÇAO (Dutch)

De Nederlandse Antillen

The Netherlands Antilles

544 GESCHIED-, taal-, land- en volkenkundig genootschap, Willemstad. Jaarlijksch verslag. 1-6 (1896-1902). Closed 1902. (U-2:1714a; B-2:287b).
In Dutch.

CYPRUS (Greek or Turkish)

Kypriaki Demokratia

ΚΥΠΡΙΑΚΗ ΔΗΜΟΚΡΑΤΙΑ

KIBRIS CUMHURİYETİ

*(545 ΓΕΩΓΡΑΦΙΚΑ Χρονικα. (Γεωγραφικος 'Ομιλος Κύπρου). Λευκωσία.

*(545 GEOGRAPHIKA chronika. Geographical chronicles. (Geographikos homilos Kyprou; Bulletin of the Cyprus geographical association). Nicosia. 1- (1971-) 2 per year. (N75-1:906c; GZS).
In Greek or English.
[P.O.Bo. 3656, Nicosia, Cyprus.]

CZECHOSLOVAKIA (Czech or Slovak)

Československá Socialistická Republika, ČSSR

*(546 ACTA geographica universitatis comenianae. (Bratislava. Univerzita Komenského). Bratislava. 15- (1978-) Irregular.
Supersedes Acta geographica universitatis comenianae: economico-geographica [547] and continues its numbering.
Supersedes also Acta geographica universitatis comenianae: geographico-physica, [548].
In English or Slovak. Abstracts in Russian and in Slovak or English.
[Slovenské Pedagogické Nakladatel'stvo, Bratislava, Czechoslovakia.]

547 ______: Economico-geographica. Bratislava. 8-14 (1969-1977) Closed 1977. (N77).
Supersedes in part Acta geologica et geographica universitatis comenianae: geographica [549] and continues its numbering.
Superseded by Acta geographica universitatis comenianae, [546].
In English or Slovak. Abstracts in English or German.

548 ______: Geographico-physica. Bratislava. 1-2 (1973-1976). Closed 1976. (N75-1:323c; N76).
In English, German, or Slovak.

549 ACTA geologica et geographica universitatis comenianae: geographica. (Bratislava. Univerzita Komenského). Bratislava. 1-7 (1959-1968). Closed 1968. (N70-1:58c; R-1:144a).
Superseded by Acta geographica universitatis comenianae: Economico-geographica [547].
In English, German, or Slovak.

*(550 ACTA universitatis carolinae: geographica. (Praha. Universita Karlova). Praha. 1- (1966-) 2 nos a year. (N70-1:71a; N70-2:2280b; B68:10b).
Supersedes in part Acta universitatis carolinae, 1954-1957 (not in this list).
In Czech, English, German, or other languages, and with abstracts in a second language.

551 BRATISLAVA. Univerzita. Přírodovědecká fakulta. Geografický ústav. Bratislava.
Sborník; Práce geografického ústavu. 1-2 (1945-1946). Closed 1946. (U-1:763c; B-4:429b; R-6:30b).
Subseries of Sborník prác Prírodovedeckej fakulty Slovenskej univerzity v Bratislave. No1 (1945) v12 of parent series; no2 (1946) v14 of parent series (not in this list.
In Slovak with abstracts in English or French.

552 BRNO (Brünn). Universita. Pedagogická fakulta. Brno.
Sborník prací: Geografie. 1- (1967-) Irregular. (N70-1:897b; N75-1:905a).
v1 of geography subseries is v23 of whole series: Geografie. Sborník prací Katedry geografie pedagogické fakulty UJEP.
In Czech with summaries in Russian and German.
[Universita Jana Evangelisty Purkyně, Brno, Czechoslovakia.]

*(553 ______. ______. Přírodovědecká fakulta.
Geographia. 1- (1964-). (N70-1:897b).
Subseries of university's Folia (not in this list): Folia facultatis scientiarum naturalium universitatis Purkynianae Brunensis: Geographia.
In Czech. Summaries in Russian, and English, or German.
[Universita Jana Evangelisty Purkyně, Brno, Czechoslovakia.]

(554 ______. ______. ______.
Geographica. 1- (1971-) Irregular. (N75-1:351a).
Subseries of its Scripta: Scripta facultatis scientiarum naturalium universitatis Purkynianae Brunensis: geographia.
In Czech or German. Summaries in English, German, or Russian.
[Universita Jana Evangelisty Purkyně, Brno, Czechoslovakia.]

555 ČESKOSLOVENSKÁ akademie věd. Geografický ústav. Brno.
Rozvojové země. (N70-1:1199b).
In Czech.

556 ______. ______.
Stručná zpráva o nejdůležitějších výsledcích, činnosti ústavu. 1967?- . (N70-1:1199b).
In Czech.

*(557 ______. ______.
Studia geographica. 1- (1969-) Irregular. (N70-4:5575c).
In Czech, English, and other languages. Abstracts in English and Russian.
[Geografický ústav ČSAV, Mendlovo nám 1, Brno, Czechoslovakia.]

*(558 ______. ______.
Zprávy. 1- (1964-) 8 nos a year. (N70-3:4399a; B73:374b; R-6:524a).
Formerly: Československá akademie věd. Slezsky ústav, Opava. Zprávy ústav geografického.
1964 had 10 nos; 1965, 8; 1966, 10; 1967- 8 nos per v and year.
1964-66 continued numbering of former series: Zprávy slezského ústavu ČSAV. Issues 1964-1968 not numbered by volumes but constitute v1-5. 1964-1967 issued by Geografický ústav, Opava.
Index: 1964-1968 in 1968 no8.
Supplementary tables of contents in Russian, English, and German.
In Czech with English abstracts.
[Vydal Geografický Ústav, ČSAV, Mendlovo nám 1, Brno, Czechoslovakia.]

559 ______. ______.
Zprávy o vědecké činnosti. 1-9 (1963-1967) Irregular. Closed 1968. (N70-1:1199b). Processed.
Superseded by the academy's Studia geographica [557].
In Czech with English abstracts.

560 ČESKOSLOVENSKÁ geografická společnost. (Until 1979: Československa společnost zeměpisná. [Czechoslovak geographical society). Praha.
Knihovna. 1-17 (1904-1949).
In Czech, some with English summaries.

*(561 ______.
Sborník (Journal). 1- (1894/95-) Quarterly. (U-2:971a; B-1:528a; BS:192b; F-4:426a; R-6:28ab).
1894/95 as Česká spolecnost zeměvědná. 1896-1978 as Československá společnost zeměpisná.
Index: 1-50 (1895-1945).
In Czech with supplementary titles in table of contents and abstracts in English, French, German, or Russian.
[Československá Společnost Zeměpisná v Academii, Nakladelstvá ČSAV, Vodičkova 40, Praha, Czechoslovakia.]

562 ______. Odbor v Brně. Brno.
Spisy.
sA. Spisy Tatranské komise. v1-16 (1929-1935). Closed 1945.
sB. Čechy a Morava. v1-13 (1930-1948). Closed 1948.
sC. Oblasti mimo Československo. v1-10 (1931-1941). Closed 1941.
In Czech.

563 ČESKOSLOVENSKÝ kras. (Československá akademie věd). Brno, Praha. 1- (1948-) Annual. (R-2:334a).
In Czech. Abstracts in English.
[Academia. Nakladatelství ČSAV, Praha, Czechoslovakia.]

564 DĚJEPIS a zeměpis ve škole. (Ministerstvo školství). Praha.
1959/60-1967/68). 10 issues a year. Closed 1968. (N70-2:1663c; R-3:22b).
Formed by a union of Dějepis ve škole (not in this list) and Zeměpis ve škole [592].
Superseded from 1968/69 to 1971 by Zeměpis ve škole [592], then by Přírodní vědy ve škole [585].
In Czech.

565 FIRGENWALD. Vierteljahrsschrift für Geologie und Erdkunde der Studetenländer. (Sudetendeutsche Anstalt für Landes- und Volksforschung). (Reichenberg) Liberec. v1-13 (1928-1942). Closed 1942. (U-2:1569b; R-3:183b).
In German.

(566 GEOGRAFICKÉ prácé. (Košice. Univerzita P. J. Šafárika. Pedagogická Fakulta, Prešov. Katedra geografie. Kabinet pre výskum krajiny). Bratislava.
1- (1970-) Irregular.
In Slovak. Abstracts in Russian, German, or English.
[Slovenské Pedagogické Nakladatel'stvo. Bratislava, Czechoslovakia.]

*(567 GEOGRAFICKÝ časopis. Geographical review. (Slovenská akadémia vied, Bratislava. Geografický ústav). Bratislava. 1- (1949-) Quarterly. (U-2:1693a; BS:341b; R-3:267b-268a).
v1 (1949) as Geographica slovaca--Hromádkov sborník; v2-4 (1950-1952) as Zemepisný sborník SAVU. 1949-1952 by the Slovenská akadémia vied a umení.
Indexes: 1-10 (1949-1958); 11-20 (1959-1968).
In Slovak. Abstracts in English or German.
[Vydavatelstvo Slovenskej akademie vied, Klemensova 27, Bratislava, Czechoslovakia.]

(568 HISTORICKÁ geografie. (Komise pro historickou geografii při historickém ústavu ČSAV). Praha. 1- (1968-). 2 nos a year. (N70-2:2551c).
In Czech with supplementary table of contents and abstracts in Russian and German.

569 KARTOGRAFICKÝ přehled. (Československá akademie věd, Praha. Kabinet pro kartografii). Praha. v1- (1946-) Quarterly. (B-2:640a; BS:478b; R-4:187a).
1946-1952 issued by Státní mapová sbírka.
In Czech with English summaries.

570 LIDÉ a země. (Academia. Nakladatelství Československá akademie věd). Praha. v1- (1952-) 12 nos a year. (N70-3:3474a; BS:511b; R-4:240a).
In Czech or Slovak.
[Nakladatelství ČSAV,Vodičkova 40, Praha 2, Czechoslovakia.]

571 LIDÉ a země. Ročenka. (Československá akademie věd, Praha). Praha. 1961- Irregular. (N70-3:3474a).
In Czech or Slovak.
[Academia. Nakladatelství ČSAV, Praha, Czechoslovakia.]

572 NÁUKA o zemi. Seria geographica. (Slovenská akadémia vied. Geografický ústav). Bratislava. 1- (1966-) Irregular. (N75-2:1545a).
Geographica 1 (1966) is No 2 in parent series Náuka o zemi; 2 (1969) is no 4; 3 (1973) is no 6.
Cover title: Nauka o zemi. Geographica.
In Slovak. Abstracts in English and Russian.
[Slovenská akadémia vied, Klemenova 27, 895 30 Bratislava, Czechoslovakia.]

573 NOVINKY literatury. Přírodní vědy. Řada geologicko-geografická. (Statní knihovna ČSSR). Praha. 1964-1972. 10 issues a year. Closed 1972.
(N70-3:4306a; B73:237a: R-5:89a).
Edited by the Universitni knihovna.
In Czech.
Superseded by Novinky literatury: geologie, geografie [574].

(574 NOVINKY literatury: geologie, geografie. (Statní knihovna ČSSR). Praha. 1973- Quarterly. (N75-2:1623b).
Supersedes Novinky literatury. Přírodní vědy. Řada geologicko-geografická [573].
In Czech.
[Státní Knihovna ČSSR, Liliova 5, Praha, Czechoslovakia.]

575 OLOMOUC, Moravia. Palackého universita. Pedagogická fakulta. Praha. Sborník prací: zeměpis. 1- (1972-). (N75-2:1650c).
In Czech. Abstracts in Russian and German.
[Státni Pedagogické Nakladatelství, Praha, Czechoslovakia.]

(576 ______. ______. Přírodovědecká fakulta. Olomouc. Geographica-geologica. 1- (1960-) Annual. (N70-3:4387c).
Subseries of its Acta: Acta universitatis palackianae olomucensis facultas rerum naturalium: geographica-geologica.
Supersedes in part Sborník: Přírodní vědy.
Obory geologie, geografie, biologie, issued by Vysoká škola pedagogická [577].
In Czech or a world language. Summaries in Russian, English, or German.
[Státní Pedagogické Nakladatelství, Praha, Czechoslovakia.]

577 ______. Vysoká škola pedagogická. Olomouc.
Sborník. Přírodní vědy. Obory geologie, geografie, biologie.
no1-3 (1956-1959). Closed 1959. (N70-3:4388b).
Subseries of its Sborník. Přírodní vědy: Sborník vysoké školy pedagogické: přírodní vědy. Obory geologie, geografie, biologie.
Superseded by Olomouc, Moravia. Palackého universita. Přírodovědecká fakulta. Geographica-geologica [576].
In Czech.

578 OSTRAVA. Pedagogická fakulta. Praha.
Sborník prací. Dějepis, zeměpis. 1- (1966-) Irregular. (N70-3:4467a).
In Czech. Abstracts in English, German, or Russian.
[Statní pedagogické nakladatelstvi, Praha, Czechoslovakia.]

(579 PLZEŇ [Pilsen]. Pedagogická fakulta. Plzeň.
Sborník. Zeměpis. 1- (1958-). (N70-3:4671c).
v1-2 (1958-1959) as Plzeň. Vyšší pedagogická škola. Pedagogický institut. Sborník. Marxismus-Leninismus a dějepis-zeměpis;
v3 (1961) as Plzeň. Pedagogický institut. Sborník. Dějepis a zeměpis; v4 (1963) as Plzeň. Pedagogická fakulta. Sborník. Zeměpis a přirodopis. Present title v5- (1965-).
In Czech. Abstracts in English, German, or Russian.
[Státní Pedagogické nakladatelství, Praha, Czechoslovakia.]

580 PRAHA. Deutsche universität. Geographisches institut. Praha.
Arbeiten. nsv1-15 (1921-1934). Closed 1934. (U-4:3420c; B-4:430a; R-1:369a).
Original series consisted of reprints from Deutscher naturwissenschaftlich- medizinischer Verein "Lotos," Praha.
Sitzungsberichte (not in this list).
In German.

581 ______. ______. ______.
Kartographische Denkmäler der Sudentenländer. no1-10 (1930-1936).
Closed 1936. (U-4:3420c).
In German.

582 ______. Universita Karlova. Geografický ústav. Praha.
Travaux géographiques tchèques. v1-17 (1901-1932); 18 (1938).
Closed 1938. (U-4:3422a; B-4:343b).
no1-10 as Institut géographique de Prague. v9 never published.
In Czech, English, French, or German.

583 ______. ______. Geologicko-geografická fakulta. Praha.
Vědecká konference. 1955- . (N70-3:4746c).
In Czech with titles in English and Russian.

584 ______. Vojenský zeměpisný ústav. Praha.
Výroční zpráva [Annual report]. no1-26 (1921-1949). Closed 1949.
(U-4:3422b).
In Czech; in early years with French table of contents and some abstracts.

(585 PŘÍRODNÍ vědy ve škole: časopis pro teorii i praxi vyučování přírodním vědám a zeměpisu. (Ministerstvo školství ve Státním pedagogickém nakladatelství). Praha. 1- (1971-). 10 nos a year.
Supersedes Zeměpis ve škole [592] and Dějepis a zeměpis ve škole [564].
Section for geography.
In Czech with supplementary table of contents in German and Russian.
[Státní Pedagogické Nakladatelství, Opletalova ul. 3, Praha 1, Czechoslovakia.]

586 ŠIRÝM světem. (Česká grafická unie A.S.). Praha. v1-21 (1924-1944).
Closed 1944.
In Czech.

587 SLOVENSKÁ zemepisná spoločnost'. (Bratislava. Slovenská univerzita. Geografický ústav). Bratislava.
Spisy. 1 (1947). Closed 1947.
In Slovak with summaries in English.

588 SLOVENSKÝ kras. (Liptovsky Mikuláš. Muzeum Slovenského krasu). Bratislava.
v1- (1956-) Annual. (N70-4:5357c; R-6:105b).
In Slovak; some abstracts in German, Russian, English, or French.
[Nakladatel'stvo Osveta, Martin, Czechoslovakia.]

589 TASCHENBUCH zur Verbreitung geographischer Kenntnisse. Praha.
1-25 (1823-1847); nsv1 (1848). Closed 1848. (U-5:4155b; B-4:287b; F-4:659b).
In German.

590 ÚSTÍ nad Labem. Pedagogický institut. Praha.
Sborník. Řada zeměpisná. 1950- (N70-4:6117b).
Issued 1950- by Vyšši pedagogická skola.
In Czech with summaries in German or Russian.

591 ZEMĚ a lidé. (Česká grafická unie A.S.). Praha. v1-109 (1921-1938).
Closed 1938.
In Czech.

592 ZEMĚPIS ve škole. (Ministerstvo školství). Praha.
v1-5 (1954-1958); 1968/69-1971. Closed 1971. (N70-4:6469a; R-6:491a).
Joined with Dějepis ve škole (not in this list) to form Dějepis a zeměpis ve škole [564], 1959-1967/68.
Superseded by Přírodní vědy ve škole [585].
In Czech.

593 ZEMĚPISNÁ knihovna. (Praha. Universita Karlova. Geografický ústav).
Praha. 1-7 (1901-1907). Closed 1907. (U-5:4624c).
In Czech.

594 ZEMĚPISNÉ aktuality. (Praha. Universita Karlova. Oddělení pro zeměpis slovanských zemí geografického ústavu). Praha. no1-4 (1947-1948).
Closed 1948. (U-5:4624c; B-4:614b).
In Czech with some English abstracts.

595 ZEMĚPISNÉ práce. (Bratislava. Universita Komenského v Bratislavě.
Geografický seminář). Bratislava. v1-13 (1930-1937). Closed 1937.
(U-5:4624c; B-4:614b; R-6:491a).
French title: Travaux géographiques.
In Czech or Slovak with French titles and some French summaries.

596 ZEMĚPISNÉ zprávy. (Praha. Universita Karlova. Geografický ústav).
2-3 (1947-1949). Closed 1949. (B-4:614b).
Mimeographed.
In Czech.

597 ZEMĚPISNÝ magazín; populární zeměpisný a cestopisný časopis.
(Ceskoslovenská společnost zeměpisná). Praha. v1-5 no3 (1945-1949).
Closed 1949. (U-5:4624c; B-4:614b; BS:985a; R-6:491a).
In Czech.

598 ZEMĚPISNÝ sborník. Praha. 1-3 (1886-1888). Closed 1888. (U-5:4625a).
In Czech.

DENMARK (Danish)

Kongeriget Danmark

(599 AARHUS. Universitet. Geografisk institut. Aarhus.
Arbejdsrapport. Working paper. 1- (1973-). (N78).
In English.

(600 ______. ______. ______.
Skrifter. 1- (1946-) Irregular. (U-1:9b; N70-1:26a; BS:3).
Partly reprints, partly processed.
English abstracts; some French or German abstracts.
Occasional article in English.

601 ______. ______. Geologisk institut. Laboratoriet for fysisk geografi.
Aarhus.
GeoKompendier.
Supersedes Laboratory's Kompendier i fysisk geografi [604].

602 ______. ______. ______. ______.
GeoRapporter. 1- (1975-).
1-4 (1975-1977) as Rapporter i fysisk geografi.

603 ______. ______. ______. ______.
GeoSkrifter. 9- (1978-).
Supersedes and continues numbering of Laboratory's Skrifter i fysisk geografi [605].

604 ______. ______. ______. ______.
Kompendier i fysisk geografi. nol-14 (1970-1978). Closed 1978.
Superseded by Laboratory's GeoKompendier [601].

605 ______. ______. ______. ______.
Skrifter i fysisk geografi. 1-8 (1870-1974). Closed 1974.
1-3 (1970) as Aarhus. Universitet. Geografisk institut. Laboratoriet for fysisk geografi. Skrifter i fysisk geografi.
Superseded by Laboratory's GeoSkrifter [603], which continues its numbering.

606 ARCHIV for historie og geographie samlet og udgivet af J. C. Riise.
København. vl-39 (1820-1829); ns vl-36 (1830-1838). Closed 1838.
(U-1:439b; B-1:192b).
Superseded by Historisk-geographisk archiv [617].

607 ATLAS over Danmark. (K. Danske geografiske selskab) København. Series 1.
(1949-) Irregular; Series 2. 1- (1976-) Irregular.
Some English abstracts.
[C. A. Reitzels Forlag, København, Denmark.]

608 BYGD. Kulturgeografisk tidsskrift. Esbjerg. 1- (1970-) 6 nos a year.
(N75-1:376b).
[Eget Forlag, 6700 Esbjerg, Denmark.]

609 COLLECTED papers Denmark. (K. Danske geografiske selskab). København.
1960- .
Every four years for the International geographical congresses.
1960-1968 issuing agency was København. Universitet. Geografiske institut.
In English.
[Komission C. A. Reitzels Forlag, København, Denmark.]

610 FAGLIGT forums kulturgeografiske haefter. København. no 1- (1973-).
[Tidsskriftet Fagligt Forum, Haraldsgade 68, DK2100 København Ø, Denmark.]

(611 FOLIA geographica danica. (K. Danske geografiske selskab). København.
1- (1940-) Irregular monographs. (U-2:1586c; B-2:200b).
Articles mainly in English, but also in Danish or German.
[Kommission C. A. Reitzels Forlag, København, Denmark.]

612 FRA alle lande. København. v1-38 (1865-1883). Closed 1883.
(U-2:1620a; B-2:218b; F-2:637b).
Index: 1865-1876.

613 GEOGRAFI for realafdelingen. (Gyldendal). København. (N70-2:2277c).

614 GEOGRAFISK orientering. (Geografforbundet).
Aabrinken. 1- (1971-). (N75-1:905b).
[Geografforlaget, 5464 Brenderup, Denmark.]

*(615 GEOGRAFISK tidsskrift. (K. Danske geografiske selskab). København.
v1- (1877-) Annual before 1964; semiannual, 1964-1969; annual since
1970. (U-2:1693a; B-2:268a; F-2:727a).
Indexes: v1-20 (1877-1910) in v20 (1910); 21-33 (1911-1930) in 36 (1933);
34-50 (1931-1950) in 54 (1955); 51-59 (1911-1960) in 60 (1961; 60-69
(1961-1970).
In Danish or English; Danish articles typically have English summaries.
Supplementary table of contents in English.
[C. A. Reitzels Forlag, København, Denmark.]

616 GEOGRAFISKE kompendier. (Gyldendal). København. no1-(1963-).
(N70-2:2279c).

617 HISTORISK-geographisk archiv. København. v1-75 (1839-1864).
Closed 1864. (B-1:192b).
Supersedes Archiv for historie og geographie samlet og udgivet af
J. C. Riise [606].

618 KØBENHAVN (Copenhagen). Skalling-laboratoriet. København.
Meddelelser. 1- (1935-) Irregular. (U-2:1198a).
In Danish, English, or German.
[C. A. Reitzels Forlag, København, Denmark.]

619 ______. Universitet. Geografisk centralinstitut. Laboratorium for
bebyggelsesgeografi, bygeografi og fysisk planlaegning. København.
Rapport. no1- (1973-).

620 ______. ______. ______. Laboratorium for geomorfologi. København.
Geo-noter. no1- (1973-) Irregular.

621 ______. ______. ______. ______.
______. Saerhefte. no1- (1974-) Irregular.

622 ______. ______. ______.
Laboratorium for kulturøkologi og landbrugsgeografi. København.
Rapport. no1- (1976-).
[Geografforlaget, 5464 Brenderup, Denmark.]

*(623 KULTURGEOGRAFI: befolkningsgeografi, bebyggelsesgeografi, erhvervsgeografi,
politisk geografi, regionalplanlaegnig, anvendt geografi [English title:
Kulturgeografi: Danish journal of cultural, social, economic, political
and applied geography]. København. 1- (1949-) Semiannual.
(U-3:2323a; BS:492b).
Supplementary English table of contents from beginning and English
abstracts from v5 no30 (1953).
Index: 1949-1975 in no124 (1976).
[Kulturgeografisk Institut, Vestergade 19, DK-8000 Århus C, Denmark.]

624 KULTURGEOGRAFISKE Skrifter. (Aarhus. Universitet. Kulturgeografisk institut). Aarhus. 1-17 (1967-) Irregular. Closed. (B73:189a). In Danish or English. Reprints from Kulturgeografi [623].

(625 KULTURGEOGRAFISKE skrifter. (K. Danske geografiske selskab). København. 1- (1936-) Irregular monographs. (U-2:1259c; N70-2:1647a; B-2:268a). Some English summaries.
[Nyt Nordisk Forlag Arnold Busck, København, Denmark.]

626 LAESNING til geografitimen. København. 1- (1951-). (N70-3:3399b).

(627 MEDDELELSER om Grønland. (Denmark. Kommissionen for videnskabelige undersøgelser i Grønland). København. 1- (1879-) Irregular monographs. (U-3:2571b; B-3:163a).
1879-1931 as Denmark, Kommissionen for ledelsen af de geologiske og geografiske undersøgelser i Grønland.
Index: 1876-1899, 1876-1912, 1876-1926, 1876-1941.
Text mainly in English but also in Danish, German, or French. Some English abstracts.
[C. A. Reitzels Forlag, København, Denmark.]

628 MUSEUM tidskrift for historie og geografi. København. 1-7 (1890-1896). Closed 1896. (U-3:2794c).
Index: 1890-1895.

629 ORION. Historisk-geografisk maanedskrift. København. 1-4 (1839-1841). Closed 1841. (U-4:3206c; B-3:463a).
Superseded by Orion; historisk quartalskrift (not in this list).

DOMINICAN REPUBLIC (Spanish)

República Dominicana

630 BIBLIOTECA dominicana de geografía y viajes. (Sociedad dominicana de geografía). Santo Domingo. 1- (1970-). (N76).
[Editora Educative Domicana, Santo Domingo, Dominican Republic.]

631 PANTA rhei. Una revista geográfica. Santo Domingo. 1 (Mr 1961). Only one issued. (N70-3:4533c).

632 SOCIEDAD dominicana de geografía. Santo Domingo.
Boletín. 1- (1970-) 3 issues a year. (N75-2:1988c; B73:48).
[Sociedad Dominicana de Geografía, Calle Mercedes 50, Santo Domingo, Dominican Republic.]

ECUADOR (Spanish)

República del Ecuador

633 CENTRO de estudios históricos y geográficos de Cuenca. Cuenca.
Revista. no1- (1921-) Irregular. (U-2:963a).

634 INSTITUTO ecuatoriano de antropología y geografía. Quito.
Informe. 1-3 (1952-1953).

635 LLACTA. (Instituto ecuatoriano de antropología y geografía). Quito.
1- (1956-). (N70-3:3509a).

*(636 REVISTA geográfica. (Ecuador. Instituto geográfico militar). Quito.
1- (Jl 1963-). (N70-3:5016c).
[Departamento geográfico, Instituto geográfico militar, Apartado 2435, Quito, Ecuador.]

637 SOCIEDAD geográfica de Quito. Quito.
Boletín. v1 no1 (O 1911). Closed 1911? (U-5:3909c).

EGYPT (Arabic)

Arab Republic of Egypt

Jumhūrīyah Miṣr al-'Arabīyah

جمهورية مصر العربية

639 الجميعة الجغرافية المصرية

640 المجلة الجغرافية العربية

644 منشورات

645 الجميعة الجغرافية المصرية

638 BIBLIOGRAPHIE géographique de l'Égypte. (Société de géographie d'Égypte).
Cairo. 1-2 (1928-1929). Closed 1929. (U-1:670c; B-1:333b).

639 al-JAM^{c}IYAH al-jughrāfīyah al-miṣrīyah. (Société de géographie d'Égypte).
Cairo.
al-Muhadarat al-ammah. 1956- Annual. (N70-2:3069a).
In Arabic.

640 al-MAJALLAH al-jughrafiya al-^{c}Arabiyah. (al-Jam'īyah al-jughrāfiya al-Miṣrīyah). Cairo. 1- (1968-) Irregular.
In Arabic.

641 REVUE d'Égypte. Recueil mensuel de documents historiques et géographique relatifs à l'Égypte. Cairo. v1-4 no4 (1894-1897). Closed 1897.
(U-4:3630c; B-3:712a; F-4:178b).

642 SOCIÉTÉ d'études historiques et géographiques de l'Isthme de Suez. Cairo.
Bulletin. 1-5 (1947-1953/54). Closed 1956. (U-5:3940b; B-4:132a; BS:802b).
In French.

643 ______.
Cahiers v1 (1955). Closed 1956. (N70-4:5383a; BS:802b).
In French.

644 ______.
Mémoires. 1-3 (1950-1956). Closed 1956. (N70-4:5383a; BS:802b).
Title in Arabic: Manshūrāt.
In French.

*(645 SOCIÉTÉ de géographie d'Égypte, Cairo. Cairo.
Bulletin. s1-7 (1876-1912); ns (s7a) no1 (1915); nsv8- (1917-) Annual. (U-5:3972a; B-4:135a; F-1:866b-867a).
s2-7 in twelve numbers each. 1875-1912 as Société khédiviale de géographie; 1917-1922 as Société sultanieh de géographie; 1923-1953 as Société royale de géographie d'Égypte.
Index: v1-15 (1875-1927), v16-30 (1928-1957).
Beginning in 1887 in French. From 1917 increasingly in English. Now mainly in English.
Abstracts and articles also in Arabic.
Society name in Arabic: al-Jamʿīyah al-jughrāfīyah al-miṣrīyah.
[rue Kasr el-Aīni, Bureau de Poste de Kasr El Doubara, Cairo, Egypt.]

646 ______.
Mémoires. 1-17 no3 (1919-1952). (U-5:3972a; B-4:135a; F-3:466b).

647 ______.
Publications spéciales. 1924-1940. (B-4:135a).
Unnumbered monographs.
In English, French, or Italian.

ኢትዮጵያ ። ETHIOPIA (Amharic)

648 የኢትዮጵያ ጂኦግራፊ መጽሔት ።

648 ETHIOPIAN geographical journal. (Addis Ababa. Mapping and geography institute). v1-5 no2 (1963-1967) Semiannual. Suspended 1968. (N70-2:1964c; B68:198a).
In Amharic and English.
[Mapping and Geography Institute, P.O. Box 597, Addis Ababa, Ethiopia.]

FINLAND (Finnish or Swedish)

Suomen Tasavalta--Republiken Finland

649 ACTA geographica. (Suomen maantieteellinen seura; Geografiska sällskapet i Finland; Geographical society of Finland). Helsinki. 1-27 (1927-1972). Closed 1972. (U-1:53c; B-1:44b; F-1:54b).
Merged into Fennia [652].
In English or German.

(650 ACTA wasaensia. Geography (Universitas economica vasaensis. Vaasa. School of economics). Vaasa. 1- (1971-) Irregular. (N75-1:26a).
In English or German.
[Vaasa School of Economics, Raastuvankatu 31, SF 65100 Vaasa, Finland.]

651 BIDRAG till kännedom av Finlands natur och folk. (Finska vetenskapssocieteten). Helsinki. 1- (1858-) Irregular. (U-1:699b; B-1:348a; F-1:632a).
In Finnish, German, or Swedish.
Indexes: 1838-1910, 1911-1926, 1858-1938.
[Snellmaninkatu 9-11, Helsinki 17, Finland.]

*(652 FENNIA. (Suomen maantieteellinen seura; Geografiska sällskapet i Finland; Geographical society of Finland). Helsinki. 1- (1889-) Irregular. From 1979 2nos a year. (U-2:15553b; B-2:179b; F-2:585a).
Index: 1-50 (1889-1928) in v52 no10 (1931).
In English.
[Academic Bookshop, Keskuskatu 1, SF 00100, Helsinki 10, Finland.]

653 GEOGRAFISKA föreningen i Finland. Helsinki.
Meddelanden. 1-11 (1892-1920). Closed 1920. (U-2:1693b; B-2:268a; F-4:903a).
1-3 as Vetenskapliga meddelanden. Society united with Sällskapet för Finlands geografi, to form the Geographical society of Finland, which publishes Fennia [652], and Terra [661].

654 HELSINKI (Helsingfors). School of economics. (Helsinki. Kauppakorkeakoulu).
Publications. Geographical studies. (Maantieteellisiä tutkielmia).
1-33 (1949-1974) Irregular. (U-3:1837a).
1-2 (1949) as Publications of the Finnish university. School of business. Geographical studies.
In 1974 merged into Acta Academiae oeconomicae Helsingiensis, series A, B, and C (not in this list).
In English, Finnish, German, or Swedish.

(655 ______. Svenska handelshögskolan. Ekonomisk-geografiska institutionen. (Swedish school of economics. Department of economic geography). Helsinki.
Meddelanden. 1- (1952-) Irregular. (N70-2:2521a).
1- (1952-) issued by the institute under its earlier name, Geografiska institutionen.
In Swedish; occasionally in English.
[Arkadiagatan 22, SF-00100 Helsingfors 10, Finland.]

656 ______. University. Department of geography.
Publications. (Publicationes instituti geographici universitatis Helsingiensis; Helsingin yliopiston maantieteen laitoksen julkaisuja).
1-57 (1939-1967). (U-3:1837b; B-1:78a).
sA. 58- (1968-). In English. Reprints.
sB. 1- (1968-). In Finnish with English summary.
[Hallituskatu 11-13, SF-00100 Helsinki 10, Finland.]

657 NORDIA. (Pohjois-Suomen maantieteilijäin seura [Society of geographers in northern Finland]). Oulu. 1- (1967-) Irregular.
In English.
[University of Oulu, Department of Geography, SF-90100 Oulu 10, Finland.]

658 NORDIA, tiedonantoja. (Pohjois-Suomen maantieteilijäin seura. [Society of geographers in Northern Finland]). Oulu.
1- (1970-) Irregular.
In Finnish.
[Department of Geography, University of Oulu, SF-90100 Oulu 10, Finland.]

659 OULU. Yliopisto. Maantieteen laitos. (University. Department of geography; Universität. Geographisches institut). Oulu.
Julkaisuja. (Publications; Schriftenreihe). 1- (1960-) Irregular. (N70-3:4471b).
Also as Publicationes instituti geographici universitatis Ouluensis.
Reprints, mainly from Fennia or Nordia.
In English or German.
[Department of Geography, University of Oulu, SF-90100 Oulu, Finland.]

(660 SUOMALAINEN tiedeakatemia. (Academia scientiarum fennica). Helsinki. Toimituksia (Annales). sA. 3. Geologica-geographica. 1- (1942-) Irregular. (U-5:4117a; B-4:254b).
In English or German.
[Snellmaninkatu 9-11, SF-00170 Helsinki 17, Finland.]

*(661 TERRA. (Suomen maantieteellinen seura; Geografiska sällskapet i Finland; Geographical society of Finland). Helsinki. 1- (1888-) Quarterly. (U-5:4183a; B-2:268a; F-4:669a; F-2:717a).
1-24 (1888/89-1912) as Geografiska föreningen i Finland. Tidskrift.
Indexes: 1-20 (1888-1908) in v21 (1909), pp93-128; 21-40 (1909-1928) in v40 (1928); 41-65 (1929-1953) in v66 (1954); 66-73 (1954-1961) in v75 (1963); 74-81 (1962-1969) in v83; 82-89 (1970-1977) in v90.
In Finnish or Swedish; English abstracts listed by title in table of contents.
[Academic Bookshop, Keskuskatu 1, SF-00100 Helsinki 10, Finland.]

662 TURKU (Åbo). Åbo akademi. Handelshögskolan. (Abo Swedish university. School of economics). Ekonomisk geografiska institutionen. Turku. Meddelanden. 1- (1961-). (N70-4:5888b).
In Swedish.

663 ______. ______. ______.
______.
Memorandum. 1- (1967-). 5 nos a year. (N75-2:2155c).
In Swedish with English summary.
[Henriksgatan 7, Turku, Finland.]

664 ______. Yliopisto [University]. Turku.
Julkaisuja. Annales. (Annales universitatis turkuensis; Turku. Turun yliopisto. Julkaisuja).
sA. II. Biologica-geographica. 23- (1957-). (N70-4:5888c).
Continued in part sA. Physico-mathematica-biologica (not in this list).
Superseded by Julkaisuja. sA. II: Biologica-geographica-geologica [665].
In English or German.

665 ______. ______.
______. sA II: Biologica-geographica-geologica.
v45- (1970-).
Continues its Julkaisuja. sA II: biologica-geographica [664].
Title also in Latin.
Text in English, German, or French.

(666 ______. ______. Maantieteen laitos. Turku.
Julkaisuja. (Publicationes instituti geographici universitatis turkuensis; Turku. University. Department of geography. Publications. 1- (1927-) Irregular. (U-1:8c; B-4:254b).
Mainly reprints. 1-12 (1927-1935) as Publicationes instituti geographici universitatis aboensis.
In English or Finnish.

667 VAASA. School of economics. Vaasa.
Proceedings: reprint series, geographica. 1- (1969-).
In English, Finnish, or Swedish with summaries in English or German.
[Raastuvankatu 31, SF-65100 Vaasa 10, Finland.]

668 ______. ______.
Proceedings: research series, geographica. 1- (1969-).
In English, Finnish, German, or Swedish.
[Raastuvankatu 31, SF-65100 Vaasa 10, Finland.]

FRANCE (French)

République Française

669 À TRAVERS le monde. Paris. 1-20 (1895-1914). Closed 1914. (U-1:8b; B4:332b; F-1:3a).
Supplement to (Le) Tour du monde [882]. Supersedes Nouvelles géographiques [794].

670 À TRAVERS le monde. (Société de géographie de Compiègne). Compiègne. v1-2 no6 (1946-1947). Closed 1947? (U-1:8b).

671 ACADÉMIE des sciences, Paris.
Comptes rendus hebdomadaires des séances. Groupe X. Géographie et navigation, physique du globe. 1964-1965. (N70-1:44c).
Superseded by sD. Sciences naturelles (not in this list).

672 ACADÉMIE internationale de géographie botanique, Le Mans. Le Mans, Agen.
Bulletin de géographie botanique. 1-29 (1891-1919). Closed 1919. (U-1:32a; B-3:241b; F-1:746a; F-3:608a).
v1-8 (no1-109) (0 1891-1898) as Monde des plantes; v8 (no112)-v29 (1899-1909) as Bulletin.

*(673 ACTA geographica. (Société de géographie de Paris). Paris. no1-79 (1947-1969); s3 v1- (1970-) Quarterly. (U-1:53c; B-1:44b; BS:23a; N75-1:23b). Succeeds Geographie [768].
no1-19 (1947-1954) carry subtitle Comptes rendus; no20-24 (1956-1957) Bulletin officiel. Includes Supplément bibliographique, later Bibliographie analytique.
Indexes: 1947-1969; 1970-1975 in no17 (Ja 1976).
[Société de Geographie, 184, Boulevard Saint-Germain, 75006 Paris, France.]

674 ______. Bibliographie analytique. 10/13-19 (1973-1977). (N75-1:23b).
10/13 (1973) one number; 14 and 15 (1974) not separately numbered as bibliographies but as nos 18 and 19 of Acta geographica. Nos16, 17, 18, 19 (Jl 1975-Mr 1977), also as nos22, 25, 28, and 30 of Acta geographica. From no31 (Jn 1977) bibliographies not separately numbered but carried as integral part of each issue of Acta geographica [673].
Supersedes Acta geographica: supplément bibliographique [675].

675 ______. Supplément bibliographique. 1949-1972: Quarterly. (N70-1:58b; B-1:44b; BS:803a).
1949-1959 no2 as Bibliographie mensuelle. Monthly. Assumes numbering of society's Acta geographica. New series 29-79 (1959-1969); s3, 1-9 (1970-1972).
Superseded by Acta geographica: Bibliographie analytique [674].

676 AMIENS. Université. Centre d'études géographiques. Groupe de recherches sur l'eau.
Travaux. 1- (1975-) Annual.
[Centre d'Études Géographiques, Université d' Amiens, Chemin du Thil, 80044 Amiens cedex, France.]

*(677 ANALYSE de l'espace. (Centre de recherche "Analyse de l'espace." Cahiers). Paris. 1972- Quarterly.
[Centre de Recherches "Analyse de l'Espace," Institut de Géographie, 191 rue St.-Jacques, 75005 Paris, France.]

(678 ANALYSE, organisation et gestion de l'espace. (Nice. Centre d'analyse de l'espace). Nice. 1- (1977-).
[Centre d'Analyse de l'Espace, 117, rue de France, 06 Nice, France.]

679 ANALYSE spatiale quantitative et appliquée. (Association d'analyse spatiale quantitative et appliquée). Nice.
1- (1976-) 2 nos a year.
Summaries in English and French.
[Laboratoire de Géographie Raoul Blanchard, UER Lettres et Sciences humaines, 98 bd. Edouard Herriot, 06000 Nice, France.]

*(680 ANNALES de géographie. Bulletin de la Société de géographie. Paris.
1- (1891-) 6 nos a year. (U-1:374c; B-1:150b; F-1:203b-204a).
v1-24(1891/92-1913/14) included v1-24 of Bibliographie géographique internationale [696], then called in turn Bibliographie de l'année... Bibliographie annuelle, and Bibliographie géographique annuelle.
Indexes: 1891-1901, 1902-1911, 1912-1921, 1922-1931, 1932-1951, 1952-1961, 1962-1971.
English abstracts of articles from v76 no413 (Ja/F 1967).
[Librairie Armand Colin, 103 Boulevard St.-Michel, 75240 Paris, Cedex 05, France.]

681 ANNALES des voyages, de la géographie, de l'histoire et de l'archéologie. Paris. 1-188 (1819-1865); s6 v45-64 (1866-1870). Closed 1870. (U-1:381a; B-1:155a; F-1:251b; F-3:777b).
1-188 (1819-1865) as Nouvelles annales des voyages, de la géographie et de l'histoire..., 1-30 (1819-1826); 31-60 also as s2 v1-30 (1826-1833); 61-84 as s3 v1-24 (1834-1839); 85-104 as s4 v1-20 (1840-1844; 105-144 as s5 v1-40 (1845-1854); 145-188 as s6 v1-44 (1855-1865).
Supersedes Annales des voyages de la géographie et de l'histoire [682].
Index: 1-s3 v24 (1819-1839).

682 ANNALES des voyages, de la géographie et de l'histoire. Paris.
1-24 (1807-1814). Closed 1814. (U-1:381a; B-1:155a; F-1:251b).
Superseded by Annales des voyages, de la géographie, de l'histoire et de l'archéologie [681].
Index: v1-20 (1808-1813).

683 ANNALES statistiques et géographiques universelles. Paris. 1 (1831).

684 ANNÉE cartographique; supplément annuel à toutes les publications de géographie et de cartographie. Paris. 1-23 (1891-1914). Closed 1914. (U-1:393a; B-1:158b; F-1:282a).

685 ANNÉE géographique. Revue annuelle des voyages de terre et de mer. Paris. 1-17 (1862-1878). Closed 1878. (U-1:393b; B-1:158b; F-1:283b).
15-17 also as s2 v1-3.

686 ANNUAIRE des géographes de la France (et de l'Afrique francophone). Paris. 1969-1973 Irregular. (N75-1:133c).
Supplement to Intergéo bulletin [778].
Superseded by Répertoire des géographes francophones [812a].

687 ANNUAIRE des voyages et de la géographie. Paris. 1-4 (1844-1847). Closed 1847. (U-1:396c; B-1:165a; F-1:307ab).

688 ANNUAIRE universel des sociétés de géographie. Paris. 1 (1892-1893). Only one issued.

*(689 ASSOCIATION de géographes français. Paris.
Bulletin. 1- (1924-) Bimonthly. (U-1:517c; B-1:238a; F-1:755a).
Indexes: 1924-1934; 1955-1975.
Summaries in English and French.
[Association de Géographes Français, 191 rue Saint-Jacques, 75005 Paris, France.]

(690 ASSOCIATION française pour l'étude du quaternaire. Paris.
Bulletin. 1- (1964-) Quarterly. (N70-1:510b; B68:99a).
[Laboratoire de Géologie I, Université Pierre et Marie Curie, 4 Place Jussieu, Tour 16, 75230 Paris Cedex 05, France.]

691 ASSOCIATION géographique d'Alsace. Strasbourg. Mulhouse.
Bulletin. 1- (1954-) Annual. Closed?

692 ASSOCIATION Marc Bloch de Toulouse. Toulouse.
Cahiers. Études géographiques. 1-2 (1964). Closed? (N70-1:512a).

693 ASSOCIATION professionnelle des géographes. Maurepas.
Bulletin. 1- (1972-) Quarterly. Closed?
[Association Professionnelle des Géographes, 14/617 place du Sancerrois, 78310 Maurepas, France.]

(694 BESANÇON. Université. Institut de géographie. Besançon.
Séminaires et notes de recherche des Cahiers de géographie de Besançon. 1- (1971-) Irregular, but 2-4 a year.
[Institut de Géographie, U.E.R. de Lettres, 25030 Besançon, France.]

695 BIBLIOGRAPHIE cartographique internationale. (Comité national français de géographie; International geographic union). Paris. 1936-1975. Closed 1975. Annual. (U-1:668a; B-1:333a; F-1:603a).
1936 as Bibliographie cartographique de la France; 1937-1938/45 as Bibliographie cartographique française. Name of issuing body varies.

*(696 BIBLIOGRAPHIE géographique internationale. International geographical bibliography. (France. Centre national de la recherche scientifique. Laboratoire d'information et de documentation en géographie [Intergéo]. Under the auspices of the International geographical union and l'Association de géographes français. With the assistance of UNESCO and Comité international d'historiens et géographes de langue française). Paris. 1- (1891-). 1-81 (1891-1975/76) Annual, some years grouped; 82- (1977-) Quarterly with annual cumulative index. (U-1:670c; B-1:150b; F-1:609b).
1-23/24 (1891-1914) in or supplement to Annales de géographie [680] entitled successively Bibliographie de l'année, Bibliographie annuelle, and Bibliographie géographique annuelle. 25-41 (1915-1931) as Bibliographie géographique.
Table of contents and headings also in English from 1972.
[BGI, Laboratoire d'Information et de Documentation en Géographie (Intergéo), CNRS, 191 rue St-Jacques, 75005 Paris, France.]

697 BIBLIOTHÈQUE d'histoire et de géographie universelle. Paris. 1-8 (1900-1903). Closed 1903. (U-1:693b).

698 BIBLIOTHÈQUE des géographes arabes. Paris. 1-2 (1927-1928). Closed 1928. (U-1:695b; B-1:345b).
2 published in 1927.

(698a BORDEAUX. Laboratoire de géographie physique appliquée. Bordeaux.
Travaux. 1- (1977-).

(699 BROUILLONS Dupont. Avignon. 1- (1976-) Annual.
[Groupe Dupont, Faculté des Lettres et Science Humaines, 35, rue Joseph Vernet. 84000 Avignon, France.]

700 BULLETIN de géographie d'Aix-Marseille. (Société de géographie de Marseille [and] Aix-en-Provence. Université. Faculté des lettres. Laboratoire de géographie). Marseille; Aix-en-Provence. nsv1-5 (66-70) (1955-1959). Closed 1959. (BS:803b). nsv1 (1955) also as v66 of Société de géographie et d'études coloniales de Marseille [859]. Superseded by Société de géographie de Marseille. Bulletin [859].

701 BULLETIN universel des sciences et de l'industrie. (Société pour la propagation des connaissances scientifiques et industrielles). Paris. Section 6, Bulletin des sciences géographiques, etc. Economie publique; voyages. 1-28 (1824-1831). Closed 1831. (U-1:844b; B-1:450a; F-1:900a).

(702 CAEN. Centre de géomorphologie. (Centre national de la recherche scientifique). Caen.
Bulletin. 1- (1967-) 2 nos a year, irregular. (N75-1:492a).
[Centre de Géomorphologie de Caen, C.N.R.S., Rue des Tilleuls, 14000 Caen, France.]

(703 ______. Université. Centre d'études régionales et d'aménagement. Caen.
Bulletin. 1- (1978-).
[Département de Géographie, CERA, Université de Caen, Caen, France.]

(704 ______. ______. Centre de recherches sur l'évolution de la vie rurale. Caen.
Publications. 1- (1971-) Irregular. (N75-1:492b).
[Faculté des Lettres et Sciences Humaines de l'Université de Caen, Caen, France.]

(705 ______. ______. ______.
Travaux. 1- (1961-) Irregular.
[Association Normande de Géographie, Université de Caen, 14032 Caen Cedex, France.]

(706 ______. ______. Département de géographie. Caen.
Cahiers: bulletin semestriel. 1- (N 1970-) 2 nos a year. (N75-1:385b).
[Association Normande de Géographie, Université de Caen, 14000 Caen, France.]

*(707 CAHIERS d'Outre-Mer; revue de géographie de Bordeaux. (Université de Bordeaux III. Institut de géographie et d'études régionales; Institut d'Outre-Mer de Bordeaux; [and] Société de géographie de Bordeaux). Bordeaux. 1- (1948-) Quarterly. (U-2:865c; B-1:470b).
Indexes: v1-10 (1948-1957); v11-20 (1958-1967).
Resumés in French and English from no89 (Ja/Mr 1970).
[Les Cahiers d'Outre Mer, Institut de Géographie, Université de Bordeaux III, 33405 Talence, France.]

(708 CAHIERS de Fontenay. (Fontenay-aux-Roses. École normale supérieure). Fontenay-aux-Roses.
Fascicule spécialisé "Géographie." 1976- Irregular.
Summaries in English, German, Spanish.
[École Normale Supérieure, 5, rue Boucicaut, 92260 Fontenay-aux-Roses, France.]

*(709 CAHIERS de géographie de Besançon. (Besançon. Université). Paris.
1- (1954-) Annual. (N70-1:989b).
Subseries of Besançon. Université. Annales littéraires (not in this list).
["Les Belles Lettres," 95 boulevard Raspail, 75006 Paris, France.]

*(710 CAHIERS de géographie de Dijon. (Dijon. Faculté des sciences humaines. Section de géographie. Groupe d'études géographiques de Dijon). Dijon.
1- (1975-) 2 nos a year.
[Groupe d'Etudes Géographiques de Dijon, Section de Géographie, Faculté des Sciences Humaines, 2, bd. Gabriel, 21000 Dijon, France.]

(712 CAHIERS de géographie physique de Lille. Lille. 1- (1972-).
[Institut de Géographie, Université des Sciences et Techniques de Lille, 9, rue Auguste-Angellier, 59000 Lille, France.]

(713 CAHIERS de l'Atlas de Franche-Comté. Besançon. 1- (1975-) 2 nos a year.
[Association pour l'Atlas de Franche-Comté, Institut de Géographie, U.E.R. Faculté des Lettres, 25030 Besançon Cedex, France.]

*(714 CAHIERS des Amériques Latines. Série sciences de l'homme. (Institut des hautes études de l'Amérique Latine). Paris. 1- (1968-) 2 nos a year. (N70-1:992b).
Summaries in English, French, and Spanish.
[Institut des Hautes Études sur l'Amérique Latine, 28, rue Saint-Guillaume, 75007 Paris, France.]

*(715 CAHIERS géographiques de Rouen. Mont-Saint-Aignan.
1- (1973-) 2 nos a year. Abstracts in English.
[U.E.R. Lettres de Rouen, rue Lavoisier, 76130 Mont Saint-Aignan, France.]

(715a CAHIERS mulhousiens de géographie (Association géographique d'Alsace).
Mulhouse. 4- (1973-). (N79).

(716 CAHIERS nantais. (Centre nantais de recherche pour l'aménagement régional).
Nantes. 1- (1969-) 2 nos a year.
[Institut de Géographie et d'Aménagement Régional, Chemin de la Sensive du Tertre, 44036 Nantes Cedex, France.]

717 CERCLE d'études géographiques du Bas Maine. Laval.
Bulletin. no1-12 (1951?-1955). Closed 1955. Mimeographed.

(718 CESSIÈRES. Centre de recherches et d'enseignement de Cessières. Cessières (Aisne).
Notes et cahiers. 1- (1974-) Irregular.
[École Normale Supérieure de Saint-Cloud, Laboratoire de Biogéographie, 2, avenue du Palais, 92120 Saint-Cloud, France.]

719 CHRONIQUE géographique des pays celtes. Rennes; Paris. 1942-1953.
Closed 1953. (B-1:565a).
Succeeded by Norois [793].

720 CHRONIQUE géographique mensuelle. Paris. 1888. Closed.

(721 CLERMONT-FERRAND. Université. Institut de géographie. Faculté des lettres de Clermont-Ferrand II. Paris.
Publications. 1- (1951-) Irregular. (N70-1:1317c).
Monographs and reprints from Revue d'Auvergne. A subseries of Clermont-Ferrand. Université. Faculté des lettres. Publications (not in this list).
[Institut de Géographie de Clermont-Ferrand, 29 bd. Gergovia, 63037 Clermont-Ferrand, France.]

*(722 COMITÉ français de cartographie. Paris.
Bulletin. 1- (Mr 1958-) Quarterly. (N70-1:1416a).
1-22 (1958-1964) as Comité français de techniques cartographiques. Bulletin.
[Comité Français de Cartographie, 39 ter, rue Gay-Lussac, 75005 Paris, France.]

723 COMITÉ français de géographie historique et d'histoire de la géographie.
Nancy.
Bulletin. 1-4 (1935-1938). Closed 1938. (B-1:610b; F-1:918a).

724 COMITÉ national français de géographie. Paris.
Bulletin. 1920/27-1939. Closed 1939? (F-1:919a; F-2:149a).
1920/27-1935 as Année, 1937 as Bulletin.

725 CONGRÈS de géographie économique et commerciale. Moulins. 3 (1900).

*(726 CONGRÈS national des sociétés savantes (earlier: Congrès des sociétés savantes de Paris et des départements). Section de géographie. Paris.
Actes. 84th- (1959-) Annual. (N70-1:1486a).
Preceded by France. Comité des travaux historiques et scientifiques. Section de géographie. Bulletin [747].
[Bibliothèque. Nationale, 58 rue de Richelieu, 75084 Paris Cedex 02, France; Comité des Travaux Historiques et Scientifiques, 110, rue de Grenelle, 75357 Paris, France.]

727 CONGRÈS national des sociétés françaises de géographie. Paris.
Compte rendu. 1-31 (1878-1913). Closed 1913. (U-2:1168c; B-1:636a; F-2:213b-214a).
31 (1913) in Géographie [768] v28 (1913).

(728 DIJON. Université. Centre de recherche de climatologie. Dijon.
Cahiers. 1- (1972-) Annual.
[Centre de Recherche de Climatologie, Université de Dijon, 36, rue Chabot Channy, 21000 Dijon, France.]

(729 DOCUMENTS de cartographie écologique. (Université scientifique et médicale de Grenoble. Laboratoire de biologie végétale). Grenoble. 11- (1973-) 2 nos a year. (N75-1:711b).
Supersedes and continues numbering of Documents pour la carte de végétation des Alpes [731].
Summaries in English, French, and occasionally Italian.
[Laboratoire de Biologie Végétale, Domaine Universitaire, 38041 Grenoble Cédex, France.]

730 DOCUMENTS historiques et géographiques relatifs à l'Indochine. Paris.
1-4 (1910-1914). Closed 1914. (U-2:1352c).

731 DOCUMENTS pour la carte de la végétation des Alpes. (Grenoble. Université. Faculté des sciences. Laboratoires de biologie végétale de Grenoble et du Lautaret). Grenoble. 1-10 (1963-1972). Closed 1972. (N70-2:1771b). Superseded by Documents de cartographie écologique [729], which continues its numbering.

732 ÉCHO du monde savant. Travaux des savants de tous les pays dans toutes les sciences. Paris. Pt2. Sciences naturelles et géographiques.
1-13 (1834-1846). Closed 1846. (U-2:1387a; B-2:86b; F-2:432a).

(733 ÉCONOMIE, géographie. (Conseil national du patronat français). Paris.
1- (1963-) Monthly. (N75-1:741a).
1963-1971 as Géographie et industrie.
[ETP, 31 avenue Pierre 1er de Serbie, 75016 Paris, France.]

(734 ESPACE, Temps. Cachan. 1- (1975-) Quarterly.
[École Normale Supérieure de l'Enseignement Technique, Section Histoire et Géographie, B. P. 10, 94230 Cachan, France.]

*(735 ESPACE géographique: régions, environnement, aménagement. Paris. 1- (1972-) Quarterly. (N75-1:789a; B73:109b; BL).
Index: v1-5 (1972-1976).
Titles in table of contents and abstracts in French and English.
[Doin, Éditeurs, 8 place de l'Odéon, 75006 Paris, France.]

(736 ESPACES et sociétés. Revue critique internationale de l'aménagement de l'architecture et de l'urbanisation. Paris. 1- (1970-) Quarterly. (N75-1:789a).
Supersedes Utopie (not in this list).
[Editions Anthropos, 12, avenue du Maine, 75015 Paris, France.]

737 ÉTUDES canadiennes. (Grenoble. Université. Institut de géographie alpine). Grenoble. no1-5 (1930-1934?); s2 no1-3 (1936-1938?); s3 no1 1939- . (U-2:1488b).

(738 ÉTUDES de géographie tropicale. (Centre d'études de géographie tropicale. Centre national de la recherche scientifique). Talence, Bordeaux. 1- (1972-). (N75-1:798b).

739 ÉTUDES de la région parisienne. (Société d'études historiques, géographiques et scientifiques de la région parisienne). Paris.
ns1- (38-) (1964-1975) Quarterly. Closed 1975. (N70-2:1969b).
Supersedes and continues numbering of society's Bulletin [837].

740 ÉTUDES et travaux de "Méditerranée." Revue géographique des pays méditerranéens. Gap.
1- (1964-) Irregular. (N70-2:1970c).
[Éditions Ophrys, 05 Gap, France.]

(741 ÉTUDES rurales: revue trimestrielle d'histoire, géographie, sociologie et économie des campagnes. (École des hautes études en sciences sociales. Laboratoire d'antropologie sociale.) Paris. 1- (Ap/Je 1961-) Quarterly. (N70-2:1972c: B68:200a).
1-56 (1961-1974) issued by École pratique des hautes études, Sorbonne. Sixième section: Sciences économiques et sociales.
[Ed. Mouton et Co., 7 rue Dujuytren, 75006 Paris, France.]

(742 ÉTUDES vauclusiennes. Avignon. 1- (1969-) 2 nos a year.
Summaries in English and French.
[Association des Études Vauclusiennes, 35 rue Joseph Vernet, 84000 Avignon, France.]

743 EXPLORATEUR. Journal géographique et commercial, sous le patronage de la commission géographique commerciale. Paris. 1-4 (no1-81) (1875-1876). Closed 1876. (U-2:1510a; B-2:158a; F-2:554a).
Superseded by Exploration [744].

744 EXPLORATION. Journal des conquêtes de la civilisation sur tous les points du globe. Paris. v1-18 no1-412 (1876-1884). Closed 1884. (U-2:1510a; B-2:158a; F-2:554a).
Supersedes Explorateur [743]. Superseded by Gazette géographique et l'exploration [764].

*(745 FRANCE. Centre national de la recherche scientifique (CNRS). Paris. Mémoires et documents de géographie. 1- (1979-) Irregular.
Supersedes France. Centre national de la recherche scientifique (CNRS). Centre de documentation cartographique et géographique, later Service de documentation et de cartographie géographiques. Mémoires et documents [746].
[Éditions du CNRS, 15 quai Anatole France, Paris 75700 France; Intergéo, Laboratoire d'Information et de Documentation en Géographie, 191, rue Saint-Jacques, 75005 Paris, France.]

746 ______. ______. Service de documentation et de cartographie géographiques. Memoires et documents. 1-10 no4 (1949-1966); nsv1-18 (1966-1978) Irregular. Closed 1978. (N75-1:880c; B-2:224a).
Superseded by Mémoires et documents de géographie [745].

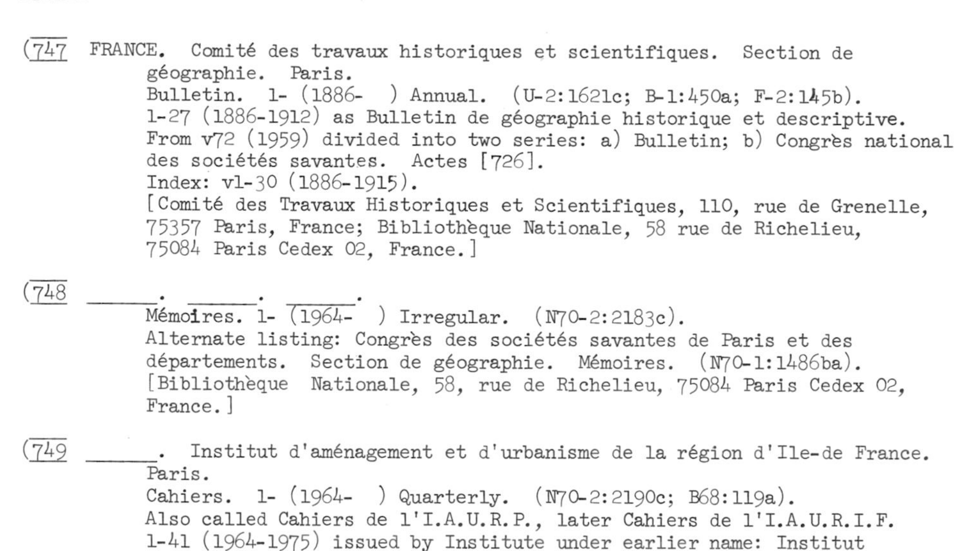

(747 FRANCE. Comité des travaux historiques et scientifiques. Section de géographie. Paris.
Bulletin. 1- (1886-) Annual. (U-2:1621c; B-1:450a; F-2:145b).
1-27 (1886-1912) as Bulletin de géographie historique et descriptive.
From v72 (1959) divided into two series: a) Bulletin; b) Congrès national des sociétés savantes. Actes [726].
Index: v1-30 (1886-1915).
[Comité des Travaux Historiques et Scientifiques, 110, rue de Grenelle, 75357 Paris, France; Bibliothèque Nationale, 58 rue de Richelieu, 75084 Paris Cedex 02, France.]

(748 ______. ______. ______.
Mémoires. 1- (1964-) Irregular. (N70-2:2183c).
Alternate listing: Congrès des sociétés savantes de Paris et des départements. Section de géographie. Mémoires. (N70-1:1486ba).
[Bibliothèque Nationale, 58, rue de Richelieu, 75084 Paris Cedex 02, France.]

(749 ______. Institut d'aménagement et d'urbanisme de la région d'Ile-de France. Paris.
Cahiers. 1- (1964-) Quarterly. (N70-2:2190c; B68:119a).
Also called Cahiers de l'I.A.U.R.P., later Cahiers de l'I.A.U.R.I.F.
1-41 (1964-1975) issued by Institute under earlier name: Institut d'aménagement et d'urbanisme de la région parisienne.
[21-23 rue Miollis, 75732 Paris Cedex 15, France.]

(750 ______. Institut géographique national. Paris.
Bulletin d'information. 1953; 1- (1964-) Irregular. (N70-2:2190c).
[Institut Géographique National, 136 bis, rue de Grenelle, 75700 Paris, France.]

(751 ______. ______.
Cahiers. 1- (1973-).
[Institut Géographique National, 136 bis rue de Grenelle, 75007 Paris, France.]

752 ______. ______.
Études de photo-interprétation. 1- (1964-) Irregular. (N70-2:2190; B73-111b).
In French with English summaries.

753 ______. ______.
Exposé des travaux. 1888-1966. Annual. Closed 1966. (ULG:223b; B-2:486b; F-4:488a; F-2:11b).
1888-1921 as France. Service géographique de l'armée. Rapport sur les travaux exécutés; 1922-1945 as France. Service géographique de l'armée. Cahiers; rapports sur les travaux exécutés. 1946-1951 as Rapport sur l'activité.

754 ______. ______.
Publications scientifiques. 1-2 (1944-). (U-2:1622b; B-2:486b).

755 ______. ______. Service de la documentation technique. Paris.
Bulletin analytique et bibliographique. 1- (1953-). Closed. Irregular. (N70-2:2191a).
Superseded by its Bulletin d'information [750] and its Bulletin bibliographique [756].

(756 ______. ______. ______.
Bulletin bibliographique. 1- (1953-). (N75-1:876a).
[Institut Géographique National, 2 Avenue Pasteur, 94160 Saint-Mandé, France.]

757 FRANCE. Ministère des colonies. Paris.
Revue coloniale: explorations...travaux historiques et géographiques. 1-7 (1895-1901); nsv1-11 no1-99 (1901-1911). Closed 1911. (ULG:219a; B1:452b; F-4:169b).

*(758 ______. Office de la recherche scientifique et technique d'Outre-mer (ORSTOM). Bondy.
Cahiers de l'O.R.S.T.O.M. Série sciences humaines. 1- (1963-) Quarterly. (N70-1:997a).
Summaries in English and French.
[O.R.S.T.O.M., 70-74 Route d'Aulnay, 93140 Bondy, France.]

(759 ______. ______. ______.
Mémoires de l'O.R.S.T.O.M. 1- (1961-) Irregular. (N70-2:2201a).
Includes monographs from Section de géographie, ORSTOM.
Summaries in English.
[O.R.S.T.O.M., 70-74 route d'Aulnay, 93140 Bondy, France.]

*(760 ______. ______.
Travaux et documents de l'O.R.S.T.O.M. 1- (1969-) Irregular.
Summaries in English.
[O.R.S.T.O.M., 70-74 route d'Aulnay, 93140 Bondy, France.]

761 ______. Service de la carte phytogéographique. Paris.
Bulletin. (N70-2:2202b; BS:328a).
sA. Carte de la végétation au 200,000e. v1-7 (1956-1962). Closed 1962.
sB. Carte des groupements végétaux au 20,000e. v1-7 (1956-1962). Closed 1962.
Continued by service's Notes et documents [763] and later by service's Mémoires [762].

762 ______. ______.
Mémoires. 1- (1969-).
Supersedes the service's Notes et documents [763].
[15, Quai Anatole France, Paris 7e, France.]

763 ______. ______.
Notes et documents. 1962-1963. Closed 1963. (N70-2:2202a).
Supersedes the service's Bulletin [761] and is superseded by its Mémoires [762].

764 GAZETTE géographique et l'exploration. Paris. nsv1-3 no1-38 (1885-1887). Closed 1887. (U-2:1674b; B-2:158a; F-2:807b).
Supersedes Exploration [744]. Also numbered 21-24 in continuation of Exploration. Merged into Revue française de l'étranger et des colonies... [823].

(765 GÉOCATALOGUE. (Paris. Universités de Paris I, IV, VII. Centre de géographie et Laboratoire Intergéo du CNRS). Paris. 1- (1966-) Irregular.
[Bibliotheque du Centre de Géographie des Universités Paris, I, IV, VII, 191, rue Saint-Jacques, 75005 Paris, France.]

(766 GÉODOC. (Toulouse. Université de Toulouse le Mirail. Institut de géographie). Toulouse. 1- (1974-) Irregular.
[Institut de Géographie, Université de Toulouse le Mirail. 109, rue Vauquelin, 31081 Toulouse, France.]

767 GEOGRAPHIA; Magazine de la géographie et de l'histoire. Paris.
no1-139 (1951-1963) Monthly. Closed 1963. (N70-2:2280a; N70-2:2546c; BS:341b). no1-96 (1951-1959) as Geographia; Revue d'informations et d'actualités géographiques; no97-99 (1959) as Geographia: Historica; 100-139 (1960-1963) as Geographia histoire; magazine de la géographie et de l'histoire. Superseded by Historama, 140- (1963-) not in this list).

768 GÉOGRAPHIE. (Société de géographie). Paris. v1-72 no3 (1900-O/D 1939). Closed 1939. (U-2:1695b; B-4:134b-135a; F-2:728a).
v1-32 (1900-1918/19) carries subtitle: Société de géographie, Paris. Bulletin. v56-58 (1931-1933) as Terre, air, mer; La géographie. 1940-1946 combined with Annales de géographie [680]. Succeeded in 1947 by Acta geographica [673].
Index: v1-72 (1900-1939).

769 GÉOGRAPHIE; revue générale des sciences géographiques. Paris. v1-11 no1-489? (1888-1898). Closed 1898. (U-2:1695c; B-2:269b; F-2:728ab).

*(770 GÉOGRAPHIE et recherche. (Association géographie et recherche). Saint Cloud. 1- (1972-) Quarterly. (N76).
Before 1978 published in Dijon.
[Association Géographie et Recherche, 11, rue des Girondins, 92210 Saint-Cloud, France.]

*(771 HÉRODOTE. Stratégies, géographies, idéologies. Paris. 1- (Ja/Mr 1976-) Quarterly.
[Editions F. Maspero, 1, Place Paul-Painlevé, 75005 Paris, France.]

(772 HISTORIENS et géographes; revue de l'Association des professeurs d'histoire et de géographie de l'enseignement public. Paris. 1- (1910-) Quarterly to 1965, bimonthly 1965- . (U-5:3954c; F-1:852a).
1-55 (1910-1965) as Société des professeurs d'histoire et de géographie de l'enseignement public. Bulletin.
[Association des Professeurs d'Histoire et de Géographie, 8, rue Nicolas-Charvet, 75015 Paris, France.]

773 HOMME; cahiers d'ethnologie, de géographie et de linguistique. (Paris. École pratique des hautes études). Paris. 1-3 (1950-1951); ns1- (1958-). (N70-2:2569bc; BS:390a).
Monographs in French or English.

*(774 HOMMES et terres du Nord. (Lille. Université I. Institut de géographie [and] Société de géographie de Lille). Lille. 1963- Semiannual. (N70-2:2570b; B68:241b).
Supersedes the geographical issue of Revue du Nord [821], and Société de géographie de Lille, Bulletin [856].
Abstracts in English and French.
[Institut de Géographie de l'Université de Lille I, Cité Scientifique, 59650 Villeneuve d'Ascq, France; Société de Géographie de Lille, 116 rue de l'Hôpital militaire, 59000 Lille, France.]

(775 IMAGES du Val de Marne. (Association géographique d'études et de recherches sur le Val de Marne). Maisons Alfort. 1- (1972-) Quarterly.
[Association Géographique d'Études et de Recherches sur le Val de Marne (AGER 94), 59, rue de Jemmapes, 94700 Maisons Alfort, France.]

*(776 INFORMATION géographique. Paris. 1- (1936-) 5 nos a year. (U-3:1987b; B-2:478b; BS:412a; F-2:882a).
[J. B. Baillière et Fils, 19 rue Hautefeuille, 75279 Paris Cedex 06, France.]

777 ______. (supplément). Cahiers. Paris. 1952-1954. Closed 1954. (N70-2:2784b; BS:412a).

*(778 INTERGÉO bulletin. Organe trimestriel des instituts et centres de recherches de géographie. Paris. 1- (1966-) Quarterly. (N75-1:1141b).
1966-1977 as Intergéo. Bulletin de liaison des instituts et centres de recherches de géographie.
Supplemented by Annuaire des géographes de la France...[686].

Index: 1966-1975.
[C.N.R.S. Laboratoire d'Information et de Documentation en Géographie "Intergéo," 191, rue Saint-Jacques, 75005 Paris, France.]

(779 INTER-NORD: Revue internationale d'études arctiques et nordiques; International journal of arctic and nordic studies. (Paris. Ecole des hautes études en sciences sociales. Centre d'études arctiques. Paris. 1- (1960-) Annual. (N70-2:2979b).
Subtitle varies: v7-9 (1965-1967) as Inter-Nord: Revue de géographie [économique et politique] des pays du Nord. Until 1974 issued by Paris. École pratique des hautes études. Centre d'études arctiques et finno-scandinaves.
In French and English. Abstracts.
[Mouton, Publishers, P.O. Box 482, The Hague, The Netherlands.]

780 JOURNAL des voyages, découvertes et navigations modernes ou Archives géographiques et statistiques du XIXe siècle. (Société de géographes français et étrangers). Paris. v1-44 no1-134 (1818-D 1829). Closed 1829. (U-3:2189b; B-2:627a; F-3:159б-160a).

(781 LETTRE d'Intergéo. (Laboratoire d'information et de documentation en géographie Intergéo). Paris. 1977- 8 nos a year.
[Laboratoire d'Information et de Documentation en Géographie Intergéo, CNRS, 191, rue Saint-Jacques, 75005 Paris, France.]

(782 LILLE. Université des sciences et techniques (Lille I). Institut de géographie de Lille. Travaux. 1- (1974-) Irregular.
[Institut de Géographie, Université des Sciences et Techniques de Lille, B.P. 36, 59650 Villeneuve d'Ascq, France.]

783 ______. ______. Laboratoire de géographie rurale de Lille. Travaux et recherches. 1- (1973-) Irregular.
[Laboratoire de Géographie Rurale, U.E.R. de Géographie, 59650 Villeneuve d'Ascq, France.]

(784 LIMOGES. Université. U.E.R. des lettres et sciences humaines de Limoges. Limoges.
Publications. Travaux et mémoires. [Collection géographie.]
1- (1973-) Irregular.
Summaries in English and French.
[U.E.R. des Lettres et Sciences Humaines, Université de Limoges, 181, rue d'Isle, 87000 Limoges, France.]

(785 LYON (Lyons). Université. Institut des études rhodaniennes. Lyon. Mémoires et documents. 1- (1942-) Irregular. (U-3:2493c; BS:526b).

(786 ______. Université II. Centre de recherches sur l'environnement géographique et social. Bron.
Cahiers. 1- (1973-) Irregular.
[Centre de Recherches sur l'Environnement, Université de Lyon II, Centre universitaire, 69500 Bron, France.]

(787 ______. ______. Laboratoire rhodanien de géomorphologie. Bron.
Bulletin. 1- (1977) Biennial.
[Laboratoire Rhodanien de Géomorphologie. Université Lyon II, Boulevard de l'Université, 69500 Bron, France.]

788 MAGASIN asiatique; ou, revue géographique et historique de l'Asie centrale et septentrionale. Paris. 1-2 (1825-1826). Closed 1826. (U-3:2508a; B-3:117a; F-3:368b).

*(789 MÉDITERRANÉE; revue géographique des pays méditerranéens. (Université d'Aix-Marseille. Institut de géographie; Université de Nice. Laboratoire Raoul Blanchard; Laboratoire de géographie d'Avignon). Aix-en-Provence. 1-10 (1960-1969); nsl- (1970-) Quarterly. (N70-3:3714a; B68:351a).
Summaries in English and French and occasionally in German, Spanish, or Italian.
Index: 1960-1969.
[Institut de Géographie de l'Université d'Aix-Marseille, 29 avenue Robert Schuman, 13100 Aix-en-Provence, France.]

(790 MÉMOIRES de géographie tropicale. (France. Centre national de la recherche scientifique. Centre d'études de géographie tropicale. Talence). Paris. 1- (1975-) Irregular.
[C.N.R.S., 15 quai Anatole France, 75700 Paris, France.]

*(791 MOSELLA. (Metz. Université. Centre d'études géographiques [Also: Faculté des lettres de Metz. Département de géographie]). 1- (1971-) Quarterly. (N75-2:1486c; B73:220b; BL).
Abstracts in French, English, and German.
[Centre d'Études Géographiques de l'Université de Metz. Faculté des Lettres, Ile de Saulcy, 57000 Metz, France.]

792 NANCY. Université. Institut de géographie. Nancy.
Publications. 1- (1957-) Irregular monographs. (N70-3:3984a).
1-3 as subseries of Annales de l'est. Mémoire (not in this list).
[23 bd. Albert ler, 54 Nancy, France.]

*(793 NOROIS. Revue géographique de l'Ouest et des pays de l'Atlantique nord. (Universités de Angers, Brest, Caen, Le Mans, Limoges, Nantes, Orléans, Poitiers, Rennes, Rouen, Tours. Instituts de géographie). Poitiers. 1- (1954-) Quarterly. (N70-3:4251a; BS:626a).
Supersedes Poitiers. Université. Groupe poitevin d'études géographiques. Bulletin [807] and Chronique géographique des pays celtes [719].
Resumés in English and French from no. 49 (Ja 1966).
Indexes: vl-10 (1954-1963), vll-20 (1964-1973).
[8, rue Descartes, 86022 Poitiers Cedex, France.]

794 NOUVELLES géographiques. Paris. 1-4 (1891-1894). Closed 1894. (U-4:3110a; B-4:332b; F-3:782b).
Supplement to (Le) Tour du monde [882]. Superseded by À travers le monde [669].

(795 PARIS. École pratique des hautes études (3ème section). Laboratoire de géomorphologie. Dinard.
Mémoires 1- (1951-) Irregular. (N70-3:4545b; BL).
Summaries in English.
[Laboratoire de Géomorphologie de l'École Pratique des Hautes Études, 15 Boulevard de la Mer, 35800 Dinard, France.]

796 ______. Université de Paris. Faculté des lettres. Paris.
Positions des mémoires (Histoire et géographie). 1895-1907. Closed 1907. (U-4:3269a).
Title varies slightly.

797 ______. ______. ______. Fédération des groupes d'études. Paris.
Bulletin de géographie. 1945-1947. Closed 1947.

*(798 ______. Université de Paris-Sorbonne). Département de Geographie. Paris.
Publications. 1- (1974-) Irregular.
[191, rue Saint-Jacques, 75005 Paris, France.]

*(799 PARIS. Université de Paris III. [France. Centre national de la recherche scientifique (CNRS)].
Institut des hautes études de l'Amérique Latine. Paris.
Travaux et mémoires. ,1- (1957-) Irregular. (N70-3:4550c).
[Institut des Hautes Etudes de l'Amérique Latine, 28 rue St-Guillaume, 75007 Paris, France.]

(800 ______. Université de Paris VII. Laboratoire de géographie physique. Paris.
Travaux. 1- (1973-) Annual.
[U.E.R. de Géographie et Sciences de la Société, Université de Paris VII, 2 Place Jussieu, 75005 Paris, France.]

(801 ______. ______. U.E.R. Géographie et sciences de la société.
Groupe de recherches sur les équilibres des paysages. Paris.
Cahiers. 1- (1971-) Irregular.
[Groupe de Recherches sur les Équilibres de Paysage, Institut de Géographie de Paris, 191 rue Saint-Jacques 75005 Paris, France.]

(802 ______. Université de Paris-Nord. Section de géographie. Villetaneuse.
Cahiers. 1- (1973-) Irregular.
[Université de Paris-Nord, Avenue J. B. Clément, 93430 Villetaneuse, France.]

803 PAYS africains. Paris. no1-6 (1951-1955). 6 nos in 2v. Closed 1955. (N70-3:4564c).

804 PETITE revue de géographie et d'histoire. Limoges. no1-7 (1886-1887). Closed 1887. (F-3:940b).

(805 PHOTO interprétation. (Rueil-Malmaison. École nationale supérieure du pétrole et des moteurs). Rueil-Malmaison. 1- (1962-) Bimonthly. (N70-3:4633bc; B68:415a).
In French with parallel text or abstracts in English and Spanish.
[Editions Technip, 27 rue Ginoux, 75747 Cedex 15, France.]

(806 POITIERS. Université. Centre géographique d'études et de recherches rurales. Poitiers.
Travaux. 1- (1972-) Irregular.
[Centre Géographique d'Études et de Recherches Rurales, Université de Poitiers, 95 avenue du Recteur Pineau, 86022 Poitiers Cedex, France.]

807 ______. ______. Faculté des lettres. Institut de géographie.
Groupe poitevin d'études géographiques. Poitiers.
Bulletin. 1-6 (1940-1953). Closed 1953. (U-4:3380a).
Suspended between Ja/F 1940 and Ja/Mr 1949. v1 complete in one issue. Succeeded by Norois. Revue géographique de l'Ouest et des pays de l'Atlantique nord [793].

*(808 RECHERCHES géographiques à Strasbourg. (Association géographique d'Alsace). Strasbourg. 1- (1976-). 3 nos a year.
Summaries in English, French, and German.
[U.E.R. de Géographie, 43 rue Goethe, 67083 Strasbourg Cedex, France.]

(809 RECHERCHES régionales. (Côte d'Azur et contrées limitrophes). Nice. 1961- Quarterly.
[Archives Départementales, 5 ter, avenue Edith Cavel, 06052 Nice Cedex, France.]

810 RECUEIL de voyages et de documents pour servir à l'histoire de la géographie, etc. Paris. 1-24 (1882-1923). Closed 1923.
(U-4:3548b; B-3:670a).

*(811 REIMS. Université. Institut de géographie de Reims. Reims.
Travaux. 1- (1969-) Quarterly.
Summaries in English and French from 18/19 (1974).
[Institut de géographie, U.E.R. Lettres et Sciences Humaines,
57, rue Pierre Taittinger, 51100 Reims, France.]

(812 RENNES. Université. Laboratoire de géographie. Rennes.
Travaux. 1-17 (1903-1954) Irregular. Closed. (U-4:3566a; F-4:736a).

(812a RÉPERTOIRE des géographes francophones (France. Centre national de la
recherche scientifique. Laboratoire d'information et de documentation
en géographie "Intergéo"). Paris. 1979- Irregular.
[Laboratoire "Intergéo," CNRS, 191 rue Saint Jacques, 75005 Paris, France.]

813 RÉUNION. Centre universitaire.
Cahier. Special géographie. 1- (1974-) Irregular.
[Laboratoire de Géographie, Centre Universitaire, 97489 Saint-Denis
de la Réunion, Réunion.]

814 REVUE de géographie. Paris. 1-55 (1877-1905). Closed 1905.
(U-4:3635b; B-3:715b: F-4:192ab).
Succeeded by Revue de géographie annuelle [816].
Index: 1-13 (1877-1883), 14-35 (1884-1894).

*(815 REVUE de géographie alpine. (Grenoble. Université. Institut de géographie
alpine). Grenoble. v1- (1913-) Quarterly. (U-4:3635b; B-2:351b-
352a; F-4:192b; F-4:48b).
1-7 (1913-1919) as Grenoble. Université. Institut de géographie
alpine. Recueil des travaux.
Indexes: v1-10 (1913-1922), 11-20 (1923-1932), 21-30 (1933-1942),
31-40 (1943-1952), 41-50 (1953-1962), 51-60 (1963-1972).
Summaries in English, French, and German.
[Institut de Géographie Alpine, rue Maurice Gignoux, 38000 Grenoble,
France.]

816 REVUE de géographie annuelle. Paris. 1-12 no2 (1906/07-F 1924). Closed
1924. (U-4:3635b; B-3:715b; F-4:192ab).
Succeeds Revue de géographie [814] as nsv1-12 no2.

*(817 REVUE de géographie de Lyon. Lyon. 1- (1925-) Quarterly. (U-4:3635b;
B-2:147a; B-3:715b; BS:526b; F-2:536b).
1 (1925) as Lyon. Université. Institut des études rhodaniennes.
Études et travaux; 2-23 (1926-1948) as Études rhodaniennes. After 1942
carries subtitle: Bulletin de la société de géographie de Lyon et
de la région lyonnaise. Absorbed Société de géographie de Lyon et
de la région lyonnaise. Bulletin [857] in 1942.
Indexes: 1-10 (1925-1934); 11-20 (1935-1945), 21-30 (1946-1955).
Summaries in English and French.
[Association des Amis de la Revue géographique de Lyon, 74, rue Pasteur,
69007 Lyon, France.]

818 REVUE de géographie humaine et d'ethnologie. Paris. v1 no1-4 (1948-1949).
Closed 1949. (U-4:3635b; B-3:715b).

*(819 REVUE de géologie dynamique et de géographie physique. Paris. v1-12 no3
(1928-1939); s2v1- (1957-) 5 nos a year. (U-4:3635c; N70-3:5027b;
B-3:715b; BS:737a; F-4:192b). v1-12 (1928-1939) and s2v1-20 (1957-1978)
as Revue de géographie physique et de géologie dynamique.
Suspended 1940-1956.
Titles in table of contents and abstracts in English and French from
s2v1 (1957).
[Masson et Cie., 120, Boulevard Saint-Germain, 75280 Paris Cedex 06,
France.]

*(820 REVUE de géomorphologie dynamique: Géodynamique externe. Étude intégrée du milieu naturel. Paris. 1- (1950-) Quarterly. (N70-3:5027b; B-3:715b; BS:737a).
In French or English, often with abstracts in both languages.
[Éditions CDU et SEDES Réunies, 88 bd Saint-Germain, 75005 Paris, France.]

821 REVUE du Nord; revue historique trimestrielle. (France. Centre national de la recherche scientifique. Direction de l'enseignement supérieur. Conseil général du Nord [and] Lille. Université). Lille. (U-4:3650c; B-3:723a).
Livraison géographique. 1-11 (1952-1962). Closed 1962.
The third issue of the years 1952-1962 of the parent series devoted to geography.
Superseded by Hommes et terres du nord [774].

(822 REVUE économique française. (Société de géographie commerciale de Paris). Paris. 1- (0 1878-) Quarterly. (U-4:3651c; B-4:135a; F-4:267b; F-1:834b). 1878-1903 as society's Bulletin. 1904-1920, 1921-1929 half-title as the society's Bulletin. Suspended 1940-1948.
Index: 1-10 (1878-1888).
[Société de Géographie Commerciale de Paris, 8 rue Roquépine, 75008 Paris, France.]

823 REVUE française de l'étranger et des colonies. Exploration et gazette géographique. Paris. v1-39 no1-428 (Ja 1885-Ag 1914). Closed 1914. (U-4:3653a; B-3:714a; F-4:274a).
1-6 (1885-1887) as Revue française de l'étranger et des colonies.

824 REVUE géographique de l'année...Paris. no1-4 (1862-1865). Closed 1865. (U-4:3655c; F-4:283b).
Reprints from Revue maritime et coloniale (not in this list).

*(825 REVUE géographique de l'Est: géographie générale; Europe rhénane, centrale et orientale; Moyen-Orient. (Universités de Besançon, Dijon, Metz, Nancy, et Strasbourg). Nancy. 1- (1961-) Quarterly. (N70-3:503b; B68:483b).
Supplementary table of contents in English and German.
Summaries in French and in English or German.
[Association de Géographes de l'Est, 23 boulevard Albert 1er, 54015 Nancy Cedex, France.]

*(826 REVUE géographique des Pyrénées et du Sud-Ouest. (Instituts de géographie de Pau, Bordeaux et Toulouse). Toulouse. 1- (1930-) Quarterly. (U-4:3655c; B-3:715b; F-4:283b).
Indexes: 1930-1939, 1940-1949, 1950-1959, 1960-1969.
Supplementary English titles in table of contents.
Summaries in English and French.
[Service des Publications de l'Université, 56, rue du Tour, 31000 Toulouse, France.]

827 REVUE géographique et industrielle de la France. Paris. 1899-1937. Nouvelle série. (F-4:283b).

828 REVUE géographique internationale; Journal illustré d'enseignement et d'émigration. Paris. v1-28 (no1-326) (1876-Ja 1903). Closed 1903. (U-4:3655c; B-3:715b; F-4:283b).

829 REVUE lyonnaise de géographie. (Organe hebdomadaire des sociétés géographiques départementales et des clubs alpins). Lyon. v1 no1-30? (1877-1878). Closed 1878.

(830 ROCCAFORTIS. (Société de géographie de Rochefort. Bulletin). Rochefort.
s2 v3 no1- (My 1972-).
Supersedes Société de géographie de Rochefort, Bulletin [862].

831 SOCIÉTÉ académique de Brest. Section de géographie. Brest.
Bulletin. 8-25 (1893-1900). Closed 1900.
v1-7 not published?

832 SOCIÉTÉ bourguignonne de géographie et d'histoire. Dijon.
Mémoires. 1-30 (1884-1914). Closed 1914. (U-5:3934a; B-4:128a; F-3:450b).
1881-1882 as Société de géographie de Dijon. Bulletin. [851].
Index: 1-30.

833 SOCIÉTÉ bretonne de géographie. Lorient.
Bulletin. no1-116 (1882-1938). Closed 1938. (U-5:3934a; B-4:128a).
Suspended 1914-1921
Index: v1-6 (1882-1887) in v6.

834 SOCIÉTÉ d'encouragement pour les études géographiques. Paris.
Bulletin. 1-2 (1875). Closed 1875. (U-5:3939b).

835 SOCIÉTÉ d'études historiques et géographiques d'Athis-Mons et de la Plaine de Longboyau. Athis-Mons.
Bulletin. 1- (1947-). (U-5:3940b).

836 SOCIÉTÉ d'études historiques et géographiques de Bretagne. Rennes.
Bulletin. v1-3 no1 (1897-1899). Closed 1899. (U-5:3940b; F-1:824a).

837 SOCIETE d'études historiques, géographiques et scientifiques de la région parisienne. Paris.
Bulletin. v1-36 (1927-1962). Closed 1962. (U-5:3940b; F-1:824a).
Superseded by Études de la région parisienne [739].
Index: 1927-1957.

(838 SOCIÉTÉ de biogéographie. Paris.
Compte rendu (des séances). 1- (1923-) Quarterly.
(U-5:3944a; B-4:127a; F-2:181b).
Summaries in English or French.
[57, rue Cuvier, 75005 Paris, France.]

839 ______.
Mémoires. 1-8 (1926-1946). Closed 1946. (U-5:3944a; B-4:127b; F-4:532b).

840 SOCIÉTÉ de géographie, Paris. Paris.
Bulletin. v1-20 (1822-33); s2 v1-20 (1834-43); s3 v1-14 (1844-50); s4 v1-20 (1851-60); s5 v1-20 (1861-70); s6 v1-20 (1871-80); s7 v1-20 (1881-99). Closed 1899. (U-5:3944c; B-4134b; F-1:834ab).
134 vols. in all. Superseded by Géographie [768].
Indexes: s1-2 (1822-1843); s3-4 (1844-1860), s5-7 (1861-1899).

841 ______.
Comptes-rendus. 1-19 (1882-1899). Closed 1899. (U-5:3945a; B-4:135a; F-2:185b).
Earlier numbers in society's Bulletin [840]. Continued in Géographie [768].
Index: 1-19 (1882-1899).

842 ______.
Rapports annuels sur les progrès de la géographie. 1-3 (1867-1892). Closed 1892. (U-5:3945a; B-4:135a; F-4:10a).
Annual reports from society's Bulletin [840], for years 1867 to 1892, reprinted in 3v.

843 ______.
Recueil de voyages et de mémoires. 1-8 (1824-1866). Closed 1866.
(U-4:3945a; B-3:670a; F-4:43a).

844 SOCIÉTÉ de géographie commerciale de Bordeaux. Bordeaux.
Bulletin. nsnol- (1958-) Annual. (N70-4:5380c).
Called new series as continuation of society's Revue de géographie commerciale [845].
[Hôtel des Sociétés Savantes, 71, rue du Loup, Bordeaux, France.]

845 ______.
Revue de géographie commerciale. no1-2 (1874-1875); s2 v1-37 (1878-1911); s3 v38-40 no7 (1912-1914); v43-58 (1917-1934); v59-63 (1935-1939). Closed 1939. (U-5:3945a; B-4:135a; F-4:533a; F-4:192b).
1874-1911 as the society's Bulletin. Suspended Ag 1914-D 1916.
Superseded by the society's Bulletin ns [844].
Index: 1874-1903.

846 SOCIÉTÉ de géographie commerciale de Nantes. Nantes.
Bulletin. v1-32 (1883-1914). Closed 1914. (U-5:3945a; B-4:135a; F-4:533a; F-1:1027a).

847 SOCIÉTÉ de géographie commerciale de Saint-Nazaire. Saint-Nazaire.
Bulletin. v1-20 (1886-1906). Closed 1906. (U-5:3945a; B-4:135b; F-1:836b). v1-10 (1885-1893) as Société de géographie et du musée commercial de St. Nazaire. Bulletin.

848 SOCIÉTÉ de géographie commerciale du Havre. Le Havre.
Annuaire. 1885-1893. Reprints. (B-4:135a; F-4:533b).
Other years in society's Bulletin [849].

849 ______.
Bulletin. 1-55 (1884-1938). Closed 1938. (U-5:3945a; B-4:135a; F-4:533b).
1896 is the first numbered volume.

850 SOCIÉTÉ de géographie de Boulogne-sur-Mer. Boulogne.
Actes. 1-13 (1899-1912). Closed 1912. (F-4:533b).

851 SOCIÉTÉ de géographie de Dijon. Dijon.
Bulletin. no1-3 (1881-1882). Closed 1882. (F-4:533b).
Superseded by Société bourguignonne de géographie et d'histoire. Mémoires [832].

852 SOCIÉTÉ de géographie de Dunkerque. Dunkerque.
Bulletin. no1-40 (1898-1912); 1927-1936. Closed 1936. (U-5:3945b; F-1:835a).
Suspended 1914-1926. 1927-1936 unnumbered.

853 SOCIÉTÉ de géographie de l'Ain, Bourg-en-Bresse.
Bulletin. Bourg. 1-22 (1882-1903). Closed 1903. (U-5:3945b; F-1:835a).
Index: 1-15 (1882-1896) in v15.

854 SOCIÉTÉ de géographie de l'Aisne, Laon. Laon.
Bulletin. Arrondissements de Laon, Vervins, Soissons et Château-Thierry. 1890-1914. Closed 1914. (U-5:3945b; F-1:835a).

855 SOCIÉTÉ de géographie de l'Est. Nancy; Paris.
Bulletin. v1-35 no1 (1879-1914). Closed 1914. (U-5:3945b; B-4:135b; F-1:835ab; F-4:534a).
Index: 1-10 (1879-1888) in 10.

856 SOCIÉTÉ de géographie de Lille. Lille.
Bulletin. 1-84 no2 (1882-1940); 1942-1953; ns no1-5 (1958-1962).
Closed 1962. (U-5:3945b; B-4:135b; F-4:534a; F-1:835b).
Suspended My 1914-1919; Jl-D 1939; 1954-1957. 1942-1953 as Publications.
Superseded by Hommes et terres du Nord [774].

857 SOCIÉTÉ de géographie de Lyon et de la région lyonnaise. Lyon.
Bulletin. 1-22 (1875-1908); s2 v1-5 (1908-1912); 1913-1914; 1921/22-1937/38. Closed 1938. (U-5:3945b; B-4:135b; F-1:835b-836a; F-4:4534b).
1875-1898 as Société de géographie de Lyon. Absorbed by Revue de géographie de Lyon [817] in 1942.
Index: v1-15 (1875-1898); 1899-1914.

858 ______.
Procès-verbaux des séances. 1-12 (1882-Ag 1883). Closed 1883.
(U-5:3945b; F-4:534a).

859 SOCIÉTÉ de géographie de Marseille. Marseille.
Bulletin. 1-65 (1887-1954); 71- (1960-) Annual or years combined.
(U-1:833b; B-4:135b; F-1:836a).
Includes society's Compte rendu. 1903-1954 society was Société de géographie et d'études coloniales de Marseille. 66-70 (1955-1959) as Bulletin de géographie d'Aix-Marseille, nsv 1-5 [700]; 71- (1960-) also as nsv1- .
Index: 1-29, (1877-1905) in v30.
[Société de Géographie, 2 rue Beauvau, 13001 Marseille, France.]

860 SOCIÉTÉ de géographie de Poitiers. Poitiers.
Bulletin. no2-3 (1899/1900-1901). Closed 1901. (F-1:836a).

861 SOCIÉTÉ de géographie de Rochefort. Rochefort.
Annuaire. 1-41 (1883?-1930). Closed 1930.

862 ______.
Bulletin. v1-41 no2 (1879-1930). Closed 1930. (U-5:3945c; B-4:135b; F-1:836a).
Index: 1879-1909 in v16, 29, and 32.

863 SOCIÉTÉ de géographie de Saint-Quentin. Saint-Quentin.
Bulletin. no1-20? (1888-1893). Closed 1893. (F-1:836a).

864 SOCIÉTÉ de géographie de Saint-Valéry-en-Caux. Saint-Valéry-en-Caux.
Bulletin. v1 (1884). Closed 1884. (U-5:3945c).

865 SOCIÉTÉ de géographie de Toulon. Toulon.
Bulletin. v1-8 (1886-1890). Closed 1890. (U-5:3945c).

866 SOCIÉTÉ de géographie de Toulouse. Toulouse.
Annuaire. 1887-1916. Closed 1916. (F-1:836b).
Published in the Society's Bulletin.

867 ______.
Bulletin [mensuel]. 1-40 (1882-1921); 42-56 (1923-1937).
(U-5:3945c; B-4:135b; F-1:836b).
41 (1922) never published. 42-43 (1923-1924) also as ns no1-15.
Index: v1-21 (1882-1902) in 22.

868 ______.
Bulletin bi-mensuel. 1-621 (18..?-1923). (U-5:3945c; B-4:135b).
Merged into society's Bulletin.

869 SOCIÉTÉ de géographie de Tours. (Union géographique du Centre). Tours.
Annuaire. 1885-1888. Closed 1888?

870 ______.
Revue. v1-31 no2 (1884-1914). Closed 1914. (U-5:3945c; B-4:135b; F-4:534b).

871 SOCIÉTÉ de géographie des Ardennes. Charleville.
Bulletin annuel. no1 (1887/88). Closed 1888.

872 SOCIÉTÉ de géographie du Cher, Bourges. Bourges.
Bulletin. v1-6 (1902/03-1913/19). Closed 1919. (U-5:3945c; B-4:135a; F-4:534b).

873 SOCIETE industrielle de Reims. Comité de géographie. Reims.
Bulletin. no1-5 (1888-1890). Closed 1890. (F-4:553b).

*(874 SOCIÉTÉ languedocienne de géographie, Montpellier. (Montpellier. Université. Institut de géographie). Montpellier.
Bulletin. 1-49 (1878-1929); s2 v1-37 (1930-1966); s3 v1 [90]- (1967-) Quarterly. (U-5:3964c; B-4:140b; F-4:554a).
Index: v1-49, s2 v1-35 (1878-1964) in s2 v36 (1965).
Abstracts in English and French, in part from s3 v21 (1950).
[Société Languedocienne de Géographie, Université Paul Valéry, B.P. 5043, 34032 Montpellier, France.]

875 ______.
Géographie générale du département de l'Hérault. 1-3 (1891-1905). Closed 1905. (U-5:3964c).

876 SOCIÉTÉ normande de géographie, Rouen. Rouen.
Bulletin de l'année. 1-49 (1879-1935/38). Closed 1938. (U-5:3968c; B-4:145a; F-4:557b).

877 SOCIÉTÉS de géographie. Paris.
Annuaire. 1890. Closed 1890.

878 STATISTIQUE annuelle de géographie humaine comparée. Paris.
1-9 (1905-1913); 10-11; (1922-23). Closed. (F-4:596b).
Title varies: Statistique annuelle de géographie comparée; Statistique générale de géographie humaine comparée.

879 STRASBOURG. Université. Faculté des lettres. Fondation Baulig.
Publications. 1-2 (1965). (N70-4:5562a).

880 ______. ______. Laboratoire de géographie. Strasbourg.
Travaux. sB. Géographie industrielle. 1 (1952). (N70-4:5562a).

(881 TOULOUSE. Université. Centre interdisciplinaire d'études urbaines. Toulouse.
Travaux et documents du CIEU. 1- (1972-) Irregular.
Summaries in English.
[Centre Interdisciplinaire d'Études Urbaines, Université de Toulouse, 109 bis rue Vauquelin, 31081 Toulouse Cedex, France.]

882 TOUR du monde. (Journal des voyages et des voyageurs). Paris.
v1-68 (1860-1894); nsv1-20 no31 (1895-Ag 1914). Closed 1914. (U-5:4243a; B-4:332b; F-4:698b-699a).
Index: 1860-1910.

*(884 TRAVAUX et documents de géographie tropicale. (France. Centre national de la recherche scientifique. Centre d'études de géographie tropicale). Talence.
1- (1971-) Irregular. (N75-1:317c). (N: Bordeaux. Université. Centre d'études de géographie tropicale. Travaux et documents).
Summaries in English and French and occasionally in German.
[Centre d'Études de Géographie Tropicale, Domaine Universitaire de Bordeaux, 33 405 Talence Cedex, France.]

885 UNION géographique du Nord de la France. Lille, Douai.
Bulletin. 1-35 (1880-1914); 36 (1932). Closed 1932. (U-5:4300b; B-4:380a; F-4:777b-778a).
Suspended 1915-1931.

886 VAR historique et géographique. (Société des études locales dans l'enseignement public. Section départementale du Var).
Draguignan. 1913-1914, 1922-1923, 1926-1937, 1944. All published? (U-5:4353c; F-4:856a).

(887 VIE urbaine. (Paris. Université du Val-de-Marne. Institut d'urbanisme de Paris XII). Créteil. 1- (1919-) Quarterly. (U-5:4394; F-4:919b).
1-16 (1919-1939); ns no55- (1950-). Suspended 1924-25, 1931, 1940-1949.
1-7 (1919-1928) also as no1-35; 8-16 (1930-39) also as ns no1-54.
1919-1923 by Institut d'histoire, de géographie et d'économie urbaine de la ville de Paris; 1925- by Institut d'urbanisme de l'Université de Paris. 1953-1955 as Urbanisme et habitation.
Summaries in English.
[Institut d'Urbanisme de Paris XII, Université du Val de Marne, Avenue du Général de Gaulle, 94010 Créteil, France.]

(888 VILLES en parallèle. (Université de Paris X. Laboratoire de géographie urbaine). Nanterre. 1- (1978-).
Published in collaboration with Association de géographes français and Revue française d'études politiques méditerranéennes.
[Laboratoire de Géographie Urbaine de l'Université de Paris X, Nanterre, France.]

FRENCH EQUATORIAL AFRICA

Afrique Équatoriale Française

(ceased to exist in 1958)

889 FRENCH Equatorial Africa. Service géographique.
Rapport annuel technique et administratif. 1920-1927. (ULG:224c).

890 SERVICE géographique de l'Afrique Équatorial Française et du Cameroun.
Brazzaville?
Exposé des travaux executés au cours de l'année. 1955. Only year published? (N70-4:5292c).

GERMANY TO 1946 (German)

Deutschland

Serials listed in this section include those closed prior to 1946, or from discontinued political units, or older serials with modest current geographic content. Geographical serials published after 1945, even if started earlier, are listed under the Federal Republic of Germany or the German Democratic Republic.

891 ABHANDLUNGEN zur badischen Landeskunde. Karlsruhe. v1-5 (1913-1916).
Closed 1916. (U-1:12b; B-1:7a; F-1:20b; D:335).

892 ABHANDLUNGEN zur Landeskunde der Provinz Westpreussen. (Danzig. Museum für Naturkunde und Vorgeschichte). Danzig. no1-15 (1890-1919).
Closed 1919. (U-1:12c; B-1:7b; F-1:21a; G).

893 ALLGEMEINE geographische Ephemeriden. Weimar. 1-51 (1798-1816). Closed 1816.
(U-1:143b; B-1:84a; F-1:119a).
Superseded by Neue allgemeine geographische und statistische Ephemeriden [1066]. Supplement: see Allgemeiner Monatsbericht für Deutschaland [895].
Index: 1-51 in v51.

894 ALLGEMEINE Länder und Völkerkunde. Stuttgart. 1-6 (1837-1844). Closed 1844.

895 ALLGEMEINER Monatsbericht für Deutschland. Weimer. 1-37 (1811-1837).
Closed 1837. (U-1:146c; F-1:123ab; G).
Supplement to Allgemeine geographische Ephemeriden [893] later to Neue allgemeine geographische und statistische Ephemeriden [1066].

896 ALLGEMEINES Archiv für die Länder-und Völkerkunde. (F.C.G. Hirsching).
Leipzig. 1-2 (1790-1791). Closed 1791. (B-1:87b).

897 ALLGEMEINES Jahrbuch der Geographie und Statistik. Weimar. 1800. (B-1:87b; G).

898 ALMANACH...den Freunden der Erdkunde gewidmet von H. Berghaus...Stuttgart.
1-5 (1837-1841). Closed 1841. (B-1:92a).

899 ANALEKTEN für Erd- und Himmelskunde. München. v1-2 no7 (1828-1931).
(B-1:137b; F-1:170b; G).
Superseded by Neue Analekten für Erd- und Himmelskunde [1067].

900 ANGEWANDTE Geographie. Hefte zur Verbreitung geographischer Kenntnisse in ihrer Beziehung zum Kultur- und Wirtschaftsleben. (Institut für Auslandkunde und Auslandsdeutschtum). Frankfurt-am-Main. 1-50 (1902-1921). Closed 1921. (U-1:363c; B-1:141a; D:337-D:341).
1-49 also numbered in series: s1 v1-12 (1902-1904); s2 v1-12 (1904-1906); s3 v1-12 (1907-1911); s4 v1-11 (1911-1916); s5 v1-2 (1920); v50 (1921).

901 ANNALEN der Erd-, Völker- und Staatenkunde. Stuttgart; Berlin; Breslau; Oppeln. v1-14 (1825-1829); s2 v1-12 (1829-1835); s3 v1-12 (1835-1841); s4 v1-4 (1842-1843). Closed 1843. (U-1:369b; B-2:394b; F-1:188b; F-2:809a; G). 1-14 (1825-1829) as Hertha, Zeitschrift für Erd-, Völker-, und Staatenkunde.

902 ANNALEN der Geographie und Statistik. Braunschweig. 1-3 (1790-1792). Closed 1792. (U-1:369c; B-1:147a; F-1:189a).

903 ANNALEN der Reisen, der Geographie und Statistik in Original-Aufsätzen und Uebersetzungen aus fremden Sprachen. Berlin. v1-2 (1809-1810).

904 ANTON Friedrich Büschings...wöchentliche Nachrichten von neuen Landcharten, geographischen, statistischen und historischen Büchern und Sachen. Berlin. 1-15 (Ja 1773-1787). Closed 1787. (U-1:412a; B-2:1; F-1:342a; G). Binder's title: v1-2, Nachrichten von neuen Landcharten; 2 as D. Anton Friedrich Büschings...wöchentliche Nachrichten von neuen Landcharten...

905 ARBEITEN zur Landes- und Volksforschung. (Jena. Universität. Anstalt für geschichtliche Landeskunde). Jena. 1-10 (1938-1941). Closed 1941. (U-1:423a; D:293).

906 ARBEITEN zur Landeskunde und Wirtschafsgeschichte Ostfrieslands. Aurich. 1-7 (1926-1931). Closed 1931. (U-1:423a; G).

907 ARCHIV für ältere und neuere, vorzüglich deutsche Geschichte, Staatsklugheit und Erdkunde. Memmingen. v1-2 (1790-1792). Closed 1792. (U-1:439b; B-1:189b; F-1:369b; G).

908 ARCHIV für alte Geographie, Geschichte und Alterthümer insonderheit der germanischen Völkerstämme. Breslau, v1 no1-3 (1821-1822). Closed 1822. (U-1:439b; B-1:189b; F-1:369b).
Second page of title: Blicke auf die alten Völker und Städte des östlichen Germaniens von dem Donau bis zum Ostsee.
Superseded by Deutsche Alterthümer, oder Archiv für alte und mittlere Geschichte...[952].

909 ARCHIV für Landes- und Volkskunde der Provinz Sachsen, nebst angrenzenden Landesteilen. Halle an der Saale. 1-29 (1891-1919). Closed 1919. (U-1:446a; B-1:193b; F-1:379b; D:107; G).
Continues the scientific part of Sächsisch-thüringischer Verein für Erdkunde. Mitteilungen [1098]. 1906- issued also as first section of each issue of Mitteilungen.
Index: 1-10 (1891-1900) in 10 (1900).

910 ARCHIV für Landes- und Volkskunde von Niedersachsen. Oldenburg. v1-5 no1-24 (1940-1944). Closed 1944. no25 (1951) is index to no1-24. (U-1:446a; B-1:193b; D:101; G).
Superseded by Neues Archiv für Niedersachsen [1304]. Forms sC of Göttingen. Universität. Niedersächsisches Institut für Landeskunde und Landesentwicklung, Göttingen. Veröffentlichungen [1256].

911 ARCHIV für Landeskunde der preussischen Monarchie. Berlin. v1-4 (1855-1856); nsv1-6 (1857-1859/64). Closed 1859. (U-1:446a; B-1:193b; F-1:379b; G).
v1 (1855/56) as Archiv für Landeskunde im Königreich Preussen.

912 ARCHIV für Landeskunde in den Grossherzogthümern Mecklenburg und Revüe der Landwirtschaft. Schwerin. 1-20 (O 1850-1870). Closed 1870. (U-1:446a; G).
no1-15 (1850-D 1851) as Mecklenburgisches gemeinnütziges Archiv.

913 ARCHIV zur neuern Geschichte, Geographie, Natur- und Menschenkenntniss. Leipzig. pt1-8 (1785-1788). Closed 1788. (F-1:388a).

914 AUS allen Weltteilen. Deutsch-national Zeitschrift für Länder- und Völkerkunde. Berlin; Leipzig. 1-29 (1870-1898). Closed 1898. (U-1:559a; B-1:257b; F-1:512a).
Sub-title varies.
Merged with Globus [1019] which later merged with Petermanns geographische Mitteilungen [1370].

915 AUSLAND, Wochenschrift für Erd- und Völkerkunde. Stuttgart, München. v1-66 (1828-1893). Closed 1893. (U-1:560b; B-1:258a; F-1:513a).
Superseded by Neue Ausland [1068], which later was merged with Globus [1019] which later merged with Petermanns geographische Mitteilungen [1370].

916 ______. Supplement. Blätter zur Kunde der Literatur des Auslandes. Stuttgart, etc. 1-5 (1836-1840). Closed 1840. (U-1:717a; B-1:362a; F-1:647b).
Also published separately.

917 AUSWAHL der besten ausländischen geographischen und statistischen Nachrichten zur Aufklärung der Völker- und Ländeskunde. Halle. 1-14 (1794-1800). Closed 1800. (F-1:515a; G).
Succeeds Neue Beiträge zur Völker- und Landerkunde [1070].

918 AUSWAHL kleiner Reisebeschreibungen und anderer statistischer und geographischer Nachrichten. Leipzig. no1-22 (1784-1785). Closed 1785?
pt8- also as Neue Beiträge zur Völker- und Länderkunde [1070].

919 BADENIA; oder das badische Land und Volk; eine Zeitschrift für vaterländische Geschichte und Landeskunde. (Verein für badische Ortsbeschreibung). Karlsruhe; Heidelberg. 1-3 (1839/40-1844); nsv1-3 (1858-1864/66). Closed 1864. (U-1:583c; B-1:279b; F-1:533a; G).

920 BADISCHE geographische Abhandlungen. (Freiburg. Universität. Geographisches Institut). Freiburg; Karlsruhe. no1-19 (1926-1938). Closed 1938. (U-1:584a; B-1:279b; D:188; G).
Succeeded by Oberrheinische geographische Abhandlungen [1080].

921 BADISCHE geographische Gesellschaft zu Karlsruhe. Karlsruhe. Verhandlungen. 1880-1886. Closed 1886. (U-1:584a; B-1:279b).

922 BAUSTEINE zur elsass-lothringischen Geschichts- und Landeskunde. Zabern. 1-15 (1896-1914). Monographs. (B-1:294a).

923 BEITRÄGE zur Geographie, Geschichte, und Staatenkunde. Nürnberg. v1-2 (1794-1796). Closed 1796. (B-1:302a).
Superseded by Magazin für die Geographie, Staatenkunde und Geschichte [1050].

924 BEITRÄGE zur Landes- und Volkeskunde von Elsass-Lothringen und den angrenzenden Gebieten. Strassburg. 1-53 (1887-1918). Closed 1918. (U-1:626a; B-1:304a: F-1:559a; G).

925 BEITRÄGE zur Landes- , Volks- und Staatskunde des Grossherzogthums Hessen. (Geographische Gesellschaft in Darmstadt). Darmstadt. 1-2 (1850-1853). Closed 1853. (U-1:626a; B-1:304a; F-1:559b; D:39).
Superseded by Verein für Erdkunde und verwandte Wissenschaften zu Darmstadt; later Verein für Erdkunde, Darmstadt. Notizblatt [1110].

926 BEITRÄGE zur Landeskunde. (Württemberg-Hohenzollern. Statistisches Landesamt). Tübingen. 6-7 (1951-1952). Closed 1952.
Supplement to Württemberg-Hohenzollern in Zahlen [1136] v6-7.

927 BEITRÄGE zur Landeskunde der Rheinlande. (Bonn. Universität.Geographisches Institut). Bonn. no1-5 (1922-1927); s2 no1-5 (1933-1936); s3 no1-5 (1939-1941). Closed 1941. (U-1:626a; B-1:304a; F-1:559a; D:177; D:178; D:179; G).
Superseded by Arbeiten zur rheinischen Landeskunde [1165]. no1-4 (1922-1923) under earlier name of institute: Geographisches Seminar.

928 BEITRÄGE zur Methodik der Erdkunde als Wissenschaft wie als Unterrichtsgegenstand. Halle. 1 (1894). Closed 1894.

929 BEITRÄGE zur nordwestdeutschen Volks- und Landeskunde. (Naturwissenschaftlicher Verein zu Bremen). Bremen. v1 no1-3 (1895-1901). Closed 1901. (U-1:627a; G).

930 BEITRÄGE zur vaterländischen Historie, Geographie, Statistik und Landwirtschaft. München. 1-10 (1788-1817). (B-1:306a; F-1:562b).
v9-10 also have title Neue Beiträge...
Superseded by Neue Beiträge zur vaterländischen Geschichte, Geographie, und Statistik, ns [1069].

931 BEITRÄGE zur Völker- und Länderkunde. Leipzig. 1-14 (1781-1790). Closed 1790. (U-1:629b; B-1:306b; F-1:563a; G).
Superseded by Neue Beiträge zur Völker- und Landerkunde [1070].

932 BERICHT über die neuere Litteratur zur deutschen Landeskunde. Berlin. 1-3 (1896/99-1902/03 [1901-1906]). Closed 1906. (U-1:639c; F-1:578b; D:9).

933 BERLIN. Universität. Institut für Meereskunde (und geographisches Institut). Berlin.
Veröffentlichungen. 1-15 (1902-1911). Closed 1911. (U-1:645c; B-2:238b; F-4:890a; D:258; G).
Continued in two series: sA.Geographisch-naturwissenschaftliche Reihe [934]; sB. Historisch-volkswirtschaftliche Reihe [935].

934 _____. _____. _____.
_____. sA. Geographisch-naturwissenschaftliche Reihe.
1-40 (1912-1942). Closed 1942. (U-1:645c; B-2:238b; F-4:890ab; D:259; G).

935 _____. _____. _____.
_____. sB. Historisch-volkswirtschaftliche Reihe.
1-16 (1911-1941). Closed 1941. (U-1:645c; B-2:238b; F-4:890b; D:260; G).
1 also as main series 16.

936 _____. _____. Seminar für Staatenkunde und historische Geographie. Veröffentlichungen. Berlin. 1-5 (1929-1938). Closed 1938. (U-1:646b; D:176).

937 _____. _____. Verein der Studierenden der Geographie. Berlin. Mitteilungen. 1-2 (1915-1918). Closed 1918. (U-1:646b; D:34).

938 BIBLIOTHECA geographica: Jahresbibliographie der gesamten geographischen Literatur. (Gesellschaft für Erdkunde zu Berlin). Berlin.
1-19 (1891/92-1911/12). [1895-1917]. Closed 1912. (U-1:685c; B-1:341b; F-1:618a; D:8; G).
1853-1890 published in Gesellschaft für Erdkunde zu Berlin. Zeitschrift [1011].

939 BIBLIOTHECA geographica. Göttingen. v10-22 (1862-1874). Closed 1874. (U-1:685c; B-1:342a; F-1:618a; F-1619b).
1-9 as part of Bibliotheca historico-geographica [941]; 10-17 (1862-1869) Bibliotheca geographico-statistica et oeconomico-politica; 18-19 (1870-1871) as Bibliotheca oeconomico-politica et statistica... then, Bibliotheca geographica oder Systematisch geordnete Übersicht der in Deutschland und dem Auslande auf dem Gebiete der gesamten Geographie neu erschienen Bücher.

940 BIBLIOTHECA geographica. Leipzig. 1-2 (1857-1858). Closed 1858. (U-1:685c).

941 BIBLIOTHECA historico-geographica. Göttingen. 1-9 (1853-1861). Closed 1861. (U-1:342a; F-1:618ab).
Continued in two parallel series: Bibliotheca geographico-statistica et oeconomico-politica, later Bibliotheca geographica [939] and Bibliotheca historica (not in this list).

942 BIBLIOTHEK arabischer Historiker und Geographen. Leipzig. 1-5 (1926-1930). Closed 1930) (U-1:689c; B-1:343b).
2, 4 not published?

943 BIBLIOTHEK der Länderkunde. Berlin. 1-11 (1898-1902). Closed 1902. (U-1:690b; B-1:344a; D:347).
Monographs.

944 BIBLIOTHEK der neuesten und wichtigsten Reisebeschreibungen und geographischen Nachrichten zur Erweiterung der Erdkunde. Weimar. 1-50 (1800-1814). Closed 1814. (U-1:690c).
Monographs.

945 BIBLIOTHEK der sächsischen Geschichte und Landeskunde. Leipzig. 1-4 (1902/09-1912/14). Closed 1914. (U-1:690c; B-1:344b; G).

946 BRESLAU. Osteuropa-Institut. Hamburg.
Quellen und Studien. 5. Geographie und Landeskunde. no1 (1926). Only one issued. (U-1:768a; B-3:471b; D:328).

947 ______. ______.
Vorträge und Aufsätze. 4. Geographie und Landeskunde. 1 (1921). Closed 1921. (U-1:768b; D:333).

948 BUCHONIA; eine Zeitschrift für vaterländische Geschichte, Alterthumskunde, Geographie, Statistik und Topographie. Fulda. 1-3 (1811-1813); nsv1-4 (1826-1829). Closed 1829. (U-1:811b; B-1:436a; F-1:717a; G).
Subtitle varies.

949 CHEFS des Kriegs-Karten- und Vermessungswesens. Berlin.
Mitteilungen. 1-3 (1942-1944). Closed 1944. (D:89).

950 CONGRESS für Handelsgeographie. Berlin.
Bericht über die Verhandlungen. 1880. (F-1:580a).

951 DANZIG. Technische Hochschule. Geographisches Seminar. Breslau.
Veröffentlichungen. (1931). Closed 1931. (D:184).

952 DEUTSCHE Alterhümer, oder Archiv für alte und mittlere Geschichte, Geographie und Alterhümer insonderheit der germanischen Völkerstämme. Halle. 1-3 (1824-1830). Closed 1830. (U-2:1294b; B-1:189b; F-2:356a).
Supersedes Archiv für alte Geographie, etc., [908]. Superseded by Neue Zeitschrift für die Geschichte der germanischen Völker (not in this list).

953 DEUTSCHE Erde; Zeitschrift für Deutschkunde; Beiträge zur Kenntnis deutschen Volkstums allerorten und allerzeiten. Gotha. v1-13 no8 (1902-1914). Closed 1914. (U-2:1299a; B-2:28b; F-2:358a; D:29a; G).

954 DEUTSCHE Landschaftskunde in Einzeldarstellungen. München. 1-5 (1934-1937). Closed 1937. (U-2:1306c; D:351; G).
v5 (2nd ed.) issued 1939.

955 DEUTSCHE Sammlung. Reihe: Geographie. Greifswald, later Karlsruhe. 1-8 (1925-1937). Closed 1937. (U-2:1315c; D:352).

956 DEUTSCHE Schriften zur Landes- und Volksforschung. Leipzig. 1-15 (1939-1944). Closed 1944. (U-2:1315c; D:353).

957 DEUTSCHER Geographen-Almanach (Adolf Miessler). Hagen i.W. 1884. Only one issued. (B-2:41b).

958 DEUTSCHES Archiv für Landes- und Volksforschung. Leipzig. v1-8 no2/4 (1937-S 1944). Closed 1944. (U-2:1329b; B-2:44b; D:95; G).

959 DRESDNER geographische Studien. (Dresden. Technische Hochschule. Geographisches Institut und geographische Arbeitsgemeinschaft). Dresden. v1-15 (1931-1940). Closed 1940. (U-2:1364c; B-2:66b; D:185; G).

960 DÜSSELDORFER geographische Vorträge und Erörterungen. Düsseldorf. no1- (1927-).

961 ERDBALL; Illustrierte Monatsschrift für das gesamte Gebiet der Länder- Menschen- und Völkerkunde [later Anthropologie, Länder- und Völkerkunde]. Berlin. v1-6 no2 (1926/27-F 1932). Closed 1932. (U-2:1469b; F-2:502b; D:18: G).

962 ERDE. Leipzig; Braunschweig. v1-4 no6 (1923-1926). Closed 1926. (U-2:1469b; G).
v1-2 (1923-1924) as Naturwissenschaftliche Korrespondenz.

963 ERDE; Illustrierte Halbmonatschrift für Länder- und Völkerkunde, Reise und Jagd. Weimar, Dresden. v1-2 no12 (Ja 1912-Je 1914). Closed 1914. (U-2:1469b; B-2:134a; F-2:502b; D:19).

964 ERDE und Wirtschaft. Vierteljahresschrift für Wirtschaftsgeographie und ihre praktische Anwendung. Braunschweig. v1-7 no3/4 (Ap 1927-Ja 1934). Closed 1934. (U-2:1469b; D:20; G).

965 ______.
Beiheft. 1 (1929). (D:355; G).

966 ERDKUNDE; eine Darstellung ihrer Wissensgebiete, ihrer Hilfswissenschaften und der Methode ihres Unterrichtes. Leipzig; Wien. 1-26 (1903-1906). Closed 1906. (U-2:1469c; G).

967 ERDKUNDLICHE Blätter. Frankfurt am Main. no1-6 (My 1925-Ag 1927). Closed 1927. (U-2:1469c; D:75).

968 EXPORT. (Zentralverein für Handelsgeographie und Förderung deutscher Interessen im Auslande). Berlin. v1-47 no32 (1879-1925). Closed 1925. (U-2:1511a; B-2:158b; F-2:554b; G).

969 FORSCHUNGEN zur bayerischen Landeskunde. München. 1-2 (1920-1921). Closed 1921. (D:358).

970 FORSCHUNGEN zur Kolonialfrage. (Leipzig. Universität. Kolonial-geographisches Institut). Würzburg. 1-12 (1939-1943). Closed 1943. (U-2:1609a; B-2:212a; G).

971 FORTSCHRITTE der Geographie und Naturgeschichte. Jahrbuch. Weimar. 1-5 (no1-75) (1846-1848). Closed 1848. (U-2:1613b; B-2:215b; F-2:632a).

972 FRÄNKISCHE Studien. (Geographische Gesellschaft zu Würzburg). Würzburg. 1-4 (1937-1942). Closed 1942. (D:245; G).
Succeeds Geographische Gesellschaft zu Würzburg. Mitteilungen [994].
Succeeded by Würzburger geographische Arbeiten [1342].

973 GAEA. Natur und Leben. (Centralorgan zur Verbreitung naturwissenschaftlicher und geographischer Kenntnisse). Leipzig; Köln.
1-45 (1865-1909). Closed 1909. (U-2:1661c; B-2:248b; F-2:678a; D-22; G).
Merged with Naturwissenschaftliche Rundschau (not in this list).

974 GEOGRAPHEN-Kalender. Gotha. 1-12 (1903/04-1914). Closed 1914. (U-2:1693b; B-2:268b; D:2; G).

975 GEOGRAPHIE des Menschen- und Völkerlebens in Geschichte und Gegenwart. Bonn. 1-3 (1921-1922). Closed 1922. (B-2:269b; D:358a).

976 GEOGRAPHISCHE Abhandlungen. (Berlin. Universität. Geographisches Institut). Berlin; Leipzig; Stuttgart; Wien. 1-10 (1886-1914/21); s2 v1-4 (1923-1928); s3 v1-12 (1929-1939). Closed 1939. (U-2:1695c; B-2:270a; F-2:728b; D:360; D:362; G).

977 ______. (Berlin. Universität. Geographisches Institut).
Veröffentlichungen. Leipzig. nsv1-3 (1912-1917). Closed 1917. (U-2:1696a; F-2:728b; D:172; G).

978 GEOGRAPHISCHE Abhandlungen aus den Reichslanden Elsass-Lothringen. Stuttgart. 1-2 (1892-1895). Closed 1895. (U-2:1696a; B-2:270a; F-2:728b).

979 GEOGRAPHISCHE Arbeiten. Stuttgart. 1-10 (1908-1914), 11-12 (1927-1929).
Suspended 1915-1926. Closed 1929. (U-2:1696a; B-2:270a; D:363).
Title in German, French, and Polish.

980 GEOGRAPHISCHE Bausteine. Schriften des Verbandes deutscher Schulgeographen. Gotha. v1-22 (1913-1936). Closed 1936. (U-2:1696a; D:230; G).
New series [1354].

981 GEOGRAPHISCHE Gesellschaft für das Land Braunschweig. Braunschweig.
Jahrbuch. v1 (1931-1933). Closed 1933. (D:35).

982 GEOGRAPHISCHE Gesellschaft (für Thüringen) zu Jena. Jena.
Mitteilungen. v1-42 (1882-1939). Closed 1939. (U-2:1696a; B-2:270a; F-3:567b; D-53; G).
Index: 1-12 (1882-1893) in 12; 13-25 (1894-1907) in 28; 26-35/36 (1908-1917/18).

983 GEOGRAPHISCHE Gesellschaft, Greifswald. Greifswald.
Excursionen. 1-8 (1885-1902). Closed 1902. (D:237).

984 GEOGRAPHISCHE Gesellschaft, Karlsruhe. Karlsruhe.
Bericht über die Tätigkeit. 1-8 (1925-1932/33). (U-2:1696a).
Continued as the society's Vortragsfolge [985]).

985 ______.
Vortragsfolge. 1934/35-1935/36. Closed 1936.
A continuation of the society's Bericht über die Tätigkeit [984].

986 GEOGRAPHISCHE Gesellschaft in Bremen. Bremen.
Berichte über die Sitzungen. 1-40 (1870-Ja 1877). (U-2:1696a; B-4:455a; D:36; G).
1870-1876 as Verein für die deutsche Nordpolarfahrt. Superseded by Deutsche geographische Blätter [1194].

987 ______.
Jahresbericht. v1-11 (1878-1891). Closed 1891. (U-2:1696a; B-4:455a; F-3:28a; D:37; G).
Issued as Beilage to Deutsche geographische Blätter. Later as a section in Deutsche geographische Blätter [1194].

988 GEOGRAPHISCHE Gesellschaft in Hamburg.
Jahresbericht. 1-2 (1873/74-1874/75). Closed 1875. (U-2:1696a; B-2:270a; F-3:28a; D:50; G).
Bound as Mitteilungen, v1-2. Superseded by society's Mitteilungen [1233].

989 GEOGRAPHISCHE Gesellschaft in München. München.
Jahresberichte. 1-20 (1869/70-1901/02). Closed 1902. (U-2:1696b; B-2:270a; F-3:28a; D:61; G).
Continued as society's Mitteilungen [1236].
Index : 1-20 in 20.

990 GEOGRAPHISCHE Gesellschaft in Nürnberg.
Mitteilungen und Jahresbericht. 1-7 (1920-1940). Closed 1940. (U-2:1696b; D:63).

991 GEOGRAPHISCHE Gesellschaft zu Hannover. Hannover.
Jahresberichte. 1-12 (1878-1911). Closed 1911. (U-2:1696b; B-2:270a; F-3:282a; G).
Subsequently published in society's Jahrbuch [ ].

992 GEOGRAPHISCHE Gesellschaft zu Rostock. Rostock.
Mitteilungen. 1-26/30 (1910-1934/39). Closed 1939. (U-2:1696b; B-2:270a; F-3:567b; D:64: G).

993 ______.
______. Beihefte. v1-12 (1934-1940). Closed 1940. (U-2:1696b; B-2:270a; D:243; G).

994 GEOGRAPHISCHE Gesellschaft zu Würzburg. Würzburg.
Mitteilungen. v1-8 (1925-1935). (U-2:1696b; D:70; D:244; G).
Continued as Fränkische Studien [972].

995 GEOGRAPHISCHE Mitteilungen aus Hessen. (Gesellschaft für Erd- und Völkerkunde zu Giessen). Giessen. v1/2-6, no6 (1900-1911). Closed 1911. (U-2:1696c; B-2:270b; F-2:728b; D:45; G).

996 GEOGRAPHISCHE Nachrichten für Welthandel und Volkswirtschaft. (Centralverein für Handelsgeographie und Förderung deutscher Interessen im Auslande). Berlin. 1-3 (1879-1881). Closed 1881.
Subtitle varies. (U-2:1696c; B-2:270b).
Superseded by Neue Zeit (not in this list).

997 GEOGRAPHISCHE Schriften. Leipzig. 1-7 (1923-1935). Closed 1935.
Monographs. (U-2:1696c; B-2:270b; D:365; G).

998 GEOGRAPHISCHE Vereinigung, Bonn. Bonn.
Veröffentlichung. 1 (1905). Only one issued. (D:233).

999 GEOGRAPHISCHE Zeitung der Hertha. Stuttgart: Tübingen. 1-4 (1825-1829). (F-2:729a).
Supplement to Hertha. Zeitschrift für Erd-, Völker- und Staatenkunde, which became Annalen der Erd-, Völker- und Staatenkunde [901].

1000 GEOGRAPHISCHER Abend der Leipziger Geographen. Leipzig.
Jahresbericht. 1-2 (1901-1903). (D:56).

1001 GEOGRAPHISCHER Anzeiger. Gotha. v1-45 no18 (1899-1944). Closed 1944.
(U-2:1696c; B-2:632a; F-2:729ab; D:77).
v1-3 (1900-1902) with Petermanns Mitteilungen, Jahrgang 46-48 (1900-1902) [1370]. In 1911 absorbed Zeitschrift für Schulgeographie [164]. Prior to 1935 carried subtitle Blätter für den geographischen Unterricht. 1912-1933 also as Zeitschrift des Verbandes deutscher Schulgeographen.
Indexes: 1-25 in v25; 26-35 in v36.

1002 GEOGRAPHISCHER Büchersaal, zum Nutzen und Vergnügen eröffnet. Chemnitz.
1-3 (1766-1778). Closed 1778. (U-2:1696c).

1002a GEOGRAPHISCHER Literaturbericht: eine Bücherschau über Geographie, Auslandskunde, Kolonialliteratur und Reisen. Hamburg. 1-2 (1925-1926). (G).

1003 GEOGRAPHISCHES Jahrbuch zur Mittheilung aller wichtigen neuen Erforschungen.
Gotha. v1-4 (1850-1852). Closed 1852. (B-2:270b; G).
Part of Berghaus' physikalischer Atlas.

1003a GEOGRAPHISCHES Lesebuch zum Nutzen und Vergnügen. Halle. 1-7 (1782-1878).
(G).

1004 GEOGRAPHISCHES Magazin. Dessau; Leipzig. 1-4 (no1-16) (1783-1785).
Closed 1785. (G).
Superseded by Neues geographisches Magazin [1077].

1004a GEOGRAPHISCHES Wochenblatt. Dresden. 1-4 (1794). (G).

1005 GESELLSCHAFT der Wissenschaften in Göttingen. Göttingen.
Mathematisch-physikalische Klasse.
Nachrichten. 5. Geographie. nsv1 no1 (1935). Only one issued.
Closed 1935. (U-1:103b; B-4:117b; F-3:675b; D:307; G).

1006 GESELLSCHAFT für Erdkunde und Kolonialwesen zu Strassburg. Strassburg.
Mitteilungen...zugleich. Abteilungen Strassburg der Deutschen Kolonialgesellschaft. 1-6 (1910-1915/17). Closed 1917. (U-2:1716a; B-2:289a: F-3:568b; G).

1007 GESELLSCHAFT für Erdkunde zu Berlin. Berlin.
Jährliche Übersicht der Thätigkeit. 1-6 (My 4 1833-My4 1839).
Closed 1839. (U-2:1716a; B-2:289a; D-32; G).
Continued as society's Monatsberichte über die Verhandlungen [1008].

1008 ______.
Monatsberichte über die Verhandlungen. 1-4 (1839-1843); nsv1-10 (1844-1853). Closed 1853. (U-2:1716a; B-2:289a; F-3:600a; D:32; G).
Transactions 1853-1872 published in society's Zeitschrift [1011].
Continuation of Jährliche Übersicht der Thätigkeit [1007]; succeeded by Zeitschrift [1011].
Index: 1839-1853.

1009 ______.
Übersicht der Aufsätze, Miscellen und Karten, welche in den Monatsberichten über die Verhandlungen [v1-4, 1839-1843 and nsv1-10, 1844-1853], so wie in die Zeitschrift fur allgemeine Erdkunde enthalten sind [v1-6, 1853-1856 and nsv1-14, 1856-1863]. 1840-1863. (B-2:289a).

1010 ______.
Verhandlungen. 1-28 (1873/74-1901). Closed 1901. (U-2:1716a; B-2:289a; F-4:872b; D:31; G).
Earlier transactions published in the Jährliche Übersicht der Thätigkeit [1007] 1833-1839; Monatsberichte über die Verhandlungen [1008] 1839-1853; Zeitschrift für allgemeine Erdkunde [1011] 1853-1865 and the Zeitschrift 1866-1872 [1011].

1011 ______.
Zeitschrift. 1-6 (1853-1856); nsv1-19 (1856-1865); (s3) v1-36 (1866-1901) [37-79]; 1902-1944 no4. Closed 1944. (U-2:1716a; B-4:598b; F-4:1006b; D:30; G).
1853-1865 as Zeitschrift für allgemeine Erdkunde (G). 1853-1890 contained annual summaries of geographic literature continued as Bibliotheca geographica (1891-1912) [938]. Superseded by Erde [1201].
Indexes: 1853-1863; 1863-1901.

1012 ______.
______. Ergänzungsheft. v1-5 (1924-1937). Closed 1937. (U-2:1716b; B-4:598b; F-4:1000b; D:232; G).
See also Bibliotheca geographica [938].

1013 GESELLSCHAFT für Erdkunde zu Köln.
Jahresbericht. 1896/97-1900/03. Closed 1903. (U-2:1716b; D:55).

1014 GESELLSCHAFT für Erdkunde zu Leipzig. Leipzig.
Jahresbericht. 1-11 (1861-1871). Closed 1871. (U-2:1716b; B-4:457a; F-3:42a; D:57).
Continued as society's Mitteilungen [1015]. Through 1871 as Verein von Freunden der Erdkunde zu Leipzig; 1872-1910 Verein für Erdkunde zu Leipzig.

1015 ______.
Mitteilungen. v1-56 (1872-1941). Closed 1941. (U-2:1716b; B-4:457a; F-3:581b; D:57).
Succeeds society's Jahresbericht 1861-1871 [1014].
Index: 1861-1909.

1016 ______.
Wissenschaftliche Veröffentlichungen. v1-11 (1891-1938). Closed 1938. (U-2:1716b; B-4:457a; F-4:978b; D:241; G).
Suspended 1922-1935.

1017 GESELLSCHAFT für Völker- und Erdkunde zu Stettin. Stettin.
Bericht. 1/2-11/12 (1897/99-1908/10). Closed 1910. (U-2:1719a; B-2:291b; D:66; G).
Supersedes Verein für Erdkunde, Stettin. Jahresbericht [1115].

1018 GIESSEN. Universität. Anstalt für hessische Landesforschung.
Arbeiten. Giessen.
Geographische Reihe. 1-15 (1930-1937). Closed 1937. (U-2:1725b; B-1:12b; D:274).

1019 GLOBUS: illustrierte Zeitschrift für Länder- und Völkerkunde.
Hildburghausen; Braunschweig. v1-98 no24 (1861-1910). Closed 1910. (U-2:1736c; B-2:304b; F-2:755b; D:26).
Merged into Petermanns geographische Mitteilungen [1370].
Absorbed Neue Ausland [1068] and Aus allen Weltteilen [914].

1020 GÖTTINGEN. Universität. Geographisches Seminar. Göttingen.
Landeskundliche Arbeiten. v1-2 (1925-1926). Closed 1926. (U-2:1740b; D:193).

1021 GRENZMÄRKISCHE Forschungen. (Grenzmärkische Gesellschaft zur Erforschung und Pflege der Heimat e.V. in Schneidemühl). Leipzig. no1-5 (1939-1941). Closed 1941. (U-2:1766a; G).

1022 HAMBURG. Museum Godeffroy. Hamburg.
Journal. (Geographische, etnographische und naturwissenschaftliche Mitteilungen). v1-6 (no1-17) (1873-1910). Closed 1910.
(U-3:1788b; F-3:148b; D:49).

1023 HANDBÜCHER zur deutschen Landes- und Volkskunde. 1-4 (1887-1896). Closed 1896. Monographs. (B-2:371a; D:167).
v2 without series name or number.

1024 HANNOVERSCHE geographische Arbeiten. (Hannover. Technische Hochschule. Geographisches Institut. Veröffentlichungen. [and] Geographische Gesellschaft, Hannover). Hannover. 1 (1942). Only one issued. (D:199).

1025 HEIMAT; Monatsschrift des Vereins zur Pflege der Natur- und Landeskunde in Schleswig-Holstein und Hamburg. Kiel, Flensburg, Neumünster. v1- (1891-). (U-3:1833c; D:109; G).
Subtitle 1891-1943: Monatsschrift des Vereins zur Pflege der Natur- und Landeskunde in Schleswig-Holstein, Hamburg, Lübeck und dem Fürstentum Lübeck.
Index: 1-53 (1891-1943) [1955/1956], 54-69 (1947-1962) [1964].

1026 HEIMATKUNDLICHE Arbeiten. (Erlangen. Universität. Geographisches Institut). Erlangen. v1-10 (1926-1941). Closed 1941. (U-3:1834a; B:186; G).

1027 Der HESSISCHE Raum. (Marburg. Universität. Geographisches Institut). Marburg. 1-3 (1938-1941). Closed 1941. (D:213).

1028 HISTORISCH-geographisch-und genealogische Anmerkungen über Verschiedene in denen neuesten Zeitungen des Jahres...vorkommende besondere Sachen. Königsberg. 1-4? (1723-1726?). Closed 1726? (F-2:821a).

1029 HISTORISCH-geographisch-statistisch-literarisches Jahrbuch für Westfalen und den Niederrhein. Coesfeld. 1-2 (1817-1818). (B-2:404a).

1030 HISTORISCH-geographisches Journal. Halle. no1-6 (1789-1790). Closed 1790?

1031 HISTORISCH-geographisches Journal. (Jena. Akademie der Wissenschaften). Jena. v1 (1812). Only 1 published?

1032 JAHRBUCH der Kartographie. (Deutsche kartographische Gesellschaft). Leipzig. 1-2 (1941-1942). Closed 1942. (U-3:2139a; D:87).

1033 JENA. Universität. Geographisches Institut. Hamburg.
Publikationen. 1 (1909). Closed 1909. (U-3:2159c; D:201).

1034 JOURNAL für die neuesten Land- und Seereisen und das Interesanteste aus der Völker- und Landeskunde... Berlin. 1-84 (1808-1836); nsv1-5 (1837-1838). Closed 1838. (U-3:2191b; B-2:618b; F-3:175a).
1-20, 68-72, 75 as Magazin der neuesten Reisebeschreibungen in unterhaltenden Auszügen, with second title Neues Journal für die neuesten Land- und Seereisen...

1035 KÖNIGSBERG. Universität. Geographisches Institut. Königsberg.
Veröffentlichungen. 1-11 (1919-1929). Closed 1929. (U-3:2303c; B-1:10a; F-4:890a; D:207).
Continued as Reihe Geographie [1037] and as Reihe Etnographie (not in this list).

1036 ______. ______. ______.
______. Ausser der Reihe. 1-5 (1925-1931). Closed 1931.
(U-3:2303c; B-1:10a; D:208).

1037 ______. ______. ______.
______. Reihe Geographie. ns1-11 (1931-1938). Closed 1938.
(U-3:2303c; B-1:10a; D:209).
no11 also as university's Schriften. Naturwissenschaftliche Reihe
no1 (not in this list).

1038 KÖNIGSBERGER geographische Gesellschaft. Königsberg.
Landeskundliche Litteratur der Provinzen Ost- und Westpreussen.
1- (1842). Closed 1842. (U-3:2304b; B-3:13a).
B reports 1 (1892).

1039 KRIEGSGEOGRAPHISCHE Zeitbilder; Land und Leute der Kriegsschauplätze.
Leipzig. no1-8 (1915). Closed 1915. (U-3:2318b; B-2:669b; F-3:264b;
D:369; G).

1040 KRIEGSSCHAUPLÄTZE. Leipzig, Berlin. 1-6 (1916-1918). Closed 1918.
(B-2:669b; D:370).

1041 KULTURGEOGRAPHISCHE Beiträge. Hamburg. 1-2 (1939). Closed 1939. (D:371).

1042 LEIPZIG. Museum für Länderkunde. Leipzig.
Deutsche Sammlung: Mensch und Volkstum. 1-2 (1934-1939). Closed
1939. (D:248).

1043 ______. ______.
Führer und Erläuterungen. 1-4 (1928-1938). Closed 1938. (D:247).

1044 ______. ______.
Schriftenreihe. 1-2 (1934-?). (U-3:2384b).

1045 ______. Staatliche Forschungsinstitute. Forschungsinstitut für Geographie.
Leipzig.
Publikationen. 1 (1923). Only one issued. (U-3:2385a; G).

1046 ______. Universität. Geographisches Seminar.
Veröffentlichungen. Berlin. 1-12 (1930-1936). Closed 1936.
(U-3:2385c; B-3:36a; D:211).

1047 ______. ______. Verein der Geographen.
Mitteilungen. 1-18 (1911-1941). Closed 1941. (U-3:2386b; B-3:36b;
F-3:581a; D:58).

1048 LIPPISCHE Mitteilungen aus Geschichte und Landeskunde. (Naturwissenschaftlicher und historischer Verein für das Land Lippe). Detmold.
1-10 (1903-1914); 11-17 (1921-1939); 18- (1949-). Suspended 1915-
1920 and 1940-1948. (U-3:2431a; B-3:230a; F-3:563a; G). 1-25 (1903-1956)
as Mitteilungen aus der lippischen Geschichte und Landeskunde.
[Meyersche Hofbuchhandlung, D-493 Detmold, BRD].

1049 MACHT und Erde. Leipzig; Berlin. 1-10? (1936-1939?). (U-3:2503a;
B-3:113a).
Monographs.

1050 MAGAZIN für die Geographie, Staatenkunde und Geschichte Nürnberg.
v1-3 (1797). Closed 1797. (B-3:118a; F-3:374a; G).
Supersedes Beiträge zur Geographie, Geschichte und- Staatenkunde
[923].

1050a MAGAZIN für die Geographie und Statistik des preussischen Staats. Braunschweig. 1 (1791). (G).

1051 MAGAZIN für die neue Historie und Geographie. Hamburg; Halle. 1-23 (1767-1793). Closed 1793. (U-3:2510a; B-3:119a; F-3:375a; G).
Suspended 1789-1792.
Index: 1-23 (1767-1793).

1052 MARBURG. Universität. Institut für Grenz- und Auslanddeutschtum. Marburg. Jahresbericht. 1-8 (1919/20-1931). Closed 1931? (U-3:2537a).

1053 ______. ______. ______.
Schriften. 1 (1922). (U-3:2537a; B-1:15; D-329).
Only Heft 1 (1922) was a geographic monograph.

1054 MEER in volkstümlichen Darstellungen. (Berlin. Universität. Institut für Meereskunde). Berlin. 1-8 (1933-1940). Closed 1940. (U-3:2596a; D:262; F-3:429a; G).
Supersedes Meereskunde [1055].

1055 MEERESKUNDE. (Berlin. Universität. Institut für Meereskunde). Berlin. v1-18 no10 (no1-206). (1907-1932). Closed 1932. (U-3:2596a; B-3:173b; F-3:429ab; D:261; G).
84 of the 206 monographs written by geographers. Superseded by Meer in volkstümlichen Darstellungen [1054].

1056 MITTEILUNGEN des Ferdinand von Richthofen-Tages. Leipzig (Teubner). 1-3 (1911-1914). Closed 1914. (F-3:575b; D:33).

1057 MITTEILUNGEN über physisch-geographische und statistische Verhältnisse von Frankfurt am Main von dem Geographischen Vereine daselbst. Frankfurt am Main. 1-3 (1839-1841). Closed 1841. (F-3:584b).

1058 MONATLICHE Correspondenz zur Beförderung der Erd- und Himmelskunde. Gotha. 1-28 (1800-1813). Closed 1813. (U-3:2724b; B-3:238b; F-3:598a; G).
Index: 1-28 (1800-1813).

1059 MONOGRAPHIEN deutscher Landschaften. Berlin. 1-4 (1927-1930). Closed 1930. (U-3:2736a; B-3:244b).
Monographs.

1060 MONOGRAPHIEN zur Erdkunde. Bielefeld; Leipzig. 1-49 (1898-1934). Closed 1934. (U-3:2736c; B-3:12a; D:372; G).
Monographs.
1-29 (1898-1914) as Land und Leute; Monographien zur Erdkunde.
Suspended 1915-1924.

1061 MÜNCHEN. Technische Hochschule. Geographisches Institut. München. Arbeiten. 1-3 (1937-1941). Closed 1941. (U-3:2785a; D:218).
no1-2 reprints from Geographische Gesellschaft in München. Mitteilungen [1236].

1062 MÜNCHENER geographisch-philologische Studien. (Münchener geoplast. Inst.). München. 1-10?

1063 MÜNCHENER geographische Studien. München. no1-29 (1896-1915). Closed 1915. (U-3:2780b; B-3:272b; F-3:653a; D:215).
Continued as Neue Münchener geographische Studies [1072].

1064 MÜNSTER. Provinzialinstitut für westfälische Landes- und Volkskunde. Geographische Kommission. Münster in Westfalen.
Arbeiten. v1-7 (1938-1942). Closed 1942. (U-3:2781c; D:284).

1065 ______. Universität. Geographisches Seminar. Emsdetten in Westfalen. Beiträge zur westfälischen Landeskunde. 1-5 (1935-1937). Closed 1937. (U-1:629c; U-3:2782a; D:219).

1066 NEUE allgemeine geographische und statistiche Ephemeriden. Weimar. 1-31 (1817-1831). Closed 1831. (U-4:2965a; B-1:84a; F-3:700a). Supersedes Allgemeine geographische Ephemeriden [893]. 1-10 as Neue allgemeine geographische Ephemeriden.

1067 NEUE Analekten für Erd- und Himmelskunde. München. v1-2 no2 (1834-1836). Closed 1836. (B-1:137b; F-3:700b; G). Supersedes Analekten für Erd- und Himmelskunde [899].

1068 Das NEUE Ausland. Wochenschrift für Länder- und Völkerkunde. Leipzig. 1 (1894). Closed 1894. (U-4:2965b; B-1:258a). Superseded Ausland [915]; merged with Globus [1019], which later merged with Petermanns geographische Mitteilungen [1370].

1069 NEUE Beiträge zur vaterländischen Geschichte, Geographie und Statistik. München. ns1 (1832). (B-3:343a; F-3:701a). Supersedes Beiträge zur vaterländischen Historie, Geographie, Statistik und Landwirtschaft [930].

1070 NEUE Beiträge zur Völker- und Länderkunde. Leipzig. 1-13 (1790-1793). Closed 1793. (U-4:2965c; B-1:306b; F-3:701a; G). Supersedes Beiträge zur Völker- und Landerkunde [931]. Succeeded by Auswahl der besten ausländischen geographischen...[917].

1071 NEUE Geographie. Braunschweig. 1-4 (no1-17) (1922-1925/26). Closed 1926. (U-4:2967c; D:27).

1072 NEUE Münchener geographische Studien. München. 1 (1920). Closed 1920. Only 1 issued. (U-4:2970a; B-3:272b; F-3:706a; D:216). Supersedes Münchener geographische Studien [1063].

1073 NEUE nordische Beyträge zur physikalischen und geographischen Erd- und Völkerbeschreibung, Naturgeschichte und Oekonomie. St. Petersburg (Leningrad); Leipzig. 1-7 (1781-1796). Closed 1796. (U-4:2970a; B-3:345a; F-3:706a). 5-7 have additional title Neueste nordische Beyträge...

1074 Der NEUE Orient; Abhandlungen zur Geographie, Kultur und Wirtschaft der Länder des Ostens. (Deutsche Vorderasien-Gesellschaft). Halle. 1-13 (1905-1918). Closed 1918. (U-4:2970b; F-3:706b; F-3:845a). 1-8 as Orient, Vorträge und Abhandlungen zur Geographie und Kulturgeschichte der Länder des Ostens. Subtitle varies.

1075 NEUE Sammlung geographisch-historisch-statistischer Schriften. Weissenburg im Nordgau. 1-17 (1783-1795). Closed 1795. (B-3:345b; F-3708a). Index: 1-8.

1076 NEUE wöchentliche Nachrichten von neuen Landkarten, geographischen statistischen, historischen, wie auch Handlungsbüchern und Sachen. Göttingen. 1-2 (Ja 2 1788-D 28 1789). Closed 1789. (U-4:2927c; B-3:346a; G).

1077 NEUES geographisches Magazin. Halle. 1-4 (1785-1787). Closed 1787. (G). Supersedes Johann Ernst Fabri's Geographisches Magazin [1004].

1078 NEUES westphälisches Magazin zur Geographie, Historie und Statistik. Bielefeld. v1-3 no1-12 (1789-1794). Closed 1794. (B-3:348a; F-3:716a; G). Successor to Westphälisches Magazin zur Geographie, Historie und Statistik [1130]. Superseded by P. F. Weddingens neues fortgestztes westphälisches Magazin zur Geographie, Historie und Statistik [1125].

1079 NORDOBERFRÄNKISCHER Verein für Natur-, Geschichts- und Landeskunde. Hof.
Bericht. 1- (1896-). (U-4:3074b; B-3:391b; F-3:573a).

1080 OBERRHEINISCHE geographische Abhandlungen (Geographische Institute der Universitäten Freiburg im Breisgau und Heidelberg). 1-5 (1939-1941). Closed 1941. (U-4:3127c; B-1:279b; D:189; G).
Supersedes Badische geographische Abhandlungen [920].

1081 OSNABRÜCKER Mitteilungen. (Verein für Geschichte und Landeskunde von Osnabrück). Osnabrück, Paderborn. v1- (1848-). (U-4:3211b; B-4:458b; N70-3:4464b; G).
Suspended 1914-1915, 1942-1946. 1-64 (1848-1950) as society's Mitteilungen.
Indexes: 1-16 (1848-1891); 17-32 (1893-1907); 33-55 (1908-1933); 56-65 (1936-1952); 66-79 (1954-1971).

1082 OSTALPINE Formenstudien. Berlin.
Abteilung 1, v1-5 (1920-1923); Abteilung 2, v1-3 (1921-1924); Abteilung 3, v1-2 (1921-1923). Closed 1924. (U-4:3212a; D:374).

1083 OTTO Hübner's geographisch-statistische Tabellen aller Länder der Erde. Wien, Frankfurt am Main. v1-73 (1851-1939). Closed 1939. (B-3:473b; G).
1-28 (1851-1879) as Statistische Tafel aller Länder der Erde, Leipzig.

1084 PETERMANNS geographische Mitteilungen. Geographischer Literatur-Bericht. 32-55 (1886-1909). Closed 1909. (U-4:3316a; B-2:632a; F-2:729b; D:7; G).
1-30 (1855-1884) as Geographische Literatur and 31 (1885) as Literaturbericht, as integral parts of Mitteilungen aus Justus Perthes' geographischer Anstalt über wichtige neue Forschungen auf dem Gesamtgebiete der Geographie, later A. Petermann's Mitteilungen aus Justus Perthes' geographischer Anstalt, later as Petermanns geographische Mitteilungen [1370]. 32-55 (1886-1909) as separately paged supplement bound at back of each volume; 56- (1910-) as Geographischer Literaturbericht again as an integral part of Petermanns geographische Mitteilungen.

1085 ______.
Kartographischer Monatsbericht (later Kartenbibliographie). (D:12).
v1-2 no7 (1908-1909) issued as separately paged supplement to Petermanns geographische Mitteilungen [1370]. Closed as a separate publication 1909. Continued as an integral part of the Mitteilungen, scattered and not separately paged.

1086 PFÄLZISCHE Gesellschaft zur Förderung der Wissenschaften in Westmark. Neustadt an der Haardt, Speyer.
Veröffentlichungen. 1-26 (1927-1936); 27- (1950-). Monographs. (U-2:1720c; B-3:532b; G). 25-26 (1936) as Schriften. Replaced 1937-1943 by Abhandlungen zur saarpfälzischen Landes- und Volksforschung 1 (1937); Saarpfälzische Abhandlungen zur Landes- und Volkforschung, 2-3 (1938-1939), and Westmärkische Abhandlungen zur Landes- und Volksforschung, 4-5 (1940-1941/42) [1128].

1087 POMMERSCHE geographische Gesellschaft. Greifswald.
Jahrbuch. v1-59/60 (1882-1942). Closed 1942. (U-4:3394a; B-2:270a; F-3:28a; F-3:6b; D:46; G).
1882-1926 as Geographische Gesellschaft zu Greifswald. 1-39 (1882-1922) as Jahresbericht; 40/42-43/44 (1922/24-1925/26) as Jahrbuch.

1088 ______.
______. Beihefte. 1917-1940. Closed 1940. (B-2:270a; D:238; G).
Volumes issued: 16 (1916/17), 17/38 (1917-20), 40/42 (1922-24), 43/44 (1925/26), nos. 1-2, 45/46 (1927-28), 47/48 (1929-30), nos. 1-2, 49/50 (1931/32), 51/52 (1933/34), 57/58 (1939-40), nos. 1-2.

1089 PREUSSEN. Grosser Generalstab. Berlin.
Registrande der geographisch-statistischen Abtheilung. v1-13 (1867/68-1883). Closed 1883. (F-4:64a; G).
Also under the title: Neues aus der Geographie, Kartographie und Statistik Europas und seiner Kolonien.

1090 QUELLEN und Forschungen zur alten Geschichte und Geographie. Berlin, Leipzig.
1-30 (1901-1909 [1918]). Closed 1909. (U-4:3508a; B-3:643b; F-3:1108a; G). 16,20 never issued. Various numbers are reprints.

1091 QUELLEN und Forschungen zur Erd- und Kulturkunde. Leipzig. 1-13 (1909-1934). Closed 1934. (U-4:3508b; B-3:643b; D:376; G). Monographs. 1 as Quellen und Forschungen zur Geschichte der Erdkunde. Suspended 1922-1929.

1092 QUELLEN und Forschungen zur Geschichte der Geographie und Völkerkunde. Leipzig.
1-7 (1938-1940). Closed 1940. (U-4:3508b; B-3:643b; D:377; G).

1093 RAUM und Volk. Erdkundliche Arbeitshefte. Berlin, Leipzig, Langensalza. 1931-1942. Closed 1942. (D:378).
Gruppe 1, v1-4; Gruppe 2, v1-5; Gruppe 3, v1-3; Gruppe 4, v2; Gruppe 5, v1-8; Gruppe 6, v1-25; Gruppe 7, v1-21.

1094 Der RHEINISCHE Bund; eine Zeitschrift historisch-politisch-statistisch-geographischen Inhalts. Frankfurt am Main. 1-23 (1806-1813), no1-69. Closed 1813. (U-4:3671b; B-3:730a; G).
Superseded by Allgemeine Staatskorrespondenz (not in this list).
Index: 1-8 in 9; 10-20 in 21.

1095 RHEINISCHE Vierteljahrsblätter. (Bonn. Universität. Institut für geschichtliche Landeskunde der Rheinlande). Bonn. v1- (1931-). (U-4:3671c; BS:740b; F-4:356a; D:104).
Formed by the union of Rheinische Neujahrsblätter and Geschichtliche Landeskunde, both not in this list.

1096 RHEINLANDE in naturwissenschaftlichen und geographischen Einzeldarstellungen. Braunschweig, Berlin. v1-12 (1912-1920). Closed 1920. (U-2:1304a; D:380; G).

1097 RUNDSCHAU auf dem Gebiete der Geographie und Naturwissenschaft. Zeitschrift für Deutschlands Lehrer. Kamenz. v1-2 (1869-1870). Closed 1870.

1098 SÄCHSISCH-thüringischer Verein für Erdkunde. Halle.
Mitteilungen. v1-62 (1877-1940). Closed 1940. (U-5:3753a; B-4:456b; F-4:581b; D:48; G).
For 1891- see also Archiv für Landes- und Volkskunde der Provinz Sachsen [909]. 1877-1907 Verein für Erdkunde zu Halle a.S.
Index: 1-20 (1877-1896) in 20 (1896).

1099 ______.
______. Beihefts. v1-12 (1930-1940). Closed 1940. (U-5:3753a; B-4:456b; F-4:581b; D:239; G).

1100 SCHLESISCHE Gesellschaft für Erdkunde, Breslau. Breslau.
Jahresbericht. 1925-1933. Closed 1933. (D:38).

1101 ______.
Veröffentlichungen. no1-31 (1922-1941). Closed 1941. (U-5:3792c; B-4:35a; D:183)
no24 never published.

1102 SCHRIFTEN zur Geopolitik. Heidelberg. 1-26 (1932-1952). (U-5:3804a; F-4:441a; G). 1-22 (1932-1943) published in Berlin. Some of early nos. reprints from Zeitschrift für Geopolitik [1145].

1103 SCHRIFTEN zur schleswig-holsteinischen Landesforschung. (Kiel. Universität. Institut für Volks- und Landesforschung). Neumünster. v1-4 (1938-1941). Closed 1941. (U-5:3804b; B-4:39b; D:292). 1 (1938), 2 (1941), 4 (1940). 3 never issued?

1104 STUDENTISCHER Verein für Erdkunde zu Halle a.S. Halle. Bericht. v1-2 (1895-1897). Closed 1897?

1105 STUTTGART. Technische Hochschule. Auslandkundliche Vorträge. 1-14/15 (1930-1937). Closed 1937? (U-5:4105a).

1106 ______. ______. Geographisches Institut. Stuttgart. Veröffentlichungen. (U-5:4105ab).
sA. see Stuttgarter geographische Studien [1325].
sB. Unterrichtsbeiträge zur Pflege der Geographie und der geographischen Landeskunde mit besondere Berücksichtigung Württembergs (title varies slightly). v1-14 (1925-1937). Closed 1937. (D:224).
sC. Geographische Exkursions-führer für Württemberg. 1 (1925). Only one issued. Closed 1925. (D:225).

1106a TAGESBERICHTE über die Fortschritte der Natur- und Heilkunde. Abtheilung für Geographie und Ethnologie. Weimar. 1 (1852). Only one issued. (G).

1107 TASCHENBIBLIOTHEK der neuesten unterhaltendsten Reisebeschreibungen. Frankfurt am Main. 1-5 (1826-1830). Closed 1830.

1108 TÜBINGER geographische und geologische Abhandlungen. (Tübingen. Universität. Geologisches und geographisches Institut). Öhringen. s1. Deutschland. 1-29 (1920-1944). Closed 1944. (U-5:4275b; D:226; G). no1-22 (1920-1936) as Erdgeschichtliche und landeskundliche Abhandlungen aus Schwaben und Franken, not designated as s1.
s2. Ausser Deutschland. no1-7 (1936-1941). Closed 1941. (U-5:4275b; D:227; G).

1109 VEGETATION der Erde. Sammlung pfanzengeographischer Monographien. Leipzig. 1-15 (1896-1928). Closed 1928. (B-4:448a; F-4:860b; D:384; G).

1110 VEREIN für Erdkunde, Darmstadt. Darmstadt. Notizblatt. no1 46 (1854-1857), nsv1-3 (no1-60) (1858-1861); s3 v1-18 (no1-216) (1862-1879); s4 no1-35 (1880-1914); s5 v1-19 (1916-1938). Closed 1938. (U-5:4368b; B-1:304a; F-4:756b; D:39). 1854-1857 as Verein für Erdkunde und verwandte Wissenschaften zu Darmstadt; 1857-1879 Verein für Erdkunde und verwandte Wissenschaften zu Darmstadt und Mittelrheinischer geologischer Verein; 1880 Verein für Erdkunde zu Darmstadt, Mittelrheinischer geologischer Verein und Naturwissenschaftlicher Verein zu Darmstadt; 1881-1891 Verein für Erdkunde zu Darmstadt und Mittelrheinischer geologischer Verein; 1892-1914 Verein für Erdkunde und grossherzogliche geologische Landesanstalt zu Darmstadt; 1936-1938 Hessische geologische Landesanstalt zu Darmstadt. Supersedes Beiträge zur Landes-, Volks-, und Staatskunde des Grossherzogtums Hessen [925].

1111 VEREIN für Erdkunde, Dresden. Dresden. Jahresberichte. 1-6 (no1-27) (1863-1901). Closed 1901. (U-5:4368b; B-4:456b; D:40).
Continued as society's Mitteilungen [1112].
Indexes: 1-5 (1863-1868); 6-9 (1868/69-1871/72); 1-27 (1863-1901).

1112 ______.
Mitteilungen. v1-3 no6 (1905-1926); ns 1926-1938. Suspended Jl 1921-F 1925. Closed 1938. (U-5:4368b; B-4:456b; F-3:581ab; D:40).
1863-1901 as society's Jahresberichte [1111].

1113 VEREIN für Erdkunde, Kassel. Kassel.
Schriften. 1-58 (1884-1941). Closed 1941. (U-5:4368c; F-3:41b; F-3:28b; D:54; G).
1911-1916 as Gesellschaft für Erd- und Völkerkunde zu Cassel.
1-34 as society's Jahresbericht.

1114 VEREIN für Erdkunde, Metz. Metz.
Jahresbericht. 1-27 (1878-1911). Closed 1911. (U-5:4368c; B-4:456b; F-3:42a).
Includes the society's Abhandlungen.

1115 VEREIN für Erdkunde, Stettin. Stettin (Szczecin).
Jahresbericht. 1883/85-1889/91. Closed 1892. (U-5:4368c: B-2:291b; F-3:42a; D:66; G).
Superseded by Gesellschaft für Völker- und Erdkunde zu Stettin. Bericht [1017].

1116 VEREIN für Geographie und Statistik, eingetragener Verein. Frankfurt am Main.
Jahresbericht. 1-87/89 (1836/37-1922/25). Closed 1925. (U-5:4368c; B-2:270b; F-3:37ab; D:42; G).
1-18 (1836-1854) as Geographischer Verein zu Frankfurt am Main.
Succeeded by Frankfurter geographische Hefte [1221].
For 90/91-92 (1927-1928) see Frankfurter geographische Hefte 1-2 (1927-1928) [1221].

1117 ______.
______. Ergänzungsblätter. v1 (1895). Closed 1895.

1118 VEREIN für hessische Geschichte und Landeskunde, Kassel. Darmstadt, Marburg, Kassel. Mitteilungen. 1848-1939; nsv 1939/51, 1952/53, no3 (1954/55). (U-5:4370a; B-4:458b; G). 1870-1874; 1921/22-24/25; 1928, 1934/35-36/37 never published. nsv1- (1939/51-) [1952-] as supplements to main series with varying numbering.

1119 ______.
Zeitschrift. 1-62 (1837-1940); 63- (1952-). (U-5:4370a; B-4:458b; F-4:1005b; G).
11- (1867)-62 (1940) also as nsv1-52.
Indexes 1-10 (in 11); 1-24 (1837-1889), 1-45 (1837-1911); 46-60 (1912-1934) in 59/60.

1120 ______.
______. Supplement. 1-10 (1840-1865); (ns) 1-18 (1866-1927). (U-5:4370a; B-4:458b; F-4:1005b; G).
Monographs.
Index: 1-20, 1840-1889 (1890) with index to v1-24 of Zeitschrift.

1121 VEREIN für Naturwissenschaft und Erdkunde, Glauchau. Glauchau.
1-2? (1898/99-1899/1900). Closed 1900?

1122 VEREIN für Sachsen-Meiningische Geschichte und Landeskunde, Meiningen. Meiningen.
Schriften. 1-94 (1888-1935). Closed 1935. (U-5:4371c: B-4:459a; F-4:440a; G).
1-19 (1888-1895) as Verein für meiningische Geschichte und Landeskunde. Schriften.
Merged into Hennebergisch-fränkischer Geschichtsverein. Jahrbuch (not in this list).
Index: 1-90 (1888-1931) as 91 (1932) and 93 (1934).

1123 VERSAMMLUNG deutscher Meister und Freunde der Erdkunde in Frankfurt, Frankfurt am Main.
Amtlicher Bericht. 1 (1865). Closed 1865.

1124 VOLKSTUMSGEOGRAPHISCHE Forschungen. Leipzig. v1-2 (1939). Closed 1939. (U-5:4425a; D:385; G).
In connection with Atlas der deutschen Volkskunde.

1125 WEDDIGENS (P.F.) neues fortgesetztes westphälisches Magazin zur Geographie, Historie und Statistik. Wesel. 1-2 (1798-1799). Closed 1799. (G).
Successor to Neues westphälisches Magazin zur Geographie, Historie und Statistik [1078].

1126 WEITE Welt. Reisen und Forschungen in allen Teilen der Erde. Ein geographisches Jahrbuch. Berlin. 1-3 (1885-1887). Closed 1887. (U-5:4467a; B-4:519b; G).

1127 WELTPOLITISCHE Bücherei. Berlin. 1-29 (1928-1932). Closed 1932. (U-5:4471b; D:387).
Divided into two subseries: Grundlegende Reihe, and Länderkundliche Reihe.

1128 WESTMÄRKISCHE Abhandlungen zur Landes- und Volksforschung. (Pfälzische Gesellschaft zur Förderung der Wissenschaften). Neustadt an der Weinstrasse, Kaiserslautern. v1-5 (1937-1943). Closed 1943. (U-5:4498a; B-3:532b; D:106; G).
Supersedes the society's Veröffentlichungen [1086]. Title varies: v1 (1937) as Abhandlungen zur saarpfälzischen Landes- und Volksforschung; v2-3 (1938-1939) as Saarpfälzische Abhandlungen zur Landes- und Volksforschung.

1129 ______.
Beihefte. 1-5 (1938-1943). Closed 1943. (U-5:4498a; G).
Monographs. v1-3 (1938-1939) as Saarpfälzische Abhandlungen zur Landes- und Volksforschung. Beihefte.

1130 WESTPHAELISCHES Magazin zur Geographie, Historie und Statistik. Dessau. 1-4 (1784-1788). Closed 1788. (B-4:534b; F-4:969b; G).
Superseded by Neues westphälisches Magazin zur Geographie, Historie und Statistik [1078].

1131 WIRTSCHAFTSGEOGRAPHIE. Berlin. 1-11 (1932-1937). Closed 1937. (U-5:4515b; D:388).

1132 WIRTSCHAFTSGEOGRAPHISCHE Abhandlungen. Breslau. 1 (1933). Only one issued. Closed 1933. (U-5:4515b; D:389).

1133 WIRTSCHAFTSGEOGRAPHISCHE Arbeiten. (Königsberg. Handelshochschule. Wirtschaftsgeographisches Institut). Breslau. 1-4 (1936-1940). Closed 1940. (U-5:4515c; D:210).

1134 WOCHENSCHRIFT für Astronomie, Meteorolgie und Geographie. Leipzig; Halle. 1-45 (Ja 2 1847-D 1891). Closed 1891. (U-5:4529c; B-4:556a; F-4:980b; F-4:982a; F-4:847a; G).
1-4 (1847-1850) as Wöchentliche Unterhaltungen für Dilettanten und Freunde der Astronomie, Geographie und Witterungskunde; 5-7 (1851-1853) Unterhaltungen für Dilettanten und Freunde der Astronomie, Geographie und Meteorologie; 8-11 (1854-1857) Unterhaltungen im Gebiete der Astronomie, Geographie und Meteorologie. 12-45 (1858-1891) also as nsv1-34.

1135 WÖCHENTLICHE Unterhaltungen über die Erde und ihre Bewohner. Berlin. 1-5 (1784-1788). Closed 1788. (B-4:556a; F-4:982b). Succeeded by Wöchentliche Unterhaltungen über die Characteristik der Menschheit (not in this list).

1136 WÜRTTEMBERG-Hohenzollern in Zahlen; Zeitschrift für Statistik und Landeskunde. (Württemberg-Hohenzollern. Statistisches Landesamt). Tübingen. v1-7 no6 (1946-1952). Closed 1952. (U-5:4556c; D:98). v1 (1946) as Württemberg in Zahlen. v6-7 (1951-1952) with supplement, Beiträge zur Landeskunde [926].

1137 WÜRTTEMBERGISCHE Jahrbücher für Statistik und Landeskunde. (Württemberg. Statistisches Landesamt). Stuttgart. 1863/64-1951/52. Closed 1952. (B-4:574b; D:96; F-4:986b; G). Supersedes Württembergische Jahrbücher für vaterländische Geschichte, Geographie, Statistik und Topographie [1138]; superseded by Baden-Württemberg. Statistisches Landesamt. Jahrbücher für Statistik und Landeskunde von Baden-Württemberg [1168].

1138 WÜRTTEMBERGISCHE Jahrbücher für vaterländische Geschichte, Geographie, Statistik und Topographie. (Stuttgart. K. Statistisch-Topographisches Bureau [and] Verein für Vaterlandskunde, Stuttgart). Stuttgart. 1818/19-1862/63. Closed 1863. (B-4:574b; F-4:986b; D:96). 1818-1822 as Württembergisches Jahrbuch. Superseded by Württembergische Jahrbücher für Statistik und Landeskunde [1137].

1139 WÜRTTEMBERGISCHER Verein für Handelsgeographie und Förderung deutscher Interessen im Auslande, (e.v.), Stuttgart. Stuttgart. Jahresbericht. 1-50 (1882/83-1932). Closed 1932. (U-5:4557b; B-4:574b; F-3:43a; D:68; G). Superseded by Tribus [1327].

1140 WÜRZBURG. Universität. Institut für Amerikaforschung. Wurzburg. Studien über Amerika und Spanien. Geographische Reihe. no1-3 (1923-1927). Closed 1927. (U-5:4557c; F-4:610a; D:330; G). Monographs. Also in Spanish: Estudios sobre América y España. Serie geográfica.

1141 ZEITSCHRIFT für die allgemeine Geographie. Breslau. 1-4 (1824-1826). Closed 1826.

1142 ZEITSCHRIFT für Erdkunde. Frankfurt am Main. v1-12 no7/8 (1933-1944). Closed 1944. (U-5:4603b; B-2:270b; F-4:1017b; F-2:729a; D:24, D:29; G). v1-3 no1-48 (1933-1935) as Geographische Wochenschrift.

1143 ______. Beiheft. no1-9 (1933-1935). Closed 1935. (U-5:4603b; D:342; G).

1144 ZEITSCHRIFT für Erdkunde. Magdeburg. 1-10 (1842-1850). Closed 1850. (U-5:4603b; B-4:612a; F-4:1035b). 1-4 (1842-1845) as Zeitschrift für vergleichende Erdkunde.

1145 ZEITSCHRIFT für Geopolitik. Berlin. v1-21 no5/6 (1924-1944); Heidelberg and Darmstadt. v22-26 (1951-1955). (B-4:603a; F-4:1019a; D:85; G). Succeeded by Zeitschrift für Geopolitik in Gemeinschaft und Politik. (Institut für Geosoziologie und Politik. Bad Godesberg). Heidelberg, v27-31 (1956-1960); and in turn by Gemeinschaft und Politik und Zeitschrift für Geopolitik. Bellnhausen. v32-39 no5/6 (1961-1968). Closed 1968. (U-5:4604c).

1146 ______. Beihefte. v1-14 (1924-1942). Closed 1942. (U-5:4605a; B-4:603a; F-1:549b).

1147 ______.
Wehrwissenschaftliche Reihe. 1 (1935). (B-4:603a).
Monograph.

1148 ZEITSCHRIFT für Geschichte und Landeskunde der Provinz Posen. Posen.
1-3 (1882-1884). Closed 1884. (U-5:4605a; F-4:1019b; G).
Superseded by Historische Gesellschaft für die Provinz Posen.
Zeitschrift (not in this list).

1149 ZEITSCHRIFT für Gletscherkunde, für Eiszeitforschung und Geschichte des
Klimas. Annales de glaciologie...Annals of glaciology...
(International Gletscherkommission). Berlin. 1-28 (1906-1942).
(U-5:4605b; B-4:604b; F-4:1020a).

1150 ZEITSCHRIFT für preussische Geschichte und Landeskunde. Berlin.
1-20 (1864-1883). Closed 1883. (U-5:4616a; B-4:610a; F-4:1030b).

1151 ZEITSCHRIFT für Touristik, Geographie und Naturkunde. Frankfurt am Main.
1883/84.
Also as Deutsche Touristenzeitung.

1152 ZEITSCHRIFT für wissenschaftliche Geographie. Lahr, Weimar; Wien.
v1-8 (1880-1891). Closed 1891. (U-5:4622a; B-4:613b; F-4:1037b; G).

1153 ______.
Ergänzungshefte. no1-3 (1889-1890). Closed 1890. (B-4:613b;
F-4:1037b; G).

1154 ZENTRALINSTITUT für Erziehung und Unterricht. Berlin.
Geographische Abende. 1-10 (1919). Closed 1919. (U-5:4632b; D:359; G).

1155 ZOOGEOGRAPHICA. Internationales Archiv für vergleichende und kausale
Tiergeographie; International review for comparative and causal animal
geography; Archives internationales de géographie zoologique comparée
et causale. Jena. v1-4 no2 (1932-1942). Closed 1942. (U-5:4640a;
B-4:624a; F-4:1050a; D:83).

1156 ZUR Wirtschaftsgeographie des deutschen Ostens. Untersuchungen und
Darstellungen. (Breslau. Technische Hochschule. Geographisches
Institut). Berlin; Breslau. no1-19 (1932-1944). Closed 1944.
(U-5:4645c; D:392; G).
Monographs.

1157 ZUR Wirtschaftsgeographie des deutschen Westens; politische- und
wirtschaftsgeographische Untersuchungen and Darstellungen. (Aachen.
Technische Hochschule. Geographisches Institut). Berlin. 1-10
(1937-1943). Closed 1943. (U-5:4645c; B-4:628a; D:393; G).

FEDERAL REPUBLIC OF GERMANY (German)

including West Berlin

Bundesrepublik Deutschland (BRD)

For serials closed prior to 1946 see under Germany, p. 135-155

*(1158 AACHENER geographische Arbeiten. (Aachen. Technische Hochschule. Geographisches Institut). 1- (1969-) Irregular. (N70-1:25b; G).
In German; summaries in English and sometimes other languages.
[Geographisches Institut der Technischen Hochschule, Templergraben 55, D-51 Aachen, BRD.]

(1159 ACTA humboldtiana. Series geographica et ethnographica. (Deutsche Ibero-Amerika Stiftung). Wiesbaden. 1- (1959-) Irregular monographs. (N70-1:60b; D:336).
In German, Portuguese, or Spanish.
[Franz Steiner Verlag, D-6200 Wiesbaden, BRD.]

(1160 AFRIKA. Informationen. Forschungsberichte deutscher Geographen. (Köln. Universität. Geographisches Institut. Abteilung für Afrikaforschung). Köln. 1- (1978-) Irregular.

1161 AKADEMIE für Raumforschung und Landesplanung, Hannover. Jahresbericht. 1950/51- (N70-1:157c; B-170b; BS:34b).
Title varies: Arbeitsbericht.

(1162 ______. Forschungs- und Sitzungsberichte. Bremen. 1- (1950-) Irregular. (N70-1:157bc; B-1:70b; BS:34b).
[Hermann Schroedel Verlag KG, D-3 Hannover, BRD.]

1163 ______. Umschaudienst des Forschungsausschusses "Landschaftspflege und Landschaftsgestaltung." Hannover. v[1]-8 no3/4 (1951-1958). Closed 1958. (N70-1:157c; BS:34b).
v numbering begins with v4 n1-2 (1954).

(1164 ALEMANNISCHES Jahrbuch. (Alemannisches Institut). Freiburg im Breisgau. 1953- . (N70-1:211c; BS:40b; D:99; G).

*(1165 ARBEITEN zur rheinischen Landeskunde. (Bonn. Universität. Geographisches Institut). Bonn. 1- (1952-) Irregular. (N70-1:408c; BS:70b; D:182; G).
Supersedes Beiträge zur Landeskunde der Rheinlande [927].
Some English summaries.
[Ferdinand Dümmlers Verlag, Kaiserstrasse 33-37, 5300 Bonn, BRD.]

(1166 ARBEITSMITTEL für Geographie: Luftbild-Interpretation. Landschaftstypen und Landschaftsräume der Bundesrepublik Deutschland. (Institut für Film und Bild in Wissenschaft und Unterricht, FWU, München). Düsseldorf. 1976- .
[Hagemann Verlag, D-4000 Düsseldorf, BRD.]

1167 ARCHAEOLOGIA geographica; Beiträge zur vergleichenden geographisch-kartographischen Methode in der Urgeschichtsforschung. (Hamburg. Museum für Völkerkunde und Vorgeschichte. Vorgeschichtliche Abteilung). Hamburg. 1-10/11 (1950-1961/63). Closed 1963. (N70-1:414b; B-1:183a; D:84; G).
Publication suspended 1953-54.

1167a AUGSBURGER sozialgeographische Hefte (Augsburg. Universität. Lehrstuhl für Sozial- und Wirtschaftsgeographie). Augsburg. 1- (1977-) Irregular.
[Paul Kieser Verlag, Augsburg, BRD.]

1168 BADEN-WÜRTTEMBERG. Statistisches Landesamt. Stuttgart.
Jahrbücher für Statistik und Landeskunde von Baden-Württemberg.
v1- (1954/55-). (N70-1:609c; D:97).
Supersedes Württembergische Jahrbücher für Statistik und Landeskunde [1137].

(1168a BAMBERGER geographische Schriften. (Bamberg. Gesamthochschule. Fach Geographie). Bamberg. 1- (1978-).

1169 BEITRÄGE zur angewandten Geographie. (Verband deutscher Berufsgeographen). Wiesbaden. 1- (1964-) Annual. Processed.
v1 (1964) unnumbered and issued as Verband deutscher Berufsgeographen: Die Geographie in Wissenschaft, Wirtschaft, und Verwaltung.
Reprinted 1968. v2- (1966-).
[Editor: Martin Schneider, Bodelschwinghstr. 8, 62 Wiesbaden-Bierstadt, BRD.]

(1170 BEITRÄGE zur Länderkunde Afrikas. Sonderfolge der Kölner geographische Arbeiten. (Köln. Universität. Geographisches Institut). Wiesbaden. 1- (1961-) Irregular. (N70-1:671b; D:205).
English summaries.
[Franz Steiner Verlag, Postfach 55 29, D-6200 Wiesbaden, BRD.]

*(1171 BERICHTE zur deutschen Landeskunde. (Zentralausschuss für deutsche Landeskunde). Bad Godesberg, Meisenheim.
v1-4 (1941-1945); v5- (1948-). 2 nos a year. (U-1:641a; B-1:317b; BS:117a; D:10; D:94).
Suspended Mr 1944-1948. 1-4 (1941-1944) published by Reichsamt für Landesaufnahme. Abteilung für Landeskunde; v5-12 no1 by Amt für Landeskunde; v12 no2-v22 by Bundesanstalt für Landeskunde; v23-48 by Bundesanstalt für Landeskunde und Raumforschung. Institut für Landeskunde, Zentralarchiv für Landeskunde von Deutschalnd. Beiheft: See Bibliotheca cartographica [1179].
Index: v1-35 (1941-1965).
[Verlag Anton Hain, D-6544 Meisenheim, BRD.]

1172 ______.
Sonderheft. 1-14 (1942-1972) Irregular. (U-1:641a; N70-1:696c; D:250).

*(1173 BERLIN. Freie Universität. Geographisches Institut. Anthropogeographie. Berlin.
Abhandlungen. 1- (1953-) Irregular. (N70-1:700c; BS:332a; D:174; G).
1-6 (1953-1960) issued by Berlin. Freie Universität. Geographisches Institut. 7-16, 18-20, 22 (1963-1976) issued by Berlin. Freie Universität. 1. Geographisches Institut. 17, 21, 23- (1976-) as currently listed.
English summaries at beginning of some articles from v5 (1957).
[Dietrich Reimer Verlag, Drakestrasse 40, D-1 Berlin 45, BRD.; Geographisches Institut, Grunewaldstr. 35, D-1000 Berlin 41, BRD.]

1174 ______. ______. ______.
Veröffentlichung. 1- (1949-) Irregular.
Reprints.

(1175 ______. ______. Institut für Anthropogeographie, Angewandte Geographie und Kartographie. Arbeitsbereich Stadt- und Regionalforschung. Berlin.
Manuskripte. Empirische, theoretische und angewandte Regionalforschung.
1- (1978-) Irregular.
[Grunewaldstrasse 35, D-1000 Berlin 41, Germany.]

*(1176 BERLINER geographische Abhandlungen. (Berlin. Freie Universität. Institut für physische Geographie). Berlin. 1- (1964-). (N70-1:706b; B:68:73a; D:175; G).
1-14 (1964-1971) issued by 2. Geographisches Institut, 5, 8, and 16 also Arbeitsberichte aus der Forschungsstation Bardai, Tibesti, 1-3 (not in this list).
[Institut für physische Geographie, Grunewaldstr. 35, D-1000 Berlin 41, BRD.]

1176a BERLINER geographische Studien. (Berlin. Technische Universität. Institut für Geographie). Berlin. 1- (1977-). (N78; G).

1177 BEWOHNTE Erde: Einführungen in die Kulturgeographie. Hannover. 1-4 (1953-1958). Monographs. (N70-1:1716b; D:343; G).

*(1178 BIBLIOGRAPHIA cartographica. Internationale Dokumentation des kartographischen Schrifttums. International documentation of cartographical literature. (Staatsbibliothek Preussischer Kulturbesitz [and] Deutsche Gesellschaft für Kartographie, e. V.) München. 1- (1974- [1975]) Annual. (N77; BL; GZS).
Supersedes Bibliotheca cartographica [1179].
[Verlag Dokumentation Saur KG, Pössenbacherstr. 2, POB 711009, D-8000 München 71, BRD.]

1179 BIBLIOTHECA cartographica. Bibliographie des kartographischen Schrifttums; Bibliography of cartographic literature; Bibliographie de la littérature cartographique. (Institut für Landeskunde [and] Deutsche Gesellschaft für Kartographie]. Bonn-Bad Godesberg. 1/2-29/30 (1957-1972) 2 nos a year. Closed 1972. (N70-1:760c; BS:125a; D:13; G).
1-23 also as Berichte zur deutschen Landeskunde. Beiheft.
Superseded by Bibliographia cartographica [1178].

1180 ______. Sonderhefte. 1-2 (1962), 3 (1969). (N70-1:760c; D:13; G).

1181 BLÄTTER für den Erdkundelehrer. München. Mitteilungens- und Aussprachenblatt der Fachschaft Erdkunde des bayerischen Philologenverbandes und des Landesverbandes Bayern im Verband deutscher Schulgeographen. v1-2 no1 (1957-1958). (D:72).

*(1182 BOCHUMER geographische Arbeiten. (Bochum. Ruhr-Universität. Geographisches Institut). Paderborn. 1- (1965-) Irregular. (N70-1:799c; B68:83a; G).
Abstracts in English.
[Ferdinand Schöningh, Jühenplatz 1-3, Postfach 2540, D-4790 Paderborn, BRD.]

(1183 ______. Sonderreihe. Paderborn. 1- (1973-) Irregular. (N76; BL; G; GZS).
[Ferdinand Schöningh Verlag. Jühenplatz 1-3, Postfach 25 40, D-4790 Paderborn, BRD.]

1184 BONN. Universität. Geographisches Institut. Bonn.
Arbeiten. sA. 1- (1949-) Irregular reprints.

1185 ______. ______. ______.
______. sB. 1- (1950-).
Reprints.

*(1186 BONNER geographische Abhandlungen. (Bonn. Universität. Geographisches Institut). Bonn. 1- (1947-) Irregular. (U-1:737c; B-1:304a; D:180; G).
Some English summaries from no24 (1958).
[Ferd. Dümmlers Verlag, Kaiserstrasse 33-37, D-5300 Bonn, BRD.]

(1187 BRAUNSCHWEIGER geographische Studien. (Braunschweig [Brunswick], Technische Hochschule. Geographisches Institut). Braunschweig. 1- (1964-). (N70-1:844b; G).
[Geographisches Institut der Technischen Hochschule, Pockelstrasse 14, D-33 Braunschweig, BRD.]

1188 ______. Sonderhefte. 1- (1976-) Irregular. (N79).

1189 BREMEN. Übersee-Museum [and] Geographische Gesellschaft in Bremen.
Veröffentlichungen (ns).
sA. Naturwissenschaften. nol- (1949-). (U-1:767a; B-1:392b).
sB. Völkerkunde. 1- (1950-) Irregular. (N70-1:857c; B-1:392b).
Formerly as Bremen. Museum für Natur-, Völker- und Handelskunde.

1190 ______. ______.
sC. Geographie. 1- (1977-) Irregular.
v1 also as Deutsche geographische Blätter, v51 (1977) [1194].
[Übersee-Museum, Bahnhofsplatz 13, 2800 Bremen, BRD.]

*(1191 CATENA: Interdisciplinary journal of pedology, hydrology, geomorphology.
Cremlingen. 1- (1973-) 4 nos a year. (N75-1:481a; G; GZS).
In English, German, or French. Abstracts in English and in French
or German. Index: v1-5 (1973-1978).
[Catena-Verlag, Margot Rohdenburg, Brockenblick 8, D-3302, Cremlingen-Destedt, BRD.]

(1192 COLLOQUIUM geographicum; Vorträge des Bonner geographischen Kolloquiums
zum Gedächtnis an Ferdinand von Richthofen. (Bonn. Universität.
Geographisches Institut). Bonn. 1- (1951-) Irregular. (N70-1:385a;
BS:215b; D:181; G). Some English abstracts.
[Ferd. Dummlers Verlag, D-5300 Bonn 1, BRD.]

(1193 DEUTSCHE Forschungsgemeinschaft. Kommission für geowissenschaftliche
Gemeinschaftsforschung. Bonn-Bad Godesberg.
Mitteilung. 1- (1973-). (G; GZS).

(1194 DEUTSCHE geographische Blätter. (Geographische Gesellschaft in Bremen).
Bremen. v1-44 (1877-1941); 45-50 (1949-1965/69); nsv1- (1977-)
Irregular. (U-2:1300c; B-2:29b; F-2:358b; D:37; G).
Supersedes Verein für deutsche Nordpolarfahrt. Berichte über die
Sitzungen [986].
Index: 1-30 (1877-1907).

*(1195 DEUTSCHER Geographentag. Wiesbaden.
Tagungsbericht und wissenschaftliche Abhandlungen. 1-26 (1881-1936);
27- (1948-) Biennial. (U-2:1323b; B-2:41b; F-4:877a; D:113-D:147; G).
1-26 (1881-1936) as Verhandlungen. Suspended 1915-1920, 1937-1947.
Index: 1-34 (1881-1963).
[Franz Steiner Verlag, Postfach 5529, D-6200 Wiesbaden, BRD.]

1196 ______. Festschrift zum...Deutschen Geographentages. 13-24 (1901-1931);
30- (1955-). (D:148-D:164).
1-2, 25-29 not published. Special volumes devoted to region or topic.

1197 DOCUMENTATIO geographica: geographische Zeitschriften- und Serien-Literatur.
Papers of geographical periodicals and serials. (Bad Godesberg.
Institut für Landeskunde). Bad Godesberg. 1966-1973. Closed 1973.
6 issues a year cumulated in annual volumes 1966-1970; biennial cumulation for 1971-1972; 4 issues for 1973 (entries not cumulated but
with index volume). (N70-2:1767b; B68:175; G).
1973 carries subtitle: Viertelsjahrshefte zur Literatur-Dokumentation
aus Landeskunde, Raumordnung, Regionalforschung und Sozialgeographie.
Superseded by Dokumentation zur Raumentwicklung [1198].

1198 DOKUMENTATION zur Raumentwicklung: Vierteljahreshefte zur Literaturdokumentation aus Raumforschung, Raumordnung, Regionalforschung,
Landeskunde und Sozialgeographie. A current and annotated bibliography of regional science, regional planning and social geography.
(Bundesforschungsanstalt für Landeskunde und Raumordnung). Bonn.
1974/75-1978 Quarterly plus annual index. Closed 1978. (N76; GZS).
Supersedes Documentatio geographica [1197].

*(1199 DÜSSELDORFER geographische Schriften. (Düsseldorf. Universität. Geographisches Institut). 1- (1974-). (N77; GZS).
Abstracts in English.
[Geographisches Institut der Universität Düsseldorf. Universitätsstrasse 1, D-4 Düsseldorf, BRD.]

(1200 DUISBURGER geographische Arbeiten. (Duisburg. Gesamthochschule. Seminar für Geographie). Duisburg. 1- (1978-) Irregular.

*(1201 ERDE. (Gesellschaft für Erdkunde zu Berlin. Zeitschrift). Berlin. 1-8 [80-87] (1949-1956); 88- (1957-) Quarterly. (U-2:1469b; B-4:598b; D:30; G).
Supersedes Gesellschaft für Erdkunde zu Berlin. Zeitschrift [1011] and continues its volume numbering from v88 (1957).
English abstracts at beginning of each article from 1957 no1.
[Arno-Holz-Strasse 14, D-1000 Berlin 41, BRD.]

*(1202 ERDKUNDE; Archiv für wissenschaftliche Geographie. Bonn. 1- (1947-) Quarterly. (U-2:1469c; B-2:134b; BS:292b; D:21).
Index: v1-17 (1947-1963).
English summaries at beginning of each major article from v7 (1953).
Some articles in English from v8 (1954).
[Ferd. Dümmlers Verlag, Kaiserstrasse 33-37, Postfach 1480, D-5300 Bonn, BRD.]

1203 ERDKUNDE in der Schule. Hagen. 1-5 no4 (1956-1960). (N70-2:1940c; D:73; G).

(1204 ERDKUNDELEHRER in Rheinland-Pfalz. Mitteilungen des Landesverbandes Rheinland-Pfalz im Verband Deutscher Schulgeographen. Stuttgart. 1- (1962-). (D:74).

(1205 ERDKUNDEUNTERRICHT: Beiträge zu seiner wissenschaftlichen und methodischen Gestalung. Stuttgart. 1- (1956-). (G).
[Ernst Klett Verlag, Rotebühlstr. 77, Postfach 809, D-7000 Stuttgart 1, BRD.]

1206 ERDKUNDLICHE Bilder. Berlin.
Ausser-Europa. 1-3 (1951-1952). (N70-2:1940c).

1207 ______.
Deutschland. 1-6 (1951-1952). (N70-2:1940c).

1208 ______.
Europa. 1-6 (1950-1951). (N70-2:1940c).

1209 ERDKUNDLICHES Unterrichtswerk für höhere Lehranstalten. München. 1- (1950). (N70-2:1940c).

*(1210 ERDKUNDLICHES Wissen; Beihefte zur geographischen Zeitschrift. Wiesbaden. 1- (1952-) Irregular. (N70-2:1940c; BS:292b; D:356).
1-13 separate, not as Beihefte zur geographischen Zeitschrift.
Early nos carried subtitle: Schriftenfolge für Forschung und Praxis.
Some English abstracts.
[Franz Steiner Verlag, Postfach 5529, D-6200 Wiesbaden, BRD.]

(1211 ERDWISSENSCHAFTLICHE Forschung. (Mainz. Akademie der Wissenschaften und der Literatur. Kommission für erdwissenschaftliche Forschung). Wiesbaden. 1- (1968-) Irregular. (G).
[Franz Steiner Verlag, D-6200 Wiesbaden, BRD.]

(1212 ERLANGER geographische Arbeiten. (Fränkische geographische Gesellschaft). Erlangen. 1- (1954-) Irregular. (D:41; GZS).
Reprints from the society's Mitteilungen [1217].
[Kochstrasse 4, D-8520 Erlangen, BRD.]

(1213 ______. Sonderband. 1- (1974-) Irregular. (G; GZS).

1214 FERNE Länder. Zeitschrift für Geographie und Gegenwartskunde. Hamburg. v1-2 (1955-1956). Closed 1956. (N70-2:2066a; D:76).

*(1215 FORSCHUNGEN zur deutschen Landeskunde. (Zentralausschuss für deutsche Landeskunde). Stuttgart, Leipzig, Bad Godesberg, Trier. (v1-45 (1885-1944); v46- (1949-)Several monographs each year. (U-2:1608a; B-2:211b; F-2:628b; D:165-D:166; G).
1-32 (1885-1939) as Forschungen zur deutschen Landes-und Volkskunde. v46-206 (1949-1975) published by or in cooperation successively with Amt für Landeskunde, Bundesanstalt für Landeskunde, Bundesanstalt für Landeskunde und Raumforschung, and Bundesforschungsanstalt für Landeskunde und Raumordnung.
English summaries from v207 (1976).
[Zentralausschuss für Deutschen Landeskunde, Universität Trier, Postfach 3825, 5500 Trier, BRD.]

(1216 FORSCHUNGEN zur Raumentwicklung. (Bundesforschungsanstalt für Landeskunde und Raumordnung). Bonn-Bad Godesberg. 1- (1975-) Irregular monographs. (GZS).
[Postfach 130, D-5300 Bonn-Bad Godesberg, BRD.]

*(1217 FRÄNKISCHE geographische Gesellschaft. Erlangen.
Mitteilungen. 1- (1954-) Annual. (N70-2:2175c; BS:330b; D:41; G).
[Geographisches Institut, Kochstrasse 4, D-8520 Erlangen, BRD.]

(1218 FRANKFURT-AM-MAIN. Universität. Geographisches Institut. Kulturgeographie. Frankfurt-am-Main.
Kulturgeographie Materialien. 1- (1974-). (GZS).

(1219 FRANKFURTER Beiträge zur Didaktik der Geographie. Frankfurt-am-Main. 1- (1977-).
[Institut für Didaktik der Geographie der Universität Frankfurt, Schumannstrasse 58, D-6000 Frankfurt am Main, BRD.]

1220 ______.
Veröffentlichungen. 1- (1971-).

*(1221 FRANKFURTER geographische Hefte. (Frankfurter geographische Gesellschaft). Frankfurt-am-Main. 1-16 (1927-1942); 17/22- (1948-). (U-2:1628a; B-2:270b; F-2:657b; D:234; G).
1-36 (1927-1961) issued under earlier name of the society: Verein für Geograohie und Statistik, and carried alternate numbering Jahrgang 1-35, variable number of Hefte to each volume. Succeeds Verein für Geographie und Statistik, eingetragener Verein, Frankfurt. Jahresbericht [1116].
Some English summaries from v44 (1967).
[Verlag Waldemar Kramer, Bornheimer Landwehr 57a, D-6 Frankfurt-am-Main, BRD.]

*(1222 FRANKFURTER wirtschafts- und sozialgeographische Schriften. (Frankfurt-am-Main. Universität. Institut für Wirtschafts- und Sozialgeographie). Frankfurt-am-Main. 1- (1967-). (N70-2:2207c; G).
Issuing agency earlier called Seminar für Wirtschaftsgeographie.
[Bockenheimer Landstrasse 140, D-6000 Frankfurt am Main, BRD.]

1223 FREIBURG im Breisgau. Universität. Geographische Fachschaft. Freiburg.
Berichte. 1-4 (1958-1959). Closed 1959. (D:235).

1224 ______. ______. ______.
Mitteilungen. 1-13/14 (1927/28-1932/33); 1-3 (My-N 1953); nsv1-3 no2 (1959-1960). (U-2:1637c; D:43; D:44; G). no5 (1929) as the Fachschaft's Wissenschaftliche Veröffentlichungen no1 (1929), the only number ever issued in that series (not in this list).
Superseded by Freiburger geographische Mitteilungen [1227].

1225 FREIBURGER geographische Arbeiten. (Freiburg. Universität. Geographisches Institut). Freiburg im Breisgau. 1- (1961-) Irregular. (N70-2:2214c; B68:219a; D:190).
[Hans Ferdinand Schulz Verlag, Freiburg im Bresigau, BRD.]

*(1226 FREIBURGER geographische Hefte. (Freiburg. Universität. Geographisches Institut). Freiburg im Breisgau. 1- (1963-) Irregular. (N70-2:2214c; B68:219b; D:191; G).
In German, occasionally in English or with English summaries.
[Geographisches Institut I, Werderring 4, D-7800, Freiburg im Breisgau, BRD.]

(1227 FREIBURGER geographische Mitteilungen. (Freiburg. Universität. Geographische Fachschaft). Freiburg im Breisgau. 1969- . 2 nos a year. (N75-1:884a; N75-1:906c; G; GZS).
1969-1971 as Universität Freiburg. Geographische Fachschaft, or Geographische Fachschaft Freiburg. Mitteilunge. (G).
Supersedes Freiburg im Breisgau: Universität. Geographische Fachschaft. Mitteilungen [1224].

1227a GEO-BÜCHERBRIEF: Neuerscheinungen aus allen Bereichen der Geo-Wissenschaften. (Geo-Center. Internationales Landeskartenhaus). Stuttgart. 1- [1976-]. (G).

1228 GEO Center. Stuttgart.
Kartenbrief. no250- (1971-). Closed 1976. (N77).
Continues RV Kartenbrief, issued by Reise- und Verkehrsverlag, Stuttgart (not in this list).
Superseded by Geo-Kartenbrief [1251].

(1229 GEOCOLLEG. Kiel. 1- (1975-) Irregular monographs.
[Verlag Ferdinand Hirt, D-2300 Kiel 1, BRD.]

1230 GEO GRAFIKER. (Berliner Geographenkreis, studentische Vereinigung an der Freien Universität). Berlin. 1-7/8 (1968-1972). (N70-2:2275a; G).
[Verlag Kiepert, 1 Berlin 12, BRD.]

1230a GEOGRAPHICA historica. (Habelt). Bonn. 1- (1976-). (G, GZS).

(1231 GEOGRAPHIE im Unterricht. Zeitschrift für die Unterrichtspraxis der Sekundärstufe 1. Köln. 1- (1976-) 10 issues a year. (G; GZS).
[Aulis Verlag Deubner und Co. KG, Antwerpener Strasse 6-12, D-5000 Köln 1, BRD.]

1231a GEOGRAPHIE in Ausbildung und Planung. Bochum. 1- (1973-). (G, GZS).

(1232 GEOGRAPHIE in der Schule. (Verband deutscher Schulgeographen. Landesverbande Hamburg, Bremen, Niedersachsen und Schleswig-Holstein. Mitteilungsblatt). Kiel. 1968- 2-3 a year.
[Hirtverlag, Kiel, BRD.]

1232a GEOGRAPHIE und ihre Didaktik (GUID). (Fachbereich 2 der Universität Osnabrück; Fachbereich 2 der Universität Paderborn). Paderborn. 1-5 (1973-1977). (G, GZS).

*(1233 GEOGRAPHISCHE Gesellschaft in Hamburg. Hamburg.
Mitteilungen. 1- (1876-) Irregular. (U-2:1696a; B-2:270a; F-3:567b; D:51; G).
Succeeds society's Jahresbericht, 1873-1875 [988].
Index: 1-60 (1873-1973).
Occasional English summaries.
[Institut für Geographie und Wirtschaftsgeographie, Universität Hamburg, Bundesstrasse 55, D-2000 Hamburg 13, BRD.]

(1234 GEOGRAPHISCHE Gesellschaft in Lübek. Lübeck.
Mitteilungen. 1-12 (1882-1889); s2 v1- (1890-) Irregular. (U-2:1696b; B-2:270a; F-3:567b; D:60; G).
Suspended 1941-1946. s2 v41-43 (1947-1951) as society's Forschungen. 1890-1957 as Geographische Gesellschaft und naturhistorisches Museum in Lübeck.
[Kuckucksruf 23, D-2400 Lübeck, BRD.]

1235 GEOGRAPHISCHE Gesellschaft in München. München.
Landeskundliche Forschungen. 1-44 (1906-1972). (U-2:1696b; F-3:283b; D:242).

*(1236 ______.
Mitteilungen. 1- (1904-) Annual. (U-2:1696b; B-2:270a; F-3:567b; D:62; G).
Continues society's Jahresberichte [989].
Index: 1-40 (1904-1953) [1960].
[Heinrich-Vogl-Strasse 7, D-8 München 71, BRD.]

x(1237 GEOGRAPHISCHE Gesellschaft zu Hannover. Hannover.
Jahrbuch. 1924- Annual. (U-2:1696b; B-2:270a; F-3:6b; D:52; G).
Suspended 1942-1952. 1963-1965 issued in association with Hannover. Technische Hochschule. Geographisches Institut.
[Schneiderberg 50, 3000 Hannover 1, BRD.]

(1238 ______.
Sonderheft. 1- (1967-). (N70-2:2282a; BL; G).

1239 ______.
Sonderveröffentlichungen. 1-3 (1935-1949) Irregular. (U-2:1696b; B-2:270a; D:240).

1240 GEOGRAPHISCHE Hochschulmanuskripte. (Göttingen. Redaktionkollektiv Göttinger Geographen). Göttingen. 1- (1973-). (GZS).
[Redaktion der GHV, Rastenburger Weg 11, 34 Göttingen, BRD.]

1241 GEOGRAPHISCHE Luftbildinterpretation; Geographical interpretation of aerial photographs; Interpretation géographique des photographies aeriennes. (Institut für Landeskunde). Bad Godesberg. 1-2 (1967); 4 (1975). (N70-2:2281b; G).
Edited in cooperation with the Commission on interpretation of aerial photographs. International geographical union. Issued as Landeskundliche Luftbildinterpretation im mitteleuropäischen Raum. Sonderfolge, 1-2 [1282].
In English or in German with English and French summaries.

*(1242 GEOGRAPHISCHE Rundschau. Braunschweig. 1- (Ja 1949-) Monthly. (U-2:1696c; B-2:270b; D:78; G; GZS).
v1-20 (1949-1968) carry subtitle Zeitschrift für Schulgeographie.
Indexes: 1-21 (1949-1969), 21-27 (1969-1975).
Supplementary English table of contents and abstracts.
[Georg Westermann, D-3300 Braunschweig, BRD.]

1243 ______.
Beiheft. 1971-1979. Closed 1979. Bimonthly. (GZS).
Superseded by Praxis Geographie [1307].

1244 GEOGRAPHISCHE Zeitfragen. Frankfurt-am-Main. 1- (1960-). (N70-2:2282a, b).
1-14 (1960-1966) as Geographische Zeitfragen in Einzelheften. 1960-1964 published in Karlsruhe.
[Hirschgraben-Verlag, Grüneburgweg 118, D6 Frankfurt-am-Main, BRD.]

*(1245 GEOGRAPHISCHE Zeitschrift. Leipzig; Wiesbaden. v1-50 no3 (1895-1944); 51- (1963-) Quarterly. (U-2:1696c; B-2:270b; B68:225b; F-2:729a; D:25). Not published 1945-1962. 1895-1944 published in Leipzig; 1963- published in Wiesbaden.
For Beiheft: Erdkundliches Wissen, see Erdkundliches Wissen [1210].
Indexes: 1-10 (1895-1904), 11-20 (1905-1914), 21-30 (1915-1924), 1-50 (1895-1944).
Supplementary titles in table of contents and summaries in English.
Some articles in English from v51 (1963).
[Franz Steiner Verlag, Friedrichstrasse 24, Postfach 5529, D-6200 Wiesbaden, BRD.]

(1246 (Das) GEOGRAPHISCHE Seminar. Braunschweig. 1957- Monographs. (D:366).
[Georg Westermann Verlag, Braunschweig, BRD].

1247 ______.
Praktische Arbeitsweisen.

*(1248 GEOGRAPHISCHES Taschenbuch und Jahrweiser für Landeskunde.
(E. Meynen, Zentralverband der deutschen Geographen; 1949-1970/72 also by Institut für Landeskunde, Bad Godesberg; also 1977/78- IGU Komittee Österreichs, Schweizerische geographische Gesellschaft).
Wiesbaden. 1949- Biennial. (B-2:270b; BS:342a; D:4).
Not published 1973-1974.
[Franz Steiner Verlag, Postfach 5529, D-6200 Wiesbaden, BRD.]

1249 ______.
Supplementband. 1960/61. Only one issued. (N70-2:2282b).

*(1250 GEOJOURNAL: international journal for physical, biological and human geosciences and their application in environmental planning and ecology.
Wiesbaden. 1- (1977-) 6 nos a year.
In English.
[Akademische Verlagsgesellschaft, Postfach 1107, D-6200 Wiesbaden, BRD.]

(1251 GEO-KARTENBRIEF. (GeoCenter, Internationales Landkartenhaus). Stuttgart.
1- (1976-) 3-4 a year.
Supersedes Geo Center. Kartenbrief [1228].

(1252 GEO-KATALOG International. Landkarten, Pläne, Atlanten, Reiseführer, Globen aus aller Welt. (Geo Center, Internationales Landkartenhaus).
München. 1972- (N75-1:903c; N77; G; GZS).
From 1973 divided into 2 series:
Band 1. Landkarten, Reiseführer, Pläne, Atlanten, Globen aus aller Welt. 1973- .
Band 2. Amtliche, geographisch-thematische Karten und Atlanten. 1976- . Looseleaf.
Supersedes Zumstein-Katalog, editions 1-8 (1964-1971) (not in this list).
[Liebherrstrasse 5, 8 München 22, BRD.]

*(1253 GEOLIT. Rezensionen: Geographie, Raumwissenschaften.
(Westermann). Braunschweig. 1977- 4 a year. (N79; G; GZS).
2 nos in 1977.
[Westermann Verlag, Postfach 3320, D-3300 Braunschweig, BRD.]

(1254 GEOSPECTURM. Berlin. 1978- Irregular monographs.
[W. de Gruyter, D-1000 Berlin, BRD.]

1254a GIESSEN. Universität. Geographisches Institut. Giessen.
Werkstattpapiere. 1- (1975-). (G; GZS).

*(1255 GIESSENER geographische Schriften. (Giessen. Justus-Liebig Universität. Geographisches Institut). 1- (1957-). (N70-2:2351a; D:192; G). Some English summaries from v2 (1962).

1256 GÖTTINGEN. Universität. Niedersachsisches Institut für Landeskunde und Landesentwicklung. Göttingen.
Veröffentlichungen.
sA. Forschungen für Landes- und Volkskunde. 1. Natur, Wirtschaft, Siedlung, und Planung. (Wirtschaftswissenschaftliche Gesellschaft zum Studium Niedersachsens. Schriften). 1-26 (1941-1945); 27- (1948-). (D:279; G).
sB. Landes- , Kreis- und Ortsbeschreibungen. 1-7 (1940-1957). (D:280).
sC. see Archiv für Landes- und Volkskunde von Niedersachsen [910].
Former names: Provianzialinstitut für Landesplanung, Landes- und Volkskunde von Niedersachsen, Hannover-Göttingen, 1940-1942; Provinzialinstitut für Landesplanung und niedersächsische Landes- und Volksforschung, Hannover-Göttingen, 1942-1946; Niedersächsisches Amt für Landesplanung und Statistik, Hannover, with special section Institut für Landesplanung und niedersächsische Landeskunde an der Universität Göttingen, 1946-1961; Niedersächsisches Institut für Landeskunde an der Universität Göttingen, 1961-1962.
[August Lax, Weinberg 56, D-32 Hildesheim, BRD.]

*(1257 GÖTTINGER geographische Abhandlungen. (Göttingen. Universitat. Geographisches Institut). Göttingen. 1- (1948-) Irregular, monographs. (U-2:1740c; B-2:316a; D:194; G).

1258 HAMBURG. Universität. Wirtschaftsgeographisches Institut. Hamburg. Veröffentlichungen. 1 (1949). Only 1 issued. Closed 1949. (U-3:1790a; D:197).

*(1259 HAMBURGER geographische Studien. (Hamburg. Universität. Institut für Geographie und Wirtschaftsgeographie). Hamburg. 1- (1952-) Irregular. (N70-2:2470a; D:198; G).
[Bundesstrasse 55, D-2000 Hamburg 13, BRD.]

1260 HARMS Handbuch der Erdkunde in entwickelnder anschaulicher Darstellung. Frankfurt-am-Main. 1- (1953-). (N70-2:2483c).

1261 HARMS Landeskunde. (Paul List Verlag). München. 1- (1961-). (N70-2:2483c).

(1262 HEFTE zur Fachdidaktik der Geographie: HFG. Kastellaun. 1- (1977-) 4 nos a year. (G).
[Aloys Henn Verlag, Postfach 1180, D-5458 Kastellaun, BRD.]

1263 HEIDELBERG. Universität. Geographisches Institut. Heidelberg. Veröffentlichungen. 1- (1963-). (B68:579a; D:200a).
[F. Steiner Verlag, Wiesbaden, BRD.]

*(1264 HEIDELBERGER geographische Arbeiten. (Heidelberg. Universität. Geographisches Institut). Heidelberg. 1- (1956-) Irregular monographs. (N70-2:2514b; BS:381a; D:200; G).
English summaries.

1265 INSTITUT für Landeskunde, Bad Godesberg. Bad Godesberg.
Rundbrief. 1-25 (1948-1972). Closed 1972. (G). 1-6 (F 1948-F/Mr 1953) as Germany (BRD). Amt für Landeskunde (1 with title Neuigkeiten); 6-8 (1953-1955) as Zentralverband der Deutschen Geographen. Rundbrief des Zentralverbandes der Deutschen Geographen und der Bundesanstalt für Landeskunde; 9-12 (1956-1959) as Bundesanstalt für Landeskunde, Remagen. Rundbrief der Bundesanstalt für Landeskunde, herausgegeben in Gemeinschaft mit dem Zentralverband der Deutschen Geographen; 12-25 (1959-Ap/Jn 1972) as Institut für Landeskunde, Bad Godesberg. Rundbrief des

Instituts für Landeskunde in der Bundesforschungsanstalt (1959-1967 as Bundesanstalt) für Landeskunde und Raumordnung. Herausgegeben in Gemeinschaft mit dem Zentralverband der Deutschen Geographen. Superseded by Marburg. Universität. Fachbereich Geographie. Rundbrief [1286].

(1266 INSTITUT für Landeskunde des Saarlandes. Saarbrucken. Veröffentlichungen. 1- (1960-) 2 nos a year. (N70-2:2806b). [Institut für Landeskunde des Saarlandes, Am Ludwigsplatz 7, D-66 Saarbrücken, BRD.]

1267 INSTITUT für Raumordnung (earlier Raumforschung), Bad Godesberg. Bibliographischer Index. Bonn. 1951-53? (N70-2:2807a; B-2:492b). Title varies. 1951 as the institute's Bibliographischer Index und Literaturbericht. After 1953 reviews of current and important literature are contained in the institute's Informationen [1268] and its Raumforschung und Raumordnung [1311].

1268 ______. Informationen. 1-23 (1950-1973). Closed 1973. (N70-2:2807a; B-2:492b; D:93; G). 1-17 (1950/51-1967) issued by Institut für Raumforschung. Superseded by Informationen zur Raumentwicklung, not in this list.

1269 ______. Mitteilungen. 1- (1950-) Irregular. (N70-2:2807a; B-2:492b; BS:420a).

1270 ______. Vorträge. no1- (1950-) Irregular. (N70-2:2807b; B-2:492b; BS:420a).

*(1271 INTERNATIONAL yearbook of cartography; Internationales Jahrbuch für Kartographie. Annuaire international de cartographie (International cartographic association). Bonn-Bad Godesberg. 1- (1961-) Annual. (N70-2:2978ab; B68:276a; G; GZS). 1-13 (1961-1973) as Internationales Jahrbuch für Kartographie... (Bertelsmann), Gütersloh. 15- (1975-) in cooperation with the International cartographic association. In English, French, or German, with summaries in other two languages. [Kirschbaum Verlag, Bonn-Bad Godesberg, BRD.]

1272 INTERNATIONALE technogeographische Gesellschaft. Berlin. Mitteilungen. (N70-2:2975b). [Verlag H. Wiganskow, Chausseestrasse 72, Berlin N65, BRD.]

(1273 INTERNATIONALES Jahrbuch für Geschichts- und Geographie-Unterricht. (Georg-Eckert-Institut für internationale Schulbuchforschung mit Unterstützung des Conseil de la coopération culturelle des Europarats). Braunschweig. 1- (1951-) Annual. (N70-2:2978a; BS:444b; BL; G). Title varies: 1-9 (1951-63/64) as Internationales Jahrbuch für Geschichts-Unterricht. Sponsoring agency Arbeits-Gemeinschaft deutscher Lehrerverbände, 1-13 (1951-1970/71).

(1274 JAHRBUCH für fränklische Landesforschung. (Erlangen. Universität. Zentral-institut für fränkische Landeskunde und allgemeine Regionalforschung). Erlangen. v1- (1935-). (U-3:2141b; B-2:576b; BS:457a; F-3:17b; D:100; G). Index: 1-17 (1935-1957). Issuing institute formerly called Institut für fränkische Landes-forschung.

*(1275 KARLSRUHER geographische Hefte. (Karlsruhe. Universität. Geographisches Institut). Karlsruhe. 1- (1968-) Irregular. (G).

1276 KARTENSAMMLUNG und Kartendokumentation. (Institut für Landeskunde). Bad Godesberg. 1-10 (1966-1975) Irregular. (N70-2:3257a; G).

*(1277 KARTOGRAPHISCHE Nachrichten. (Deutsche Gesellschaft für Kartographie; Schweizerische Gesellschaft für Kartographie; Österreichische kartographische Kommission in der Österreichischen geographischen Gesellschaft). Bonn.
1951- Bimonthly. (N70-2:3257b; B-2:640a; BS:478b; D:90; G).
Index: 1951-1955 (1957).
[Kirschbaum Verlag, Siegfriedstrasse 28, Postfach 9109, D-5300 Bonn-Bad Godesberg 10, BRD.]

1278 KARTOGRAPHISCHE Schriftenreihe. Lahr/Schwarzwald. v. 1-6 (1954-58). (N70-2:3257b; BS:478b).

*(1279 KIELER geographische Schriften. (Kiel. Universität. Geographisches Institut). Kiel. v1-12 (1932-1942); 13- (1950-). (U-3:2288a; B-1:10b; D:203; G).
v1-43 (1932-1975) as Kiel. Universität. Geographisches Institut. Schriften.
v16 no2 (1956) in English.

*(1280 KÖLNER Forschungen zur Wirtschafts- und Sozialgeographie. (Köln. Universität. Wirtschafts- und Sozialgeographisches Institut. Veröffentlichungen). Wiesbaden, Köln.
1- (1963-) Irregular monographs. (N70-2:3323a; D:206).
[Wirtschafts- und Sozialgeographisches Institut der Universität zu Köln, Albertus-Magnus Platz, 5000 Köln 41, BRD.]

*(1281 KÖLNER geographische Arbeiten. (Köln. Universität. Geographisches Institut). Köln. 1- (1952-) Irregular. (N70-2:3323a; BS:487a; D:204; G). For Sonderfolge see [1170].
English summaries from no6 (1955).
[Franz Steiner Verlag, Friedrichstr. 24, Postfach 55 29, D-6200 Wiesbaden, BRD.]

(1282 LANDESKUNDLICHE Luftbild [earlier -auswertung] im mitteleuropäischen Raum. (Bundesforschungsanstalt für Landeskunde und Raumordnung). Bonn-Bad Godesberg. 1- (1952-) Irregular. (N70-3:3406a; BS:498b; D:249; G).
v1 issued by Amt für Landeskunde; v2, Bundesanstalt für Landeskunde; 3-9 by Institut für Landeskunde, Bundesanstalt für Landeskunde und Raumforschung.

(1283 LANDSCHAFTSGENESE und Landschaftsökologie. (Braunschweig. Technische Universität. Lehrstuhl physische Geographie und Landschaftsökologie). Cremlingen-Destedt. 1- (1978-) Irregular monographs.
[Catena-Verlag, Margot Rohdenburg, Brockenblick 8, D-3302 Cremlingen-Destedt, BRD.]

1284 LEHRBUCH der allgemeinen Geographie. (Walter de Gruyter und Co.). Berlin. 1- (1960-). (N70-3:3437a).

*(1285 MAINZER geographische Studien. (Mainz. Universität. Geographisches Institut). Mainz. 1 (1961), 2- (1969-).
[Geographisches Institut der Johannes-Gutenberg-Universität, Saarstr. 21, Postfach 3980, D-6500 Mainz, BRD.]

(1286 MARBURG. Universität. Fachbereich Geographie. Marburg.
Rundbrief. 1- (1973-) 6 nos a year.
Im Auftrag des Zentralverbandes der Deutschen Geographen.
Supersedes Institut für Landeskunde. Rundbrief [1265].
[Fachbereich Geographie, Universität Marburg, Deutschherrenstr. 10, D-3550 Marburg, BRD.]

*(1287 MARBURGER geographische Schriften. (Marburg. Universität. Geographisches Institut). Marburg. 1- (1949-) Irregular. Monographs. (U-3:2537b; D:214; G; GZS).
[Deutschhausstr. 10, D-3550 Marburg/Lahn, BRD.]

1288 ______.
Sonderband. 1- (1971-). (N76; GZS).

(1289 MATERIAL zum Beruf des Geographen. (Verband deutscher Berufsgeographen). Hamburg. 1- (1978-).
[Postfach 106 325, D-2000 Hamburg 1, BRD.]

1290 MATERIAL zur angewandten Geographie. (Verband deutscher Berufsgeographen e. V.). Hamburg. 1- (1978-).

1291 MATERIALIEN zur Stadtgeographie. Kallmünz/Regensburg. 1-2 (1964-1966). (N70-3:3676a; G).
1 and 2 also as Münchner geographische Hefte 25, 28 [1296].
[M. Lassleben, Kallmünz/Regensburg, BRD.]

(1292 MEDIZINISCHE Länderkunde: Beiträge zur geographischen Medizin. Geomedical monograph series: regional studies in geographical medicine. (Heidelberger Akademie der Wissenschaften. Mathematisch-naturwissenschaftliche Klasse). Heidelberg. 1- (1967-) Irregular. (N70-3:3716a; B73:208b; G).
In German or English.
[Springer Verlag, D-6900 Heidelberg, BRD.]

1293 MITTEILUNGSBLATT zur rheinhessischen Landeskunde. (Arbeitsgemeinschaft rheinhessischer Heimatforscher). Mainz. v1- (1952-). (D:105).
Index: 1-10 (1952-1961) [1962].

*(1295 MÜNCHENER geographische Abhandlungen (München. Universität. Fachbereich Geowissenschaften. Institut für Geographie). München. 1- (1970-). (N75-2:1491a; G; GZS).
1-14 (1970-1974) under former name: Geographisches Institut.
In German with abstracts in German and in English.
[Institut für Geographie der Ludwig-Maximilians-Universität München, Luisenstrasse 37/II/III, D-8000 München 2, BRD.]

*(1296 MÜNCHENER geographische Hefte. (München. Technische Universität. Geographisches Institut). Regensburg. 1- (1953-) Irregular. Monographs. (N70-3:3929c; BS:578a; D:217; G).
11-14 and 16-19 also as Materialien zur Agrargeographie;
1-8, 25, 28 also as Materialien zur Stadtgeographie;
1-2 and 30 also as Materialien zur angewandten Geographie;
1, 34 also as Arbeitstagung des Verbandes deutscher Berufsgeographen.
[Verlag Michael Lassleben, D-8411 Kallmünz, Regensburg, BRD.]

*(1297 MÜNCHNER Studien zur Sozial- und Wirtschaftsgeographie. (München. Universität. Wirtschaftsgeographisches Institut). München. 1- (1966-). (N70-3:3929c; G).
Summaries in English and other languages.
[Verlag Michael Lassleben, Kallmünz/Regensburg.]

1298 MÜNSTER. Geographische Kommission für Westfalen. Münster in Westfalen. Landeskundliche Karten und Hefte.
Reihe: Bodenplastik und Naturräume Westfalens. 1-5 (1953-1968) Irregular.
[Institut für Geographie und Länderkunde der Universität Munster i.W., Robert-Koch-Str. 26-28, D-4400 Münster in Westfalen, BRD.]

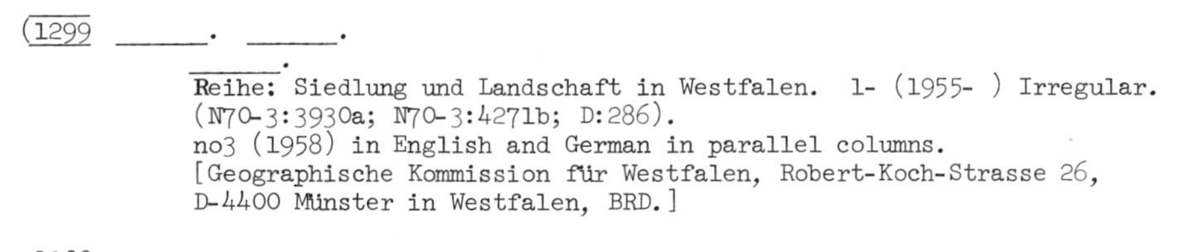

1299 ______. ______.
______.
Reihe: Siedlung und Landschaft in Westfalen. 1- (1955-) Irregular. (N70-3:3930a; N70-3:4271b; D:286).
no3 (1958) in English and German in parallel columns.
[Geographische Kommission für Westfalen, Robert-Koch-Strasse 26, D-4400 Münster in Westfalen, BRD.]

1300 ______. ______.
Landkreise in Westfalen, Die. 1-5 (1953-1969) Irregular. (D:287).

1301 ______.
Provinzialinstitut für Westfälische Landes- und Volkskunde. Veröffentlichungen.

1302 MUNDUS. A quarterly review of German research contributions on Asia, Africa, and Latin America. Stuttgart. 1- (1965-) Quarterly. (N70-3:3937a; B68:370a; G).
English language abstracts arranged in two parts in each issue: arts and economics; science and geography. Each part divided into Book notes and abstracts and Selected bibliography.
[Wissenschaftliche Verlagsgesellschaft, Postfach 40, D-7000 Stuttgart 1, BRD.]

1303 NACHRICHTEN aus dem Karten- und Vermessungswesen. (Institut für angewandte Geodäsie). Frankfurt-am-Main.
Reihe I, Deutsche Beiträge und Informationen. 1- (1951-) Irregular. (N70-3:3975c).

1304 NEUES Archiv für Niedersachsen. (Göttingen. Universität. Institut für Landeskunde und Landesentwicklung). Göttingen.
v1-5 no1-26 (1947-1951); v6- (1953-). (U-4:2974b; B-1:193b; D:102; G).
Supersedes Archiv für Landes- und Volkskunde von Niedersachsen [910]. no1-8 (1947-1948) as Neues Archiv für Landes- und Volkskunde von Niedersachsen. v1-9 (1947-1958) issued by Niedersachsen (state). Amt für Landesplanung und Statistik. Also as Niedersachsen (state). Amt für Landesplanung und Statistik. Veröffentlichungen. Reihe C.
Issuing agency formerly named Niedersächsisches Institut für Landeskunde und Landesentwicklung.
Indexes: v1-5 (n1-26 [1947-52]) as n27; v6-8 (1953-1956) as v9 n6 (1958).

*1305 NÜRNBERGER wirtschafts- und sozialgeographische Arbeiten. (Universität Erlangen-Nürnberg. Wirtschafts- und sozialgeographisches Institut. Veröffentlichungen). Nürnberg. 1- (1957-) Irregular. (N70-3:4318a; B8:637a; D:221).
Issuing agency earlier: Nürnberg. Hochschule für Wirtschafts- und Sozialwissenschaften. Institut für Wirtschaftsgeographie.
[Findelgasse 7-9, D-8500 Nürnberg, BRD.]

*1306 ORBIS geographicus; World directory of geography; Adressar géographique du monde; Geographisches Weltadressbuch. Wiesbaden. 1960- . About every four years. (N70-3:4411c; N70-2:2282b; D:5).
1960, 1964/66 as supplements to Geographisches Taschenbuch [1248]. 1968/1972, part 1. Societies, institutes, agencies, 1970 and 1968/74, part 2. Geographers by countries, 1974, separately issued. Sponsored by the International geographical union.
Preceded by World directory of geographers (International Geographical Union) New York, 1952.
In English, French, and German.
[Franz Steiner Verlag, Postfach 5529, 62 Wiesbaden, BRD.]

(1307 PRAXIS Geographie. Braunschweig. 1980- 6 nos a year.
Supersedes Geographische Rundschau. Beiheft [1243].
[Georg Westermann Verlag, D-3300 Braunschweig, BRD.]

(1308 QUELLEN und Forschungen zur Geschichte der Geographie und der Reisen. (Hanno Beck). Stuttgart. 1- (1964-). (N70-3:4884b).
[F. A. Brockhaus, D-7000 Stuttgart, BRD.]

1309 RAUM und Siedlung; Fachblatt für Raumordnung, Städtebau, Gemeindeentwicklung, Planungsrecht. Köln. 1967-1971. Monthly. Closed 1971. (N70-3:4924b; G; GZS).
Superseded by Structur: Zeitschrift für Planung, Entwicklung und Umwelt (not in this list).

(1310 RAUMFORSCHUNG und Landesplanung. Abhandlungen. (Akademie für Raumforschung und Landesplanung). Hannover. 1- (1937-) Irregular. (U-1:628b; B-1:305b).
v1-15 as Beiträge zur Raumforschung und Raumordnung; v16-36 (1950-1959) as Akademie für Raumforschung und Landesplanung. Veröffentlichungen. 37- (1963-) as Veröffentlichungen der Akademie für Raumforschung und Landesplanung. Abhandlungen. 1-6 (1937-1940) issued by Reichsarbeitsgemeinschaft für Raumforschung und Landesplanung. Heidelberg.
[Gebrüder Jänecke Verlag, Osterstrasse 22, D-3 Hannover, BRD.]

(1311 RAUMFORSCHUNG und Raumordnung. (Bundesforschungsanstalt für Landeskunde und Raumordnung, Bonn-Bad Godesberg [and] Akademie für Raumforschung und Landesplanung, Hannover). 1-8 (1936-1944); 9- (1948-) 6 nos a year. (U-4:3535b; B-3:661b; D:91; G).
1-8 (1936-1944) issued by Reichsarbeitsgemeinschaft für Raumforschung. Heidelberg. 9 (1948) as Raumforschung-Raumordnung. Hefte der Akademie für Raumforschung und Landesplanung. Berichte über Raumforschung, Raumordnung und Landesplanung, Regionalplanung und angewandte Landeskunde; 10 (1950) as Raumforschung und Raumordnung. Organ der Akademie für Raumforschung und Landesplanung. Zeitschrift für Raumforschung, Raumordnung und Landesplanung, Regionalplanung, angewandte Landes- und Auslandskunde; 11- (1953-) present title. Suspended 1945-1947, 1949, 1951-1952. Absorbed Zeitschrift für Raumforschung [1345], 1950.
[Carl Heymanns Verlag, Köln, D-5300 Köln, BRD.]

(1312 REFERATEBLATT zur Raumentwicklung. (Bundesforschungsanstalt für Landeskunde und Raumordnung). Bonn-Bad Godesberg. 1- (1975-) Quarterly. (G; GZS).
Supersedes Referateblatt zur Raumordnung [1313].
[Postfach 130, 5300 Bonn-Bad Godesberg, BRD.]

1313 REFERATEBLATT zur Raumordnung. (Dokumentation zur Raumordnung. Reihe II). (Bundesforschungsanstalt für Landeskunde und Raumordnung.Institut zur Raumordnung). Köln. 1-6 (1969-1974) Quarterly. (G).
Superseded by Referateblatt zur Raumentwicklung [1312].

(1314 REGENSBURG. Universität. Geographisches Institut. Regensburg.
Exkursionen in Ostbayern. 1- (1976-).
[Geographisches Institut der Universität Regensburg, Universitätsstr. 31, Postfach 397, D-8400 Regensburg, BRD.]

*(1315 REGENSBURGER geographische Schriften. (Regensburg. Universität. Geographisches Institut). Regensburg. 1- (1971-) Irregular. (G; GZS).
[Geographisches Institut an der Universität Regensburg, Universitätsstr. 31, Postfach 397, D-8400 Regensburg, BRD.]

*(1316 RHEIN-Mainische Forschungen. (Frankfurt am Main. Universität. Geographisches Institut). Frankfurt am Main. no1-26 (1927-1941); 27- (1949-). (U-4:3671a; D:187; G).
Monographs. Abstracts in German, English, and French from v84 (1977).
[Geographisches Institut der Universität, Senckenberganlage 36, D-6000 Frankfurt, BRD; or Verlag Waldemar Kramer, Frankfurt am Main, BRD.]

1317 ROTER Globus Sondernummer: Zeitschrift zur Kritik bürgerlicher Geographie. 1- (1973-) Irregular.
[Arndt Thomas, Weidenhäuserstr. 60, Marburg/Lahn, BRD.]

(1318 RUHR-UNIVERSITÄT, Bochum. Geographisches Institut. Forschungsabteilung für Raumordnung.
Materialien für Raumordnung. 1- (1969-) Irregular.
[Geographisches Institut der Ruhr Universität, Universistätsstrasse 150, D-4630 Bochum, BRD.]

*(1319 SAARBRÜCKEN. Universität des Saarlandes. Geographisches Institut. Arbeiten. 1- (1956-) Irregular. (N70-4:5140c; BS:757b; D:222).

(1320 SAMMLUNG geographischer Führer. Berlin, Stuttgart. v1-4 (1924-1930); 5- (1969-). D:381; G).
Not published 1931-1968.
[Gebrüder Bornträger, Johannesstr. 3A, D-7000 Stuttgart 1, BRD.]

(1321 SCHULGEOGRAPH, Der. (Verband deutscher Schulgeographen. Landesverband Hessen. Mitteilungsblatt). Frankfurt. 1967- .
[Verlag Moritz Diesterweg, Hochstr. 31, D-6000 Frankfurt am Main, BRD.]

*(1322 SPIEKER. Landeskundliche Beiträge und Berichte. (Geographische Kommission für Westfalen). Münster. 1- (1950-) Irregular, about 1 a year. (N65-2:2522b; N68-2:2057b; D:285; G).
[Geographische Kommission für Westfalen, Robert-Koch-Str. 26, D-4400 Münster in Westfalen, BRD.]

(1323 STANDORT. (Verband deutscher Berufsgeographen e.V.). Hamburg. 1- (1977-) 3 nos a year.
[H. Fangohr, Postfach 106 325, D-2000 Hamburg 1, BRD.]

1324 STUDIEN zur Kartographie. Berlin-Dahlem. 1-2 (1957-1959). (N70-4:5585b; G).

*(1325 STUTTGARTER geographische Studien. (Stuttgart. Universität [formerly Technische Hochschule.] Geographisches Institut [until 1936: Geographisches Seminar]. Stuttgart.
v1-68 (1924-1940); v69- (1957-) Irregular. (U-5:4105b; B-4:245a; BS:846b; D:223).
v1-68 as Institut's Veröffentlichungen, sA. Suspended 1941-1956.
Summaries in English and other languages from v71.
[Silcherstrasse 9, D-7000 Stuttgart 1, BRD.]

(1326 TEUBNER Studienbücher der Geographie. Stuttgart. 1975- Irregular.
[B. G. Teubner, D-7000 Stuttgart, BRD.]

1327 TRIBUS. Veröffentlichungen des Linden-Museums, Stuttgart. nsv1- (1951-). (N70-4:5856c; B-4:574b; D:69; G).
1- (1951) as Jahrbuch des Linden-Museums für Länder- und Völkerkunde, Württembergischer Verein für Handelsgeographie; 2/3 (1953) as Tribus, Jahrbuch des Linden-Museums; 4/5 (1956-1957) as Tribus. Zeitschrift für Ethnologie und ihre Nachbarwissenschaften.
Supersedes Württembergischer Verein für Handeslgeographie und Förderung deutscher Interessen im Auslande. Jahresbericht [1139].
Publisher varies among a combination of Linden Museum, Gesellschaft für Erd- und Völkerkunde Stuttgart e.V., and Württembergischer Verein für Handelsgeographie e.V.

(1328 TRIERER geographische Studien. (Geographische Gesellschaft Trier [and] Trier. Universität. Fachgruppe Geographie). Trier. 1- (1976-). (G).

(1328a TÜBINGEN. Universität. Geographisches Institut. Tübingen. Kleinere Arbeiten. 1- (1973-). (GZS).

*(1329 TÜBINGER geographische Studien. (Tübingen. Universität. Geographisches Institut). 1- (1958-) Irregular. (N70-4:5874c; D:228; G).
English abstracts.
[Geographisches Institut der Universität Tübingen, Hölderlinstrasse 12, D-7400 Tübingen, BRD.]

1330 ______. Sonderband. 1-2 (1962-1966). (N70-4:5874c; D:228; G).
Others are part of regular series 3 (34), 4 (36), 5 (44), 6 (45), 7 (46), 8 (58), 9 (60), 10 (61), 11 (68), 12 (71).

1331 UNTERLAGEN und Beiträge zur allgemeinen Theorie der Landschaft. (Hamburg. Universität. Institut für Geographie und Wirtschaftsgeographie). Hamburg. 1- (1968-). (N70-4:6104a).

(1332 URBANISIERUNG der Erde. Stuttgart. 1- (1978-) Irregular monographs.
[Gebrüder Borntraeger, Johannesstr. 3A, D-7000 Stuttgart 1, BRD.]

(1333 URBS et Regio. Kasseler Schriftum zur Geographie und Planung. (Kassel. Gesamthochschule. Geographisches Institut. Veröffentlichungen). Kassel. 1- (1976-) Irregular.

1334 VERBAND deutscher Schulgeographen. Landesverband Nordrhein-Westfalen. Paderborn.
Mitteilungsblatt. v1- (1953-) Irregular. (D:80).

(1335 WGI - Berichte zur Regionalforschung. (München. Universität. Wirtschaftsgeographisches Institut). München. 1- (1970-). (N75-2:1493b; G).
[Verlag Michael Lassleben, D-8411 Kallmünz, Regensburg, BRD.]

1335a ______.
Sonderheft. 1971. (G).

1336 WESTERMANNS geographische Bibliographie. Braunschweig. v1-11 (1955-1965) 10 nos a year. Closed 1965. (N70-4:6315c; D:16).
Perforated list of titles of international geographical publications for pasting onto library cards.

(1337 WESTFÄLISCHE Forschungen. (Münster. Provinzialinstitut für westfälische Landes- und Volkskunde. Mitteilungen). Münster. 1- (1938-) Annual. (U-5:4497b; B-4:534a; BS:961a; D:103; G).

*(1338 WESTFÄLISCHE geographische Studien. (Münster. Universität. Institut für Geographie). Münster. 1- (1949-) Irregular. (U-5:4497b; D:220).
Varying titles of sponsoring institutions: Münster. Universität. Geographisches Institut [in combination with] Geographische Kommission im Provinzialinstitut Münster, or Geographische Kommission für Westfalen, Münster, or Geographische Kommission im Provinzialinstitut für westfälischen Landes- und Volkskunde, or Geographische Kommission für westfälische Landes- und Volkskunde.
Some English summaries.
[Institut für Geographie, Robert-Koch-Strasse 26-28, D-4400 Münster/ Westfalen, BRD.]

1339 ______. Beihefte. Münster. 1- (1970-). (N75-2:2286c).

(1340 WISSENSCHAFTLICHE Länderkunden. Darmstadt. 1- (1968-). (G).
Monographs.
[Wissenschaftliche Buchgesellschaft, Hindenburgstrasse 40, Postfach 11 11 29, D-6100 Darmstadt 11, BRD.]

1341 WISSENSCHAFTLICHE Paperbacks: Geographie. Wiesbaden. 1973- Irregular. (N76).
[Franz Steiner Verlag, Friedrichstr. 24, Postfach 55 29, D-6200 Wiesbaden, BRD.]

*(1342 WÜRZBURGER geographische Arbeiten. (Geographische Gesellschaft zu Würzburg. Mitteilungen). Würzburg. 1- (1953-) Irregular. (N70-4:6405a; BS:972b; D:229; G).
Supersedes Fränkische Studien [972].
Includes a subseries: Arbeiten aus der Kommission für Geomorphologie der Bayerischen Akademie der Wissenschaften.
Some English summaries.
[Geographisches Institut der Universität Würzburg, Am Hubland, D-8700 Würzburg, BRD.]

*(1343 ZEITSCHRIFT für Geomorphologie. Annals of geomorphology. Annales de géomorphologie. Berlin, Stuttgart. v1-11 no5 (1925-1940); ns1- (1957-) 4 nos a year. (U-5460c; B-4:603a; BS:983a; F-4:1019b; D:81).
Articles in German, French, or English with abstracts in other two languages.
Index: 1925-1976 [v1-11 (1925/26-1939/43); nsv1-20 (1957-1976)].
[Gebrüder Borntraeger, Johannesstrasse 3a, D-7000 Stuttgart 1, BRD.]

*(1344 ______.
Supplementband. 1- (1960-). Irregular. (N70-4:6464b; BS:983a).
Index: v1-26 (1960-1976).
In German, English, or French.

1345 ZEITSCHRIFT für Raumforschung. (Institut für Raumforschung, Bad Godesberg). Bielefeld. v1 no1-12 (1950). Closed 1950. (N70-4:6466c; B-4:610b; D:92; G).
Absorbed by Raumforschung und Raumordnung [1311].

*(1346 ZEITSCHRIFT für Wirtschaftsgeographie. Angewandte- und Sozial- Geographie. Hagen. 1- (1957-) 8 nos a year. (N70-4:6467c; BS:985a; D:86).
1-3 (1957-1959) as Zeitschrift für Wirtschaftsgeographie (Sozialgeographie) unter Berücksichtigung der Verkehrs- und Handelsgeographie.
[Pick-Verlag, Postfach 2723, D-5800 Hagen, Westfalen, BDR.]

GERMAN DEMOCRATIC REPUBLIC (German)

Deutsche Demokratische Republick (DDR)

For serials closed prior to 1946 see under Germany, p. 135-155

1347 AKADEMIE der Wissenschaften der DDR, Berlin. Institut für Geographie und Geoökologie.
Beiträge zur Geographie. no1-13 (1896-1914); nsno1-10 (1932-1942); 11-27/28 (1952-1970). 29- (1979-). Suspended 1943-1951. (U-3:2384c; B-3:36a; D:59; D:246; N75-2:2408b; N75-2:1910b; N78; R-6: 406a).
1-3 (1896-1901) as Museum für Völkerkunde zu Leipzig. Mitteilungen; 4-9 (1903-1906) as Grassi-Museum zu Leipzig. Vulkanologische Abteilung. Veröffentlichungen; 10-13 (1909-1914) as Leipzig. Städtisches Museum für Länderkunde. Veröffentlichungen; nsno1-10 (1932-1942) as Leipzig. Deutsches Museum für Länderkunde. Wissenschaftliche Veröffentlichungen; nsno11-26 (1952-1968) as Leipzig. Deutsches Institut für Landerkunde. Wissenschaftliche Veröffentlichungen; 27/28 (1970) as Akademie der Wissenschaften der DDR, Berlin. Geographisches Institut. Leipzig. Wissenschaftliche Veröffenlichungen. Present title from 29 (1979).
[Institut für Geographie und Geoökologie, Akademie der Wissenschaften der DDR, Georgi-Dimitroff-Platz 1, 701 Leipzig, DDR.]

1348 ______. ______.
ZWL-Dokumentationsdienst. Geographie (later IfD-Referaterkartei. Geographie; ZIID-Referatekartei. Geographie; Titelkartei. Geographie). 1-15 (1957-1971) Monthly. About 200 title cards a month. (D:14).
1957-1963 as Regionale Geographie.

1349 BEIHEFTE für Erdkunde (zu "die neue Schule") Berlin; Leipzig.
no1-2 (1949). Closed 1949. (D:71; G).
Superseded by Zeitschrift für den Erdkundeunterricht [1374].

1350 BERLIN. Hochschule für Ökonomie. Institut für ökonomische Geographie und Regionalplanung. Berlin.
Wissenschaftliche Beiträge. 1-13 (1961-1969) Annual. Closed 1969.
Reprints, partly hektographed. (N70-1:701a).

1351 BERLINER geographische Arbeiten. (Berlin. Humboldt-Universität). Berlin.
no1-21 (1932-1942); no22 (1949); no23- (1955-). (U-1:647c; B-1:321b; D:173; G; R-2:53b).
no1-20 published in Stuttgart. 1-22 (1932-1949) by Berlin. Universität. Geographisches Institut; no 23- (1955-) as subseries of Berlin. Humboldt-Universität. Wissenschaftliche Zeitschrift. Mathematisch-naturwissenschaftliche Reihe. 5- (1956-) (not in this list)
Abstracts in English, French, and Russian from no 23(1955).
[Sektion Geographie, Humboldt Universität zu Berlin, Universitätsstr. 3b, 108 Berlin, DDR.]

1353 DRESDEN. Technische Universität. Institut für Geographie, later Sektion Geodäsie und Kartographie. Dresden. Arbeiten. 1- (1964-).
Reprints from university's Wissenschaftliche Zeitschrift. Mathematisch-naturwissenschaftliche Reihe (not in this list).
[Sektion Geodäsie und Kartographie, Technische Universität, Mommsenstr. 13, 8027 Dresden, DDR.]

1354 GEOGRAPHISCHE Bausteine. Gotha. no1- (1967-) Irregular. (G).
Old series, 1-22 (1913-1936) [980].
[VEB Hermann Haack, Geographisch-Kartographische Anstalt, Gotha, DDR.]

*(1355 GEOGRAPHISCHE Berichte. (Geographische Gesellschaft der Deutschen Demokratischen Republik. Mitteilungen). Berlin. 1- (1956-) 4 nos a year. (N70-2:2782a; BS:342a; D:23; G; R-3:272b-273a).
Supplementary table of contents in English and Russian.
English and Russian summaries from 1960.
[VEB Hermann Haack, Geographisch-Kartographische Anstalt, Justus-Perthes-Strasse 3/9, 58 Gotha, DDR.]

*(1357 GEOGRAPHISCHE Gesellschaft der Deutschen Demotratischen Republik.
Wissenschaftliche Abhandlungen. Berlin. 1- (1957-) Irregular. (N70-2:2282a; D:231; G; R-6:404a).
Title varies: 1 (1957) as Abhandlungen.
[VEB Hermann Haack, Geografisch-Kartographische Anstalt, Justus-Perthes-Strasse 3/9, 58 Gotha, DDR.]

(1358 GEOGRAPHISCHES Jahrbuch. Gotha. v1-58 no1 (1866-1943); v58 no2- v62 (1943/47-1967); 63-65 (1976-1979) Irregular. (U-2:1697a; B-2:270b; F-2:729b; D:6; G).
Index: v1-40, 1866-1925 in 40.
[VEB Hermann Haack Geographisch-Kartographische Anstalt, Justus-Perthes-Strasse 3/9, 58 Gotha, DDR.]

1359 HALLE. Universität. Halle.
Wissenschaftliche Beiträge. Reihe Q: Geowissenschaftliche Beiträge. 1- (1966-). (N 70-2:2465bc; B68:591b; G).

1360 _____. _____. Geographisches Institut. Halle.
Mitteilungen. 1-10 (1954/55-1968). Closed 1968. (N70-2:2464c; D:47; D:196; G).
Reprinted from Halle. Universität. Wissenschaftliche Zeitschrift. Mathematisch-naturwissenschaftliche Reihe (not in this list).

1361 ______. ______. Sektion Geographie. 1-3 (1970-1971). (GZS).

*(1362 HALLESCHES Jahrbuch für Geowissenschaften. Gotha. 1- (1977-) Annual.
[VEB Hermann Haack Geographisch-Kartographische Anstalt, Justus-Perthes-Strasse 3/9, 58 Gotha, DDR.]

(1363 HERCYNIA. Für die Sektionen Biowissenschaften, Geographie und Pflanzenproduktion. (Halle-Wittenberg. Martin-Luther Universität). Leipzig. ns1- (1963-). (N70-2:2525a; B68:237b; D:108; G; R-3:328b-329a).
Old series 1-3 (1937-1940) issued by Botanische Vereinigung Mitteldeutschlands (not in this list).
ns1-10 as Hercynia: für die Fachgebiete Botanik, Geographie, Geologie, Geophysik, Paläontologie, Zoologie. Present title from ns11 (1974).
[Akademische Verlagsgesellschaft Geest und Portig, Leipzig, DDR.]

1364 KARTENTECHNIK-Kartengestaltung. (Deutsche Demokratische Republik. Kammer der Technik der DDR). Berlin. 1- (1966-) Irregular.
[VEB Fachverlag für Bauwesen, Berlin, DDR.]

1365 LEIPZIG. Universität. Geographisches Institut. Leipzig.
Arbeiten. 1/2- (1951/53-) Irregular. (N70-3:3439a; BS:505a; D:212).
Reprinted from the university's Wissenschaftliche Zeitschrift.

*(1367 LITERATURINFORMATION. Territorialforschung. Territorialplanung. (Akademie der Wissenschaften der DDR. Institut für Geographie und Geoökologie). Leipzig. 1971- Monthly.
[Institut für Geographie und Geoökologie, Akademie der Wissenschaften der DDR, Georgi-Dimitroff-Platz 1, 701 Leipzig, DDR.]

1368 MITTEILUNGEN für Agrargeographie, landwirtschaftliche Regionalplanung und ausländische Landwirtschaft. (Halle. Universität). Halle. 1-56 (1963-1973). Reprints from Halle-Wittenberg. Universität. Wissenschaftliche Zeitschrift (not in this list).

1369 NEUERE Arbeiten zur mecklenburgischen Küstenforschung. (Greifswald. Universität. Geographisches Institut). Berlin. 1-5 (1954-1960). (N70-3:4117c; D:195).

*(1370 PETERMANNS geographische Mitteilungen. (Geographische Gesellschaft der Deutschen Demokratischen Republik). Gotha. 1-91 no3 (1855-1945) Monthly; 92- (1948-) Quarterly. (U-4:3315c; B-2:632a; F-3:397a; F-3:564b-565a; D:28; R-5:167ab-168a).
v1-24 (1855-1878) as Mitteilungen aus Justus Perthes' geographischer Anstalt. v25-83 no9 (1879-1937) as Petermanns Mitteilungen aus Justus Perthes' geographischer Anstalt. Combined with Ausland (1828-1893) [915] Aus allen Weltteilen (1870-1898) [914] and Globus (1862-1910) [1019].
Indexes: 1855-1864, 1865-1874, 1875-1884, 1885-1894, 1895-1904, 1905-1934.
In German with supplementary table of contents and abstracts in English and Russian from 1961.
[VEB Hermann Haack Geographische-Kartographische Anstalt, Justus-Perthes-Strasse 3-9, 58 Gotha, DDR.]

*(1371 ______.
Ergänzungsheft. v1-55 (no1-242) (1860-1944); no243- (1951-) Irregular. (U-4:3316a; B-2:632a; F-2:397a; F-3:565a; D:357). Monographs.
Index: 1860-1934.
[VEB Hermann Haack Geographisch-Kartographische Anstalt, Justus-Perthes-Strasse 3/9, 58 Gotha, DDR.]

1372 ROSTOCK. Universität. Geographisches Institut. Rostock.
Mitteilungen. 1- (1964-).
Reprints from university's Wissenschaftliche Zeitschrift (not in this list).
Summaries in English, French, and Russian.

(1373 STUDIENBÜCHEREI. Geographie für Lehrer. Gotha. 1- (1976-) Irregular.
[VEB Hermann Haack Geographisch-Kartographische Anstalt, Justus-Perthes-Strasse 3/9, 58 Gotha, DDR.]

(1374 ZEITSCHRIFT für den Erdkundeunterricht. (Volk und Wissen Volkseigener Verlag). Berlin. v1- (1949-) Monthly. (U-5:4599a; D:79; G; R-6442b).
Supersedes Beihefte für Erdkunde [1349].
v14-16, no5/6 (1962-1964) divided into two parts: Ausgabe A and Ausgabe B.
Originally published by Ministerium für Volksbildung der Deutschen Demokratischen Republik.
[Deutscher Buch-Export und -Import GmbH, Leninstrasse 16, 701 Leipzig, DDR.]

GHANA (English)

Republic of Ghana

*(1375 GHANA Geographical association. Legon.
Bulletin. 1- (1956-). (N70-2:2347ab; BS:355b).
v1-2 no1 (1956-1957) as Gold Coast geographical association.
In English.
[University of Ghana, Accra, Ghana.]

GREECE (Greek)

Elliniki Dimotratia

ΕΛΛΗΝΙΚΗ ΔΗΜΟΚΡΑΤΙΑ

1377 ΕΛΛΑΣ. Στρατος. Γεωγραφικη Υπηρεσία. Δελτίον.

1378 Ελληνικη Γεωγραφικη Εταιρεία. Αθήναι. Δελτίον.

(1376 EKISTICS; the problems and science of human settlements. Athens.
1- (1955-) 10 nos a year. (N70-2:1886c).
1-3 (1955-1957) as Tropical housing and planning monthly bulletin.
Subtitle varies.
Index: 1961-1967 in v26 no4 (S 1968).
In English.
[The Athens Center of Ekistics, P.O. Box 471, Athens, Greece.]

1377 GREECE. Stratos. Geōgraphikē hyperesia. Athinai.
Deltion. v1- (1935-) Semiannual.
English title: Army. Geographical service. Bulletin.
[Pedion Areos, Athens, Greece.]

1378 HELLENKIE geōgraphikē hetaireia, Athinai. Athinai.
Deltion. 1- (1925-) Irregular. (U-3:1835c).
English title: Hellenic geographic society. Bulletin.
v1 (1919-1923); v2 (1925); v3 (1926-1932); v4 (1932-1951); I; v5 (1952-1959), II; v6 Panegyric (1919-1959), III; v7 (1963) IV, special congress.
In English, French, or Greek.

GUATEMALA (Spanish)

República de Guatemala

1379 GUATEMALA. Instituto geográfico nacional. Guatemala.
Boletín trimestrial. 1- (1968-) Quarterly.
[Instituto geográfico nacional, Avenida de las Américas 5-76, Zona 13, Guatemala, Guatemala.]

(1380 SOCIEDAD de geografía e historia de Guatemala.
Anales. v1- (1924-) Annual (earlier, quarterly). (U-5:3906b; B-4:106a; F-1:176a).
Index: v1-5 (1924-1929).
[3A Avenida 8-35, Zona 1, Guatemala 1, Guatemala.]

GUYANA (English)

*(1381 GEOGRAPHICAL association of Guyana. Georgetown.
Journal. 1- (Jl 1973-). (N77).

*(1382 UNIVERSITY of Guyana. Department of geography. Georgetown.
Occasional paper. 1- (1970-) Irregular. Mimeographed. (N75-2:2223c; GZS).
[Department of Geography, University of Guyana, Box 841, Georgetown, Guyana.]

HAITI (French)

République d'Haiti

1383 SOCIÉTÉ haïtienne d'histoire, de géographie et de géologie. Port-au-Prince.
Revue. 1- (1925-) Quarterly. (U-5:3962a; B-4:137a; F-4:210b).
1 (My 1925) as Bulletin. Earlier names: Sociéte d'historie et de géographie d'Haïti, 1923-1947; Société haïtenne d'histoire et de géographie, 1947-1950?
[Avenue Magloire Ambroise no. 112, Port-au-Prince, Haiti.]

HONDURAS (Spanish)

República de Honduras

1384 INSTITUTO geográfico nacional (Honduras). Tegucigalpa. Boletín. 1- (1966?-). (N70-2:2572a; N75-1:1130c).

1385 ______. ______.
Informe anual. (N70-2:2572a; N75-1:1130c).
Subseries of institute's Boletín [1384].

1386 SOCIEDAD de geografía e historia de Honduras. Tegucigalpa.
Memorias. 1926-1934. (U-5:3906c).

1387 ______.
Revista. 1- (1904-) 12 nos a year. (U-5:3906c; B-2:416b: F-4:135ab).
Contains "Sección de geografía." Suspended 1910-My 1927, F-Je 1938.
1904-1955 by the Archivo nacional and the Biblioteca nacional.

香港 HONG KONG (Chinese or English)

Colony of Hong Kong

(1388 GEOGRAPHY bulletin. (Hong Kong. Education department. Curriculum development editorial board). 1- (1958-) Annual.
no1-2 mimeographed; no3- printed.
In Chinese and English.

1389 GEOGRAPHY teacher. (Grantham college of education. Geography society).
Hong Kong. 1-3 (1970- ?) Annual. Closed.
In Chinese and English.

1390 HONG KONG. Chinese university. Ch'ung-chi college. Hong Kong.
Geographical research paper. 1- (1966-) Irregular. Closed.
Reprints from the Chung-chi journal (Chung-chi hsüeh pao) (N70-1:1281c) (not in this list).
[Shatin, New Territories, Hong Kong.]

(1391 ______. ______.
Geography society. Hong Kong.
Geographical journal. 1- (1976-) Irregular.
In Chinese and English.

1392 ______. ______. Graduate school. Geographical research centre.
Cultural and economic geography. 1-5 (1970-1974). Closed.

1393 ______. ______. ______. ______.
Research report. 1-96 (1967-1977) Irregular. Mimeographed. Closed.
Usually in Chinese with English summary but may be entirely in Chinese or English.

1394 ______. ______. Research institute of Far Eastern studies.
Geography research paper. 1- (1966-). (N70-2:2598b; N70-2:2574b).
Also as: Hsin ya shu yuan, Kowloon.

(1395 ______. University. Geographical, geological, and archaeological society.
Hong Kong.
Annals. 1- (1972-).
In English.

1396 ______. ______. ______.
Journal. 1 (Mr 1961). Only one issued. (N70-2:2576c; B68:302b).
In English.

*(1397 HONG KONG geographical association. Hong Kong.
Bulletin. 1- (1971-) Annual.
[Hong Kong Geographical Association, Department of Geography and Geology, Hong Kong University, Hong Kong.]

1398 SHIH-TI chi k'an. (Hsiang-kang Chung wen ta hsüeh Ch'ung-chi hsüeh yuan shi ti hsi). Hong Kong. 1 (1964). Only one issued. (N70-4:5310b; C:391a).
Title also in English: History and geography magazine. (Hong Kong. Chinese University. Ch'ung-chi college. History and geography society).

1398a CHINA studies series. (International house for China studies). Hong Kong.
1- (1976-) Irregular, about 5 a year.
Usually in Chinese with English summary.
[International House for China Studies, Villa Cassia, Hang Tau, Sheung Shui, N. T., Hong Kong.]

1398b HISTORICAL and cultural geography. (International house for China studies).
Hong Kong. 1- (1980-) Annual.
In Chinese or Japanese. English summaries.
[International House for China Studies. Villa Cassia, Hang Tau, Sheung Shui, N. T., Hong Kong.]

HUNGARY (Hungarian)

Magyar Népköztársaság

*(1399 ACTA geographica. (Szeged. Tudományegyetem. Acta universitatis szegediensis. Pars geographica scientiarum naturalium). Szeged. vl- (1955-). (N70-1:58b; BS:530b; R-1:143a).
Called ns in continuation of Szeged. Tudományegyetem. Acta universitatis szegediensis. Sectio geographico-historica.
In English, German, or Russian [1448].

*(1400 ACTA geographica debrecina. (Debrecen. Tudományegyetem). Budapest. 8- (sl-) (1962-). (N70-1:58b; R-1:172a).
Subseries of Acta universitatis debreceniensis de Ludovico Kossuth nominatae. Called also Acta universitatis debreceniensis de Ludovico Kossuth nominatae. Series geographica, geologica et meteorologica.
In English, French, German, Hungarian, Latin, or Russian.

1401 BORSODI földrajzi évkönyv. (Magyar földrajzi társaság. Miskolci osztálya). Miskolc. 1- (1958-). (N70-1:829c).
Subseries of Borsodi szemle könyvtára (not in this list).

*(1402 BUDAPEST. Tudományegyetem. Budapest.
Annales. Sectio geographica. 1- (1965-) Annual. (N70-1:913c; R-1:337a).
(Annales universitatis scientiarum budapestinensis de Rolando Eötvös nominatae).
In English, German, or Russian.

1403 _____. _____. Földrajzi intézet. Budapest.
Hungarian geographical essays. no1 (1921). Closed 1921.

1404 _____. _____. _____.
Kiadványai. no1-2 (1943-1944). Suspended 1944.

1405 _____. _____. _____.
Magyar földrajzi értekezések. no1-4 (1921-1922). Closed 1922. (U-3:2517a).

1406 _____. _____. Térképtudományi tanszék. Budapest.
Térképtudományi tanulmányok; Studia cartologica. vl- (1956/58-).

*(1407 CARTACTUAL: map service. Budapest. 1- (1965-) 6 per annum. (N70-1:1150c; B68:127a; R-2:317a).
In English, French, German, and Hungarian.
Indexes: 1-20 (1965-1969)?; 21-50 (1970-1974); 51-56 (1975).
[Cartactual, P.O.B. 72, H-1367, Budapest, Hungary.]

*(1408 CARTINFORM. Publishers' selfreviews. Supplement to Cartactual. (Hungary. National office of lands and mapping. Division of cartography). Budapest. 1- (1971-) 6 per annum.
Supplement to Cartactual, 27- (1971-).
In English, French, and German.
[Cartactual, P.O.B. 76, H-1367 Budapest, Hungary.]

1409 DERBECEN. Kossuth Lajos tudományegyetem. Földrajzi intézet. Debrecen.
Közlemények. no1- (1934-) Irregular.
To 1949 as Tisza István tudományegyetem.
In Hungarian with some summaries in other languages.

1410 DEBRECENI Tisza István tudományos társaság. Honismertető bizottsága. Debrecen.
Kiadványai. v1-8 (no1-32) (1924-1932). Closed 1932. (U-2:1269b under society's series 3. Osztályának munkai).
In Hungarian or German.

(1411 FÖLD és ég. (Tudományos ismeretterjesztő társulat, Csillagászati űrhajózási és földrajz-földtan-geofizikai szakosztálya). Budapest. 1- (1966-). (N70-2:2125a).

1412 FÖLD és ember. (Magyar néprajzi társaság). Budapest. v1-10 no6 (1921-1930). Closed 1930. (U-2:1585c; B-2:200a; F-2:615b).

1413 FÖLDGÖMB. (Magyar földrajzi társaság). Budapest. v1-15 (1930-1944). Closed 1944. (U-2:1585c; B-2:200a; F-2:615b; R-3:193b).

1414 FÖLDRAJZ tañitása.(Művelődésügyi minisztérium). Budapest. v1- (1958-). (N70-2:2125a).

*(1415 FÖLDRAJZI értesítő. Geographical bulletin. (Magyar tudományos akadémia, Budapest. Földrajztudományi kutató intézet). Budapest. v1 no1- (1952-) Quarterly. (N70-2:2125a; BS:318a; R-3:193).
1-3 Mimeographed. Supersedes Földrajzi könyv-és térképtar értesítöje [1418]. Before 1968 title of issuing institute was Földrajztudományi kutatócsoport. Subtitle, Geographical bulletin, from 1969.
In Hungarian. Table of contents also in French, German, and Russian from v1 (1952). Summaries in English, German, or Russian from v4 (1955).
[Akadémiai Kiado, Budapest V, Hungary.]

1416 FÖLDRAJZI kézikönyv. (Kozoktatásügyi minisztérium [and] Országos pedagogiai intezet). Budapest.
Tankönyvkiadó. (N70-2:2125ab).

1417 FÖLDRAJZI kiskönyvtár. Budapest. 1- (1952-). (N70-2:2125b).

1418 FÖLDRAJZI könyv-és térképtár értesítője. Budapest. v1-2 (1950-1951). Closed 1951.
Mimeographed. Superseded by Földrajzi értesítő [1415].

*(1419 FÖLDRAJZI közlemények. Geographical review. (Magyar földrajzi társaság). Budapest. 1-76 (1873-1948); nsv1- (77-) (1953-) Quarterly. (U-2:1585c; B-2:200a; BS:318a; F-2:615b; R-3:193b-194a).
None published 1949-1952.
In Hungarian. Table of contents generally in Russian and English; abstracts in English, French, German, or Russian.
See also the Bulletin international edition [1420].
Index: 1-15 (1873-1888).
[Akadémiai Kiado, Budapest V, Hungary.]

1420 ______. Bulletin international. v10-46 (1882-1919); 65-71 (1937-1943). Suspended 1944. (U-21586a; B-2:200b; F-2:615b; F-1:22b).
v10-36 (1882-1908) as Abrégé; v37-46 (1909-1919) as International edition.
English, French, German, or Italian abstracts or text of articles in Földrajzi közlemények [1419] to which volume numbers correspond.
Abstracts in v47-64 (1920-1936) and nsv1- (1953-) published as integral part of Földrajzi közlemények.

(1421 FÖLDRAJZI monográfiák. (Magyar tudományos akadémia). Budapest. 1- (1955-) Irregular. (N70-2:2125b; R-3:194a).
In Hungarian. Summaries in Russian and German.
[Akademiai kiado, Publishing house of the Hungarian academy of sciences, Box 24, H-1363 Budapest, Hungary.]

(1422 FÖLDRAJZI tanulmányok. (Magyar tudományos akadémia, Budapest. Földrajztudományi kutató intézet). Budapest. 1- (1964-) Irregular book series. (N70-2:2125b; R-3:194a).
In Hungarian.
[Akademiai Kiadó Publishing House of the Hungarian Academy of Sciences, Box 24, H-1363 Budapest, Hungary.]

1423 FÖLDRAJZI zsebkönyv. (Magyar földrajzi társaság). Budapest. 1-14 (1939-1963). (N70-2:2125b).
None published 1946, 1949-1957.

1424 GAZDASÁGFÖLDRAJZI gyütemény. (Budapest. Müegyetem. Közgazdaságtudományi kar. Földrajzi intézet). Budapest.
no1-8 (1930-1934). Closed 1934. (U-2:1671c).
In English, German, or Hungarian.

(1425 GEODÉZIA és kartográfia. (Állami földmérési és térképészeti hivatal). Budapest. 1- (1949-). (BS:341a; R-3:266b).
1 (1949) as Az állami földmérés közleményei; 2-6 (1950-1954) as Földméréstani közlemények.

*(1426 GEOGRAPHIA medica; international journal of geography of health; journal international de la géographie de santé (Hungarian geographical society. Medical geographical section and International geographical union. Working group on geography of health). Budapest. (N75-1:905b).
1- (1969/1970-). 1-4/5 (1969/1970-1973/1975) as Geographia medica: international journal on medical geography. Journal internationale de la géographie medicale (Sectio medico-geographica societatis geographicae hungaricae et consilium geographia medica unionis internationalis geographiae).
Supersedes Geographia medica hungarica [1427].
In English.
[Kultura, Box 149, H-1389 Budapest, Hungary.]

1427 GEOGRAPHIA medica hungarica. (Societatis geographicae hungaricae. Sectio medico-geographica). Budapest. 1-3 (1966-1968) Irregular. (N70-2:2280a).
Superseded by Geographia medica [1426].
In English, French, and German.

1428 GEOGRAPHICA actualis. (1970?-) Quarterly.

1429 GEOGRAPHICA hungarica. Budapest. v1-2 (1930-1931). Closed 1931. (U-2:1694a; B-2:268b; F-2:727b; R-3:269b).
In German.

*(1430 GEOGRAPHY of world agriculture. (Magyar tudományos akadémia. Földrajztudományi kutató intézet). Budapest. 1- (1972-) Irregular. (N75-2:1380b).
In English.
[Akadémia Kiadó, Publishing House of the Hungarian Academy of Sciences, Budapest. Distributors: Kultura, H-1389 Budapest, P.O.B. 149, Hungary.]

(1431 GEONÓMIA és bányászat. (Magyar tudományos akadémia. Föld- és bányászati tudományok osztályána közleményei). Budapest. 4- (1971-). (N75-1:910c).
Supersedes and continues number of Magyar tudományos akadémia föld-és bányászati tudományok osztályána (X. Osztály) Közleményei [1439].

1432 MAGYAR földrajzi évkönyv. (Magyar földrajzi intézet). Budapest. 1909-1930. Closed 1930.
To 1924 as Zsebatlasz. In 1925 as Magyar földrajzi évkonyv és Zsebatlasz. None published 1920-1921.
In Hungarian with English, French, and German table of contents.

1433 MAGYAR földrajzi irodalom. Bibliographie géographique hongroise. Budapest.
1936- (F-3:380a).
Supplement to Földrajzi közlemények [1419], in which it is included.

1434 MAGYAR földrajzi társaság. Gazdaságföldrajzi szakosztály. Budapest.
Kiadványai. 1 (1917). Closed 1917. (U-3:2517a).

*(1435 MAGYAR tudományos akadémia. Földrajztudományi kutató intézet.
(Academia scientiarum hungarica. Institutum geographicum), Budapest.
Abstracts. 1- (1964-) Irregular. (N70-3:3587a).
1-11 (1964-1967) issued under former name of institute: Földrajztudományi kutatócsoport.
English abstracts of major research monographs originally in Hungarian.
[Magyar Tudományos Akadémia Földrajztudományi Kutató Intézet,
Népköztársaság utja 62, Budapest 6, Hungary.]

(1436 ______. ______.
Elmélet-módszer-gyakorlat. 1- (1978-).
Supersedes institute's Elméleti és módszertani vitaanyagok, munkajelentések [1437].

1437 ______. ______.
Elméleti és módszertani vitaanyagok, munkajelentések. 1-6 (1963-1977).
Closed 1977. Mimeographed.
Superseded by institute's Elmélet-módszer-gyakorlat [1436].

(1438 ______. ______.
Közlemények. Publicationes. 1- (1952-). (N70-3:3587a).
Reprints mainly from Földrajzi értesítő [1415] or Földrajzi közlemények [1419].

1439 MAGYAR tudományos akadémia föld- és bányászati tudományok osztályának
[X. Osztály] közleményei. (Magyar tudományos akadémia Föld- es bányászati tudományok osztályának közleményei). Budapest. 1-3 (1967-1970). Closed 1970.
Superseded by Geonómia es banyaszat [1431].

1440 MAGYARORSZÁG képekben. Pest.
Honismertető folyóirat. v1-2 (1867-1868). Closed 1868. (B-3:128a).

(1441 MAGYARORSZÁG tájföldrajza. Budapest. 1- (1967-).
[Landscape geography of Hungary].
In Hungarian.
[Akadémiai Kiadó, Box 24, H-1363 Budapest, Hungary.]

1442 NEUES ungrisches Magazin oder Beyträge zur ungrischen Geschichte, Geographie,
Naturwissenschaft...Pressburg, Wien. v1 no1-4 (1791-1792). (B-3:348a;
F-3:716a).

(1443 PÉCS. Dunántúli tudományos intézet. Pécs.
Dunántúli tudományos gyüjtemény. Series geographica. 1- (1955-)
Irregular. (N70-3:4567c).
In Hungarian with some German and Russian summaries.

1444 ______. ______.
Közlemények. 1- (1967-).
In Hungarian or German.

1444a ______. Tanarkepzo foiskola.
Pecsi tanarkepzo foiskola tudományos közleményei. Seria 5. Geographica.
1966- Annual. (N79).
Subseries of Acta academiae paedagogicae in civitate Pecs.
In Hungarian. Abstracts in German.

1445 ______. Tudományegyetem. Földrajzi intézet. Pécs.
Geographia pannonica. no1-44 (1926-1941). Closed 1941.

*(1446 STUDIES in geography in Hungary. (Geographical research institute of the Hungarian academy of sciences). Budapest. 1- (1964-) Irregular. (N70-4:5593bc; R-6:162a).
n1- carried title as Studies in geography.
In English.
[Akadémia kiadó, Budapest, Hungary.]

*(1447 SZEGED. Tudományegyetem. Szeged.
Acta climatologica. (Acta universitatis szegediensis. Pars climatologica scientiarum naturalium). 1- (1959-). (N70-4:5667c; R-1:138b).
In English or German with both English and German summaries.

1448 ______. ______.
Acta universitatis szegediensis. Sectio geographico-historica. v1-7 no2 (1932-1939); nsv1 (1943). Closed 1943. (U-1:60b; B-3:123b; F-1:56b).
Succeeded by Acta geographica [1399].
In Hungarian with abstracts in other languages.

1449 ______. ______. Földrajzi intézet. Szeged.
Értekezések. no1-5 (1942-1943). Suspended 1943.

1450 ______. ______. Földrajzi szeminárium. v1 (1935-1936).
Closed 1936.

(1451 TERMÉSZET világa. Természettudományi közlöny. (Tudományos ismeretterjesztő társulat). Budapest. 1- (108-) (1949-).
Supersedes and continues numbering in parentheses of Természettudományi közlöny [1452]. (R-6:231a).

1452 TERMÉSZETTUDOMÁNYI közlöny. (A királyi Magyar természettudományi társulat; later Természettudományi ismeretterjesztő társulat).
Budapest. 1-107 (1869-1948). Closed 1948. (U-5:4182c; B-3:126a; F-4:669a; R-6:231a).
Indexes: 1-15 (1869-1883); 1-36 (1869-1904).
Superseded by Természet világa [1451].

1453 UNGRISCHES Magazin, oder Beyträge zur ungrischen Geschichte, Geographie, Naturwissenschaft und der dahin einschlagenden Literatur.
Pressburg; Pozsony, Hungary; [now Bratislava, Czechoslovakia].
1-4 (1781-1787). Closed 1787. (U-5:4297a; F-4:767b-768a; G-2:1399b; R-6:300b).

ICELAND (Icelandic)

Lýðveldið Ísland

1454 REYKJAVIK. Náttúrugripasafnid. [Museum of natural history].
Department of geology and geography. Reykjavik.
Miscellaneous papers. 1- (1950-) Irregular reprints. (N70-3:5042a).
In Icelandic with English summaries, or in English.

भारत

INDIA

(English, Hindi, or 14 other languages)

Bharat

(1461 भूदर्शन.

(1506 उत्तर भारत भूगोल पत्रिका

(1455 AJMER geographer. Ajmer. 1- (1969-) Irregular.
In English and Hindi.
[Geographical Society, Dayanand College, Ajmer, Rajasthan, India.]

1456 ALIGARH. Muslim university. Department of geography. Aligarh.
Research publication. 1- (1964-) Irregular.

1457 ALIGARH geographical journal. (Aligarh. D. S. college. Department of geography). vl nol- (Mr 1959-). Closed 1960.

1458 ASSOCIATION of Indian geographers. New Delhi.
Bulletin. 1956, nol-1957 nol. Closed 1957. (N70-1:516b; BS:89a).

(1458a ______.
Annals. 1- (1979-) Quarterly. In English.
[Center for the Study of Regional Development, Jawaharlal Nehru University, New Mehrauli Road, New Delhi 110067, India.]

1459 AVADH geographer. (Avadh geographical society. Journal). Lucknow.
1- (1975-) Irregular. (N77).
[The Avadh Geographer, 12, Ashok Nagar, Lucknow 226001, India.]

1460 BHOOGYAN darpan. (Banaras Hindu university. Department of geography).
Varanasi.
1 (1971). Only one issued. Closed 1971.
In Hindi.

(1461 BHUDARSHAN. Ajmer, Uddaipur, Jaipur. 1- (1967-) Quarterly.
[Geography Office, Guru Kripa, Palace Road, P.O., Uddaipur, Rajasthan, India.]

1462 BHUGOL; Hindi journal of geography. Allahabad. 1924-Apr 1939. 10 nos a year. Closed. (U-1:657c; B-1:326b).
In Hindi.

1463 BOMBAY geographical magazine. (Bombay geographical association). Bombay.
1- (1953-) Annual to 1971; then irregular. (N70-1:820c; BS:133a)
Index 1-10 (1953-1962).
In English.
[The Bombay Geographical Association, c/o Geography Department, Parle College, Bombay 57, India.]

1464 BOMBAY geographical society. Bombay.
Transactions. vl-19 no2 (1836/38-1868/71). Closed 1871.
(U-1:736a; B-1:371b; F-4:712a; F-3:1029a).
v2-5 (1838-1840) as the society's Proceedings; merged in 1873 into Royal asiatic society, Bombay branch. Journal (not in this list).
In English.
Index: 1-17 (1836-1865).

1465 CALCUTTA geographical society. Calcutta.
Publication. 1-6 (1939-1944). no6 repeated 1946. Closed 1946.
(U-2:1695a; B-1:472a).
In English.

*(1466 DECCAN geographer. (Deccan geographical society). Secunderabad.
1- (1962-) Semiannual. (N70-2:1660a; B68:167b).
1-3 nol (Jl 1962-Ja 1965) issued by the society under its earlier name, Hyderabad geography association, in collaboration with the Department of geography, Osmania university.
Index: 1-5 (1962-1967); 11-13 (1973-1975).
In English.
["Subhadra Bhavan", 120/A Nehru Nagar East, Secunderabad 500026, Andhra Pradesh, India.]

(1467 DEVELOPING habitat. (International centre for rural habitat studies).
Varanasi. 1- (1980-) Biannual.
[International Centre for Rural Habitat Studies, Department of Geography, Banaras Hindu University, Varanasi 221005, India.]

*(1468 GEOGRAPHER. (Aligarh. Muslin university. Geographical society. Journal).
Aligarh. 1- (1948-) Annual. (N70-2:2280a; BS:341b).
Not published 1960-1964.
In English.
[The Geographer, Department of Geography, Aligarh Muslim University, Aligarh, Uttar Pradesh, India.]

1469 GEOGRAPHER. (Curzon geographical society). Aligarh. Ja 1926-Mr 1927.
Closed Mr 1927. (U-2:1693c; B-2:268b).
In English.

*(1470 GEOGRAPHICAL bulletin of India. (Patna university. Association of geographers). Patna. 1- (1977-) 2 nos a year.
In English.
[Association of Geographers, Patna University, Patna 800 005, India.]

*(1471 GEOGRAPHICAL knowledge. (Society for geographical studies). Kanpur.
1- (1968-) Irregular. (N70-2:2281b; B73:128a).
In English.
[The Secretary, The Society for Geographical Studies, 7/125 Swarup Nagar, Kanpur-2, Uttar Pradesh, India.]

(1472 GEOGRAPHICAL observer. (Meerut college geographical society). Meerut.
1- (1965-) Irregular. (N70-2:2281c; B68:225a).
In English or Hindi with abstracts in other language.
[Meerut College Geographical Society, Department of Geography, Meerut College, Meerut, Uttar Pradesh, India.]

*(1473 GEOGRAPHICAL outlook. (Ranchi university. Department of geography)
Ranchi. 1- (Ja 1956-) Irregular. (N70-2:2281c).
In English.

1474 GEOGRAPHICAL research. (J. V. Jain college). Saharanpur. 1- (1977-)
Annual.
[Department of Geography, J. V. Jain College, Saharanpur, U. P., India.]

*(1475 GEOGRAPHICAL review of India. (Geographical society of India). Calcutta.
1- (1936-) Quarterly. (U-2:1695a; B-1:472a).
v1-12 (1936-1950) as Calcutta geographical review.
Index: v1-29 (1936-1967).
In English.
[Geographical Society of India,35 Ballygunge Road, Calcutta 700019), West Bangal, India.]

1476 GEOGRAPHICAL series. (Kitab mahal). Allahabad. 1- (1964-). (N70-2:2281c).
[56A Zero Road, Allahabad 3, Uttar Pradesh, India.]

1477 GEOGRAPHICAL thought. (Association of North Indian geographers. Journal). Gorakhpur. 1-3 (D 1965-1967) Semiannual. Closed 1968. (N70-2:2281c). Superseded by Uttar Bharat bhoogol patrika [1506].

*(1478 GEOGRAPHICAL viewpoint. (Agra geographical society). Agra. v1- (1970-) Irregular. (N75-1:906c; N75-1:906b).
v1 no1 (Ap 1970) as Geographical outlook.
In English.
[Agra Geographical Society, Raja Balwant Singh, Department of Geography, Agra 282002, India.]

*(1479 GEOGRAPHY teacher (India). (Society for the promotion of education in India). Madras. 1- (1965-) Bimonthly. (N70-2:2282b).
In English.
[3, Thiruvengadam St. Raja Annamalaipuram, Madras, India.]

(1480 GORAKHPUR. University. Department of geography. North India geographical society. Uttar Bharat bhoogol parishad. Gorakhpur.
U.B.B.P. research series. 1- (1973-) Irregular.

1481 INDIAN council of social science research. New Delhi.
ICSSR journal of abstracts and reviews: geography. 1- (1975-) Irregular.

1482 INDIAN geographer. (Association of Indian geographers). New Delhi.
1-12 (1956-1967). Closed 1968. (N70-2:2731a; BS:407b).
In English.

*(1483 INDIAN geographical journal. (Indian geographical society). Madras.
1- (1926-) Quarterly. (U-3:1957a; B-3:115b; F-3:206a).
1 (1926) as Madras geographical association. Bulletin; 2-15 (1927-1940) as Madras geographical association. Journal.
Index: 1926-1959 in v34 no1/2 (1959), p. 67-85.
In English.
[Department of geography, University Centenary Buildings, Chepauk, Madras 600005, India.]

1484 INDIAN geographical review. (J. V. Jain college). Saharanpur. 1- (1977-) Annual.
[Department of Geography, J. V. Jain College, Saharanpur, U.P., India.]

1485 INDIAN geographical society. Madras.
Monograph. 1-3 (1945-1951). (U-3:1957a; B-2:465b).
1 (1945), 2 (1946), 3 (1951).
In English.

*(1486 INDIAN geographical studies: research bulletin (Geographical research center). Patna. 1- (1973-) 2 nos a year. (N77).
In English.
[Department of Geography, Patna University, Patna 800005, India.]

(1487 INDIAN geography. (Indian geographical association). Raipur.
1- (1973-) Annual.
In English. Abstracts in Hindi from 1976.
[Indian Geographical Association, Department of Geography, Ravishankar University, Raipur, Madhya Pradesh, 492002, India.]

1488 INDIAN journal of geography. (Jodhpur. University. Department of geography. Association of geographers. Journal). 1-12 (1966-1977) Annual. Closed 1977. (N70-2:2734a; B68:256a).
In English or in Hindi with English abstracts.

(1489 INDIAN journal of regional science. (Regional science association of India). Kharagpur. 1- (1968-). (N75-1:1080a; B73-146b).
[Indian Institute of Technology, Kharagpur, West Bengal, India.]

(1490 INSTITUTE of economic geography, India. Calcutta. 1- (Ja 1970-)
Irregular. (N75-1:1118a).
In English.
[4/1, Ashton Road, Calcutta 20, India.]

1491 INSTITUTE of Indian geographers, Patna. Patna.
Publication. no1 (1954). Closed.
In English.

1492 INTERNATIONAL council for the study of Afro-Asian geography, Aligarh.
Bulletin. 1 (1956). No more published. (N70-2:2903b).
In English.

1493 JOURNAL of geography. (Jabalpur. University. Department of studies in
geography. Geography academic society). Jabalpur. 1-2 (1959-1960).
Closed 1968. (N70-2:3169a).

(1494 MADRAS. University.
Madras university geographical series. 1- (1972-) Irregular.
(N75-2:1377a).

1495 MAHARAJA Sayajirao university of Baroda. Baroda.
Research paper series: geography. 1- (1965-) Irregular. (N70-3:3590b;
N75-2:1380b).
In English.

1496 ______. Faculty of science. Baroda.
Geographical series. 1 (1962). (N70-3:3590a).
In English.
[University of Baroda Press, Palace Road, near Palace Gate, Baroda,
Gujarat, India.]

1497 NAGPUR university geographical journal. Nagpur. v1 no1 (Ap 1956) Irregular.
Closed.
All issued?

*(1498 NATIONAL geographer. (Allahabad. Geographical society). Allahabad.
1-5 (1958-1962); 6- (1971-). v1-10 Annual; v11- (1976-) 2 nos a
year. (N70-3:4036a). Not published 1963-1970.
In English.
[The Allahabad Geographical Society, Department of Geography, University of Allahabad, Allahabad-211002, India.]

*(1499 NATIONAL geographical journal of India. (National geographical society of
India). Varanasi. 1- (1955-) Quarterly. (N70-3:4036a; BS:595b).
In English.
[National Geographical Society of India, Department of Geography,
Banaras Hindu University, Varanasi 221005, Uttar Pradesh, India.]

*(1500 NATIONAL geographical society of India. Varanasi.
NGSI Research publication series. 1- (1964-) Irregular, but recently
2 a year.
In English. (N70-3:4036a).
[National Geographical Society of India, Banaras Hindu University,
Varanasi 221005, India.]

*(1501 National geographical society of India, Varanasi.
Research bulletin series. 1-21 (1946-1954); 22- (1974-) Irregular.
(U-4:2879a; BS:595b).
1-21 (1946-1955) as Bulletin. Not published 1956-1973.
In English.
[Department of Geography, Banaras Hindu University, Varanasi 221005,
India.]

(1502 NORTH east India geographical society. Gauhati.
Journal. 1- (1969-) Semiannual. (N75-2:1611b).
In English.
[Science Block No. IV, Gauhati University, Gauhati-14, Assam, India.]

1503 OBSERVER. (Calcutta. University. Student's geographical association. Magazine). Calcutta. Annual.
v3 (1957), 4 (1958). Closed 1967.
In English.

1504 OSMANIA university, Hyderabad. Department of geography. Centre for urban research. Hyderabad.
Occasional papers.

1505 STUDIES in geography. (Rajasthan. University. Department of geography). Jaipur. 1- (1968/69-) Annual. (N78).
In English.

(1506 UTTAR Bharat bhoogol patrika. North India geographical journal. (Uttar Bharat bhoogol parishad. North India geographical society). Gorakhpur. 4- (1968-) Quarterly. Issues often combined.
Continues numbering of predecessor: Geographical thought [1477].
In Hindi with abstracts in English.
[Jagdish Singh, editor, Uttar Bharat Bhoogol Parishad, Department of Geography, Gorakhpur University, Gorakhpur 273001, India.]

1506a YUVA bhoogol vidya patrika. Young geographers journal. (Yuva bhoogol vidya parishad), Gorakhpur. 1 (1977).
In Hindi.
[Yuva Bhoogol Vidya Parishad, Gorakhpur University, Gorakhpur 273001, India.]

INDONESIA (Bahasa Indonesia)

Republik Indonesia

1507 BERITA topografi. Topographic news. (Indonesia. Angkatan darat. Direktorat topografi). Djakarta. 1- (1965-) Quarterly. (N75-2:2136a). Cover title: Berita topografi. Compiled by Dinas penelitian dan pengembangan direktorat topografi angkatan darat.
[Jawatan dan pusat topografi TNI-AD, Jl. Gunungsahari 90, Jakarta, Indonesia.]

1508 IKATAN geograf Indonesia. Djakarta.
Laporan I.G.I. 1- (1969-). (N75-1:1044c).

1509 INDISCH aardrijkskundig genootschap. Semarang.
Tijdschrift. v1 no1-4 (1880-1883). Closed 1883. (B-2:471b).
In Dutch.

1510 INDONESIA. Direktorat topografi angkatan darat. Dinas geografi. Djakarta.
Publikasi. 1- (1950-) Irregular. (N70-2:2759c).
Also listed as: Indonesia. Angkatan darat. Direktorat topografi. Dinas geografi. Publikasi; Indonesia. Departemen angkatan darat. Direktorat topografi angkatan darat; Indonesia. Direktorat topografi angkatan darat. Institute of geography. Publication; Indonesia. Departemen angkatan darat. Dinas geografi. Publikasi; Indonesia. Dinas geografi. Publication; Indonesia. Instituut geografi; Indonesia. Balai geografi.
Indonesian and Dutch to 1954. Indonesian and English from 1956.
[Dj. Dr. Wahidin Satu 11, Djakarta, Indonesia.]

1511 ______. Kementerian pertahanan. Direktorat topografi angkatan.
Balai geografi. (Ministry of defense. Topographical service of the army. Geographical institute). Djakarta.
Lapuran (Report). 1954- Irregular.
In Bahasa Indonesia.

*1512 INDONESIAN journal of geography. (Yogyakarta. Universitas Gadjah Mada. Faculty of geography). Yogyakarta. v1-5 (no1-9) (1960-Jn 1965); 6- (no10/31-) (Jn 1976-) 2 nos a year.
v1-5 (1960-1965) as English edition of Madjalah geografi Indonesia. Edisi inggeris [1513].
In English.
[The Indonesian Journal of Geography, Faculty of Geography, Gadjah Mada University, Bulaksumur, Yogyakarta, Indonesia.]

1513 MADJALAH geografi Indonesia. Edisi inggeris. (Yogyakarta. Universitas Gadjah Mada. Begian ilmu bumi). Yogyakarta. v1-5 (no1-9) (1960-1965); v6 (1976). Closed 1976. (N70-3:3570c; B68:339b).
English edition entitled: Indonesian journal of geography [1512].
In Bahasa Indonesia.

1514 MADJALAH untuk ilmu bahasa, ilmu bumi dan kebudajaan Indonesia. (Lembaga kebudajaan Indonesia). Djakarta.
1-86 (1852/53-1958). (U-3:2504b; B-4:319a; F-4:688a).
Suspended 1943-1947, 1953-1954. v4-6, 7-9, 10-14, 15-17, and 18-20 called also new series, v1-3; s3 v1-3; s4 v1-5; s5 v1-3; s6 v1-3. v1-85 no3 as Tijdschrift voor Indische taal-, land- en volkenkunde. In Dutch. v85- no4-v86 (1955/57-1958) under present name. In English or Bahasa Indonesia. v1-83 (1852/53-1949) by the society under its earlier name: Bataviaasch genootschap van kunsten en wetenschappen.
Indexes: 1-50 (1852/53-1907; 51-58 (1909-1919) with indexes to society's Verhandelingen in 2 v.

1515 NEDERLANDSCH-Indische geografische mededeelingen. (K. Nederlandsch aardrijkskundig genootschap, Amsterdam). Djakarta [Batavia]. v1 no1-6 (1941). Closed 1941. (U-4:2953a; B-3:333a).
In Dutch with English abstracts.

1516 PUSTAKA djurusan geografi. (Djurusan geografi FKIP universitas Tjenderawasih). Djajapura [Sukarnapura]. 1-13 (1964-1967). Mimeographed.
At head of title: Universitas Negeri Tjenderawasih.
In Bahasa Indonesia.

IRAN (Persian [Fārsī])

جمهوری اسلامی ایران

1517 نشریهٔ انجمن دبیران جغرافیا و تاریخ و تعلیمات اجتماعی تهران .

1518 انتشارات بنیاد فرهنگ ایران : منابع تاریخ و جغرافیای ایران .

1519 فرهنگ آبادیهای ایران .

1520 فهرست مقالات جغرافیایی ،تهران .

1522 جهان شناسی ، مشهد .

1523 جغرافیـــا

1525 انتشارات موسسهٔ جغرافیایی دانشگاه تهران ، تهران .

1517 ANJUMAN-i dabīrān-i jughrāfiyā va tārīkh va ta^c^līmāt-i ijtimā^c^ī. Tehran. Nashrīayh. 1963- Irregular.
English title: Publication of the Association of teachers of geography, history and social sciences.

1518 BUNYĀD-i farhang-i Īrān. Manābi^c^-i tārīkh va jughrāfiyā-yi Iran. Tehran. (N70-1:951a).
Subseries of its Intishârât (not in this list).

1519 FARHANG-i ābādīhā-yi Īran. Tehran. 1967- .
English title: Iranian village gazetteer.
[Iranian Statistical Center, Tehran, Iran.]

1520 FIHRIST-i maqālāt-i jughrāfiyāyi. Tehran. 1963- Irregular.
English title: Index of geographical titles [in Persian periodicals].
[M. H. Ganji, Faculty of Arts and Humanities, University of Tehran, Tehran, Iran.]

1521 INSTITUT scientifique de recherches géographiques et géochimiques en Asie. Tehran.
Bulletin. 1-4? (1930-1933). Closed 1933? (U-3:2003b).
In French.

1522 JIHĀN'SHINĀSĪ. Mashhad.
English title: Geography; quarterly review of geographical studies.

1523 JUGHRĀFIYĀ (Nashrīyah-'i Anjuman-i Jughrāfī dānān-i Īrān). Tehrân. 1976- .
English title: Geography (Organ of the Association of Iranian geographers).

1524 SOCIÉTÉ géographique de Khorâsân. Mashhad.
Publications. 1- (1969?-) Irregular. (N70-4:5384c).
In French.

1525 TEHRAN. Dānishgāh. Tehran.
Intishārāt-i mu'assisah-'i jughrāfiyā. 1968- Irregular.
English title: Tehran. University. Geographical institute. Occasional publications.

1526 ______. ______. Markaz-i taḥqīqāt-i ᶜilmī-i manāṭiqi-i khushk (Arid zone research centre). Tehran.
Publication 1- (1959? -) Irregular. (N70-4:5725b).
In English or French.

IRAQ (Arabic)

al-Jumhūrīyah al-ᶜIrāqīyah

الجمهورية العراقية

1527 الجميعة الجغرافية العراقية

1527 JAMᶜĪYAH al-jughrāfīyah al-ᶜIraqīyah. Baghdad.
al-Majallah. v1-3 (1962-1965). (N70-2:3069a; B68:279a).
English title: Iraqi geographical journal.
In Arabic or English.
[Geographical Section, College of Arts, University of Baghdad.]

IRISH REPUBLIC (English or Irish)

Éire

1528 BAILE. (Dublin. University college. Department of geography. Geographical society. Magazine). Dublin. 1- (1966-) Annual.
In English.
[Department of Geography, University College, Dublin 4, Ireland.]

*1529 GEOGRAPHICAL viewpoint. (Association of geography teachers of Ireland. Journal). Dublin. v1 no1- (1964-) Annual. (N70-2:2281c-2282a; B73:225b).
v1 no1 (1964) as Association's Journal.
In English. Abstracts.

*(1530 IRISH geography. (Geographical society of Ireland). Dublin. v1- (1944-) Annual. (U-3:2113c; B-2:269a).
v1 no1-3 as Geographical society of Ireland. Bulletin.
Index every five years.
In English. Abstracts.
[Geographical society of Ireland, Department of Geography, Trinity College, Dublin 2, Ireland.]

1531 MILIEU. (Maynooth geographical society). Maynooth. 1975- Annual.
In English, occasionally Irish.
[Department of Geography, St. Patrick's College, Maynooth, County Kildare, Ireland.]

ISRAEL (Hebrew or Arabic)

Medinat Yisra'el; Isrāil

מדינת-ישראל

(1532 ארץ-ישראל

(1536 מחקרים בגיאוגרפיה של ארץ-ישראל.

1537 מרחבים

1538 כופים

1539 אופקים

(1532 ERETS-YISRAEL. (Jewish Palestine exploration society. Jerusalem). Jerusalem. 1- (1951-) Biennial. (N70-2:1941a).
Added title page. Erets-Israel: Archaeological, historical and geographical studies.
Text in Hebrew or English.
[Israel Exploration Society, P. O. Box 7041, Jerusalem, Israel.]

(1533 ISRAEL exploration journal. (Israel exploration society). Jerusalem. 1- (1950/51-). (N70-2:3019c; BS:449a).
Index: v1-10 (1950-1960) in v11 (1961), p. 209-264.
[POB 7041, Jerusalem, Israel.]

1534 JERUSALEM studies in geography. (Jerusalem. Hebrew university. Department of geography). Jerusalem. 1-2 (1970-1971). Closed 1971. (N75-1:1231c; B73:166a).
In English.

1535 JERUSALEM urban studies. (Jerusalem. Hebrew university. Institute of urban and regional studies). Jerusalem. 1- (1970-). (N75-1:1231c).
In English.

*(1536 MEHKARIM ba-ge'ografyah shel érets Yisrael. Jerusalem. 1- (1959-) Irregular. (N70-3:3718a).
English title: Studies in geography of Israel. v1-5 (1959-1965) as Studies in the geography of Éretz-Israel. 1-2 (1959-1960) as offprint from Bulletin of the Israel Exploration Society (Yediot); 3-5 (1962-1965) from Yediot Bahakirat Eretz-Israel Weatikoteha, continuing the Bulletin of the Israel Exploration Society, which ceased publication in 1967. New series 6- (1968-) continues this series but as an independent publication.
In Hebrew with table of contents and abstracts in English.
[Department of Geography, The Hebrew University of Jerusalem, Jerusalem, Israel.]

1537 MERHAVIM. (Universitat Tel Aviv. Ha-hug le-ge'ografyah). Jerusalem. 1- (1974-). (N76).
"Kovets mekharim ba-ge'ografyah shel erets Yisra'el veha-mizrah ha-tikhon."
Added title page: Spaces: collected studies on the geography of Israel and the Middle East (Tel Aviv. University. Geography department).
In Hebrew with supplementary table of contents and abstracts in English.

1538 NOPHIM. (Aveshalom institute for homeland studies). Tel-Aviv. 1- (1975-).
Added title page: Iyunim beyediat haaretz (Homeland studies).
In Hebrew.
[Department of Geography, Tel-Aviv University, Tel-Aviv, Israel.]

1539 OFAKIM. (Haifa. University. Department of geography). Haifa. 1- (1975-) Irregular.
Added title page: Ofakim ba-ge'ografyah - kovets le-zekher Amiasaf Hochman. Horizons: Studies in geography in memory of Amiasaf Hochman.
In Hebrew with titles in table of contents and abstracts in English.
Some articles in English.

1540 STUDIES in geography. (Jeruslaem. Hebrew university). Jerusalem. 1-3 Irregular. (N70-4:5593c).
Subseries of university's Scripta hierosolymitana, 1954-1968 (N70-2:3113a) (not in this list).

ITALY (Italian)

Repubblica Italiana

1541 ALMANACCO geografico. Bergamo. v1-2 (1892-1893). Closed 1893.
Supplement to Universo. Geografia per tutti [1657].

1542 ANNALI di geografia e di statistica. Genova. v1-2 no4 (1802). Closed 1802.
(U-1:385b; B-1:155b; F-1:274b).

1543 ANNALI di ricerche e studi di geografia. (Genova. Università. Istituto di geografia). Genova. 1- (1945-) Quarterly. (U-1:386a; B-1:156a).
no1-2 (1945-1946) as Istituto di geografia dell'Ateneo Genovese. Annali.
Index: 1-10 (1945-1954) in v11 no2 (1955); 11-20 (1955-1964) in v21 no1 (1965), p. 27-35.
English, French, and German abstracts.
[Direttore Prof. Emilio Scarin, Via Molfino 32, Ruta di Camogli, Genova 16100, Italy.]

1544 ANNALI universali di statistica, economia pubblica, geografia, storia e viaggi. Milano. 1-80 (1824-1844); s2 v1-36 (1844-1853); s3 v1-24 (1854-1859); s4 v1-48 (1860-1871). Closed 1871. (U-1:386b; B-1:156b; F-1:276a). 1 as Annali universali di viaggi, geografia, storia, economia pubblica e statistica.

1545 ANNUARIO geografico italiano. Bologna. 1-2 (1844-1845). Closed 1845. (U-1:402b; B-1:169b; F-1:334b).

(1546 ASSOCIAZIONE italiana di cartografia. Firenze, later Novara. Bollettino. 1- (1964-) 3 nos a year. (N70-1:522b). [Corso della Vittoria 91, 28100 Novara, Italy.]

(1547 ATLANTE: Alla scoperta del mondo. (Istituto geografico de Agostini. Mensile). Novara. 1- (1960-) Monthly. (N70-1:532c). [Corso della Vittoria 91, 28100 Novara, Italy.]

(1548 BARI. Università degli studi. Facoltà di economia e commercio. Istituto di geografia. Bari. Memorie. nsl- (1963-) Irregular. [Largo Fraccacreta 1, 70122 Bari, Italy.]

(1549 ______. ______. Facoltà di lettere e filosofia. Istituto di geografia. Quaderni di geografia. nsl- (1976-). [Palazzo Ateneo, 70121 Bari, Italy.]

(1550 ______. ______. Facoltà di magistero. Istituto di geografia. Bari. Pubblicazioni. sl. 1- (1971-) Irregular. s2. 1- (1974-) Irregular. [Via Quintino Sella 268, 70122 Bari, Italy.]

1551 ______. ______. Università. Istituto di geografia. Bari. Memorie. 1-18 (1936-1959). Closed 1959. (U-1:605b).

1552 ______. ______. ______. Sussidi didattici e scientifici. 1-8 (1936-1940); nsvl-8 (1941-1953). Closed 1953. (U-1:605b).

1553 BIBLIOTECA di geografia storica. Roma. 1-3 (1906-1907). Closed 1907. (U-1:679b). Monographs.

1554 BOLLETTINO nautico e geografico di Roma. Roma. vl-2 (1861-1863). Closed 1863. (B-1:457a).

1555 BOLOGNA. Università. Istituto di geografia economica. Bologna. Contributi. 1- (1974-) Irregular. [Largo Scaravilli 2, 40126 Bologna, Italy.]

1556 CAGLIARI. Università. Facoltà di economia e commercio. Istituto di geografia. Cagliari. Ricerche di geografia regionale. 1 (1965). [Viale Sant'Ignazio da Laconi 15, 09100 Cagliari, Italy.]

1557 ______. ______. Istituto di geografia. Contributi alla geografia della Sardegna. 1- (1951-) Irregular. (N70-1:987a). [Istituto di Geografia, Piazza d'Armi, 09100 Cagliari, Italy.]

(1558 CALENDARIO-atlante de Agostini. (Istituto geografico de Agostini). Novara. 1- (1903-) Annual. (B-1:473a).

1559 CAMERINO. Università. Istituto di mineralogia, geologia, e geografia. Camerino. Pubblicazioni. (N60-1:375a). sA. Studi e ricerche originali. 1 (1954). Closed 1954. sB. Sussidi didattici e metodologici. 1 (1955). Closed 1955. (N70-1:1050b).

1560 CATANIA. Università. Istituto di geografia fisica e vulcanologia. Pubblicazioni. no1-4 (1914-1915). Closed 1915. (U-2:943a; B-1:514a).

1561 CIRCOLO geografico italiano, Torino. Torino. Pubblicazioni. Periodico bimestrale di geografia, etnografia e scienze affini. 1-4 (1872-1875). Closed 1875. (U-2:1058a; B-3:526b).

(1562 COLLANA di bibliografie geografiche delle regioni italiane. (Italy Consiglio nazionale delle ricerche. Comitato per le scienze storiche, filologiche e filosofiche). 1- (1959-) Irregular.
1959-1970 issued by Comitato per la geografia, geologia e mineralogia. Some have English table of contents.

1563 COLTURA geografia. (Trieste. Università. Istituto de geografia). Trieste. 1-3 (F 1930-Ap 1932). Closed 1932. (U-2:1115a).
Continued as Rivista di geografia e cultura geografica [1629].

(1564 COMITATO dei geografi italiani. Commissione per la geografia storica delle sedi umane in Italia. Firenze. 1- (1974-) Irregular.
[Olschki, Casella Postale 295, Firenze, Italy.]

(1565 ______. Commissione porti. (Venezia. Istituto di geografia economica). Venezia. Documenti. 1-4 (1973-1977).
[Campo San Polo 2169, 30125 Venezia, Italy.]

1566 ______. Commissioni di ricerca. Studi su: Città, sistemi metropolitani, sviluppo regionale. (Bologna. Istituto di geografia economica). Bologna. Quaderni. 1-2 (1973-1974).

1567 COMUNICAZIONI di un collega. Rivista illustrata di geografia e storia. Organo di communicazioni fra gli insegnanti delle scuole secondarie. Cremona; Bergamo. 1-17 (1894-1911). Closed 1911.

*(1568 CONGRESSO geografico italiano. Roma.
Atti. 1- (1892-) Irregular. About every 3 years. (U-2:1172a; B-1:639a; F-1:494a).
No congresses held between 1910 and 1921 and 1937 and 1947.
Indexes: 1-10 (1892-1927), 11-20 (1930-1967).

1569 CORRESPONDANCE astronomique, géographique, hydrographique et statistique du baron de Zach. Genova. v1-15 no1 (1818-1826). Closed 1826. (U-2:1210a; B-1:658b; F-2:266b).

1570 COSMOS. Comunicazioni sui progressi più recenti e notevoli della geografia e delle scienze affini di Guido Cora. (Società di geografia ed etnografia). Torino; Roma. s1 v1-10 (1873-1891); v11-13 (1892-1913) as s2 v1-3. Closed 1913. (U-2:1213b; B-1:660b; F-2:277b).

1571 CULTURA geografica. Rassegna quindicinale illustrata di geografia. Firenze. v1 no1-10 (1899). Closed 1899.

1572 EROSIONE del suolo in Italia. (Consiglio nazionale delle ricerche. Centro di studio per le geografia fisica). Padua. 1-2 (1962). Closed 1962. (N70-2:1946a).

1573 ESPLORATORE. Giornale di viaggi e geografia commerciali. Organo della Società d'esplorazione commerciali in Africa. Milano. 1-10 (1876-1886). Closed 1886. (U-2:1478b; B-2:139b; F-2:515a).
Superseded by Esplorazione commerciale [1575].

1574 ESPLORAZIONE Rassegna quindicinale delle conquiste geografiche e degli'interessi italiani in tutti i punti del globo. Napoli. v1 no1-6 (1883). Closed 1883. (U-2:1487b; F-2:515a).

1575 ESPLORAZIONE commerciale. (Società italiana di geografia commerciale). Milano. v1-43 no1/3 (1886-1928). Closed 1928. (U-2:1478b; B-2:139b; F-2:515a).
vl as Esplorazione commerciale. Giornale di viaggi e di geografia commerciale; v2-13 as Esplorazione commerciale e l'esploratore. Giornale di viaggi e di geografia commerciale; v14-25 as Esplorazione commerciale (già l'esploratore).
1886-1898 as Società di esplorazione commerciale in Africa; 1899-1923 as Società italiana d'esplorazioni geografiche e commerciali.
Supersedes Esploratore [1573]. v7 of Espansione commerciale d'Italia (not in this list) substituted for v38.

(1576 FIRENZE (Florence). Università. Facoltà di magistero. Firenze.
Atti. Quaderno. 1-6 (1971-1977).
Quaderno 6 issued under name: Istituto di geografia.
[Via S. Gallo 10, 50129 Firenze, Italy.]

1577 ______. ______. Istituto di geografia. Firenze.
Pubblicazioni. 1-2 (1945-1948). Closed 1948. (U-2:1576c).

1578 ______. ______. Istituto di geografia economica. Firenze.
Memorie. 1-2 (1973-1974).
No2 has summaries in English, French, German, and Polish.

1579 ______. ______. Istituto di geografia economica.
Pubblicazioni. 1- (1955-) Irregular.
In part reprints.
[Via Curtatone 1, Firenze, Italy.]

(1580 GENOVA (Genoa). Università. Facoltà di economia e commercio. Istituto di geografia economica. Genova.
Quaderno di studi e ricerche. 1- (1963-) Irregular.
1-4 (1963/64-1966/69 as Quaderno di studi e ricerche di geografia economica e regionale.
[Via A. Bertani 1, 16125 Genova, Italy.]

(1581 ______. ______. Facoltà di magistero. Istituto di scienze geografiche. Genova.
Pubblicazioni. 1- (1966-) Irregular, but several a year.
1-2 (1966) issued by Institute under earlier name: Istituto di geografia.
[Lungoparco Gropallo 3, 16122 Genova, Italy.]

1582 GEOGRAFIA. (Istituto geografico de Agostini, Novara). Novara.
1-18 (Ja 1912-D 1930). Closed 1930. (U-2:1692a; B-2:268a; F-2:727a).
Title also as: Rivista di propaganda geografica.
1914-1921 contains: Le bibliografie dell'Istituto geografico de Agostini.

(1583 GEOGRAFIA: trimestrale di ricerca scientifica e di programmazione regionale. Roma. 1- (1978-) 3 nos a year.
[Via Giorgio Baglivi 3, 00161 Roma, Italy.]

(1583a GEOGRAFIA fisica e dinamica quaternaria. (Bollettino del Comitato glaciologico italiano, ser. 3). Torino. 1- (1978-) Semiannual.

(1584 GEOGRAFIA nelle scuole. (Associazione italiana degli insegnanti di geografia. Notiziario). Napoli. 1- (1955-) Bimonthly. (N70-2:2278a).
[Via Vitelleschi 26, 00193 Roma, Italy.]

1585 GEOPOLITICA: Rassegna mensile di geografia politica, economica, sociale. Milano. 1-4 (1939-1942). Closed 1942. (U-2:1701c; F-2:733b).

(1585a GEOS: la terra che vive. Bimestrale geografico culturale. Milano.
1- (1979-) Bimonthly.
[Via Craducci 13, 20123 Milano, Italy.]

1586 HÉRODOTE/Italia: strategie, geografie, ideologie. Verona.
1- (1978-) Quarterly.
[Giorgio Bertani editore, Lungadige Panvino 37, 37100 Verona, Italy.]

1587 IN giro pel mondo. Bologna. 1-3 (1899-1901). Closed 1901.

1588 ISTITUTO coloniale italiano. Sezione studi e propaganda. Roma.
Memorie e monografie coloniali. Serie geografica. 1-3 (1918-1920).
Closed 1920. (U-3:2126a; B-3:178b).

1589 ISTITUTO geografico de Agostini, Novara. Novara.
Biblioteca geografica. 1-2 (1913-1914). Closed 1914. (U-3:2127b;
B-1:339a).

1590 ITALY. Comitato geografico nazionale italiano.
Pubblicazione. no1-3 (1922-1928). Closed 1928. (U-2:1127b).
Superseded by Italy. Consiglio nazionale delle ricerche.
Comitato per la geografia. Pubblicazioni [1591].

1591 ______. Consiglio nazionale delle ricerche. Comitato per la geografia.
Roma; etc.
Pubblicazioni. 1932-1948 in nine series. (U-3:2133b; B-1:609a;
BS:218ab).
s1. Ricerche sulle variazioni delle spiagge italiane. v1-5 (1933-1940).
For continuation see Ricerche sulle variazioni delle spiagge italiane
[1627]. Issued by Bologna. Università. Istituto di geografia, in
cooperation with the consiglio's Comitato per l'ingegneria. Preceded
by an unnumbered introductory vol. dated 1936.
s2. Ricerche sulle variazioni storiche del clima italiano. no1-9 (1934-
1942). For continuation see Ricerche sulle variazione storiche del clima
italiano [1628]. Issued by Bologna. Univeristà. Istituto di geografia.
s3. Studi geografici sulle terre redente e nuovi territori.
v1-9 (1934-1943). Closed 1943. First part of series issued by Roma.
Università. Istituto di geografia; later by Bologna. Università.
Istituto di geografia. (U-5:4093c).
s4. Ricerche sui terrazzi fluviali e marini d'Italia. v1-3 (1934-1939).
Closed 1939. Published by Bologna. Università. Istituto di geologia.
s5. Ricerche sulla distribuzione altimetrica della vegetazione in
Italia. Bologna. v1-5 (1934-1942). Closed 1942.
s6. Ricerche di geografia economica sui porti italiani. no1-5 (1936-
1942). Closed 1942. Published by Napoli. Istituto superiore navale.
Laboratoria di geografia economica.
s7. Lo Spopolamento montano in Italia. no1-8 (1932-1938). Closed 1938.
Also numbered as Italy. Istituto nazionale di economia agraria.
s8. Ricerche sulle dimore rurali in Italia. no1-4 (1938-1943). For
continuation see Ricerche sulle dimore rurali in Italia [1626]. Published by Firenze. Università. Istituto de geografia.
s9. Studi geografici sulle città italiane. v1-3 (1941-1948). Closed
1948. Issued by Firenze. Università. Istituto di geografia.
[s10. See Ricerche sulla morfologia e idrografia carsica [1625].

(1591a ______. ______. Comitato per le scienze storiche, filosofiche e filologiche,
Roma.
Glossario di termini geografici dialettali della regione italiana.
Roma, 1- (1978-).
[Istituto di Geografia dell'Università, Roma, Italy.]

1592 ______. Istituto geografico militare, Firenze. Firenze.
Annuario. 1-3 (1913-1915); 1955. (U-3:2134a; B-2:563b).
Older series superseded by Universo [1656].

1593 LECCE. Università. Facoltà di magistero. Istituto di geografia.
Quaderni. 1-4 (1974-1976). (N78).
[Palazzo Castro, 73100 Lecce, Italy.]

1594 ______. Università Salentina. Laboratorio di geografia.
Pubblicazioni. Bari. 1- (1958-). Closed 1958. (N70-3:341a).
At head of title: Consorzio universitario Salentino, Lecce.

1595 MEMORIE di geografia antropica. (Italy. Consiglio nazionale delle ricerche. Centro di studi per la geografia antropica [and] Roma. Università. Istituto di geografia). 1-16 (1946-1959). Closed 1959. (B-3:178b).
Superseded by Memorie di geografia economica e antropica [1597].

1596 MEMORIE di geografia economica. (Italy. Consiglio nazionale delle ricerche. Centro di studi per la geografia economica [and] Napoli. Università. Istituto di geografia). Napoli. 1-20 (1949-1959). Closed 1959.
Superseded by Memorie di geografia economica e antropica [1597].

1597 MEMORIE di geografia economica e antropica. (Napoli. Università. Istituto di geografia). Napoli. nsv1- (1963-) Annual. (B68:353b).
Successor to Memorie di geografia economica [1596] and Memorie di geografia antropica [1595].
In Italian with supplementary table of contents in English, French, or German.
[Istituto di geografia dell' Università di Napoli, Largo S. Marcellino 10, Napoli, Italy.]

1598 MEMORIE geografiche. Firenze. 1-13 (no1-39) (1907-1919). Closed 1919. (U-3:2601c; B-3:737b; F-3:503a).
Supplement to Rivista geografica italiana [1630].

1599 MEMORIE geografiche e geologiche. Firenze. 1-4 (1929-1934). Closed 1934).

1600 MESSINA. Università. Istituto di geografia. Messina.
Memorie. 1-7 (1930-1932). Closed 1932. (U-3:2617c).

1601 MILANO. (Milan). Università. Istituto di geografia umana. Milano.
Pubblicazioni. 1- (1971-). (N75-2:1454c).
Subseries of Milano. Università. Facolta di lettere e filosofia. Pubblicazioni (not in this list).

1602 ______. ______. Istituto di geologia, paleontologia e geografia fisica.
Pubblicazione. Ser. Ge(ografia). (N70-3:3806b).

1603 ______. Università cattolica del sacro cuore. Milano.
Saggi e ricerche. Serie terza: Scienze geografiche. 1-4 (1962-1971) Irregular. Monographs. (N70-3:3805c).
Continues in part university's Saggi e ricerche, nuova serie (1944-1960) (not in this list).
[Università cattolica del Sacro Cuore, Piazza San Ambrogio 9, Milano, Italy.]

1604 ______. Università commerciale Luigi Bocconi. Istituto de geografia economica. Milano.
[Pubblicazione]. v1-12 (1955-1960) Irregular. Closed. (N70-3:3806a).

1605 NAPOLI (Naples). Istituto universitario navale. Laboratorio di geografia economica. Napoli.
Quaderni. 1 (1968). Closed 1968.

1606 ______. Istituto universitario orientale. Napoli.
Contributi geografici. 1- (1976-) Irregular.
[Via S. Pietro a Maiella 6, 80134 Napoli, Italy.]

1607 ______. ______. Centro di studi geografici. Napoli.
Pubblicazioni. 1-2 (1938). Closed 1938. (U-4:2819c).
Earlier name: Istituto superiore orientale.

1608 ______. Università. Istituto di geografia economica.
Pubblicazioni. 1- (1964-) Irregular. (N70-3:3986a).
[Via Tommaseo 4, 80121 Napoli, Italy.]

1609 NOTIZIARIO di geografia economica. (Roma. Università. Facoltà di economia e commercio. Istituto di geografia economica). Roma. 1- (1970-) Quarterly. (N75-2:1897c; B73:236a).
[Istituto di Geografia Economica, Facoltà di Economia e Commercio, V. del Castro Laurenziano 9, 00161 Roma, Italy.]

1610 OPINIONE geografica. Rivista di geografica didattica. Firenze.
v1-10 (1905-1914). Closed 1914.
1911-1914 with subtitle: Rassegna dell'insegnamento di geografia.

1611 PADOVA (Padua). Università. Istituto di geografia. Padova.
Pubblicazioni. 1- (1948-) Irregular. (BS:659b).
Mostly, but not entirely, reprints.
[Via del Santo 26, 35100 Padova, Italy.]

1612 PAESI d'attualità. Collezione di monografie geografiche. Roma.
no1-17 (1942-1943). Closed 1943. (U-4:3240b).
Monographs.

1613 PALERMO. Università. Istituto di geografia economica. [Collana] Bologna.
[1] (1964-). (N70-3:4515b).

1614 PARMA. Università. Facoltà di magistero. Istituto di scienze geografiche. Parma.
Pubblicazioni. 1-2 (1970).

1615 ______. ______. ______. ______.
Quaderni di geografia comparata. 1 (1971).

1616 PERUGIA. Università. Istituto di geografia economica. Perugia.
Ricerche di geoeconomia in Umbria. 1-3 (1967).

1616a ______. ______. Istituto policattedra di geografia. Perugia.
Quaderni. 1- (1979-).
[Via del Fagiano 6, 56100 Perugia, Italy.]

1617 PISA. Università. Istituto di geografia generale. Pisa.
Pubblicazioni. 1- (1954-) Annual. (N70-3:4656b; BS:680a).
[Via Trieste 20, 56100 Pisa, Italy.]

1618 POLO. (Istituto geografico polare). Civitanova Marche (Macerata).
1- (1945-) 2 nos a year. (U-4:3388c; B-2:563b).
1 (1945) as institute's Bolletino mensile d'informazione.

1619 QUADERNI di geografia umana per la Sicilia e la Calabria. (Messina. Università. Istituto di geografia). Messina. 1-5 (1956-1960). Closed 1960. (N70-3:4861c).

1620 QUADERNI geografici. (Istituto geografico de Agostini). Novara.
v1 no1-11 (1918-1919). Closed 1919. (U-4:3496a; F-3:1101b).

1621 QUADERNI geografici d'attualità. (Società geografica italiana). Roma.
s1. I confini d'Italia. no1-8 (1945-1948). Closed 1948. (U-4:3496a).
s2. I territori italiani d'Africa. no1/5 (1948). Closed 1948.
(U-4:3496a).

1622 RASSEGNA della letteratura geografica. (Società di studi geografici e coloniali). Firenze. 1914-1919. Closed 1919. (B-3:737b: F-4:371a). Supplement to Rivista geografica italiana. Continued as integral part of Rivista [1630].

1623 RICERCHE limnologiche. (Italy. Consiglio nazionale delle ricerche. Centro di studi per la geografia fisica [and] Centro di studi alpini). Bologna. 1-2 (1952). (N70-3:5058a; BS:741b).

1624 RICERCHE sugli aspetti morfologici di territori italiani. (Italy. Consiglio nazionale delle ricerche. Centro di studi per la geografia fisica). Bologna. 1 (1953). (N70-2:3034b).

1625 RICERCHE sulla morfologia e idrografia carsica. (Italy. Consiglio nazionale delle ricerche. Centro di studi per la geografia fisica [and] Bologna. Università. Istituto di geografia). Bologna. 1-6 (1948-1955). Closed 1955. (B-1:525a; BS:218b).
vl-3 carry note as s10 of Italy. Consiglio nazionale delle ricerche. Comitato per la geografia. Pubblicazioni [1591].

(1626 RICERCHE sulle dimore rurali in Italia. (Italy. Consiglio nazionale delle ricerche [and] Firenze. Università. Centro di studi per la geografia etnologica). Firenze. 1- (1938-) Irregular. (U-3:2133c; N70-2:3036c; BS:218b).
1-4 (1938-1943) as Italy. Consiglio nazionale delle ricerche. Comitato per la geografia. Pubblicazioni. s8 [1591].
[Leo S. Olschki Editore, Casella Postale 295, 50100 Firenze, Italy.]

(1627 RICERCHE sulle variazioni delle spiagge italiane. (Italy. Consiglio nazionale delle ricerche. Centro di studi per la geografia fisica). Bologna and Padova. 1- (1933-) Irregular. (U-3:2133b; B-1:609a: BS:218a).
1-5 (1933-1940) as Italy. Consiglio nazionale delle ricerche. Comitato per la geografia. Pubblicazioni. s1 [1591].

1628 RICERCHE sulle variazioni storiche del clima italiano. (Italy. Consiglio nazionale delle ricerche. Centro di studi per la geografia fisica [and] Bologna. Università. Istituto di geografia). Bologna. 1- (1934-) Irregular. (U-3:2133b; B-1:609a).
1-9 (1934-1942) as Italy. Consiglio nazionale delle ricerche. Comitato per la geografia. Pubblicazioni. s2 [1591]; v10- (1949-) under present heading.

1629 RIVISTA di geografia. Firenze. vl-13 no6 (1917-1933). Closed 1933. (U-4:3688a; B-3:737b; F-4:368b).
vl-11 (1917-1927) as Rivista di geografia didattica; v12 no4-9 (1932) as Rivista di geografia e cultura geografica.
v12 (1932) published in Roma.
Not published 1928-1931.

*(1630 RIVISTA geografica italiana. (Società di studi geografici). Firenze. 1- (1893-) Quarterly. (U-4:3690a; B-3:737b; F-4:371ab).
1893-1935 as Società di studi geografici e coloniali.
Indexes: vl-50 (1894-1943); v51-60 (1944-1953).
English titles in annual table of contents and abstracts in English from v55 (1948).
[Via Laura 48, 50121 Firenze, Italy.]

1631 ROMA (Rome). Istituto universitario pareggiato di magistero "Maria SS. Assunta." Istituto di geografia. Roma.
Pubblicazioni. 1- (1968-).
[Via della Traspontina 21, Roma, Italy.]

(1632 ______. Università. Facoltà di economia. Istituto di geografia economica.
Pubblicazioni. Collana di geografia economica. 1- (1964-).
(N70-3:5095a?).
[V. del Castro Laurenziano 9, 00161 Roma, Italy.]

1633 ______. ______. Facolta di magistero. Istituto di scienze geografiche e cartografiche.
s1. Memorie geografiche. 1-9 (1954-1964). Closed 1964. (N70-3:5095a).
[Via delle Terme di Diocleziano 10, Roma, Italy.]

(1634 ______. ______. Istituto di geografia.
Pubblicazioni. (U-4:3706b; B-3:746a).
sA. Ricerche originali. 1-9 (1931-1951). Antropica e fisica. 10-22 (1970-1976).
sB. Scritti di storia della cartografia. 1-3 (1932-1955); Geostorica. 1-6 (1969-1978).
sC. Testi e sussidi per le università. 1 (1943). Miscellanea. 2-5 (1969-1975).

1635 ______. ______. ______.
______. nsno1-25 (1961-1968). Closed 1968. (N70-3:5095a). Reprints.

1636 SALERNO Università. Istituto di geografia.
Pubblicazioni. 1 (1974) Irregular.
[Via Irno, 84100 Salerno, Italy.]

1637 ______. ______. Istituto universitario di magistero. Gabinetto di geografia. Salerno.
Quaderni. 1-3 (1950). Closed 1950.

1638 SCRITTI di geografia umana ed economica. Milano. 1-8 (1964-1977).
[Cisalpino-Goliardica, Milano, Italy.]

1639 SOCIETÀ di geografia ed etnografia. Torino.
Atti. 1 (1884/86).

(1640 SOCIETÀ geografica italiana, Roma.
Bibliografia geografica della regione italiana. Firenze, Roma. 1- (1925-) Annual. (U-5:3918b; B-1:329b; F-1:597a).
no1-3 (1925-1927) as Comitato geografico nazionale italiano. Bibliografia geografica dell'Italia; no4-13 (1928-1937) issued as December issue of society's Bollettino (1641); no14- (1938-) separately issued.
Indexes: 1-10 (1925-1934), 11-19/20 (1935-1943/44), 21-30 (1945-1954), 31-40 (1955-1964).

*(1641 ______. Firenze (1868-1871); Roma (1872-).
Bollettino. 1- (1868-) 12 nos a year, often combined.
(U-5:3918b; B-4:111a; B68:87b; F-1:681b).
Also in series of 12v each.
Indexes: v1-12 (1868-1875); s2, v13-24 (1876-1887); s3, v25-36 (1888-1899); s4, v37-48 (1900-1911); s5, v49-60 (1912-1923); s6, v61-72 (1924-1935).
English titles in table of contents and summaries of articles from s8 v2 no1 (1949).
[Via della Navicella, 12, Roma, Italy.]

1642 ______.
Memorie. 1- (1878-) Irregular. (U-5:3918b; B-4:111a; F-3:502a).

1643 STRUMENTI di lavoro. Milano. 1- (1971-).
1-2 (1971-1974) as Trieste. Università. Laboratorio di geografia.
3-4 (1977) as Strumenti di lavoro: geografia.
Translated from Russian.
[Via Petitti 19, 20149 Milano, Italy.]

1644 TERRA e la vita. (Società geografica italiana). Roma. v1-2 no11 (1922-1923). Closed 1923. (U-5:4183b; B-4:302b).

(1645 TORINO (Turin). Università. Facoltà di economia e commercio. Laboratorio di geografia economica. Torino.
Pubblicazioni. 1- (1965-) Irregular.
[Laboratorio di geografia economica "Piero Gribaudi," Facoltà di economia e comercio, Università di Torino, Piazza Arbarello 8, Torino, Italy.]

1646 ______. ______. ______. ______.
Studi geografici su Torino e il Piemonte. 1 (1954), 2 (1975).

1647 ______. ______. Facoltà di magistero. Istituto di geografia.
sA. Memorie e studi geografici. 1-10 (1950-1976) Irregular. (N70-4:5882a).

(1648 ______. ______. Istituto di geografia alpina.
Pubblicazioni. 1- (1962-). (N70-2:3027b).
Subseries: Studi sul Manto Nevosco 1- (1962-). (N70-4:5573c).
Studi sulle Valanghe 1- (1967-). (N70-4:5573c).
Summaries in English and French.
[Via S. Ottavio 20, 10124 Torino, Italy.]

1649 TRIESTE. Università. Facoltà di economia e commercio. Istituto di geografia. Trieste.
Notiziario ad uso degli studenti. 1-3 (1950-1953). Closed. (N70-4:5858b).

1650 ______. ______. ______. ______.
Trieste.
Pubblicazioni. 1-9 (1957-1975) Irregular.

1651 ______. ______.
Facoltà di lingue e letterature straniere. Sede staccata di Udine.
Pubblicazioni. 1-2 (1974). Atti IV incontro geografico italo-sloveno.
In Italian. Summaries in Slovenian and English.

1652 ______. ______. Facoltà di magistero. Laboratorio di geografia.
Pubblicazioni. 1-4 (1963-1964).

1653 ______. ______. Istituto di geografia.
Collana di monografie. sII. Le superpotenze economiche. Milano. 1- (1955-). (N70-4:5858b).

1654 ______. ______. ______.
Pubblicazioni. 1-16 (1942-1956). (U-5:4263b; BS:880b).
s3. Stati nuovi. 1-2 (1956-1958). Closed. (N70-4:5858b).

1655 UDINE. Università. Istituto di geografia. Udine.
Pubblicazioni. 3 (1978).
[Via Antonini 8, 33100 Udine, Italy.]

*(1656 UNIVERSO: rivista di divulgazione geografica. (Italy. Istituto geografico militare). Firenze. 1- (1920-) 6 nos a year. (U-5:4336a; B-4:435b; F-4:844b).
Supersedes Italy. Istituto geografico militare, Firenze. Annuario. [1592].
Index: 1920-1960.
[Istituto Geografico Militare, Via Cesare Battisti, 10, 50100 Firenze, Italy.]

1657 UNIVERSO. Geografia per tutti. Bergamo, Milano. 1-9 (1891-1899). Closed 1899. (U-5:4336a; B-2:268a; F-4:844b; F-2:727a).
v1-5 (1891-1895) as Geografia per tutti. Rivista quindicinale per la diffusione delle cognizioni geografiche.

1658 VERONA. Università di Padova. Facoltà di economia e commercio di Verona. Istituto di geografia. Verona.
Ricerche di geografia applicata. 2 (1975).
[Via dell'Artigliere 19, 37100 Verona, Italy.]

IVORY COAST (French)

République de Côte d'Ivoire

(1659 ABIDJAN. Université nationale de Côte d'Ivoire. Abidjan.
Annales. Série G: Géographie. 1 (1965); 1-2 (1969-1970); nsl- (1974-)
Irregular. (N75-1:12c; B73:20b).
Formerly Abidjan. Université.

1660 ______. ______.
Centre universitaire de recherches de développement. Abidjan.
Bulletin de liaison. 1967- 1976. (N70-1:29a; B68:105b).
Title varies. 1967 nosl-2 and 1968 nol, Bulletin d'information et de liaison des instituts d'ethno-sociologie et de géographie tropicale; 1968 no2-, Bulletin des instituts de recherche. 1967-1968 nol issued by the university's Institut d'ethno-sociologie and Institut de géographie tropicale.

JAMAICA (English)

1661 CARIBBEAN geographer. (Mona, Jamaica. University of the West Indies. Department of geography). Mona. 1-2 (1971-1972). Closed. (N75-1:474c).

1662 JAMAICAN geographical society. Kingston.
Newsletter. 1- (1972-) Monthly.
[Jamaican Geographical Society, Geography Department, University of the West Indies, Mona, Kingston 7, Jamaica.]

1663 MONA, Jamaica. University of the West Indies. Department of geography.
Occasional papers. 1-9 (1966-1972). Closed 1973. Mimeographed. (N70-3:3867b; B73:240b; DP).
Title varies: 1 (1966) as Department of geology and geography. Occasional papers in geography; others as Occasional publication.

1664 ______. ______. ______.
Research notes. 1-7 (1970-1972). Closed 1973. (DP).

JAPAN 日本 (Japanese)

(Nippon)

(1665 秋田地理

1669 日本生物地理学会報

1670 地学論叢

*(1671 地学雑誌

1672 地域

(1673 地域研究

1674 地人教養地理叢書

1675 地球

1676 地理
1677 地理
1678 地理研究
1679 地理教育
1680 地理論叢
1681 地理誌叢
1682 地理と経済
1683 地理学
1684 地理学報告
1685 地理学評論
1686 地理学研究
1687 地理学年報
1688 地理学报
1689 地理学史研究
1690 地理科学
1691 地図
1693 学芸地理
1694 学術研究 - 地理学.歴史学 社会科学編
1696 弘大地理
1699 北海道地理
1700 法政大学地理学集报
1703 人文地理
1704 人文研究
1706 鹿児島地理学会紀要
1707 経済地理学年報
1708 国土地理院時報
1709 駒沢地理
1710 駒沢大学大学院地理学研究
1711 教育地理
1712 京都帝国大学文学部地理学研究報告
1713 奈良大学地理学研究報告

(1714 日本大学研究紀要(地理)
(1715 日本地図センター. 地図センターニュース
(1717 岡山大学地理学研究報告. 都市と農村
1718 大塚地理学会論文集
(1719 歴史地理
(1720 歴史地理学紀要
1721 歴史と地理
(1722 歴史学・地理学年報
1723 琉球大学文理学部紀要: 社会編
1724 琉球大学法文学部紀要: 社会編
(1725 政治地理
1726 社会地理
1727 資源科学研究所報告
1728 資源科学研究所彙報
1729 資源科学研究所欧文報告
1730 資源科学研究叢書
1731 資源科学研究所特別報告
(1732 新地理
1733 地理学研究
*(1734 東北地理
*(1735 東北大学理科報告第七輯(地理学)
1736 東京地学協会報告
(1737 東京大学理学部紀要
(1738 東京大学教養学部人文科学科紀要(人文地理学)
1740 東京大学理学部地理学文献
1741 東京大学地理学研究
1744 東京教育大学人文地理学研究所報
1745 東京教育大学理学部地理学研究報告
1747 鳥大地理
*(1748 筑波大学人文地理学研究
(1749 筑波大学水理実験センター報告

(1665 AKITA chiri. (Akita chiri gakkai). Akita. 1- (1965-) Annual.
Title in English: Annals of the Akita geographical association.
In Japanese.
[Association of Akita Geographers, c/o Department of Geography, Faculty of Education, Akita University, 1-1, Tegata Gakuencho, Akita 010, Japan.]

1666 APPLIED geography. (Association of applied geographers). Tokyo.
1-5 (1960-1964). (N70-1:39c; B68:50b).
In English or French.

1667 BIOGEOGRAPHICA. (Biogeographical society of Japan. Transactions).
Tokyo. 1-4 no1 (1935-1942). Closed 1942. (U-1:705b).
In Japanese or English.

1668 BIOGEOGRAPHICAL society of Japan. Tokyo.
Bulletin. 1-31 (1929-1943). Closed 1943. (U-1:705b; B-1:351b).
Succeeded by society's Proceedings [1669].
Text in Japanese, English, and German.

1669 ______.
Proceedings. 1- (1948-) Irregular.
Succeeds the society's Bulletin [1668].

1670 CHIGAKU ronso. Collected papers of earth science. (Tokyo chigaku kyokai. Tokyo geographical society). Tokyo. 1-5 (1908-1913). Closed 1913.

*(1671 CHIGAKU zasshi. Journal of geography. (Tokyo chigaku kyokai. Tokyo geographical society). Tokyo. 1- (1889-) Quarterly. (1949-1960); 6 nos a year (since 1961). (U-2:1011b; B-1:546b; F-285a).
Supersedes Tokyo chigaku-kyokai hokoku [1736].
Indexes: no1-500 (1889-1930); 491-755 (1930-1966).
In Japanese. Supplementary English table of contents back to at least v9 (1897) and English abstracts from v58 (1949).
[Tokyo Chigaku Kyokai, Niban-cho, Chiyoda-ku, Tokyo 102, Japan.]

1672 CHIIKI. Geographical region. (Nihon shoin). no1-7 (1952). Closed 1952.

(1673 CHIIKI kenkyu. (Rissho chiri gakkai). Tokyo. 1- (1955-) 2 nos a year; irregular 1-13 (1955-1972).
Title in English: Regional study. (Rissho geographical association).
In Japanese.
[Rissho Chiri Gakkai, c/o Department of Geography, Faculty of Literature, Rissho University, 2-16, Osaki 4-chome, Shinagawa-ku, Tokyo 141, Japan.]

1674 CHIJIN kyoyo chiri sosho. (Chijin shobo). Kyoto. 1- (Ag 1965-).
(N70-1:1247b).

1675 CHIKYU. Globe. (Kyoto. Chikyu gakudan. Association of earth scientists).
Kyoto. 1-27 (1924-1937) Monthly. Closed 1937. (U-2:1011c).
Table of contents in English.

(1676 CHIRI. Geography. (Kokon-shoin). Tokyo. 1- (1956-) Monthly.
(N70-1:1266c).
In Japanese.
[2-10, Surugadai Kanda, Chiyoda-ku, Tokyo 101, Japan.]

1677 CHIRI. Geography. (Otsuka chiri gakkai. Association of Otsuka geographers).
Tokyo. v1-5 no4 (1938-1943). Closed 1943.
Merged into Chirigaku hyoron [1685].

1678 CHIRI kenkyu. Geographical studies. (Chiri kenkyukai). Tokyo.
no1-3 (1941). Closed 1941.
Succeeds Chiri kyoiku [1679].

1679 CHIRI kyoiku. Geographical education. (Chiri kyoiku kenkyukai).
Tokyo. Chukokan. 1924-1941. Closed 1941.
Succeeded by Chiri kenkyu [1678].

1680 CHIRI ronso. Geographical bulletin. (Kyoto. [Chirigaku kyoshitsu]
Imperial university. Faculty of literature. Geographic institute).
Kyoto. 1-13 (1932-1943). Closed 1943.

(1681 CHIRI shiso. (Nihon daigaku chiri gakkai). Tokyo. 1- (1966-) Annual.
English title of issuing agency: Geographical society of Nihon university.
In Japanese.
[Nihon Daigaku Chiri Gakkai, c/o Department of Geography, Faculty of Literature and Sciences, Nihon University, 3-25-40, Sakurajosui, Setagaya-ku, Tokyo, 156, Japan.]

1682 CHIRI to keizai. Geography and economics. (Nihon keizai chiri gakkai. Japanese society of economic geography). Tokyo.
no1-16 (1936-1937). Closed 1937.
In Japanese with English titles.

1683 CHIRIGAKU. Geography. (Kokon shoin). Tokyo. v1-12 no3 (1933-1944).
Closed 1944.

(1684 CHIRIGAKU hokoku. (Aichi kyoiku daigaku. Chirigakkai). Kariya.
1- (1952-) Irregular, 1 or 2 a year. (N70-1:1266c).
1-24 under former name of issuing agency: Aichi gakugei daigaku chirigakkai.
Title in English: Geographical report. (Aichi university of education. Geographical society).
In Japanese.
[Geographical Society of Aichi University of Education, 1 Hirosawa, Igayacho, Kariya 448, Japan.]

x(1685 CHIRIGAKU hyoron. Geographical review of Japan. (Nippon chiri gakkai. Association of Japanese geographers). Tokyo. 1- (1925-) 12 a year. (U-2:1027a).
Indexes: 1-10 (1925-1934); 11-20 (1935-1944); 21-30 (1947-1957); 31-40 (1958-1967); 41-50 (1968-1977).
In Japanese with English table of contents from v1 (1925); occasional English abstracts from v8 (1932) and regular abstracts from v23 (1950).
[Nippon Chiri-Gakkai, Building of Japan Academic Societies Center, 2-4-16, Yayoi, Bunkyo-ku, Tokyo 113, Japan.]

1686 CHIRIGAKU kenkyu. Geographical studies. Tokyo (Chukokan). v1-3 (1942-1944).
Closed 1944.
Succeeds Chiri kenkyu [1678].

1687 CHIRIGAKU nempo. Geographical yearbook. Tokyo. 1933-1935. Closed 1935.

(1688 CHIRIGAKUHO. (Osaka kyoiku daigaku. Chirigaku kyoshitsu). Osaka.
1- (1950-) Annual. (Irregular 1950-1972). (N75-1:521c).
Title in English: Geographical reports. (Osaka kyoiku university. Department of geography).
1-10 issued by Osaka gakugei daigaku.
In Japanese. Occasional article in English.
[Chirigaku Kyoshitsu, Osaka Kyoiku Daigaku, 43 Minami Kawahori-machi, Tennoji-ku, Osaka, 543, Japan.]

1689 CHIRIGAKUSHI kenkyu. Researches in the history of geography. (Chirigakushi kenkyukai. Society of researches in the history of geography). Kyoto. 1- (1957-) Irregular. (N70-1:1266c).
In Japanese; abstracts in English or German.
[c/o Nobuo Muroga, 36 Koyama-Kitabayashi-cho, Yamashina, Higashiyama-ku, Japan.]

*(1690 CHIRIKAGAKU. Geographical sciences. (Chiri kagaku gakkai. Hiroshima geographical association). Hiroshima. 1- (1961-) 2 nos a year.
In Japanese with English abstracts.
[Hiroshima Geographical Association, c/o Department of Geography, Faculty of Literature, University of Hiroshima, Higashi-senda-machi, Hiroshima, 730, Japan.]

(1691 CHIZU. Map. (Nihon kokusai chizu gakkai. The Japan cartographers association). Tokyo. 1- (1963-) Quarterly. (N70-1:1268a; B68:344a).
In Japanese. Abstracts in English.
[Japan Cartographers Association, c/o The Japan Map Center, Kudan Ponpian Bldg., 4-8-8, Kudan-minami, Chiyoda-ku, Tokyo 102, Japan.]

*(1692 CLIMATOLOGICAL notes. (Tsukuba university. Institute of geosciences). Tsukuba. 1- (1969-) Irregular, 1 or 2 a year. (B73:75a).
1-15 (1969-1973) published by Department of geography, Hosei university.
In English.
[Institute of Geosciences, University of Tsukuba, Ibaraki-ken, 300-31, Japan.]

(1693 GAKUGEI chiri. (Tokyo gakugei daigaku chiri gakkai). Tokyo. 1- (1947-) Annual.
Title in English: Journal of geography. (Tokyo gakugei university. Geographical society).
In Japanese.
[c/o Department of Geography, Tokyo Gakugei University, Koganei, Tokyo 184, Japan.]

1694 GAKUJUTSU kenkyu: chirigaku, rekishigaku, shakaikagaku hen. Tokyo. Annual.
Title in English: Scientific researches: geography...
In Japanese.
[School of Education, Waseda University, 1-6-1, Nishi-waseda, Shinjuku-ku, Tokyo 162, Japan.]

*(1695 GEOGRAPHICAL survey institute. (Kokudo chiriin; before 1960: Chiri chosasho hokoku). Tokyo, now Tsukuba.
Bulletin. 1- (1948-) Irregular. (BS:459a).
In English.
[Geographical Survey Institute, Ministry of Construction, 1 Kitago, Yatabemachi, Tsukuba-gun, Ibaraki-ken, 300-31, Japan.]

(1696 HIRODAI chiri. (Hirosaki daigaku. Kyoiku gakabu. Chirigaku kenkyushitsu). Hirosaki. 1- (1965-) Annual.
Title in English: Hirosaki university geography. (Hirosaki university. Faculty of education. Geographical institute).
In Japanese.
[Chirigaku Kenkyushitsu, Kyoiku Gakubu, Hirosaki Daigaku, Hirosaki, 036, Japan.]

1697 HIROSHIMA. University. Department of geography. Hiroshima.
Special publication. 1-5 (1972-1974) Irregular.
In English.

*(1698 ______. ______.
Research and sources unit in regional geography. Hiroshima.
Special publication. 1- (1975-) Irregular.
In English.
[Research and Sources Unit in Regional Geography, University of Hiroshima, 1-1-89, Higashi Sendamachi, Hiroshima 730, Japan.]

(1699 HOKKAIDO chiri. (Hokkaido chiri gakkai). Sapporo. Irregular; 1 or 2 nos a year.
Title in English: Hokkaido geography. (Hokkaido geographical society).
In Japanese.
[Hokkaido Chiri Gakkai, c/o Hokkaido Kyoiku Daigaku, Sapporo bunkō, Sapporo 060, Japan.]

(1700 HOSEI daigaku chirigaku shuho (Hosei daigaku). Tokyo. 1- (1972-) Annual. (N78).
Title in English: Geographical reports of Hosei university.
In Japanese. Abstracts in English.
[Hosei Daigaku Chirigaku Shuho Kankokai, c/o Graduate School, Hosei University, 17-1, 2-chome, Fujimi, Chiyoda-ku, Tokyo 102, Japan.]

1701 JAPANESE journal of geology and geography; transactions, titles and abstracts. (Science council of Japan). Tokyo. 1-45 (1922-1975). Closed 1975. (U-3:2155a; B-2:589b; F-3:56b).
1922-1947 name of issuing agency was National research council of Japan.
Index: 1-20 (1922-1947).
In English.

*(1702 JAPANESE progress in climatology. (Hosei university. Japan climatology seminar. 1964-1977 at Tokyo university of education. Laboratory of climatology). Tokyo. 1964- Annual. (N70-2:3105a).
In English.

*(1703 JIMBUN chiri. Human geography. (Jimbun chiri gakkai. Human geographical society of Japan). Kyoto. 1- (My 1948-) 6 nos a year. (BS:395a).
In Japanese with supplementary English table of contents, titles, and abstracts from v2 (1950).
Index: v1-20 (1948-1968) as v22 (1970), p. 603-736.
[Jimbun Chiri Gakkai, c/o Geographical Institute, Faculty of Letters, Kyoto University, Kyoto 606, Japan.]

(1704 JIMBUN kenkyu. Osaka. 1- (1949-) Annual. (U-3:2169a; B-2:596a).
Title in English: Studies in the humanities. (Osaka. City university. Faculty of literature).
In Japanese.
[Department of Geography, Faculty of Literature, Osaka City University, Sugimotocho, Sumiyoshi-ku, Osaka 558, Japan.]

1705 JOURNAL of earth science. (Nagoya. University. Institute of earth sciences). Nagoya. 1- (1953-). (N70-2:3159c: BS:468a).

(1706 KAGOSHIMA chirigakkai kiyo. Kagoshima. 1- (1950-) Irregular.
Title in English. Memoirs of the Kagoshima geographical society.
In Japanese.
[Kagoshima Chirigakkai, c/o Department of Geography, Faculty of Education, Kagoshima University, Kagoshima 890, Japan.]

*(1707 KEIZAI chirigaku nempo. Annals. (Keizai Chiri Gakkai. Association of economic geographers). Tokyo. 1- (1954-) 3 nos a year from 1977; 2 nos a year (1966-1976); Annual (1954-1965).
In Japanese with supplementary English subtitles, table of contents, and abstracts.
[Keizai Chiri Gakkai, c/o Graduate School, Meiji University, Kanda-Surugadai, Chiyoda-ku, Tokyo 101, Japan.]

(1708 KOKUDO chiriin jiho. (Kokudo chiriin. Geographical survey institute). Tokyo, now Tsukuba. 1- (1946-) Annual. (N70-1:1266c).
1-30 (1946-1965) as Chiri chisasho jiho. GSI Journal.
In Japanese.
[Geographical Survey Institute, Ministry of Construction, 1 Kitago, Yatabemachi, Tsukuba-gun, Ibaraki-ken, 300-31, Japan.]

(1709 KOMAZAWA chiri. (Komazawa daigaku chirigakkai). Tokyo.
1- (1958-) Annual. (N70-2:3330b).
Nol-3 published by Komazawa chiri gakkai.
Title in English: Komazawa geography (Komazawa university. Department of geography).
In Japanese.
[Department of Geography, Komazawa University, 1 Komazawa, Setagaya-ku, Tokyo, 154, Japan.]

(1710 KOMAZAWA daigaku daigakuin chirigaku kenkyu. Tokyo. 1- (1971-) Annual.
Title in English: Geographical research of graduate students, Komazawa university.
In Japanese.
[Geographical Research Association of Graduate School Students, c/o Department of Geography, Komazawa University, 23-1, 1-chome, Komazawa, Setagaya-ku, Tokyo, 154, Japan.]

1711 KYOIKU chiri. (Nihon shoin). Tokyo. 1967?- . (N77).
Some months are issued together.

1712 KYOTO teikoku daigaku bungakubu chirigaku. (Kyoto. Imperial university. College of letters. Institute of geography). Kyoto.
Kenkyu hōkoku. Geographic reports. nol-2 (1937-1938). Closed 1938.

(1713 NARA daigaku. Chirigaku kenkyu hokoku. Nara. 1- (1976-) Annual.
In Japanese. Abstracts in English.
[Department of Geography, Faculty of Literature, University of Nara, 1230, Horaicho, Nara, 631, Japan.]

(1714 NIHON daigaku. Tokyo.
Kenkyu-kiyo (Chiri). 1- (1966-) Annual.
Title in English: Nihon university. Institute of natural sciences-geography. Proceedings.
In Japanese. Abstracts in English.
[Institute of Natural Sciences, Nihon University, Sakurajosui, Setagaya-ku, Tokyo 156, Japan.]

(1715 NIHON chizu senta. Chizu senta nyusu. Tokyo. 1- (1972-) Monthly. (N75-2:1597a).
Title in English: Japan map center news.
In Japanese.
[Nihon chizu senta, 4-8-8 Kudanminami, Chiyoda-ku, Tokyo 102, Japan.]

*(1716 NIPPON chiri-gakkai. Association of Japanese geographers. Tokyo.
Special publication. 1- (1966-) Irregular. (N70-1:516c).
In English.
[Nippon Chiri-Gakkai, Building of Japan Academic Societies Center, 2-4-16, Yayoi, Bunkyo-ku, Tokyo 113, Japan.]

(1717 OKAYAMA daigaku. Okayama.
Chirigaku kenkyu hokoku. Toshi to noshon. 1- (1975-) 2 nos a year.
Title in English: Publications in geography. City and country. (Okayama. University. Faculty of law and literature. Institute of geography).
In Japanese.
[Institute of Geography, Faculty of Law and Literature, Okayama University, 1-1, 3-chome, Naka Tsushima, Okayama 700, Japan.]

1718 OTSUKA chiri gakkai rombunshu. (Association of Otsuka geographers. Collected papers). Tokyo. vl-6 (1933-1935). Closed 1935.

(1719 REKISHI chiri. Historical geography. (Nippon rekishi chiri gakkai. Japanese historical and geographical society). Tokyo. vl- (1897-) Quarterly. (U-4:3561b; F-4:66a).
Cumulative index: 1-100 in 1978 no3, p. 150-184.
In Japanese.

(1720 REKISHI chirigaku kiyo. (Nippon rekishi-chirigaku kenkyukai). Tokyo. 1- (1959-) Annual. (N70-3:4962c).
Title in English: Historical geographical report. (Japanese historical-geographical society).
In Japanese.
[Nippon Rekishi-chirigaku Kenkyukai, c/o Department of Geography, Faculty of Literature, Rikkyo University, Toshima-ku , Tokyo, 171, Japan.]

1721 REKISHI to chiri. History and geography. (Shigaku chiragaku dōkō kai). Kyoto. 1-34 (1917-1934). Closed 1934. (U-4:35616b; F-4:66a).

(1722 REKISHIGAKU-CHIRIGAKU nenpoh. Fukuoka. 1- (1977-) Annual.
Title in English: Annals of historical and geographical studies. (Kyushu university. College of general education).
In Japanese.
[College of General Education, Kyushu University, 4-2-1, Ropponmatsu, Chuo-ku, Fukuoka 810, Japan.]

1723 RYUKYU daigaku, Naha. Bunrigakubu. Ryukyu daigaku bunrigakubu kiyo: shakai hen. 8-11 (Jn 1964-1967). (N70-3:5126c).
Title also in English: University of the Ryukyus. Arts and Science Division.
Supersedes in part and continues the numbering of its Ryukyu daigaku bunrigakubu kiyo. Jimbum shakai (not in this list).

1724 ______. Ho-bungakubu. Ryukyu daigaku ho-bungakubu kiyo: shakaihen. Naha. 12- (1968-) Annual. (N75-2:1905b).
Continues Ryukyu daigaku bunrigakubu kiyo: shakai hen, and assumes its numbering [1723].
Title also in English: University of the Ryukyus. College of law and literature. Bulletin: Geography, history, and sociology.

(1725 SEIJI chiri. (Nippon seiji chiri gakkai). Tokyo. 1- (1960-) Irregular. (N70-4:5260c).
Title in English: Political geography. (Japanese association of political geographers).
In Japanese.

1726 SHAKAI chiri. Geography for social life. (Nippon shakai chiri kyōkai. Japanese association of geography for social life). Tokyo. 1-30 (1947-1950). Closed 1950. (U-5:3855c).
Indexes: 1-10 in no11, 11-20 in no21.
In Japanese with English subtitle and table of contents.

1727 SHIGEN kagaku kenkyūsho. Research institute for natural resources. Tokyo. Hōkoku. Bulletin. 1- (1943-). (U-5:4230c).

1728 ______.
Ihō. Miscellaneous reports. 1-75 (1943-1971). Closed 1971. (U-5:4230c).

1729 ______.
Ōbun hōkoku. [Reports in European languages.] 1- (1943-). (U-5:4230c).
Publication subtitled the society's Journal.

1730 ______.
Shigen kagaku kenkyū sōsho. Natural resources research series. no1- (1943-). (U-5:4230c).

1731 ______.
Tokubetsu hōkoku. Transactions. no1- (1943-). (U-5:4230c).

(1732 SHIN-CHIRI. (Nippon chirikyoiku gakkai). Tokyo. 1-5 (1947-1951); nsv1- (1952-) Quarterly. (U-5:3862c; N70-4:5311c).
Title in English: The new geography. (Association for geographical education in Japan).
In Japanese with English table of contents from v2 (1953), and English summaries.
[Nippon Chirikyoiku Gakkai, c/o Department of Geography, Tokyo Gakugei University, Koganei, Tokyo 184, Japan.]

1733 SHIRIGAKU kenkyu. Takamatsu. Irregular.
In Japanese.
[Department of Geography, Faculty of Education, University of Kagawa, Takamatsu 760, Japan.]

*(1734 TŌHOKU chiri. (Tōhoku chiri gakukai). Annals of the Tohoku geographical association. Sendai. 1- (1948-) Quarterly. (U-5:4228a).
Indexes: 1-20 (1948-1968), 21-30 (1969-1978) as v31, special no. (S 1979).
In Japanese with English table of contents and abstracts.
[c/o Tōhoku Daigaku Rigakubu Chirigaku Kyoshitsu. Katahira-cho, Sendai, Japan.]

*(1735 TŌHOKU daigaku rika hōkoku. Chirigaku. (Tohoku university. Faculty of science. Department of geography). Sendai.
Science reports. s7. Geography. 1- (1952-). v1-17 (1952-1968) annual; v 18- (1969-) 2 nos a year. (N70-4:5801b; BS:872b).
Formerly Chirigaku hōkoku.
Index: 1-20 (1952-1971).
In English.
[Department of Geography, Faculty of Science, Tohoku University, Aobayama, Sendai 980, Japan.]

1736 TOKYO chigaku-kyokai hokoku. (Journal of the Tokyo geographical society). Tokyo. v1-18 no4 (1879-1897). Closed 1897.
(U-5:4233a; F-4:695a).
Superseded by Chigaku zasshi [1671].
In Japanese with English table of contents v14-18 (1892-1897).

(1737 TOKYO daigaku. Faculty of science. Tokyo.
Journal. pt2: Geology, mineralogy, geography, seismology. 1- (1925-) Irregular. (U-5:4231c; B-4:326b; F-3:203b).
In English.

(1738 ______. Kyoyogakubu. Jimbun kagakuka kiyo. Jimbun chirigaku. Proceedings of the Department of humanities, series of human geography. (University of Tokyo. College of general education. Institute of human geography).
1- (1965-) Irregular. (N70-4:5809a).
1 (1965) is no34 of parent series; 3 (1971), 52; 4 (1973), 56.
In Japanese with table of contents and abstracts in English or other languages.
[Institute of Human Geography, College of General Education, University of Tokyo, 3-8-1, Komaba, Meguro-ku, Tokyo 153, Japan.]

*(1739 ______. Rigakubu. Chirigaku kyoshitsu. (University of Tokyo. Faculty of science. Department of geography). Tokyo.
Bulletin. 1- (1969-) Annual. (N70-4:5808a; N78; B73:55a).
In English.
[Chairman, Department of Geography, Faculty of Science, University of Tokyo, Hongo, Tokyo, 113, Japan.]

1740 ______. ______. ______.
Contributions to geographical literature. 1-4 (1959-1965) Irregular.
Closed 1965. (N70-4:5809c).
Reprints bound together.
In English or Japanese.

1741 ______. ______. ______.
Tokyo daigaku chirigaku kenkyu. Tokyo university geographical studies.
1-3 (1950-1954). Closed 1954. (N70-4:5809c).
Added title: Bulletin.
In Japanese with English subtitle, table of contents, summaries, and some titles of maps and graphs.

1742 TOKYO journal of climatology.
(Tokyo university of education. Laboratory of climatology).
Tokyo. 1-3 (1964-66)? (N70-4:5811a).
In English.

1743 TOKYO kyoiku daigaku (Tokyo university of education). Tokyo.
Science reports. Section C. 1-13 (1932-1977). Closed 1977.
Irregular. (U-5:4232c; B-4:325b; BS:873b; F-4:453b).
v1-2 (1932-1952) as Tokyo bunrika daigaku (University of literature and science). Science reports. Section C. Contributions from the Geological and mineralogical institute or from the Geographical institute.
Index: v1-3 (1932-1955).
In English.

1744 ______. Institute of human geography. Tokyo.
Geographical research paper. 1- (1965-). (N75-1:906c).

1745 ______. Rigakubu. Chirigaku kyoshitsu. (Faculty of science. Department of geography). Tokyo.
Chirigaku kenkyu hōkoku. Tokyo geography papers. 1-21 (1957-1977).
Closed 1977. Annual. (N70-4:5811c).
In Japanese with supplementary English title, table of contents, and summaries of articles.

*(1746 TOKYO toritsu daigaku. Rigakubu. Chiri-gakka. (Tokyo metropolitan university. Faculty of science. Department of geography). Tokyo.
Geographical reports. 1- (1966-) Annual. (N76; B68:225b).
In English.
[Department of Geography, Faculty of Science, Tokyo Metropolitan University, 2-1-1 Fukazawa, Setagaya-ku, Tokyo, 158, Japan.]

1747 TORIDAI chiri. (Tottori daigaku kyoikugakubu chirigaku kenkyukai).
Tottori. 1- (1965-). (N78).

*(1748 TSUKUBA daigaku Jimbun chirigaku kenkyu. Tsukuba.
1- (1977-) Annual.
Title in English: Tsukuba studies in human geography (Tsukuba university. Institute of geosciences).
In Japanese. Supplementary table of contents and abstracts in English or other languages.
[Institute of Geosciences, University of Tsukuba, Ibaraki-ken 300-31, Japan.]

(1749 ______.
Suiri jikken senta hokoku. 1- (1977-) Annual.
In Japanese.
[Environmental Research Center, University of Tsukuba, Ibaraki-ken, 300-31, Japan.]

*(1750 TSUKUBA university. (Tsukuba daigaku). Institute of geoscience. Tsukuba.
Annual report. 1- (1975-) Annual. (N78).
In English.
[Institute of Geoscience, University of Tsukuba, Ibaraki-ken 300-31, Japan.]

KENYA (Swahili, English)

Jamhuri ya Kenya

1751 AFRICAN geographical studies. Nairobi. 1- (1971-). (B74).
In English.

(1752 KENYAN geographer. (Geographical society of Kenya. Journal). Nairobi.
1- (1975-) 2 nos a year. (N77; B78).
In English. Abstracts in English.
[East African Literature Bureau, P.O. Box 30022, Nairobi, Kenya.]

KOREA (Korean)

Taehan Min'guk

南朝鮮

*(1753 地理學

(1754 地理学科 地理教育

*(1755 地理學論叢

*(1756 應用地理

*(1753 CHIRI HAK. Geography. (Taehan chiri hakhoe. Korean geographical society).
Seoul. 1- (1963-) Irregular 1963-1967, Annual 1968-1973, Semiannual
1974- . (N79).
In Korean. Supplementary English table of contents and abstracts.
[Korean Geographical Society, c/o Department of Geography, Seoul
National University, Sinlim-dong, Kwanak-ku, Seoul 151, Korea.]

(1754 CHIRIHAK kwa chiri-kyoyuk. Geography-education. (Seoul national university. College of education. Department of geography education).
Seoul. 1- (1973-) Annual.
In Korean. Summaries in English.

*(1755 CHIRIHAK nonchong. Journal of geography. (Seoul national university. College of social sciences. Department of geography). Seoul. 1- (1970-)
Irregular. 1970-1977 as Naksan jiri [chiri]. (N79).
In Korean. Summaries in English or German.

*(1756 UNGYONG chiri. Applied geography. (Korean institute of geographical research). Seoul. 1- (1975-) Annual.
In Korean. Summaries in English.

KOREA, NORTH 北朝鮮 (Korean)

Chosŏn Minjujuŭi In'min Kong-hwaguk

1757 CHIJIL kwa chiri. P'yongyang. (Kwahagwon ch'ul'ansa). 1- (1961-).
(N70-1:1247b; R-1:103b).
Russian title: Geologiia i geografiia.

KUWAIT (Arabic)

Dawlat al-Kuwayt
دولة الكويت

1758 KUWAIT geographical society.
Magazine. 1.
Geographical lectures delivered to the society.

LAOS (Lao)

Lao People's Democratic Republic

1759 CYCLE d'études sur la géographie du Laos. (Royaume du Laos. Ministère de l'éducation nationale). Rapport. Vientiane. 1- (1970-).
(N75-1:643c).
In French.

LEBANON (Arabic)

al-Jumhūrīyah al-Lubnānīyah
الجمهورية اللبنانية

*1760 HANNON: revue libanaise de géographie. (Université libanaise. Faculté des lettres et des sciences humaines. Département de géographie. Recueil des travaux). Beirut. 1-7 (1966-1972); 8- (1977-)
Annual. (N70-2:2480a; B73:134b). Not published 1973-1976.
In French. Arabic table of contents and resumés.
[B.P. 2691, Beirut, Lebanon.]

LIBYA (Arabic)

al-Jumhūrīyah al-ᶜArabīyah al-Lībīyah

الجمهورية العربية الليبية

1761 CIRENAICA. Ufficio studi. Benghazi.
Bollettino geografico. 1-16 (1926-1933). Closed 1933. (B-1:456b).
In July 1933 combined with Tripolitania. Ufficio studi. Bollettino geografico [1761].
In Italian.

1762 LIBYA. Ufficio studi. Tripoli.
Bollettino geografico. 1-10 (1931-1936). Closed 1936. (B-1:456b; F-1:685a).
Ufficio studi initially under Tripolitania; later Tripolitania e Cirenaica.
In Italian.

MADAGASCAR (Malagasy, French)

République Malgache

*(1763 MADAGASCAR: revue de géographie. (Tananarive. Université de Madagascar. Laboratoire de géographie). Tananarive. 1- (1962-) 2 nos a year. (N70-3:3566bc; B68:339b).
Index: no1-30 (1962-1977) in no30 (Ja-Jn 1977), p. 145-162.
In French. Abstracts in English and French.

1764 MADAGASCAR. Service géographique.
Exposé des travaux sur le territoire de la République Malgache. Tananarive. 1958-1967 Annual. (N70-3:3609b).
Also listed as France. Institut géographique national.
Service géographique. Tananarive. Exposé des travaux sur le territoire de la République Malgache. (N70-2:2191a).

MALAWI (English)

Republic of Malawi

1765 DZIKO. (Chancellor college geographical society). Limbe. 1- (1968-)
Annual. (N70-2:1803bc).

(1766 MALAWI. University. Chancellor college. Department of geography and earth sciences. Limbe.
Occasional paper. 1- (1975-). (DP).
[Department of Geography and Earth Sciences, Chancellor College, University of Malawi, P.O. Box 280, Zomba, Malawi.]

*(1766a MALAWIAN geographer (Geography teachers association). Zomba. 1- (1968-)
Semiannual.
[Mr. E. D. Kadzombe, Department of Geography and Earth Sciences, Chancellor College, P. O. Box 280, Zomba, Malawi.]

MALAYSIA (Malay, English)

*(1767 GEOGRAPHICA. (University of Malaya. Department of geography [and] Geographical society). Kuala Lumpur. 1- (1965-) Annual. (N75-1:905b; N78).
Index: v1-9 (1965-1974) in v10 (1975), p. 1-8.
In English.
[Editor, Geographica, c/o Department of Geography, University of Malaya, Kuala Lumpur, Malaysia.]

1768 GEOGRAPHICAL bulletin. (National geographical association of Malaysia). Kuala Lumpur. 1- (1975-) Irregular.
In English and Malay.
[Department of Geography, University of Malaya, Kuala Lumpur, Malaysia].

1769 ILMU alam. (Jabatan ilmu alam, Universiti kebangsaan Malaysia). Kuala Lumpur. 1- (1972-) Annual. (N75-1:1055a).
Journal of the Department of geography and the Geographical society of the National university of Malaysia.
In English and Malay.
[Department of Geography, National University of Malaysia, Bangi, Selangor, Malaysia].

1770 MALAYSIAN geographers. (National geographical association of Malaysia). Kuala Lumpur. 1- (1978-) Annual.
In English.
[Department of Geography, University of Malaya, Kuala Lumpur, Malaysia].

*(1771 UNIVERSITY of Malaya. Department of geography. Kuala Lumpur. Occasional papers. 1- (1975-) Irregular. (N78; DP).
In English.
[Department of Geography, University of Malaya, Kuala Lumpur, Malaysia.]

MALTA (Maltese, English)

Repubblika Ta Malta

1772 SOCIETÀ geografica maltese. Valletta.
Monitore geografico e scientifico de Malta. ns no1 (1887).
Closed 1887?

MEXICO (Spanish)

Estados Unidos Mexicanos

1773 ACADEMIA nacional de historia y geografía, México. (Universidad nacional autónoma de México). México, D.F.
Memoria. s2 v1- (1945-) Irregular no. of issues each year.
(U-1:22a; B-1:16a).

1774 ______.
Publicaciones. 1. (N70-1:37b).

1775 ______.
Publicaciones. Serie Divulgación cultural. 1- (1962-). (N70-1:137b).

1776 ANALES de geografía. (Universidad nacional autónoma de México. Facultad de filosofía y letras. Centro de investigaciones geográficas).
México, D.F. 1-2 (1975-1976). (79).
Reprints from Anuario de geografía. 14-15 (1974-1975) [1777].
Note: Anuario de geografía 3 (1977) a continuation of this entry consists of reprints from Anuario de geografía, v16 (1976) [1777].

*(1777 ANUARIO de geografía. (Universidad nacional autónoma de México. Facultad de filosofía y letras). México, D.F. 1- (1961-) Annual.
(N70-1:391c; B68:49b).
[Editor, Anuario de Geografía, Facultad de Filosofía y Letras, Ciudad Universitaria, México 20, D.F.]

1778 CUADERNOS geográficos. (Chiapas. Dirección general de educación pública).
Tuxtla Gutierrez, Chiapas. 1- (1955-). (N70-1:1594b).

1779 DIORAMA; semanario histórico, geográfico y literario. México, D.F.
v1 no1-10 (Ag 14-O 16, 1837). Closed 1837? (U-2:1341c).

1780 GUADALAJARA. Universidad. Instituto de geografía y estadística.
Boletín.

1781 MAGAZINE de geografía nacional. México, D.F. v1-2 (Jl 1925-O 1926).
Closed 1926. (U-3:2511a).

(1782 MÉXICO (city). Universidad nacional autónoma de México. Instituto de geografía. México, D.F.
Boletín. 1- (1969-) Irregular. (N70-3:3756a; B73:47b).
v4 (1971) has abstracts in Spanish and English.

1783 ______. ______. ______.
Publicaciones. 1 (1965) Only one published. (N70-3:3756ab).

1784 ______. ______. ______.
Serie varia. 1- (1977-) Irregular.

1785 REVISTA mexicana de geografía. (México. Universidad nacional autónoma de México. Instituto de geografía). México, D.F.
1-4 (1940-1943). (U-4:3618b).

1786 SOCIEDAD de historia, geografía y estadística de Aguascalientes.
Aguascalientes.
Boletín. v1 no1-16 (1934-1935). Closed 1935? (U-5:3906c).

(1787 SOCIEDAD mexicana de geografía y estadística. México, D.F.
Boletín. v1- (1839-) 2 vols. a year, each of 3 nos. often combined. (U-5:3911a; B-2:507b; F-1:669b).
J1-D 1849 as Boletín de geografía y estadística de la República Méxicana. 1839-1959 also issued in series: 1-12 (1839-1868); s2 v1-4 (1869-1872); s3 v1-6 (1873-1887); s4 v1-4 (1888-1901); s5 v1-88 (1902-1959).
Issuing agency through O 1850 was Instituto nacional de la geografía y estadística de la República mexicana; N 1850-Mr 1851 as Sociedad de geografía y estadística de la República mexicana.
Recent volumes have consisted of monographs, v126 (1978): Memorie del VII Congreso nacional de geografía aplicada, 1978. 454 p.
Index: v1-63 (1839-1947).
[Calle del Maestro Justo Sierra 19, (Apartado Postal No.M-10739), México, D.F., Mexico.]

1788 ______.
Gaceta. v1- (1974-) 3 nos. a year.

1789 ______.
Selección de estudios y conferencias. 1- (1964-) Irregular. (N70-4:5375).

1790 ______.
Temas de México.
Serie geográfica. 1-4 (1946-1952). (U-5:3911a).
Monographs.

1791 ______. Junta auxiliar jalisciense, Guadalajara.
Boletín. 1-6 (1919-1940). (U-5:3911b).
Suspended J1 1920-F 1923; Ap 1923-D 1924; Mr 1925-Ja 1933.

1792 SOCIEDAD michoacana de geografía y estadística, Morelia. Morelia.
Boletín. v1-8 (1905-1912). Closed 1912. (U-5:3911b).

MOÇAMBIQUE

1793 MAPUTO. Instituto de investigação científica de Moçambique. Maputo.
Memórias. sB: Ciências geográficas-geológicas. v7- ? (1965- ?). Closed 1975. (N70-3:3539a).
Supersedes in part institute's Memórias (not in this list).

1794 SOCIEDADE de geographia de Moçambique. Moçambique.
Boletim. no1-6 (1881). Closed 1883. (U-5:3914b).

MONGOLIA (Khalkha Mongolian)

Mongolian People's Republic

Bügd Nayramdakh Mongol Ard Uls

БҮГД НАЙРАМДАХ МОНГОЛ АРД УЛС

(1795 МОНГОЛ орны газарзүйн асуудлууд. (Шинжлэх ухааны академи газарзүй-цэвдэг судлалын хүрээлэн). Улаанбаатар.
ВОПРОСЫ географии Монголии. (Академия наук Монгольской Народной Республики. Институт географии и геокриологии).

(1795 MONGOL orny gazarzuin asuudluud. (Shinzhlekh ukhaany akademi. Gazarzui-tsevdeg sudlalyn khureelen). Ulaanbaatar. 1- (1963-) Irregular. (N70-4:6255c; R-1:61b).
Russian title: Voprosy geografii Mongolii. (Akademiia nauk Mongol'skoi Narodnoi Respubliki. Institut geografii i geokriologii).
In Mongolian; Russian table of contents and summaries.

MOROCCO (Arabic, French)

al-Mamlakah al-Maghribīyah

المملكة المغربية

1796 NOTES marocaines. (Société de géographie du Maroc. Institut des hautes études marocaines). Rabat. 1-16 (1951-1961). Closed 1961. (N70-3:4291a; B68:483a).
Superseded by Revue de géographie du Maroc [1798].
In French.

1797 RABAT. Institut scientifique Chérifien.
Travaux. Série géologie et géographie physique. Tangier. 1- (1951-) Irregular. (N70-4899a; BS:805b).
In French; summaries in English.
1 (1951) published as v. 3 of earlier series Travaux.

*(1798 REVUE de géographie du Maroc. (Société de géographie du Maroc). Rabat. 1-22 (1962-1972); ns1- (1977-). Not published 1973-1976. 2 nos. a year. (N70-3:5027b; B68:483a).
Supersedes Notes marocaines [1796] and Revue de géographie marocaine [1799].
In French with English and Arabic summaries.
[Société de Géographie du Maroc, Faculté des Lettres, Rabat, Morocco.]

1799 REVUE de géographie marocaine. (Société de géographie du Maroc). Casablanca. v1-33 (1916-1949). Closed 1949. (U-4:3635b; B-4:135b; F-4:192b; F-1:836b). 1-4 (1916-1924/25) as Société de géographie du Maroc. Bulletin.
Superseded by Notes marocaines [1796] and by Société de géographie du Maroc. Travaux [1800]. v33 also as nsv1 of the Travaux.

1800 SOCIÉTÉ de géographie du Maroc. Institut des hautes études marocaines. Rabat.
Travaux. 1- (1949-) Irregular.
Monographs. Succeeds Revue de géographie marocaine [1799].

NEPÁL (Nepali)

Nepāl Adhirājya

नेपाल

*(1800a GEOGRAPHICAL journal of Nepal. (Tribhuvan university. Institute of humanities and social sciences. Geography instruction committee). Kirtipur. 1- (1978-) Annual. In English.

*(1801 HIMALAYAN review. (Nepal geographical society). Kirtipur. 1- (1968-) Annual. (N75-1:1001b).
In English.
[Department of Geography, Tribhuvan University, Kirtipur, Nepal.]

NETHERLANDS (Dutch)

Koninkrijk der Nederlanden

1802 AARDRIJKSKUNDIG weekblad. Amsterdam. 1-4 (1877-1881). Closed 1881. (U-1:9b).
1879-1881 as ns.

(1803 ACTA cartographica. (Theatrum orbis terrarum). Amsterdam. 1- (1968-) 3 vols a year. (N70-1:55c; B73:4a).
Reprints of monographs and articles published since 1801 in about 50 leading journals.
Index: v1-15 (1968-1972), v16-21 (1973-1975).
[O. Z. Voorburgwal 85, Amsterdam, The Netherlands.]

(1804 AMSTERDAM. Universiteit. Economisch geografisch instituut.
EGI-Paper. 1- (1974-) Irregular.
In Dutch or English.
[Economisch Geografisch Instituut, Universiteit van Amsterdam, Jodenbreestraat 23, 1011 NH Amsterdam, The Netherlands.]

(1805 ______. ______.
Fysisch-geografisch en bodemkundig laboratorium, Amsterdam.
Publicaties. no1- (1958-) Irregular. (N70-1:334a).
In English.
[Fysisch-Geografisch en Bodemkundig Laboratorium, Dapperstraat 115, 1093 BS Amsterdam, The Netherlands.]

1806 ______. ______. Geografisch en sociografisch seminarium. Amsterdam.
Publicaties. nol-3 (1936-1937). Closed 1937. (U-1:352a; B-1:252b).

1807 ______. ______.
Planologisch en demografisch instituut. Amsterdam.
Mededelingen. 1-10 (1968-1977). Closed 1977.
Superseded by Rooilijn [1834].
In Dutch.

(1808 ______. ______. Sociaal-geografisch instituut. Amsterdam.
Publicatie. 1- (1975-) Irregular.
[Sociaal Geografisch Instituut, Universiteit van Amsterdam,
Jodenbreestraat 23, 1011 NH Amsterdam, The Netherlands.]

1809 BIBLIOTHECA geographorum arabicorum. Leiden. 1-8 (1870-1894).
Closed 1894. (U-1:685c; B-1:341b).
From 1938 (s2 v1), some reprints.

(1810 BIJDRAGEN tot de sociale geografie. (Amsterdam. Vrije universiteit).
Amsterdam. 1- (1970-) Irregular.
In Dutch.
[Geografisch en Planologisch Instituut, Vrije Universiteit,
De Boelelaan 1105, Amsterdam, The Netherlands.]

*(1811 BIOGEOGRAPHICA. The Hague. 1- (1972-) Irregular monographs.
(N75-1:296b; B73:43b; GSZ).
In English or German.
[Dr. W. Junk b.v., Publishers, 13 van Stolkweg, The Hague, The
Netherlands.]

1812 BLUMEA; tijdschrift voor de systematiek en de geografie der planten.
A journal of plant taxonomy and plant geography.
(Leiden. Bijksherbarium). Leiden. 1- (1934/35-). (U-1:702b;
B-1:15a; BS:7a; F-1:649a).
Supersedes the herbarium's Mededeeling (not in this list).
Supplement. 1- (1937-).
In English or Dutch.

1813 CYBELE. Tijdschrift ter bevordering van land-en volkenkunde. Amsterdam.
1-7 (1826-1830). Closed 1830. (U-2:1248b; B-1:684b).

(1814 GEODERMA. An international journal of soil science. (Elsevier publishing
company). Amsterdam.
1- (1967-) Quarterly. (N70-2:2276b).
In English.
[Elsevier Scientific Publishing Company, P. O. Box 330, 1000 AH
Amsterdam, The Netherlands.]

*(1815 GEOGRAFISCH tijdschrift. (Koninklijk Nederlands aardrijkskundig
genootschap). Groningen. 1-19 (1948-1966), (U-2:1693a); nsv1-
(1967-) 5 nos. a year. (N70-2:2279b; N70:3:4091c; N75-1:905b;
B73:128a).
Supersedes K. Nederlandsch aardrijkskundig genootschap. Tijdschrift
[1826].
In Dutch with some English summaries.
[Koninklijk Nederlands Aardrijkskundig Genootschap, p/a Koninklijk
Instituut voor de Tropen, Mauritskade 63, 1092 AD Amsterdam, The
Netherlands.]

1816 GLOBE. (Album van buitenlandsche lettervruchten). 's Gravennage. 1841-1848; s2 v1-28 (1849-1862); 1863-1889. Closed 1889. (U-2:1736b; F-2:755a).
1841-48, 1863-89 have four vols. each. Subtitle varies. s2 as Globe: Schetsen van landen en volken, bijeenverzameld door H. Picard; also Keur van reisverhalen en merkwaardigheden van vreemde landen en volken.

(1817 GRONINGEN. Rijksuniversiteit van Groningen. Geografisch instituut. Groningen. Sociaal-geografisch reeks. 1- (1972-) Irregular.
In Dutch.
[Geografisch Instituut, Rijksuniversiteit Groningen, P.O. Box 800, Groningen, The Netherlands.]

(1818 ITC journal. (International Institute for Aerial Survey and Earth Sciences ITC). Enschede. 1973- Quarterly. (N75-1:1160b; B73-162b).
Supersedes ITC Publications. Series A, Photogrammetry, 1-53 (1960-1972) and Series B, Photo-interpretation, 1-67 (1960-1972) [1819].
Mostly in English.
[350 Boulevard 1945, P.O. Box 6, Enschede, The Netherlands.]

1819 ITC publications. (Institute for Aerial Survey and Earth Sciences). Delft. sB: (Photo-interpretation). 1-67 (1960-1972). Closed 1972. (N70-2:2965; B68:445a).
Superseded by ITC journal [1818].
In English.

1822 KARTOGRAFIE. (Koninklijk Nederlandsch aardrijkskundig genootschap. Kartografische sectie). Groningen. 1-17? (1958-1974). (N70-2:3257b; N70-3:4091c).
Reprinted from the cartographic section of K. Nederlandsch aardrijkskundig genootschap. Tijdschrift 1958-1966 [1826], and Geografisch tijdschrift 1967-1974 [1815].
Superseded by Kartografisch tijdschrift [1823].

(1823 KARTOGRAFISCH tijdschrift. (Nederlandse vereniging voor kartografie). Amersfoort. 1- (1975-) Quarterly. (N77).
Supersedes(Kartografie Koninklijk Nederlandsch aardrijkskundig genootschap. Kartografische sectie. Mededelingen). 1959-1974 [1822] and (b) section's Kaartbulletin, 1959-1964.
In Dutch. Some English abstracts.
[c/o Drs P. W. Geudeke, Seringenstraat 3, 2636 Schipluiden, The Netherlands.]

1824 LAND en volk; moderne geographische beschrijvingen. Zutphen. v1-6 (1924-1933). Closed 1933. (U-3:2348b).

1825 K. NEDERLANDS aardrijkskundig genootschap. Amsterdam. Verhandelingen. 1- (1973-) Irregular monographs. (N75-2:1549b).
[Wolters-Noordhoff, Groningen, Netherlands.]

1826 K. NEDERLANDSCH aardrijkskundig genootschap, Amsterdam. Amsterdam. Tijdschrift. 1-7 (1876-1883); s2 v1-83 (1884-1966) Quarterly. Closed 1966. (U-4:2952a; B-1:4a; F-4:687ab). s2 v1-6 (1884-1889) in two sections.
Superseded by Geografisch tijdschrift [1815].
Indexes: 1876-1904; 1905-1922; 1923-1940; 1941-1960; 1961-1966.
In Dutch with English abstracts at beginning of major articles from s2 v64 (1947) and occasional article in English or French.

1827 ______.
______. Bijbladen. v1-3, no1-12 (1876-1884). Closed 1884. (U-4:2952a; B-1:4a; F-1:636a). One unnumbered issue.

(1828 NIEUWE geografenkrant. (Koninklijk Nederlands aardrijkskundig genootschap). Utrecht. 1- (1977-) 10 nos. a year.
In Dutch.
[KNAG, Mauritskade 63, 1092 Amsterdam, The Netherlands.]

(1829 NIJMEEGSE geografische cahiers. (Nijmegen. Katholieke universiteit Nijmegen. Geografisch en planologisch instituut). Nijmegen.
1- (1971-) Irregular.
In Dutch or English.
[Geografisch en Planologisch Instituut, Katholieke Universiteit Nijmegen, Berg en Dalseweg 122, 6522 BW Nijmegen, The Netherlands.]

1830 NOMINA geographica Neerlandica. Geschiedkundig onderzoek der Nederlandse aardrijkskundige namen. (K. Nederlandsch aardrijkskundige genootschap). Amsterdam. v1-5 (1894-1901); 6-14 (1928-1954). (U-4:3070a).
Index: 1-10.
In Dutch.

(1831 PALAEOGEOGRAPHY, palaeoclimatology, palaeocology. (Elsevier publishing company). Amsterdam. 1- (1965-) 2 v of 4 issues annually. (N70-3:4512c; B68:407b).
Subtitle: An international journal of the geosciences.
Basically in English. French or German articles have English summaries.
[Elsevier Publishing Company, P.O. Box 211, Amsterdam, The Netherlands.]

1832 PRACTISCHE onderzoekingen op sociaal-economisch en planologisch gebied. (Series) D. Publicaties van de afdeling sociale en economische geographie van de Landbouwhogeschool in Wageningen. Assen. 1- (1950-) Irregular. (N70-3:4742c).

1833 RIJKS planologische dienst. Den Haag.
Publikatie. 1- (1970-) Irregular.
In Dutch.
[Rijks Planologische Dienst, Willem Witsenplein 6, Den Haag, The Netherlands.]

(1834 ROOILIJN; mededelingen van het planologisch en demografisch instituut (Amsterdam. Universiteit). 1- (1977-) Irregular.
Supersedes institute's Mededelingen [1807].
In Dutch.
[Subfaculteit Planologie en Demografie, Universiteit van Amsterdam, Jodenbreestraat 23, 1011 NH Amsterdam, The Netherlands.]

(1835 ROTTERDAM. Erasmus universiteit Rotterdam. Economisch-geografisch instituut. Working papers. sA. 1- (1975-) Irregular.
In English.
[Erasmus universiteit Rotterdam was formed in 1973 incorporating the Nederlandse economische hogeschool and the Medische Faculteit Rotterdam].
[Economisch Geografisch Instituut Erasmus Universiteit, Burg. Oudlaan 50, 3062 PA Rotterdam, The Netherlands.]

(1836 STEDEBOUW en volkshuisvesting. (Nederlands instituut voor ruimtelijke ordening en volkshuisvesting). Alphen aan de Rijn.
1- (1919-) Monthly. (U-5:4074b; B-4:319b).
1920-1957 as Tijdschrift voor volkshuisvesting en stedebouw.
Name of issuing agency earlier: Nederlandsch instituut voor volkshuisvesting en stedebouw.
In Dutch.
[Van Speijkstraat 25, 2518 EV Den Haag, The Netherlands.]

(1837 TERRAE incognitae; the annals of the society for the history of discoveries. Amsterdam. 1- (1969-) Annual. (N70-4:5744c; B73:342b).
In English.
[N. Israel, Publishing Department, Keizersgracht 539, Amsterdam C-3, The Netherlands.]

1838 THEATRUM orbus terrarum: a series of atlases in facsimile. (Meridian publishing co.). Amsterdam.
Series 1. 1- (1961-) Irregular. (N70-4:5773b).

1839 ______.
2nd series. 1- (1965-) Irregular. (N70-4:5773b).
[Orbis Terrarum, Amsterdam, The Netherlands.]

*(1840 TIJDSCHRIFT voor economische en sociale geografie. Journal of economic and social geography. (Nederlandsche vereeniging voor economische en sociale geografie 1910-1966; Koninklijk Nederlands aardrijkskundig genootschap, Amsterdam, 1967-). Rotterdam, Amsterdam.
1- (Ja 1910-) Bimonthly. (U-5:4219a; B-4:318b; BS:871a).
v1-39 no3 (Ja 1910-Mr 1948) as Tijdschrift voor economische geographie; v35 no8-v36 (Ag 1944-D 1945) not published.
Indexes: 1910-1934; 1935-1952.
In English or Dutch.
[Koninklijk Nederlands Aardrijkskundig Genootschap, p/a Koninklijk Instituut voor de Tropen, Mauritskade 63, 1092 AD Amsterdam, The Netherlands.]

1841 TIJDSCHRIFT voor het onderwijs in de aardrijkskunde. Rotterdam.
v1-20 no1 (1923-1942). Closed Ja 1942. (U-5:4219b).

1842 TROPICAL and geographical medicine. Amsterdam. 1- (1949-) Quarterly. (U-5:4266a; B-2:57a; BS:263b).
Earlier as Documenta de medicina geographica et tropica, a continuation of Documenta neerlandica et indonesica de morbis tropicis (not in this list).
In English with summaries in Spanish.
[57 Mauritskade, Amsterdam, The Netherlands.]

1843 UTRECHT. Rijksuniversiteit. Geografisch instituut. Utrecht.
Bulletin.
s1. Algemene sociale geografie van Europa. 9-10 (1972-1974). (N75-2:2235c).
Supersedes and continues numbering of Afdeling algemene sociale geografie van Europa. Bulletin [1851].
s2. Sociale geografie ontwikkelingslanden. 1-4 (1971-1975). (N75-2:2236a; B73:56a).
In Dutch with English summaries.
s3. Historische geografie. 1 (1971). Only one issued. (N75-2:2235c; B73:56a).
Supersedes Historisch geografische afdeling. Mededelingen.
In Dutch or English. Some articles have English summaries.

1844 ______. ______. ______.
Geographische en geologische mededeelingen.
Anthropo-geographische reeks. 1-4 (1929-1932). Closed 1932. (U-5:4347b; B-1:28b).
no1 (1929) as Anthropologische reeks.
In Dutch or English.

1845 ______. ______. ______.
______.
Physiographisch-geologische reeks. 1-18 (1927-1945); s2 no1-9 (1947). Closed 1947. (U-5:4347b; B-1:28b).
In Dutch, English, French, or German.

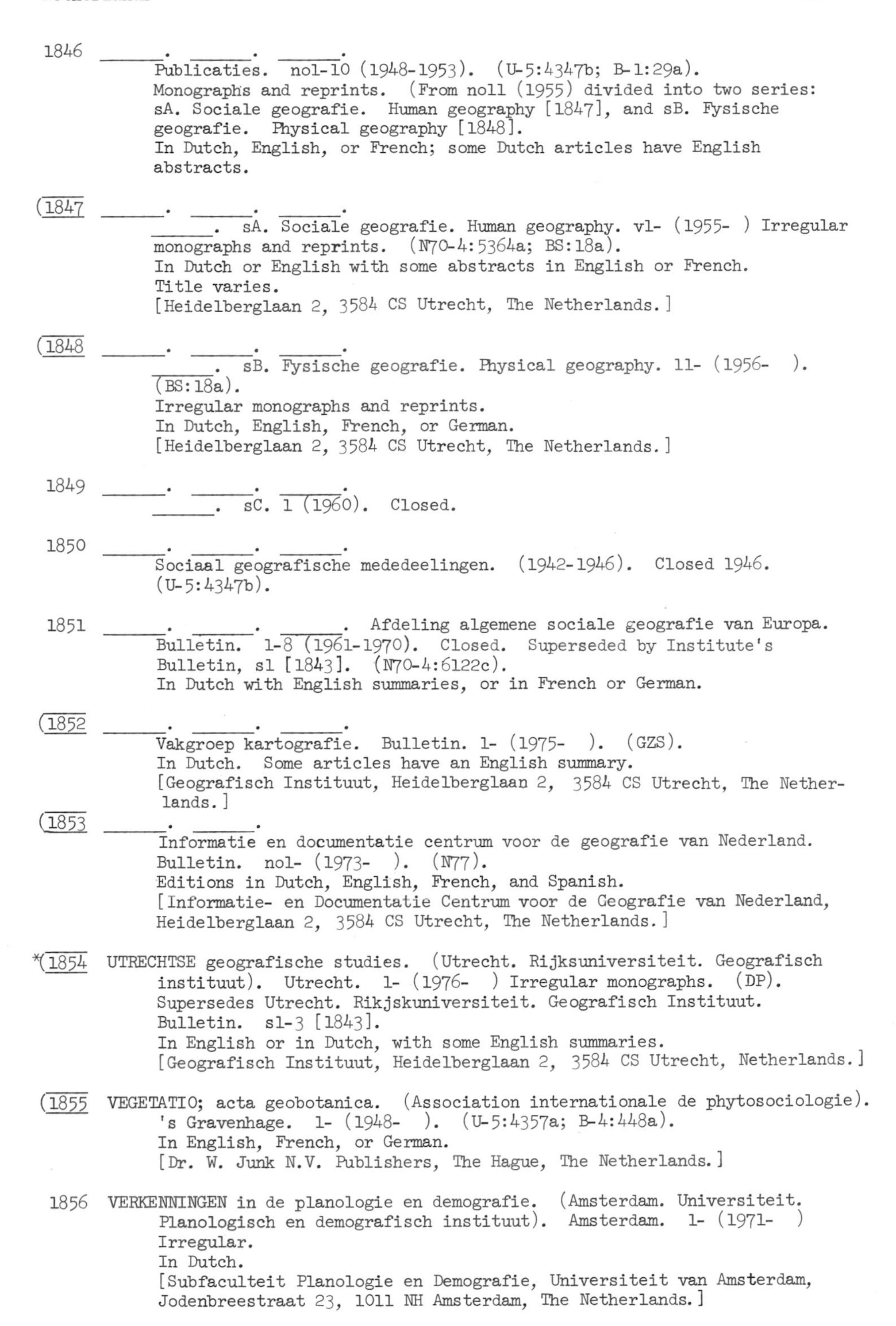

1846 ______. ______. ______.
Publicaties. no1-10 (1948-1953). (U-5:4347b; B-1:29a).
Monographs and reprints. (From no11 (1955) divided into two series:
sA. Sociale geografie. Human geography [1847], and sB. Fysische
geografie. Physical geography [1848].
In Dutch, English, or French; some Dutch articles have English
abstracts.

(1847 ______. ______. ______.
______. sA. Sociale geografie. Human geography. v1- (1955-) Irregular
monographs and reprints. (N70-4:5364a; BS:18a).
In Dutch or English with some abstracts in English or French.
Title varies.
[Heidelberglaan 2, 3584 CS Utrecht, The Netherlands.]

(1848 ______. ______. ______.
______. sB. Fysische geografie. Physical geography. 11- (1956-).
(BS:18a).
Irregular monographs and reprints.
In Dutch, English, French, or German.
[Heidelberglaan 2, 3584 CS Utrecht, The Netherlands.]

1849 ______. ______. ______.
______. sC. 1 (1960). Closed.

1850 ______. ______. ______.
Sociaal geografische mededeelingen. (1942-1946). Closed 1946.
(U-5:4347b).

1851 ______. ______. ______. Afdeling algemene sociale geografie van Europa.
Bulletin. 1-8 (1961-1970). Closed. Superseded by Institute's
Bulletin, s1 [1843]. (N70-4:6122c).
In Dutch with English summaries, or in French or German.

(1852 ______. ______. ______.
Vakgroep kartografie. Bulletin. 1- (1975-). (GZS).
In Dutch. Some articles have an English summary.
[Geografisch Instituut, Heidelberglaan 2, 3584 CS Utrecht, The Nether-
lands.]

(1853 ______. ______.
Informatie en documentatie centrum voor de geografie van Nederland.
Bulletin. no1- (1973-). (N77).
Editions in Dutch, English, French, and Spanish.
[Informatie- en Documentatie Centrum voor de Geografie van Nederland,
Heidelberglaan 2, 3584 CS Utrecht, The Netherlands.]

*(1854 UTRECHTSE geografische studies. (Utrecht. Rijksuniversiteit. Geografisch
instituut). Utrecht. 1- (1976-) Irregular monographs. (DP).
Supersedes Utrecht. Rikjskuniversiteit. Geografisch Instituut.
Bulletin. s1-3 [1843].
In English or in Dutch, with some English summaries.
[Geografisch Instituut, Heidelberglaan 2, 3584 CS Utrecht, Netherlands.]

(1855 VEGETATIO; acta geobotanica. (Association internationale de phytosociologie).
's Gravenhage. 1- (1948-). (U-5:4357a; B-4:448a).
In English, French, or German.
[Dr. W. Junk N.V. Publishers, The Hague, The Netherlands.]

1856 VERKENNINGEN in de planologie en demografie. (Amsterdam. Universiteit.
Planologisch en demografisch instituut). Amsterdam. 1- (1971-)
Irregular.
In Dutch.
[Subfaculteit Planologie en Demografie, Universiteit van Amsterdam,
Jodenbreestraat 23, 1011 NH Amsterdam, The Netherlands.]

NEW ZEALAND (English)

1857 AUCKLAND. University. Auckland.
Geography series. 1 (1956). (N70-1:543b; BS:94b).
A subseries of the university's Bulletin (not in this list); Geography series no1 (1956) was Bulletin no49, issued by university under its earlier name, Auckland university college.

(1858 ______. ______.
Department of geography. Auckland.
Occasional paper. 1-10 (1961-1977). (DP).

(1859 AUCKLAND student geographer. (University of Auckland geographical association. Journal). Auckland. 1-7 (1964-1970); 8- (1977-). (N70-1:543c; B:68;62b).

(1860 CARTOGRAM. (New Zealand cartographic society). Wellington. 1- (1975-)
Irregular. Newsletter.

1861 COMPASS. (Massey university student geographical association).
Palmerston North. 1-6 (1966-1972) Irregular. Closed 1972.

1862 DUNEDIN. University of Otago. Department of geography. Dunedin.
Publications in geography. 1- (1945-) Irregular, partly reprints.
Closed. (U-2:1373a).

1863 EARTH science journal. (Waikato geological society). Hamilton.
v1-5 (1967-1971). Closed 1971. (N70-2:1812c; B68:179a; N75-2:2377b).

1864 GEOGRAPHY of New Zealand. (Massey university. Geography department).
Palmerston North. 1 (1961), 2 (1968) Irregular. (N70-2:2282b).

(1865 NEW ZEALAND cartographic journal. (New Zealand cartographic society).
Wellington. 1- (1970-) Semiannual. (N77).
[New Zealand cartographic society, Inc., P.O. 9331 Courtenay Place, Wellington, New Zealand.]

*(1866 NEW ZEALAND geographer. (New Zealand geographical society). Christchurch.
1- (1945-) 2 times a year. (U-4:3053c; B-3:376a; BS:618b).
Index: 1-25 (1945-1969).
[Secretary, New Zealand Geographical Society, Department of Geography, University of Canterbury, Christchurch 1, New Zealand.]

(1867 NEW ZEALAND geographical society. Christchurch.
Annual report. 1956- . (BS:618b).

*(1868 ______.
Conference series. (Proceedings). Christchurch. 1- (1955-) Triennial to 1970, then biennial. (N70-3:4200b; N77).
1 (1955) Auckland; 2 (1958) Christchurch; 3 (1961) Palmerston North; 4 (1964) Dunedin; 5 (1967) Auckland; 6 (1970) Christchurch; 7 (1972) Hamilton; 8 (1974) Palmerston North; 9 (1977) Dunedin; 10 (1979) Auckland.
[Secretary, New Zealand Geographical Society, Department of Geography, University of Canterbury, Christchurch, New Zealand.]

1869 ______.
Miscellaneous publications. (N70-3:4200b).

1870 ______.
Miscellaneous series. 1-11 (1950-1970) Irregular monographs. (N70-3:4200b).
[University of Canterbury, Christchurch, New Zealand.]

1871 ______.
Record of proceedings of the society and its branches. no1-46 (1946-1968). Closed 1968. (U-4:3053c; B-3:376a; BS:618b; B73:286a).
Superseded by New Zealand journal of geography [1875].

(1872 ______. Auckland branch.
Annual lectures. 1971- (N70-3:4200b).

(1873 ______. Canterbury branch. Christchurch.
Publication. 1- (1958-) Irregular.

1874 ______. Manawatu branch. Palmerston North.
Publication. Closed. (N70-3:4200b).

1874a ______. Waikato branch. Hamilton.
Publication. 1-2 (1970-1972).

*(1875 NEW ZEALAND journal of geography, (New Zealand geographical society). Dunedin. 47- (1969-) 2 nos. a year.
Superseded New Zealand geographical society. Record of proceedings [1871], and continues its numbering. (N70-4:6635a; B73:229b).
[University of Canterbury, Christchurch, New Zealand.]

(1876 NEW ZEALAND map keepers circle. Palmerston North, Auckland. Newsletter. 1- (1977-) Irregular.

(1877 OTAGO geographer. (Otago university students' geographical association). Dunedin. 1- (1967-). (N70-3:4468a).

*(1878 PACIFIC viewpoint. (Victoria university of Wellington. Department of geography). Wellington. 1- (1960-) 2 nos. a year. (N70-3:4495a; B68:406b).
Index: 1-5 (1960-1964).
[Department of Geography, Victoria University of Wellington, Wellington, New Zealand.]

1879 PACIFIC viewpoint monograph. (Victoria university of Wellington. Department of geography). Wellington. 1-4 (1967-1969) Irregular. Closed 1969. (N70-3:4495a; B73:248b).

(1880 PERSPECTIVE. (New Zealand geographical society, Manawatu branch). Palmerston North. 1- (1965-) Quarterly. (N70-3:4599c; B73:254b).
1-8 published by New Zealand geographical society, Christchurch.

(1881 J. T. STEWART lecture in planning. (New Zealand geographical society. Manawatu branch, [and] Palmerston North city council). Palmerston North. 1- (1967-) Annual. (N75-1:1215c).
[Geography Department, Massey University, Palmerston North, New Zealand.]

(1882 UNIVERSITY of Waikato, Hamilton. Department of earth sciences. Hamilton. Occasional reports. 1- (1976-) Irregular.

NICARAGUA (Spanish)

República de Nicaragua

1883 ACADEMIA de geografía e historia de Nicaragua. Managua.
Revista. 1-41 (1936-1972); 42- (1977-) Irregular. (U-1:19b; B-1:12b; F-4:127a).
Not published 1973-1976.

NIGERIA (English)

Federal Republic of Nigeria

1884 AHMADU Bello university, Zaria. Department of geography. Zaria.
Occasional paper. 1- (1965-) Irregular. (N70-1:141c; B68:396b).
[Department of Geography, Ahmadu Bello University, Zaria, Northern Nigeria.]

1885 GEOGRAPHER. (Lagos. University. Geographical society). Yaba. (N77).
Continues Unilag geographers (not in this list).

1886 GEOGRAPHICA. (Ife. University. Geographical society). Ile-Ife.
1- (1967-) Annual. (N75-1:905b).

1887 GEOGRAPHICAL society, Ibadan.
Annual magazine. 1- (1965-). (N70-2:2281c).
At head of title: University college, Ibadan.
Also called: university geographer.

1888 IBADAN. University. Department of geography. Ibadan.
Research notes. no1-12 (Je 1952-F 1959). Closed. (N70-2:2650a; B-2:436b; BS:399b).
Issued under former name: Ibadan university college.
Mimeographed.

x(1889 NIGERIAN geographical journal. (Nigerian geographical asssociation). Ibadan.
1- (1957-) Semiannual. (N70 3:4222b; DO:623a).
Indexes: 1-4 (1957-1961), 5-7 (1962-1964), 10-12 (1967-1969), 13-16 (1970- 1973) in v18 no1 (Jn 1975), p. 83-87.
[Department of Geography, University of Ibadan, Ibadan, Nigeria.]

*(1890 SAVANNA: a journal of environmental and social sciences.
(Ahmadu Bello university). Zaria. v1- (Jn 1972-) 2 nos a year. (N75-2:1929a; B73:306a).
Abstracts.
[Department of Geography, Ahmadu Bello University, Zaria, Nigeria.]

NORWAY (Norwegian)

Kongeriket Norge

(1891 AD NOVAS. Norwegian geographical studies. Bergen; Oslo. 6- (1968-)
Irregular monographs.
Supersedes Bergen. Norges handelshøyskole. Skrifter i rekken geografiske avhandlinger [1894] and Ad novas [1892], whose numbering it continues.
In Norwegian with English summary, or in English.
[Universitetsforlaget, Oslo, Norway.]

1892 AD NOVAS. (Norske geografiske selskab. Skrifter). Oslo. 1-2 (1944); 3-5 (1960). Closed. Irregular monographs. (N70-1:78c).
1-2 lacked the series title. Superseded by Ad novas. Norwegian geographical studies [1891].

1893 BERGEN. Norges handelshøyskole. Geografisk institutt. Bergen.
Meddelelser 36-37 (1977) Irregular.
Title in English: Norwegian school of economics. Department of geography.
Occasional paper.
In English or Norwegian.

1894 ______. ______. ______.
Skrifter i rekken geografiske avhandlinger. 1-3, 5-9 (1947-1967).
Closed. Irregular monographs. (U-1:638c; B-3:409a).
Superseded by Ad novas. Norwegian geographical studies [1891].
English subtitle: Norwegian school of economics and business administration. Publication. Geographical series.
Some monographs in English; others in Norwegian with English summary.

1895 GEOGRAFISKE studier av utkantstrøk i Norge: Arbeidsrapport. (Bergen. Norges handelhøyskole og Universitet i Bergen. Geografisk institutt). Bergen. 2 (1974). (GZS).

1896 NORGES almenvitenskapelige forskningraad. Gruppe: Geografi. (Universitetsforlaget). Oslo.
Skrifter. Irregular. (N70-3:4249a).

1897 NORGES land og folk. Oslo. 1-20 (1884-1906). Closed 1906. (U-4:3075a; B-3:393a).
A second v12 issued in 1920 (12a). Subtitle varies.

*(1898 NORSK geografisk tidsskrift. (Norske geografiske selskab). Oslo. 1- (1926/27-) 4 nos. a year, 8 to a v. until v20 (1966); v21- (1967-) 4 nos. to a v. (U-4:3076c; B-3:394b; F-3:745a).
Supersedes Norske geografiske selskab, Oslo. Aarbok [1899].
Index: 1-10 (1926-1945); 11-24 (1946-1970).
In English or Norwegian.
[Universitetsforlaget, Postboks, 7508, Skillebekk, Oslo 2, Norway, or P.O. Box 142, Boston, Massachusetts 02113, U.S.A.]

1899 NORSKE geografiske selskab, Oslo. Oslo.
Aarbok. 1-32 (1889/90-1921). Closed 1921. (U-4:3078b; B-3:294a; F-3:746a).
1-4 as Arbog; 5-20 as Aarbog. Superseded by Norsk geografisk tidsskrift [1898].
Index: 1-17 (1889-1906) in 17.

(1900 NORWAY. Norges geografiske oppmåling. Oslo.
Beretning om Norges geografiske oppmålings virksomhet i tidsrummet. 1885- Annual. (ULG:467c; B-3:409a).
In Norwegian with English abstracts.

(1901 OSLO. Norsk polarinstitutt (prior to 1948 as: Norway. Norges Svalbard- og Ishavs-undersøkelser). Oslo.
Årbok. 1960- (1962-) Annual. (N70-3:4462c; B68:52b).
In English with a few articles in Norwegian.
[Norsk Polarinstituut, Postboks 158, 1330 Oslo Lufthavn, Norway. Orders through Universitetsforlaget, Postboks 307, Blindern, Oslo 3, Norway; 16 Pall Mall, London S.W.1, England; P.O. Box 142, Boston, Massachusetts 02113, U.S.A.]

(1902 ______. ______.
Meddelelse. 1- (1926-). (U-4:3077b: B-3:407b; F-3:743a).
In Norwegian with English summaries.
[Address: see under Årbok [1901].]

1903 ______. ______.
Polarhandbok. 1- (1964-). (N70-3:4462c; B68:421b).
[Address: see under Årbok [1901].]

(1904 ______. ______.
Skrifter. 1- (1929-). (U-4:3077b; B-3:407b; BS:631a).
Title varies: 1-11, Resultater av de Norske statsunderstøttede Spitsbergenekspedisjoner; 12, Skrifter om Svalbard og Nordishavet; 13-81, Skrifter om Svalbard og Ishavet; 82-89, Norges Svalbard- og Ishavsundersøkelser. Skrifter; 90- (1948-), Norsk polarinstitutt. Skrifter.
In English.
[Address: see under Årbok [1901].]

پاکستان PAKISTAN (Urdu, English)

Islāmī Jamhūrīya-e-Pākistān, Islamic Republic of Pakistan

1905 DIRECTORY of Pakistan geographers. Karachi. 1 (1961). (N70-2:1747b).
Issued by the Pakistan institute of geography for the Department of geography, University of Sind.

1906 GEOGRAFIA: a research journal of geography. (Pakistan institute of geography). Karachi. 1-5 (1962-1966). Publication suspended 1966. (N70-2:2277c-2278a; B68:224b).
Successor to Karachi geographical society. Bulletin [1908].

1907 KARACHI geographers association. Karachi.
Monograph. 1- (1964-). (N70-2:3254a).
no6 (1968) as Publication.

1908 KARACHI geographical society. Karachi.
Bulletin. no1-2 (1949-1951). Closed 1951. (U-3:2272a).
In English.

*(1909 PAKISTAN geographical review. (Lahore. University of the Punjab. Department of geography). Lahore. 1- (1942-) 2 nos. a year. (U-4:3242c; BS:712a).
v1-3 (1942-1948) as Punjab geographical review. Suspended 1943-1946. Indexes: v1-10 (1942-1955) as v11 no2 (1956); 11-17 (1956-1962) in v17 no2 (1962); 18-22 (1963-1967) in v22 no2 (1967); 23-33 (1968-1978). In English.

1910 PAKISTAN institute of geography. Karachi.
Monograph. 1 (1965). Only one published. (N70-3:4508c-4509a).

PANAMA (Spanish)

República de Panamá

1911 PANAMÁ (city). Universidad. Departamento de geografía. Panamá.
Publicación. 1- (1953-) Irregular. (N70-3:4527a).
[Apartado 289, Caudad de Panamá, Panama.]

PAPUA NEW GUINEA (English)

*(1912 PAPUA New Guinea. University. Geography department.
Occasional paper. Port Moresby. 1- (1971-) Irregular. (N78).
Processed. In English.

PERU (Spanish)

1913 ANUARIO geográfico del Perú. (Sociedad geográfica de Lima). Lima.
1962. Only one issued. (N70-1:393b).

1914 ASOCIACIÓN de geógrafos del Perú. Lima.
Publicación. 1- (1960-). (N70-1:493b).

1915 ASOCIACIÓN nacional de geógrafos peruanos. Lima.
Serie "Educación y geografía." 1- (1960-). (N70-1:496c).

1916 ______.
[Publicaciones]. (N70-1:496c).

1917 BOLETÍN geográfico. (Lima. Universidad de San Marcos. Instituto superior de geografía. Asocación de geógrafos egresados).
1- (1956-). (N70-1:811a).

1918 CENTRO geográfico del Cuzco. Cuzco.
Boletín. v1 no1 (1942).

1919 CONGRESO nacional de geografía. (Asociación nacional de geógrafos peruanos). Lima.
Anales. 1- (1961-) Irregular. (N70-1:1491a).
1, 1961, 3 vols.; 2, 1964, 3 vols.; 3, 1965, 4 vols.

1920 LIMA. Colegio nacional de Nuestra Señora de Guadalupe. Departamento de geografía. Lima.
Boletín mensual. no1-2 (1937-1938). All issued? (U-3:2423b).

1921 ______. Pontificia universidad católica del Perú. Lima.
Ensayos geográfocos. 1- (1938-). (U-3:2424a).

1922 ______. Universidad de San Marcos. Instituto de geografía. Lima.
Publicaciones.
sI. Monografías y ensayos geográficos. 1-4 (1958-1960).
(N70-3:3484a).

1923 ______. ______. ______.
______.
sII. Viajes de estudios y trabajos de campo. 1 (1958).
(N70-3:3484c).

1924 ______. ______. ______.
Revista. 1949; 1-6 (1954-1960) Irregular. (N70-3:3484b; BS:924b).
1949 as Viajes de estudio.

1925 ______. ______. ______.
Serie conferencias. 1- (1957-) Irregular. (N70-3:3484b).

1926 MÉTODOS de investigación geográfica. (Lima. Universidad de San Marcos. Instituto de geografía. Asociación de geógrafos egresados).
1 (1960). Only one issued? (N70-3:3748a).

1927 REVISTA geográfica del Perú. (Lima. Universidad de San Marcos. Instituto do geografía. Asociación de geógrafos egresados). Lima. 1-2 (1956-1961). All issued? (N70-3:5017a; B68:479a).

*1928 SOCIEDAD geográfica de Lima. Lima.
Boletín. 1- (Ap 1891-) Quarterly. (U-5:3909b; B-4:106a; F-1:664b-665a).
Index: v1-58 (1891-1941).

1929 ______.
Publicaciones. no1-6 (1920-1929). Closed 1929. (U-5:3909c).

1930 UNIVERSIDAD nacional Federico Villarreal. Departmento de ciencias históricas y geográficas. Lima.
Serie B. Documentos y reimpresiones. 1- (1966-). (N70-4:6093c).

PHILIPPINES (Pilipino, English, Spanish)

Republic of the Philippines,

Republika ñg Pilipinas, República de Filipinas

*(1931 PHILIPPINE geographical journal. (Philippine geographical society [and] National committee on geographical sciences, NSDB). Manila. 1- (1953-) Quarterly. (N70-3:4622c; BS:675a). None published 1958-1962.
Index: v1-20 (1953-1976) in v21, nos. 3-4 (1977), p. 99-151.
In English.
[Philippine Geographical Journal, P. O. Box 2116, Manila, Philippines.]

POLAND (Polish)

Polska Rzeczpospolita Ludowa

*(1932 ACTA geographica lodziensia. (Łódzkie towarzystwo naukowe. Wydział 3 [and] Łódź. Uniwersytet. Instytut geograficzny). Łódź. no1- (1948-) Irregular. (BS:23a; R-1:143a).
1-10 (1948-1962) as Acta geographica universitatis lodzensis.
In Polish with English or French abstracts.
[ul. M. Curie-Sklodwskiej 11, Łódź, Poland.]

1933 BADANIA fizjograficzne nad Polską zachodnią. (Poznańskie towarzystwo przyjaciół nauk. Wydział matematyczno-przyrodniczy. Komitet fizjograficzny). Poznań. 1-22 (1948-1969). (U-1:583b; N75-2:2355b; BS:104b; R-2:16a).
With v23 (1969) divided into two subseries: Seria A. Geografia fizyczna [1934], Seria B. Biologia, Seria C. Zoologia (not in this list).
In Polish with English titles in table of contents. Abstracts chiefly in English.

*(1934 BADANIA fizjograficzne nad polską zachodnią. Seria A: Geografia fizyczna. (Poznańskie towarzystwo przyjaciół nauk. Komitet fizjograficzny). Poznań. 23- (1969-) Irregular. (N75-1:242b).
Supersedes in part Badania fizjograficzne nad Polską zachodnią [1933] and continues its numbering.
In Polish with summaries chiefly in English.

*(1935 BIBLIOGRAFIA geografii polskiej. (Polska akademia nauk. Instytut geografii i przestrzennego zagospodarowania). Warszawa. 1- (1956-) Irregular. (N70-1:722a; BS:120a).
A bibliography of the geography of Poland from 1918.
[Krakowskie Przedmieście 30, Warszawa, 64, Poland.]

1936 BIULETYN geograficzny. (Polskie towarzystwo geograficzne). Warszawa. 1-3 (1952-1954). Closed 1954. (N70-1:790a; N70-3:4702b; R-2:87a).
Called also the society's Wydawnictwo s2. v3 (1954) as Polska akademia nauk. Instytut geografii. Biuletyn geograficzny.
Superseded by Dokumentacja geograficzna [1965].

*(1937 BIULETYN peryglacjalny. (Łódzkie towarzystwo naukowe. Wydział 3; Societas scientiarum lodziensis, sectio 3). Łódź. 1- (1954-) Irregular. (N70-1:790b; BS:129a; R-2:89b).
Predominantly in English or French but articles also in German, Polish, and Russian.
[Ars Polona, Krakowskie Przedmieście 7, Warszawa, Poland.]

*(1938 CZASOPISMO geograficzne. (Polskie towarzystwo geograficzne). Łódź, Warszawa. Wrocław. 1- (1923-) Quarterly. (U-2:1250a; B-1:687b; BS:245a; F-2:322b; R-2:450b).
Until 1939 by Zrzeszenie polskich nauczycieli geografii, Łódź and Warszawa. v17 no3-4 covered the years 1939-1946.
English title: geographical journal.
Index: 1-50 (1923-1979) as v50 no4 (1979).
In Polish. Supplementary English titles in table of contents and abstracts.
[Ars Polona-Ruch, Krakowskie Przedmieście 7, Warzawa, Poland.]

*(1939 FOLIA geographica. (Polska akademia nauk. Oddział w Krakowie. Komisja nauk geograficznych). Kraków.
Series geographica-oeconomica. 1- (1968-). (N70-2:2126b; B73:120b).
In Polish with English summaries, or in English.

*(1940 ______.
Series geographica-physica. 1- (1967-). (N70-2:2126b; B73:121a).
In Polish with English summaries, or in English.

1941 FOTOINTERPRETACJA w geografii. (Polskie towarzystwo geograficzne. Komisja fotointerpretacji). Warszawa. Warszawa. 1-10 (1964-1977) Irregular. (N70-2:2172a).
In Polish; supplementary table of contents and abstracts in English.
Beginning with 11 (1977) subseries of Prace naukowe uniwersytetu śląskiego w Katowicach (not in this list) as Fotointerpretacja w geografii. Katowice. 1- (11-) (1977-) Annual.
Contents and abstracts in English and French.

(1942 GDAŃSK. Uniwersytet. Wydział biologii i nauk o ziemi. Gdańsk.
Zeszyty naukowe: geografia. 1- (1970-). (N75-1:653a).
Supersedes Zeszyty geograficzne [2016].

(1943 GEOGRAFIA w szkole; czasopismo dla nauczycieli. (Polskie towarzystwo geograficzne. Organ Ministerstwa oświaty). Warszawa.
1- (1948-) 6 nos. a year. (U-2:1692a; R-3:267b).
[Plac Dąbrowskiego 8, Warszawa, Poland.]

*(1944 GEOGRAPHIA polonica. (Polska akademia nauk. Instytut geografii i przestrzennego zagospodarowania. Polish academy of sciences. Institute of geography and spatial organization). Warszawa. 1- (1964-) Irregular. (N70-2:2280a; B68:224b; R-3:269a).
Index: v1-32 (1964-1975).
In English and other major languages.
[Foreign Trade Enterprise, Ars Polona, Krakowskie Przedmieście 7, Warszawa, Poland.]

1945 GÓRNOŚLĄSKIE prace i materiały geograficzne. (Katowice. Śląski instytut naukowy). Katowice. 1- (1962-) Irregular. (N70-2:2371b; R-3:302a).
In Polish. English and Russian summaries.
[Wydawnictwo "Śląsk," Katowice, Poland.]

1946 INSTYTUT zachodnio-Pomorski, Szczecin. Sekcja przestrzenno-geograficzna. Prace. 1- (1965-). (N70-2:2849a).

1947 KOSMOS. (Polskie towarzystwo przyrodników imienia Kopernika). Lwów.
1-52 (1875-1927). Closed 1927. (U-3:2314c; B-2:667b; F-3:260b: R-4:206a).
Continued in two series, of which A is included here [1948].
Index: 1-20 (1875-1895).

1948 ______.
sA. Rozprawy 53-64 (1928-1939). Closed 1939. (U-3:2314c; B-2:667b; F-3:260b: R-4:206a).

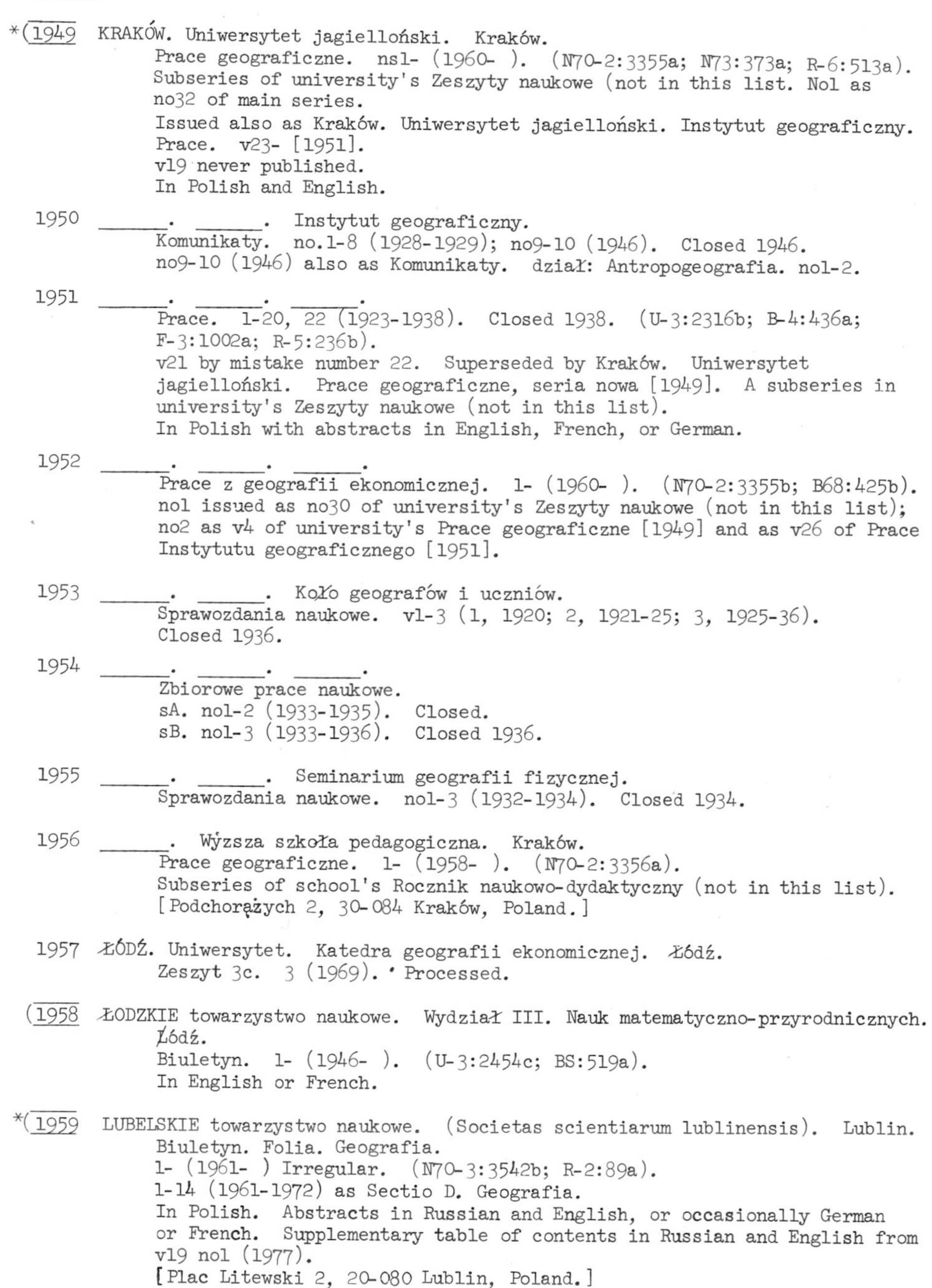

*(1949 KRAKÓW. Uniwersytet jagielloński. Kraków.
Prace geograficzne. nsl- (1960-). (N70-2:3355a; N73:373a; R-6:513a).
Subseries of university's Zeszyty naukowe (not in this list. Nol as no32 of main series.
Issued also as Kraków. Uniwersytet jagielloński. Instytut geograficzny. Prace. v23- [1951].
v19 never published.
In Polish and English.

1950 ______. ______. Instytut geograficzny.
Komunikaty. no.1-8 (1928-1929); no9-10 (1946). Closed 1946.
no9-10 (1946) also as Komunikaty. dział: Antropogeografia. nol-2.

1951 ______. ______. ______.
Prace. 1-20, 22 (1923-1938). Closed 1938. (U-3:2316b; B-4:436a; F-3:1002a; R-5:236b).
v21 by mistake number 22. Superseded by Kraków. Uniwersytet jagielloński. Prace geograficzne, seria nowa [1949]. A subseries in university's Zeszyty naukowe (not in this list).
In Polish with abstracts in English, French, or German.

1952 ______. ______. ______.
Prace z geografii ekonomicznej. 1- (1960-). (N70-2:3355b; B68:425b).
nol issued as no30 of university's Zeszyty naukowe (not in this list); no2 as v4 of university's Prace geograficzne [1949] and as v26 of Prace Instytutu geograficznego [1951].

1953 ______. ______. Koło geografów i uczniów.
Sprawozdania naukowe. v1-3 (1, 1920; 2, 1921-25; 3, 1925-36).
Closed 1936.

1954 ______. ______. ______.
Zbiorowe prace naukowe.
sA. nol-2 (1933-1935). Closed.
sB. nol-3 (1933-1936). Closed 1936.

1955 ______. ______. Seminarium geografii fizycznej.
Sprawozdania naukowe. nol-3 (1932-1934). Closed 1934.

1956 ______. Wyższa szkoła pedagogiczna. Kraków.
Prace geograficzne. 1- (1958-). (N70-2:3356a).
Subseries of school's Rocznik naukowo-dydaktyczny (not in this list).
[Podchorążych 2, 30-084 Kraków, Poland.]

1957 ŁÓDŹ. Uniwersytet. Katedra geografii ekonomicznej. Łódź.
Zeszyt 3c. 3 (1969). Processed.

(1958 ŁODZKIE towarzystwo naukowe. Wydział III. Nauk matematyczno-przyrodniczych. Łódź.
Biuletyn. 1- (1946-). (U-3:2454c; BS:519a).
In English or French.

*(1959 LUBELSKIE towarzystwo naukowe. (Societas scientiarum lublinensis). Lublin.
Biuletyn. Folia. Geografia.
1- (1961-) Irregular. (N70-3:3542b; R-2:89a).
1-14 (1961-1972) as Sectio D. Geografia.
In Polish. Abstracts in Russian and English, or occasionally German or French. Supplementary table of contents in Russian and English from v19 nol (1977).
[Plac Litewski 2, 20-080 Lublin, Poland.]

(1960 LUBLIN. Uniwersytet Marii Curie-Skłodowskiej. Lublin.
Roczniki. Annales. Dział B. Geografia, geologia, mineralogia i petrografia. nol- (1946-) Annual. (U-3:2482c; B-4:431a; BS:926a; R-1:336a).
In Polish. Abstracts in Russian and in English, French, or German.
[ul. Narutowicza 30, Lublin, Poland.]

1961 LUBUSKIE towarzystwo naukowe. Wydział nauk przyrodniczych. Komisja geograficzno-geologiczna. Zielona Góra.
nol (1973).
nol (1973) of commission is nol3 of parent series of section (not in this list).

1962 LWÓW (Lvov). Uniwersytet. Instytut geograficzny. Lwów. Prace.
1-16 (1932-1939). Closed 1939. (U-3:2491c).
Reprints.
In Polish with English, French, or German abstracts.

1963 POLAND. Służba geograficzna. (Wojskowy instytut geograficzny). Warszawa. Wiadomości. v1-13 (1927-1939); v14 (1948). Closed 1948.
(ULG-488b; B-3:568b; F-4:971a; R-6:393a).
In Polish with French or German abstracts.

(1964 POLSKA akademia nauk. Warszawa. Bulletin. Série des sciences de la terre.
19- (1971-). (N75-2:1753c).
Continues academy's Bulletin. Serie des sciences géologiques et géographiques [1972].
In English. Abstracts in English.

(1965 ______. Instytut geografii i przestrzennego zagospodarowania. Warszawa.
Dokumentacja geograficzna. 1955- Bimonthly. (N70-3:4702b; B68:176b).
Issues numbered 1-6 within each year. Supersedes Biuletyn geograficzny [1936].
In Polish. Titles also in English. Abstracts in English and Russian.
[Krakowskie Przedmieście 30, Warszawa 64, Poland.]

1966 ______. ______.
Streszczenia prac habilitacyjnych i doktorskich. 1970- (N75-2:1754a).
Subseries of Documentacja geograficzna [1965]. Continues its Abstrakty prac habilitacyjnych i doktorskich (not in this list).

(1967 ______. Komitet przestrzennego zagospodarowania kraju. Committee for space economy and regional planning. Warszawa.
Biuletyn. 1- (1960-). (N70-3:4704c; B73:45b; R-2:88b). Mimeographed.
[Pałac Kultury i Nauki, p. 2409, Warszawa, Poland.]

*(1968 ______. ______.
Studia. 1- (1961-). (N70-3:4704c; B73:331b).
In Polish with summaries in English and Russian.
[Pałac Kultury i Nauki, p. 2409, Warszawa, Poland.]

1969 ______.
Oddział w Krakowie.
Ośrodek dokumentacji fizjograficznej studia. 1- (1972-) Irregular.

1970 ______. Wydział III. Nauk matematyczno-fizycznych, chemicznych i geologo-geograficznych.
Biuletyn, v1-5 (1953-1957). (N70-3:4706b; BS:688a; R-2:167b).
Title in French: Bulletin de l'Académie polonaise des sciences. Classe 3. Mathématique, astronomie, physique, chimie, géologie et géographie.
Superseded in part by the academy's Série des sciences chimiques, géologiques et géographiques [1971] and its Série des sciences géologiques et géographiques [1972].
In English, French, or German.

1971 ______. ______.
Bulletin. Série des sciences chimiques, géologiques et géographiques.
v6-7 (1958-1959). (N70-3:4706a; BS:688a; R-2:168b).
Superseded in part by the academy's Série des sciences géologiques et géographiques [1972].
In English, French, or German with Russian summaries.

1972 _____. _____.
Bulletin. Série des sciences géologiques et géographiques.
v8-18 (1960-1970). Closed 1970. (N70-3:4706ab; R-2:169a).
Superseded by the academy's Bulletin. Série des sciences de la terre [1964].
In English, French, German, Polish, or Russian.

1973 POLSKA akademia umiejętności. Kraków. Komisja fizyograficzna. Kraków.
Prace monograficzne. 1-7 (1925-1931). Closed 1931. (U-4:3389c; B-1:69a; F-3:986a; R-5:242a).

1974 _____. _____.
Sprawozdania. 1-73 (1866-1939). Closed 1939. (U-4:3389c; B-4:334a; F-4:987b; F-3:987b; R-6:136b).

1975 _____. Komisja geograficzna. Kraków.
Prace. no1-4 (1931-1934). Closed 1934. (U-4:3389c; B-1:69a; F-3:986a; R-5:239b).
In Polish; some French abstracts.

1976 POLSKA bibliografia analityczna. Geografia. (Polska akademia nauk. Ośrodek bibliografii i dokumentacji naukowej). Warszawa.
1-5 (1956-1960). Closed 1960. (N70-3:4707c; BS:690a; R-5:214b-215a).
In Polish, with all titles of works cited also translated into English.

1977 POLSKI przegląd kartograficzny. (Instytut kartograficzny imienia E. Romera). Lwów; Warszawa. v1-12, no1-48 (1923-1934). Closed 1934. (U-4:3391b; B-3:576a; F-3:987b; R-5:216a).
In Polish with English, French or German abstracts. English title: Polish cartographical review. French title: Revue cartographique polonaise.

*(1978 POLSKI przegląd kartograficzny. Polish cartographical review. (Polskie towarzystwo geograficzne. Komisja Kartograficzna. Państwowe przedsiębiorstwo wydawnictw kartograficznych). Warszawa. 1- (1969-)
Quarterly. (N75-2:1754b; B73:262a).
In Polish. Table of contents and summaries in English and Russian.
[Redakcja Polskiego Przeglądu Kartograficznego PPWK, ul. Solec 18, 00-410 Warszawa, Poland.]

1979 POLSKIE towarzystwo geograficzne, Kraków. Krakowskie odczyty geograficzne. Kraków.
Wydawnictwa. no1-14 (1924-1929). Closed 1929. (U-4:3392a).

1980 _____.
Oddział w Lublinie. Lublin.
Ogólnopolski zjazd polskiego towarzystwo geograficznego; przewodnik wycieczkowy. 8 (1964). (N70-3:4709c).

(1981 POZNAJ świat: Magazyn geograficzny. (Polskie towarzystwo geograficzne). Warszawa. 1- (1948-) Monthly. (U-4:3416a; R-5:233b-234a).
Index: 1955-1965.
[Nowy Świat 49, Warszawa, Poland.]

(1982 POZNAŃ. Uniwersytet. Poznań.
Geografia. no1- (1957-) Irregular. (N70-3:4730a; R-6:512b).
A subseries of university's Zeszyty naukowe (not in this list).
[ul. Ratajczaka 38/40, Poznań, Poland.]

1983 _____. _____. Instytut geograficzny.
Badania geograficzne. 1-20 (1926-1939). Closed 1939.
(U-4:3407b; B-1:278b; F-1:533a; R-2:16b).
Title varies: Wydawnictwo (1926-1931); Badania geograficzne nad Polską północno-zachodnią; Prace Instytutu geograficznego Uniwersytetu poznańskiego.
In Polish with English, French, or German abstracts.

*(1984 ______. ______. Seria geografia. 1- (1962-) Irregular. (N70-3:4730a; B73:265b).
Early numbers as Wydzial biologii i nauk o ziemi. Prace. Seria geografia.
Abstracts in English.

1985 POZNAŃSKIE towarzystwo przyjaciół nauk. Poznań.
Wydawnictwo popularno-naukowe z zakresu nauk o ziemi. 1- (1955-). (N70-3:4740b; BS:972b).

1986 ______. Komisja geograficzno-geologiczna.
Prace. v1 no1-6 (1936-1939); v2- (1956-) Irregular. (U-4:3416b; B-4:335b; R-5:239b).
v1 (1936-1939) as Komisja geograficzna.
Abstracts in English, French, German, or Russian.
[ul. Fredry 10, Poznań, Poland.]

1987 ______. Komisja matematyczno-przyrodnicza. Prace.
sA. Nauki abiologiczne. v1-4 no4 (1921-1939), 5 (1947). Closed. (U-4:3416b; F-3:1001a; R-5:240a).

1988 PRACE geograficzne (E. Romer). Lwów; Warszawa. 1-19 (1918-1938). Closed 1938. (U-4:3416c; B-3:590b; F-3:1001b; R-5:235a).
In Polish with English, French, or German abstracts.

*(1989 PRACE geograficzne. Geographical studies. (Polska akademia nauk. Instytut geografii i przestrzennego zagospodarowania). Warszawa. 1- (1954-) Irregular. (N70-3:4702b; B-3:590b; BS:697a; R-5:235a).
In Polish. Supplementary table of contents and abstracts of major articles in English and Russian from the beginning. Some monographs in English.
[Instytut Geografii PAN. Krakowskie Przedmieście 30, 00-927 Warszawa 64, Poland.]

1990 PRACE geograficzno-ekonomiczne. (Instytut zachodni). Poznań. 1- (1961-). (N70-3:4740c).

1991 PROBLEMS of applied geography. Warszawa. 1- (1961-). (N70-3:4782b).
1 comprises the Proceedings of the Anglo-Polish (geographical) seminar.

*(1992 PRZEGLĄD geograficzny. Polish geographical review. (Polska akademia nauk. Instytut geografii i przestrzennego zagospodarowania). Warszawa. 1- (1918-) Quarterly. (U-4:3466a; B-3:622b; BS:709b; F-3:1072b; R-318b-319a).
v1-25 (1918-1953) issued by Polskie towarzystwo geograficzne.
Index: 1-10 (1918-1930); 11-25 (1931-1953); 26-35 (1954-1963); 36-50 (1964-1978).
In Polish. English and Russian table of contents and abstracts.
[Ars Polona-Ruch, Krakowskie Przedmieście 7, Warszawa 1, Poland.]

(1993 PRZEGLĄD zagranicznej literatury geograficznej. (Polska akademia nauk. Instytut geografii). Warszawa. no1- (1956-) Quarterly. (N70-3:4821c; R-5:324b). Mimeographed.
[Krakowskie Przedmieście 30, Warszawa 64, Poland.]

*(1994 QUAESTIONES geographicae (Poznań. Uniwersytet. Instytut geograficzny). Poznań. 1- (1974-) Annual. (N75-2:1808a).
English title of issuing agency: Adam Mickiewicz University. Institute of Geography.
In English.
[Ul. Fredry 10, 61-701 Poznań, Poland.]

(1994a QUATERNARY studies in Poland. (Polish academy of sciences. Branch in Poznań. Committee of Quaternary research). Warszawa-Poznań.
1- (1979-) Irregular.
In English. Abstracts in English.

1995 STUDIA geograficzno-fizyczne z obszaru opolszczyzny. 1- (1968-) Irregular. (N70-4:5575c).
[Instytut Śląski, Instytut Naukowo-Badawczy, Luboszycka 3, Opole, Poland.]

*(1996 STUDIA geomorphologica Carpatho-Balcanica. (Polska akademia nauk. Oddział w Krakowie. Komisja nauk geograficznych). Kraków. 1- (1967-) Annual. (N70-4:5575c).
Principally in English with abstracts in Polish and Russian. Articles also in French, Polish, or Russian.

(1997 TORUŃ. Uniwersytet. Toruń.
Geografia. 1- (1958-) Irregular. (N70-4:5823c).
Subseries of the university's Zeszyty naukowe: nauki matematyczno-przyrodnicze, 1 (4), 2(5), 3 (10), 4 (11), 5 (14), 6 (19), 7 (24), 8 (26), 9 (31), 10 (32), 11 (35), 12 (41), 13 (43), 14 (46).
Also as Acta universitatis Nicolai Copernici: geografia; Zeszyty naukowe Uniwersytetu Mikołaja Kopernika w Toruniu: geografia.
In Polish with English table of contents and summaries, or in English.

1998 TOWARZYSTWO naukowe katolickiego uniwersytetu lubelskiego. Instytut geografii historycznej kościoła w Polsce. Lublin.
Materiały zródlowe do dziejów kościoła w Polsce.
1- (1965-). Monographs. (B73:205b).
Title on cover: Materiały do dziejów kościoła w Polsce.

1999 ______. ______.
Prace. 1- (1958-). (N70-4:5828b).

(2000 TOWARZYSTWO naukowe w Toruniu. Toruń.
Studia. Sectio C. Geographia et geologia. 1- (1953-) Irregular. (N70-4:5828c; BS:876b; R-6:159a).
(Studia societatis scientiarum torunensis. Sectio C. Geographia et geologia).
In Polish. English or Russian summaries.

(2000a UNIWERSYTET śląski. Katowice.
Geographia: Studia et dissertationes. 1- (1976-) Irregular.
Subseries of Prace naukowe uniwersytetu śląskiego w Katowicach (not in this list).
In Polish. Supplementary table of contents and abstracts in English and Russian.
[Uniwersytet Śląski, Bankowa 14, 40-007 Katowice, Poland.]

2001 WARSZAWA [Warsaw]. Uniwersytet. Instytut geograficzny. Warszawa.
Prace i studia. 1- (1967-). (N70-4:6267b; B73:265b).
Consists in part but not wholly of three subseries: Geografia economiczna [2002], Geografia fizyczna [2003], and Klimatologia [2004].
Supplementary titles in table of contents in English and abstracts in English.

2002 ______. ______. ______.
______.
Geografia economiczna. 1- (1973-). (N76).
In Polish. Added title and summaries in English.

2003 ______. ______. ______.
______. Geografia fizyczna. 1- (1967-). (N70-4:6267c; N79).
In Polish. Added title and summaries in English.

2004 ______. ______. ______.
Klimatologia. 1- (1964-). (N70-4:6267c; B73:265b).
In Polish. Added title and summaries in English.

*(2004a WARSZAWA. Uniwersytet. Wydział geografii i studiów regionalnych. Warszawa.
Prace i studia geograficzne. 1- (1979-) Irregular.
Supersedes Warszawa. Uniwersytet. Instytut geograficzny. Prace i studia [2001-2004].
In Polish. Supplementary titles in table of contents in English and abstracts in English.
[Wydawnictwa Uniwersytetu Warszawskiego, Krakowskie Przedmieście 30, Warszawa, Poland.]

2005 ______. ______. Zakład geograficzny.
Prace wykonane. 1-25 (1922-1938); 26-27 (1946-1947). Closed 1947. (U-5:4445a; F-3:1004b; R-5:245b). Reprints.
In Polish with French or German abstracts.

2006 ______. ______. ______.
Sprawozdania. no1-3 (1924-1946). Reprints.

2007 WIADOMOŚCI geograficzne. (Polskie towarzystwo geograficzne). Kraków.
v1-17 no1 (1923-1939). Closed 1939. (U-5:4503b; B-4:542a; F-4:970b).
v1-14 with subtitle: Revue mensuelle de géographie; v15-17 with subtitle: Bulletin trimestriel de géographie.
In Polish with English, French, or German abstracts.

2008 WILNO (Vilna). Uniwersytet. Zakład geograficzny. Wilno. Prace.
1-6 (1936-1938). Closed 1938. (U-5:4401a).
In Polish with abstracts in French or German.

2009 ______. ______. Zaklady geologiczny i geograficzny. Prace.
no1-27 (1924-1936). Closed 1936. (B-1:29a).
In Polish with abstracts in French or German.

2010 WISŁA; miesięcznik geograficzno-etnograficzny. Warszawa.
1-21 (1887-1922). Closed 1922. (U-5:4526b; B4:552b; R-6:402a).
Suspended 1918-1921. v21 (1922) as Wisła. (Polskie towarzystwo etnologiczne).

*(2011 WROCŁAW. Uniwersytet. Wrocław.
Studia geograficzne. 1- (1963-) Irregular. (N70-1:859c; B68:535a).
Subseries of Acta universitatis wratislaviensis: 1 (9), 2 (10), 3 (29), 4 (34), 5 (39), 6 (41), 7 (44), 8 (47), 9 (61), 10 (68), 11 (71), 12 (92), 13 (124), 14 (127), 15 (113), 16 (161), 17 (168), 18 (173), 19 (191), 20 (219), 21 (220), 22 (302), 23 (263), 24 (311), 25 (329), 26 (356), 28 (358), 29 (359), 31 (393).
Supersedes in part: Wrocław. Uniwersytet. Zeszyty naukowe. Nauka o ziemi [2012].
In Polish. Abstracts in English, French, or German.
[Instytut Geograficzne, Uniwersytet Wrocławski, Place Uniwersytecki 1, Wrocław, Poland.]

2012 ______. ______.
Zeszyty naukowe. Nauka o ziemi. 1-5 (1960-1961). Closed 1961. (N70-1:859c-860a, B68:375b).
Superseded in part by Wrocław. Uniwersytet. Studia geograficzne [2011].

(2013 ______. ______. Instytut geograficzny. Prace.
Seria A: Geografia fizyczna. 1- (1974-) Irregular. (N76).
Subseries of Acta universitatis wratislaviensis. no1 (1974), no236 of whole series; 2 (1978), 340.
Supplementary English table of contents and abstracts.
[Instytut Geograficzne, Uniwersytet Wrocławski, Plac Uniwersytecki 1, Wrocław, Poland.]

(2014 ______. ______. ______.
Prace. Seria B. Geografia społeczna i ekonomiczna. 1- (1975-)
Irregular. (N76).
Supplementary English table of contents and abstracts.
[Instytut Geograficzny, Uniwersytet Wroclawski, Plac Uniwersytecki 1, Wrocław, Poland.]

2015 ______. ______. Obserwatorium meteorologii i klimatologii. Wrocław.
Prace. 1- (1947-). (U-1:768a; BS:972a; R-5:243a).
Prace Zakładu i Obserwatorium meteorologii i klimatologii Uniwersytetu i politechniki we Wrocławiu.

2016 ZESZYTY geograficzne. Geographical papers. (Gdańsk. Wyższa szkoła pedagogiczna. Wydział biologii i nauk o ziemi). Gdańsk.
1-11 (1959-1969). (N70-4:6474a; N70-2:1648b; R-6:509a).
(Zeszyty geograficzne Wyzszej szkoły pedagogicznej w Gdańsku).
In Polish with English title, table of contents, and summaries.

2017 ZIEMIA. (Polskie towarzystwo krajoznawcze). Warszawa. 1-29 (1910-1939); 1946-1950; 1956-1958 Irregular. Closed 1958. (U-5:4636c; F-4:1048).
18-22 omitted in numbering.
Index: 1910-1929.

2018 ZIEMIA. Parce i materiały krajoznawcze. (Polskie towarzystwo turystyczno-krajoznawcze). Warszawa. 1- (1965-) Annual. (N70-4:6477a).

PORTUGAL (Portuguese)

República Portuguesa

2019 ACADÉMIA das ciências de Lisboa. Lisboa.
Collecção de notícias para a história e geografia das nações ultramarinas. 1-7 (1812-1856). Closed 1856. (U-1:17b; B-1:594b).

2020 AGRUPAMENTO de estudos de cartografia antiga. (Portugal. Junta de investigações do ultramar). Coimbra; Lisboa.
Série memorias (publicaçao). 1- (1961-) Irregular. (N70-1:141a; B68:25a).
In two series: Secção de Coimbra, Secção de Lisboa.
Also as Portugal. Junta das missões geográficas e de investigações do ultramar. Estudos de cartografia antiga. (N70-3:4726b).

2021 COIMBRA. Universidade. Faculdade de letras. Centro de estudos geográficos. Coimbra.
Boletim. v1-3, nos1-22/23 (1950-1966/1967). Closed 1967. (N70-1:1330b).

2022 ______. ______. ______. ______.
Publicações. 1-5 (1946-1951).
Many unnumbered monographs also.

*(2023 FINISTERRA; revista portuguesa de geografia. (Lisboa. Universidade. Centro de estudos geográficos). Lisboa. 1- (1966-) 2 nos. a year.
Index: 1966-1975. (N70-2:2086a).
In Portuguese with abstracts in French and English.
[Livraria Portugal, Rua do Carmo, 70, 1200 Lisboa, Portugal.]

2024 GARCIA de ORTA. (Portugal. Junta das missões geográficas e de investigações do ultramar. Revista). Lisboa.
1-19 (1953-1972) Quarterly. Closed 1972. (N70-2:2254a; BS:337b).
Abstracts in English and French.
Superseded in part by: Garcia de Orta: série de geografia [2025].

*(2025 GARCIA de ORTA. Série de geografia. (Portugal. Junta de investigacões do ultramar). Lisboa. 1- (1973-) Irregular. (N75-1:895a).
Supersedes in part, Garcia de Orta [2024].
Abstracts in English and Portuguese.
[Rua de Jau 54, Lisboa 3, Portugal.]

2026 GEOGRAPHICA. (Sociedade de geografia de Lisboa. Revista). Lisboa. v1-9, nos1-36 (1965-1973). Closed 1973. Quarterly. (N70-2:2280b).
In Portuguese with some articles and summaries in English or French.

(2027 LISBOA (Lisbon). Universidade. Faculdade de letras. Centro de estudos geográficos. Lisboa.
Estudos de geografia física. 1- (1972-) Irregular.

(2028 ______. ______. ______. ______.
Estudos de geografia humana e regional. 1- (1973-) Irregular.

(2029 ______. ______. ______. ______.
Estudos de planeamento regional e urbano. 1- (1971-) Irregular.

*(2030 ______. ______. ______. ______.
Memorias. 1- (1972-) Irregular.
In Portuguese. Summaries in English or French or both.

2031 PORTUGAL. Commissão central permanente de geographia. (Ministério dos negócios da marinha e ultrammar; Ministério da marinha e ultramar). Lisboa.
Annaes. no1-2 (1876-1877). Closed 1877. (ULG:489c; B-3:585a; F-1:167b).

2032 ______. Junta das missões geográficas e de investigações do ultramar. Lisboa.
Anais. 1946-1953. (U-4:3406c; B-3:585a; BS:695a).
1 vol. each year with 1 to 8 nos.; one no. of each vol. typically devoted to geography.
Issued under a variant name: Portugal. Junta das missões geográficas e investigações coloniais.

2033 ______. ______.
Estudos, ensaios e documentos. 1- (1950-). (N70-3:4726b; B-3:585a; BS:695a).

2034 ______. ______.
Memórias.
Série geográfica. 1 (1954). Closed. Only one issued. (N70-3:4726c).
United with other series of Junta's Memórias to form Junta's Memórias 2. Ser. (not in this list).

2035 ______. Ministério da economia. Instituto geográfico e cadastral. Lisboa.
Boletim. 1-3 (1934-1943) Irregular.

2036 ______. Ministério das obras públicas, commércio, e indústria. Lisboa.
Relatório dos trabalhos executados no Instituto geográfico. 1867/68-1897/98. Closed 1898? (ULG:494a; F-4:67a).

2037 SOCIEDADE de geografia commercial do Pôrto. Pôrto.
Boletim. no1-5/6 (1880-1882); s2 no1-10 (1883-1885); s3 no1-9/10 (1886-1888). Closed 1888. (U-5:3914b).

2038 SOCIEDADE de geografia de Lisboa. Lisboa.
Actas das sessões. 1-19 (1876-1899). Closed 1899. (U-5:3914b; B-4:108b; F-1:65b).

*(2039 ______.
Boletim. 1- (1876-) 12 nos. a year, often grouped into quarterly issues. (U-5:3914b; B-4:108b; F-1:653b).
Index: 1-42 (1876-1924).
English abstracts and occasional English articles since v70 (1952).
[Rua das Portas de Santo Antão, Lisboa, Portugal.]

2040 ______.
Semana das colónias de 1945. 1-7 (1945).

PUERTO RICO (Spanish, English)

Commonwealth of Puerto Rico,

Estado Libre Asociado de Puerto Rico

2041 REVISTA geográfica de Puerto Rico. San Juan. v1 (no1-10) (Mr-D 1923).
Closed 1923. (U-4:3613a).
Superseded by Revista de obras públicas de Puerto Rico (not in this list).

ROMANIA (Romanian)

Republica Socialistă România

2042 ACADEMIA Republicii Socialiste România. Bucureşti.
Buletin ştiinţific. Secţia de geologie, geografie şi biologie.
Analele. 1-2 (1949). Closed 1949. (U-1:23a; R-1:261a).
In Romanian. Table of contents and summaries in French and Russian.

2043 ______.
______. ______.
sA. Memoriul. 1-3 (1949-1950). Closed 1950.
In Romanian. Table of contents and summaries in French and Russian.

2044 ______.
______. Secţia de geologie şi geografie. 1-2 (1956-1957). Closed 1957. Quarterly, 2 nos. often combined into one issue. (N70-1:39a; BS:800b).
Supersedes part of academy's former Buletin ştiinţific.
Seçtiunea de ştiinţe biologice, agronomice, geologice şi geografice [2045].
Supplementary title, table of contents, and summaries in French and Russian.

2045 ______.
______. Secţiunea de ştiinţe biologice, agronomice, geologice şi geografice. 1-7 (1949-1955). Closed 1955. (U-1:23a; B-4:119a; BS:800b; R-1:261a).
Title varies. 1 as Buletin ştiinţific. A. Matematică, fizică, chimie, geologie, geografie, biologie, ştiinţe technice şi agricole; later as Buletin ştiinţific. Seria: Geologie, geografie, biologie, ştiinţe technice şi agricole. Superseded by various of the academy's series of which only that of the Secţia de geologie şi geografie [2044] is here included.

*(2046 ______.
Studii şi cercetări de geologie, geofizică, geografie: Geografie. 11- (1964-) 2 a year. (N68-1:18b; N75-2:2066b; N79; R-6:169a).
Supersedes and continues number of Probleme de geografie [2066].
Index: 1944-1969.
In Romanian. Summaries in English, French, German, or Russian, from v14 no2 (1967).

2047 ______. Centrul de documentare ştiinţifică.
Buletin de informare ştiinţifică. Seria: geologie-geografie.
1- (1964-) Monthly. (N70-1:43a).

2048 ______. Filiala Cluj. Cluj.
Studii şi cercetări de geologie-geografie. 7-8 (1956-1957).
Closed 1957. (N70-1:39c; BS:845b; R-6:169b).
Supersedes in part and continues numbering of Academia Republicii Socialiste România. Filiala Cluj. Studii şi cercetări ştiinţifice 1-6 (1950-1955) (not in this list).
Summaries in French and Russian.

2049 ANALELE româno-sovietice. (Academia Republicii Populare Romine. Institutul di studii româno-sovietic).
Seria biologie, geografie, geologie.
Bucureşti. 4-5 (O-D 1949-O-D 1950). (U-1:355a; R-1:263b).
Title also in Russian: Rumyno-sovetskie zapiski.
Superseded for geography by Seria geologie-geografie [2050].

2050 ______.
Seria geologie-geografie. 6-15 (Ja-F 1951-1963). Closed 1963.
(N70-1:337a; R-1:264a).
Continues numbering of and supersedes Seria biologie-geografic [2049].
In Romanian; table of contents in Russian.

2051 ARCHIVELE Basarabiei. Revista de istorie şi geografie a Moldovei dintre Prut şi Nistru. Chişinău. 1-2 (1929-1930).

2052 ASOCIAŢIUNEA transilvană pentru literatura română şi cultura poporului român, Hermannstadt. Secţiune geografico-etnografică. Cluj.
Biblioteca. no1 (1927). Closed 1927. (U-1:512a).

2053 BIBLIOTECA geografului. (Societatea de ştiinţe naturale şi geografie).
Bucureşti. 1- (1964-). (N70-1:746b).

*(2054 BUCUREŞTI (Bucharest). Universitatea. Bucureşti.
Analele. Seria: [Ştiinţele naturii.] Geografie. 1- (1952-) Semi-annual or annual. (N70-1:906ab; N75-1:353ab).
1-12 (1952-1963) Stiintele-naturi. (N70-1:906a). Title and handling of geography vary, usually a separate section or part of an issue.
Separate subseries, Geologie-geografie, 13-17 (1964-1968).
Geografie 18- (1969-). (N70-1:906b; N75-1:353ab; R-1:266a, 267a)
In Romanian with supplementary table of contents and abstracts in Russian, English, or French.

2055 ______. ______. Cursuri de vară de limbă, literatură, istorie şi artă a poporului român.
Seria 3. Istorie şi geografie. 1- (1964-). (N70-1:906b).
French title: Université de Bucarest. Cours d'été et colloques scientifiques de langue, littérature, histoire et art du peuple roumain.
Série 3. Histoire et géographie.
In Romanian and French.

2056 ______. ______. Semiarul geografic.
Anuar de geografie şi antropogeografie. Anul...sl. 1909/10-1917.
Closed 1917. (F-1:342a).

2057 ______. ______. ______.
Cercetări şi studii geografice. Seria 2. v1-2 (1937-1941). Closed 1941.

2058 CAIET de informare: geografie. (Centrul de informare şi documentare al invatamintului). Bucureşti. 2 (1973) Quarterly. (N75-1:388b).
[Rompresfilatelia, Serviciul Export-Import Presa, Calea Grivitei nr. 64-66. POB 20001. Bucuresti. Romania].

2059 CLUJ. Universitatea. Institutul de geografie. Cluj.
Lucrările..Travaux. 1-7 (1922-1942). Closed 1942. (U-2:1080c; B-1:589a; F-3:350b; R-4:252a).
In Romanian and French.

(2060 ______. Universitatea Babeş-Bolyai. Cluj.
Studia universitatis Babeş-Boylai. Series geologia-geographia. 3- (1958-) 2 nos. a year. (N70-1:325b; N70-1:1324c; N75-1:538c; R-6:160a).
v3-6 (1958-1961) as s2, including also biology. Series geologia-geographia 7-14 (1962-1969); separate Series geographia 15-19 (1970-1974), 21 (1976).
In Romanian with supplementary table of contents and abstracts in Russian and in English, French, or German.

(2061 IAŞI (Jassy). Universitatea "Al. I. Cuza." Iaşi.
Analele ştiinţifice. Secţiunea 2. Ştiinţe naturale. c. Geografie. nsl- (1955-) Annual. (N70-2:3107a; BS:926b; R-1:265b).
Supersedes in part university's Analele ştiinţifice (general series, not in this list).
Titles vary: nsl-9 (1955-1963) as Secţinuea 2. Ştiinţe naturale-geografie; 10-14 (1964-1968) as Secţiunea 2. Ştiinţe naturale: b. Geologie-geografie; 15- (1969-). as c. Geografie.
In Romanian; supplementary abstracts and titles in table of contents usually in French.

2062 LECTURI geografice. (Societatea di ştiinţe geografice din Republica socialistă Romania). Bucureşti. 1- (1967-). (N76).
Subseries of Biblioteca geografului [2053].

2063 NATURA. (Societatea de ştiinţe naturale şi geografie din Republica Populară Romînă). ns. 1-12 (1949-1960). (U-4:2926a; B:600b; R-5:31a).
Superseded in part by Natura: seria geografie-geologie [2064].
Includes a section on geography.
In Romanian with tables of contents and summaries in English and Russian.

2064 NATURA: seria geografie-geologie. (Societatea de ştiinţe naturale şi geografie din Republica Socialistă România). Bucureşti. 13-20 (1961-1968). Closed 1968. (N70-3:4072c-4073a; R-5:31b).
Supersedes in part Natura [2063] and continues its numbering.
Superseded by Terra [2082] which continues its numbering.
In Romanian with tables of contents and summaries in English and Russian.

(2065 ORADEA. Institutul pedagogic. Oradea.
Lucrări ştiinţifice. Seria geografie. 1971- Annual. (N75-2:1660c).
Supersedes in part its Lucrări ştiinţifice, seria A, and seria B, (1969-1970), and its Lucrari ştiinţifice (1967-1968) (not in this list).
In Romanian. Summaries in English, French, or German.
[Institutul pedagogic Orades, Callea Armatei Rossi Nr. 5, Oradea, Romania.]

2066 PROBLEME de geografie. (Academia Republicii Populare Romîne. Institutul de geologie şi geografie). Bucureşti. 1-10 (1954-1963). Closed 1963. (N75-2:1778a; BS:703a; R-169a).
Superseded by Academia Republicii Socialiste România. Studii şi cercetări de geologie, geofizică şi geografie, seria geografie [2046] which continues its numbering.
In Romanian with supplementary table of contents and summaries in Russian and French v3-8 (1956-1961) or English v9-10 (1962-1963).

2067 REVISTA de referate şi recenzii: Seria geologie, geografie. (Academia Republicii Socialiste România. Centrul de documentare ştinţifică). Bucureşti. 1- (1964-). (N70-3:5013a; R-5:454a).

2068 REVISTA geografică. (Romania. Institutul de cercetări geografice al României). Bucureşti. 1-3 (1944-1946). Closed 1946. (U-4:3613a; R-5:444b).
Reported also as Revista geografică română.
In Romanian with supplementary French abstracts and titles in tables of contents.

2069 REVISTA geografică. (Societatea studenţilor de geografie "Soveja"). Bucureşti. 1933, 1938, 1944/1948. Closed 1948. All published?

2070 REVISTA geografică română. Bucureşti. v1-6 (1938-1943). Closed 1943. (F-4:140b; R-5:444b).
In Romanian with English, French, or German abstracts.

2071 REVUE de géologie et de géographie. (Académie de la République populaire Roumaine). Bucureşti. 1-7 (1957-1963). Closed 1963. (N70-3:5027b; BS:737a R-5:494b).
Superseded by Revue roumaine de géologie, géophysique et géographie: géographie [2072], which continues it numbering.
Articles in English, French, German, or Russian. Tables of contents and summaries in French and Russian.

*(2072 REVUE roumaine de géologie, géophysique et géographie: géographie. Bucureşti. Série de géographie. 8- (1964-) 2 a year. (N70-3:5039b; N75-2:1880a; B68:485a; R-5:494b-495a).
Supersedes in part and continues numbering of Revue de géologie et de géographie [2071].
Index: geographic articles in v1-10 (1957-1966) in v10 no2 (1966), p. 203-209.
In English, French, Russian, or German with abstract in a second language.
[Institutul de Geografie, Str. Dimitrie Racoviţă 12, R-70307 Bucureşti 20, sectorul 3, România.]

2073 ROMANIA. Institutul cartografic. Braşov.
Revista geografică şi cartografica română. 1925?

2074 ______. Institutul de cercetări geografice al României. Bucureşti. Biblioteca. sA. 1-5 (1945-1946). Closed 1946.

2075 SIEBENBÜRGISCHE Vierteljahrsschrift. (Verein für siebenbürgische Landeskunde). Hermannstadt [Sibiu]. 1-64 (1878-1941). Closed 1941. (U-5:3869a; B-4:460b; F-3:259a).
1-53 (1878-1930) as the society's Korrespondenzblatt.
In German.

(2076 SOCIETATEA de ştiinţe geografice din Republica Socialistă România. Bucureşti. Buletinul. ns1- (71-) (1971-) Irregular. (N75-2:1990b).
In Romanian. Supplementary table of contents in French and Russian.
Abstracts in French.
Number in parentheses in continuation of Societatea româna de geografie. Buletinul, 1-61 (1876-1942) [2080] and Comunicări de geografie, 1-9 (1960-1969) [2077].

2077 ______.
Comunicări de geografie. 1-9 (1960-1969) Annual. (N70-4:5379b; B68:150a; R-2:415a).
Supersedes Comunicări de geologie-geografie [2078].
In Romanian with supplementary table of contents and summaries in English, French, German, or Russian.

2078 SOCIETATEA de ştiinţe naturale şi geografie din Republica Socialistă România. Bucureşti.
Comunicari de geologie-geografie. 1957-1959. Closed 1959. (N70-4:5379b; R-2:415a).
Superseded by society's Comunicări de geografie [2077].

2079 SOCIETATEA geografică "Dimitrie Cantemir," Iaşi.
Lucarările. v1-4 (1937-1943). Closed 1943. (F-3:350b).
In Romanian with French abstracts.

2080 SOCIETATEA română de geografie, Bucureşti. Bucureşti.
Buletinul. 1-61 (1876-1942). Closed 1942. (U-5:3924a; B-4:119a; F-1:721b; F-4:517a; R-2:165b).
Through 1911 as Societatea geografică română.
Index: 1-40 (1876-1921).
In Romanian; supplementary French abstracts and titles in table of contents v37-61 (1916/1918-1942).

2080a STUDII de geografie. (Bucureşti. Universitatea. Facultatea de geologie-geografie. Secţia de geografie).
Annual. Litho.
In Romanian.

2081 STUDII şi cercetări geografice. (Societatea regală română de geografie). Bucureşti. 1-3 (1937-1939). Closed 1939. (U-5:4103c; R-6:169a).
In Romanian with French summary.

*2082 TERRA; revista ocrotirea mediului înconjurator, natura, terra. (Societatea de ştiinţe geografice din Republica Socialistă România). Bucureşti. 1- (21-) (Ja 1969-). 2 nos. a year. (N70-4:5744c).
Subtitle 1969-1975, Revistă de informare geografică.
Supersedes Natura [2063] and Natura; seria geografie-geologie [2064] and continues their numbering in parentheses.
In Romanian; supplementary table of contents in English, Russian, or French.

2083 VEREIN für siebenbürgische Landeskunde. Hermannstadt (Sibiu) later Kronstadt (Braşov).
Archiv. 1 (1840-1841); 1-4 (1843/45-1850/51); nsv1-49 no1 (1853-1936). Closed 1936. (U-5:4372a; B-1:193a; F-1:374a).
Indexes: 1-nsv20 (1843/45-1885/86); nsv21-40 (1887/88-1916/21).
In German.

2084 ______.
Bericht über die Arbeiten. 1849-1914. Closed 1914? (U-5:4372a).
In German.

2085 ______.
Jahresbericht. 1853-1914. Closed 1914. (U-5:4372a; F-3:42b; G-2:1453b).
1905-1906, 1908 in Society's Archiv [2083]. 1907 not published.
In German.

SENEGAL (French)

République du Sénégal

2086 DAKAR. Université. Département de géographie. Dakar.
Travaux. 1-10 (1953-1963). Closed 1963. (N70-2:1640c).
Reprints from various sources.
Superseded by Revue de géographie de l'Afrique occidentale [2088].
no1-4 (1953-58?) issued by the department as a part of the Institut des hautes études.

2087 INSTITUT fondamental d'Afrique noire. Dakar.
Bulletin. 1-15 (1939-1953) Quarterly. (U-3:1998a; B-1:610a).
16- (1954-) in 2 series.
sA. Sciences naturelles. 16- (1954-) Quarterly. (N70-2:2804a).
sB. Sciences humaines. 16- (1954-) 2 nos. a year. (N70-2:2804a).
Supersedes Comité d'études historiques et scientifiques de l'Afrique occidentale française (not in this list).
v1-28 (1939-1966) issued by the institute under its earlier name: Institut français d'Afrique noire.
Index: 1939-1949.
In French.
[Institut Fondamental d'Afrique Noire, Boîte postale 206, Dakar, Senegal.]

2088 REVUE de géographie de l'Afrique occidentale. (Dakar. Université. Département de géographie). Dakar. 1-3 (1965-1966). Closed. (N70-3:5027ab; B68:483a).
Supersedes Dakar. Université. Département de géographie. Travaux [2086].

2089 SOCIÉTÉ de géographie de l'Afrique occidentale française. Dakar.
Bulletin. no1-7 (1907-1908). (B-4:135a; F-1:835a).

SIERRA LEONE (English)

Republic of Sierra Leone

2090 NJALA university college. Department of geography and environmental studies. Freetown.
Geographical bulletin. 1-3 (1971-1973). Mimeographed.

2090a ______. ______. Occasional paper. 1 (F 1972). Mimeographed.

2091 SIERRA Leone geographical association. Freetown.
Occasional paper. no1 (1965), no2 (1970) Irregular. (N70-4:5326b; B68:397b). Mimeographed.
[Department of Geography, Fourah Bay College, University of Sierra Leone, Freetown, Sierra Leone.]

2092 SIERRA Leone geographical journal. Freetown. 1-15 (1957-1971). (N70-4:5326b; B68:111b).
Title varies: no1-10 (1957-1966) as Bulletin: the journal of the Sierra Leone geographical association, or Bulletin of the Sierra Leone geographical association.
Index: 1-9 (1957-1965) in 9.

星架坡 SINGAPORE (English, Chinese, Malay, Tamil)

Republic of Singapore, Hsin-chia-p'o Kung-ho-kuo,

Republik Singapura, Singapore Kudiyarasu

(2093 GEOGRAPHICAL journal. (Nanyang university, Singapore. Geographical society). Singapore. 1- (1968-) Annual. (C:445a).
Chinese title: Ti li chi k'an (Nan-yang ta hsüeh ti li hsüeh hui).
In Chinese or English.
[The Geographical Society, Nanyang University, Upper Jurong Road, Singapore 22.]

2094 HISTORY and geography. (Nanyang university, Singapore. History and geography society). Singapore. 1-2 (1960-1961) Irregular. Closed 1961. (N70-2:2553c; C:391a).
Title in Chinese: Nan-yang ta hsueh, Singapore. Shih ti hsueh hui. Shi ti.
In Chinese.

*(2095 JOURNAL of tropical geography. (University of Singapore [and] University of Malaya. Department of geography). Singapore; Kuala Lumpur. v1- (1953-) 2 nos a year. (N70-2:3211a; BS:536a).
v1-10 (1953-1957) as Malayan journal of tropical geography.
In English.
Indexes: 1-29 (1953-1969), 30-39 (1970-1974).
[Department of Geography, University of Singapore, Singapore 10.]

(2095a NANYANG university, Singapore. Department of geography. Working report series. Singapore. 1- (1977-) Irregular.
[Department of Geography, Nanyang University, Upper Jurong Road, Singapore 2263, Republic of Singapore.]

SOUTH AFRICA (Afrikaans, English)

Republiek van Suid-Afrika, Republic of South Africa

2096 ACTA geographica. Stellenbosch. v1 (1967), v2 (1975) Irregular. (N70-1:58b).
In Afrikaans.
[Universiteits-uitgewers en Boekhandelaars, Stellenbosch, South Africa.]

2096a CAPE TOWN. University. Department of geography. Cape Town Publication. 1- (1978-).

2097 ISIZWE. (Natal. University. Students geographical society). Durban. 1- (1974-) Annual. (N76).
In English.
[Students Geographical Society, University of Natal, Durban 4001, South Africa.]

*(2098 JOHANNESBURG. University of the Witwatersrand. Department of geography and environmental studies. Johannesburg.
Environmental studies. Occasional paper. 1- (1969-) Irregular. (DP).

2100 ______. ______. ______. Urban and regional research unit.
Occasional paper. 1-12 (1972-1976). Closed 1978. (DP).

*(2101 SOUTH African geographical journal. Suid-Afrikaanse geografiese tydskrif. (South African geographical society. Suid-Afrikaanse geografiese vereniging). Braamfontein. 1- (1917-) Annual. (U-5:4000c; B-4:179a; F-4:572a).
Articles in English from v1 (1917) or in Afrikaans from v24 (1943).
Abstracts in English.
[P.O. Box 31201, Braamfontein 2017, Transvaal, South Africa.]

(2102 SOUTH African landscape monograph. (South African geographical society). Braamfontein. 1- (1970-) Irregular.
In English or in Afrikaans with English abstract.
[P. O. Box 31201, Braamfontein 2017, Transvaal, South Africa.]

*(2103 SUID-AFRIKAANSE geograaf. South African geographer. (Vereeniging vir geografie. Society for geography). Dennesig. 1- (1957-) 2 nos a year. (N70-4:5892a; N75-2:2070a).
v1-3, 10 nos in each vol. v4 had 5 nos, v5 had 6 nos. v1-3 (1957-Ap 1972) as Tydskrif vir aardrykskunde. Journal for geography (Vereeniging vir aardrykskunde-onderwys. Society for the teaching of geography).
In English and Afrikaans. Abstracts in English.
[Department of Geography, University of Stellenbosch, Stellenbosch, Cape Province, South Africa.]

2104 VEREENIGING vir aardrykskunde-onderwys. Society for the teaching of geography. Stellenbosch.
Special publication. 1- (1961-) Irregular. (N70-4:6160b).

SPAIN (Spanish)

Estado Español

2105 ANUARIO español de geografía aplicada. (Spain. Consejo superior de investigaciones científicas. Instituto Juan Sebastián Elcano. Departamento de geografía aplicada). Madrid.

2106 ARCHIVO geográfico de la península ibérica. Barcelona. 1916.
Closed 1916. (U-1:468b; B-1:207b; F-1:421a).

2107 BARCELONA. Universidad. Instituto de geografía. Barcelona.
Publicación. 7 (1968).
Separately published monographs or reprints.
[Facultad de Geografía e Historia, Ciudad Universitaria, Barcelona 28, Spain.]

2108 BIBLIOTHECA hispana; revista de información y orientación bibliográficas. (Spain. Consejo superior de investigaciones científicas. Instituto Nicolás Antonio). Madrid.
Geografía. 1-4 (1943-1946) in Sección 3; 5- (1956-) in Sección 1a. Irregular. (U-1:686a [U-1:339a]).

2109 BOLETÍN de cartografía. (Madrid. Seminario de estudios cartográficos. Publicación). Madrid. 1-7 (1961-1965). Closed 1965.

2110 BOLETÍN de historia y geografía del Bajo Aragón. Tortosa. 1907-1909? (U-1:728b).

2111 CARTOGRAFIA de ultramar. (Spain. Servicio geográfico e histórico del ejército. Estado mayor). Madrid. 1-6 (1949-1955). Closed 1955. (BS:184b).

2112 COLECCIÓN Tierra. Zaragoza. 1- (1954-) Irregular. (N70-1:1357a).
No1 published jointly by the Departamento de geografía aplicada del Instituto Juan Sebastián Elcano del Consejo superior de investigaciones científicas, and the Institución Fernando el Católico, Sección de geografía.

2113 COMPENDIOS de investigación. Serie de geografía e historia. (Editorial Alpina). Granollers. no1-3 (1951-1954). (N70-1:1447c).

2114 CONGRESO español de geografía colonial y mercantil. 1884.

*(2115 CUADERNOS de geografía. (Valencia. Universidad. Facultad de filosofía y letras. Departamento de geografía). Valencia. 1- (1964-)
Annual 1964-1970; Semiannual 1971- . v1-7 (1964-1970)
Reprints from Saitabi (not in this list).
Index: v1-20 (1964-1977) in v20 (1977), p. 139-[149].
[Departamento de Geografía, Apartado 2.005, Valencia, Spain.]

(2116 CUADERNOS de investigación de geografía e historia. (Logroño. Colegio universitario.) Logroño. 1- (1975-) 2 nos. a year.

(2117 DIDÁCTICA geográfica. (Murcia. Universidad. Departamento de geografía). Murcia. 1- (1977-) Irregular.

(2118 DOCUMENTS d'anàlisi metodològic en geografia. (Universitat autònoma de Barcelona, Bellaterra. Departament de geografía). Barcelona. 1- (1977-) Irregular.

2119 DOCUMENTS d'anàlisi territorial. (Universitat autònoma de Barcelona, Bellaterra. Departament de geografia). Barcelona. 1- (1975-) Annual.

2120 DOCUMENTS d'anàlisi urbana. (Universitat autònoma de Barcelona, Bellaterra. Departament de geografia). Barcelona. 1- (1974-) Annual.

*(2121 ESTUDIOS geográficos. (Spain. Consejo superior de investigaciones científicas. Instituto "Juan Sebastián Elcano"). Madrid. v1- (1940-) Quarterly. (U-5:4038b; B-2:142b; BS:295a).
Indexes: v1-10 (no1-37) (1940-1949). no78-118 (1960-1969).
[Libería Científica, Duque de Medinaceli, 4, Madrid 14, Spain.]

2122 EUSKO lurra. Geografía del país vasco. (Editorial Itxaropena ETOR). San Sebastián. 1- (1974-). (N78).
Subseries of ETOR Bidean kultura (not in this list).

2123 FUENTES cartográficas españolas. (Spain. Consejo superior de investigaciones científicas. Instituto de geografía aplicada). Madrid. Irregular.
[Instituto de Geografía Aplicada, Serrano 115 bis, 9 planta, Madrid 6, Spain.]

2124 GEOCRÍTICA. (Barcelona. Universidad. Facultad de geografía e historia. Departamento de geografía). Barcelona. 1- (1976-) Irregular.

2125 GEOGRAFÍA. (Sociedad castellonese de cultura). Castellón de la Plana. 1- (1961-) Irregular (N70-2:2278a).

2126 GEOGRAFÍA mística de España. Madrid. 1 (1945).

2127 GEOGRAPHIA. Navarra. ?1912, 1913; 11, 13-18 (1924-1930).

*(2128 GEOGRÁPHICA. (Spain. Consejo superior de investigaciones científicas. Instituto Juan Sebastián Elcano. Instituto de geografía aplicada). Zaragoza; Madrid. 1-12 (1954-1965); s2. v13- (1971-), continues numbering. Quarterly. (N70-2:2280b).
Some issues accompanied by a separately numbered Suplemento bibliográfico [2129].
Substitle in early numbers: Revista de información y enseñanza.
[Instituto de Geografia Aplicada Patronato "Alonso de Herrera," Serrano 115 bis 9· planta, Madrid 6, Spain.]

2129 ______.
Suplemento bibliográfico. Títulos registrados en la biblioteca del departamento de geografía aplicada de Zaragoza. 1- (1954-). (N70-2:2280b).

(2130 GEOGRAPHICALIA. (Zaragoza. Universidad. Departamento de geografía). Zaragoza. 1- (1977-) Annual.

2131 GRANADA (city). Universidad. Sección de geografía. Granada.
Cuadernos de geografía de la Universidad de Granada. 1- (1970-) Annual. (N77).
Anejo del Boletín de la Universidad de Granada.
In Spanish. Summaried in English or French.
[Secretario de Publicaciones, Universidad de Granada, Granada, Spain.]

2132 ______. ______. ______.
Cuadernos geográficos de la Universidad de Granada. Serie monográfica. 1- (1975-). (N77).
Anejo del Boletín de la Universidad de Granada.
In Spanish. Summarics in English or French.

2133 MADRID. Universidad. Publicaciones. Serie de ciencias naturales y geográficas. Madrid.
Memoria. v1 no1-6 (1916). Closed 1916. (U-3:2507a).

2134 MUNTANYA. (Centre excursioniste de Catalunya). Barcelona. 1- (1976-) 12 nos. a year.

(2135 MURCIA. Universidad. Facultad de filosofía y letras. Murcia.
Papeles del departamento de geografía. 1- (1968/1969-) Irregular.

2136 OASIS; tierras, pueblos, costumbres, arte, geografía, viajes. Madrid. 1-15 (1934-1936). Closed 1936. (U-4:3126b).

2137 PARALELO 37° Revista de estudios geográficos. (Granada. Universidad. Departamento de geografía. Colegio universitario de Almeria). Almeria. 1- (1977-) Annual.

2138 PIRINEOS. (Spain. Consejo superior de investigaciones científicas. Instituto de estudios pirenaicos). Jaca. 1- (1945-). (U-4:3363c; B-3:557b; BS:680b).
Issuing body varies slightly.
[Instituto de Estudios Pirenaicos, Apartado 64, Jaca (provincia de Huesca) Spain.]

2139 ______.
Suplemento bibliográfico. 1- (1951-). (N70-3:4656a; BS:680b).

2140 REAL sociedad geográfica. Madrid.
Anuario. 1910-1936. (U-5:3909c; B-4:106a).
Formerly Sociedad geográfica de Madrid, later Sociedad geográfica nacional.

*(2141 ______.
Boletín. 1- (1876-) 12 nos. a year but often combined into quarterly or annual issues. (U-5:3909c; B-4:106a; F-1:665a).
Suspended 1937-1940. Included numerous separately paged supplements.
Indexes: v1-42 (1876-1900) in v43, 43-52 (1901-1910), 53-62 (1911-1920) in 63, 63-70 (1921-1930) in v70, 1931-1940, 1941-1950, 1951-1960.
[Calle de Valverde 22, Madrid 13, Spain.]

2142 ______. Colección geográfica. 1-? (1878-?). (U-5:3909c).

2143 ______. Hoja informativa. 1- (1975-) 10 nos. a year.

2144 ______. Publicaciones. sB. 1- (1932-). (U-5:3909c; B-4:106a).
Reprints from society's Boletin [2141].

2145 ______. Sección de la ciencia del suelo. Madrid.
Memoria. 1 (1935). Closed 1935. (U-5:3909c; B-4:106a).

*(2146 REVISTA de geografía. (Barcelona. Universidad. Departamento de geografía).
Barcelona. 1- (1967-) 2 nos. a year. (N76).
In Spanish. Abstracts in English and French.
[Departamento de Geografía, Facultad de Letras, Universidad, Barcelona 7, Spain.]

2147 REVISTA de geografía colonial y mercantil. (Real sociedad geográfica, Madrid. Sección de geografía comercial). Madrid. 1-21 (1897/1900-1924). Closed 1924. (U-4:3602b; B-3:699a; F-4:1262).
Indexes: 1876-1900 in society's Boletín, v43; 1901-1910; 1911-1920 in Boletín, v63; 1921-1924 in Boletín, v70.

2148 REVISTA de geografía comercial. (Sociedad de geografía comercial).
Madrid. v1-5 no36 (no1-160) (1885-1896). Closed 1896. (U-4:3602b; F-4:126a)
Index: v1-5 (1885-1896) in v5 no36.

2149 REVISTA de geografía universal. Madrid. 1- (1977-) Monthly.

2150 REVISTA estadística y geográfica. Barcelona. v1-4 (1878-1881). (F-4:139b).
Title varies: later Revista geográfica y estadística.

2151 REVISTA geográfica. Madrid. 1-4 (1878-1881). Closed 1881?

2152 REVISTA geográfica española. San Sebastián; Madrid. no1-43 (1939-)
Irregular. (U-4:3613a).

2153 SOCIEDAD de geografía comercial, Barcelona. Barcelona.
Publicaciones. no1-14 (1911-1919). Closed 1919. (U-5:3906b; B-4:105b).

2154 SPAIN. Consejo superior de investigaciones científicas. Institución Alfonso el Magnánimo. Instituto de geografía. Valencia.
Publicación. 1- (1966-) Irregular. (N70-4:5480a; B73:277a).
Also bears imprint of Valencia. Universidad. Facultad de filosofía y letras. Departamento de geografía.

2155 ______. ______. Instituto de estudios canarios en la Universidad de La Laguna.
Monografías. Sección 1. Ciencias historicas y geograficas. La Laguna de Tenerife. (N70-4:5480b).

2156 ______. ______. Instituto de estudios pirenaicos. Zaragoza.
Publicación (Monografía): Geografía. 1943- Irregular. (N70-4:5480c).
Sometimes with English, French, and German abstracts.

2157 ______. ______. Instituto Juan Sebastián Elcano. Departamento de geografía Madrid (formerly Zaragoza).
Publicación. 1- (1952-) Irregular. (N70-4:5481c).
Includes subseries: Serie regional, Estudios regionales.

2158 ______. Consejo superior geográfico. Madrid.
Memoria general. (N70-4:5483a).

2159 ______. Instituto geográfico y catastral. Madrid.
Memorias. 1-22 (1875-1952) Irregular. (U-5:4039b; B-4:199a; F-3:497a).
Name of agency varies: 1-15 (1875-1927) as Instituto geográfico y estadístico; 16-22 (1938-1952) under present title.

2160 ______. ______.
Reseña geográfica y estadística. 1888. (ULG:516b).
1912. v1-3 (1912). Closed 1912.

(2161 ______. Servicio geográfico del ejército. Madrid.
Boletín de información. 1- (1968-) Quarterly. (N70-5486b).

(2162 TRABAJOS de geografía. (Palma de Mallorca. Facultad de filosofía y letras. Departamento de geografía). Palma de Mallorca. 1- (1976-) Irregular.

ශ්‍රී ලංකා — SRI LANKA (formerly CEYLON) — (Sinhala; English; Tamil)

Sri Lankā Janarajaya

2163 BHŪGŌLA vidyā pravēśaya.
In Sinhalese.

2164 CEYLON geographer. (Ceylon geographical society). Colombo. 1-20 (1945-1966/1970) Quarterly, later annual. (U-2:972b; BS:196a).
v1-2 mimeographed. v1-10 as Ceylon geographical society. Bulletin.
Indexes to v3, 4-7, 8-10.
In English.
[Ceylon Geographical Society, 61 Abdul Caffoor Mawatha, Colombo 3, Sri Lanka.]

2165 NORTHERN geographer. Journal of the Northern geographical society [of Ceylon]. 1- (1960-). (N70-3:4275a).
In English and Tamil.

2166 SRI LANKA. University. Geographical society. Colombo.
Journal. Peradeniya. 1- (1958-) Annual. (N70-1:1208c).
Articles mainly by students.
In English, Sinhala, or Tamil.
[c/o Department of Geography, University of Sri Lanka, Colombo 3, Sri Lanka.]

SUDAN (Arabic)

Jumhūrīyat as-Sūdān ad-Dīmuqrātīyah

جمهورية السودان الديموقراطية

2167 GEOGRAPHICAL magazine. (Khartum. University. Geographical society). Khartum. 1- (1963-) Annual. (N70-2:2281c).
In English.
[Department of Geography, University of Khartum, Khartum, Sudan.]

SURINAM (Dutch)

Suriname

2168 SURINAM. Centraal bureau luchtkartering. Paramaibo.
Jaarverslag. 1971- Irregular. (N75-2:2075a).
[Centraal Bureau Luchtkartering, Paramaibo, Surinam.]

SWEDEN (Swedish)

Konungariket Sverige

2169 K. FYSIOGRAFISKA sällskapet i Lund. Lund.
Årsbok. 1964- (N70-2:2241c).
Supersedes society's Förhandlingar [2170].

2170 ______.
Förhandlingar. 1-33. (1931-1963). Closed 1963. (U-2:1658b; B-3:551a).
Superseded by society's Årsbok [2169].

2171 ______.
Handlingar. nsvl- (1889/90-). (U-2:1658b).
In Lund. Universitet. Acta universitatis lundensis (not in this list).

2172 GEOGRAFILÄRARNAS riksförening. Lund.
Skrifter. Irregular. (N70-2:2279b).
[C.W.K. Gleerup Bokförlag, Lund, Sweden.]

2173 GEOGRAFISKA annaler. (Svenska sällskapet för antropologi och geografi). Stockholm. 1-46 (1919-1964) Quarterly. (U-2:1693; B-2:268a; BS:341b; F-2:727a).
47- (1965-) continued in 2 series [2174, 2175].
Index: 1919-1951.
Articles mainly in English.

*(2174 ______.
sA. Physical geography. Stockholm. 47A- (1965-) Quarterly. (N70-2:2279c; B68:224b).
Continues numbering of undivided series [2173].
Articles mainly in English.
[The Almqvist and Wiksell Periodical Company, Box 62, 101 20 Stockholm, Sweden.]

*(2175 ______.
sB. Human geography. Stockholm.
47B- (1965-) 2 nos. a year. (N70-2:2279c; B68:224b).
Continues numbering of undivided series [2173].
Articles mainly in English.
[The Almqvist and Wiksell Periodical Company, Box 62, 101 20 Stockholm, Sweden.]

2176 GEOGRAFISKA notiser. (Geografilärarnas riksförening). Stockholm.
1- (1943-) Quarterly. (U-2:1693b).
[Geografiska institutionen, Sölvegatan 13, 223 62 Lund, Sweden.]

(2177 GEOGRAFISKA regionstudier. (Uppsala. Universitet. Kulturgeografiska institutionen). Uppsala. 1- (1958-) Irregular. (N70-2:2279c; BS:341b).
In Swedish. Abstracts in English. Occasional monograph in English.
1-2 (1958-1963) issued by university's Geografiska institutionen.

2178 GEOGRAPHICA. (Uppsala. Universitet. Geografiska institutionen). Uppsala.
1-38 (1936-1968). Closed 1968. Irregular monographs. (U-2:1694a; B-2:727b).
In Swedish with abstracts principally in English; occasional monograph in English.

(2179 GLOBEN. (Meddelanden utgivna av Generalstabens litografiska anstalt).
Stockholm. 1- (1922-) Quarterly. (U-2:1736b; B-2:304b; F-2:755b).
Indexes: 1-10 (1922-1931), 11-20 (1932-1941), 21-30 (1942-1951), 31-41 (1952-1962).
In Swedish.

2180 GÖTEBORG. (Gothenburg). Göteborgs högskola-handelshögskola in Göteborg.
Geografiska institutet.
Meddelanden. 1-20 (1929-1937). Closed 1937. (U-2:1751a).
Reprints and monographs. In 1937 separate geographical institutes were established as (1) Handelshögskolans i Göteborg geografiska institutionen [2181] and (2) Göteborg. Högskolas (which in 1954 became Universitet). Geografiska institutionen [2182].
Swedish with some English abstracts.

(2181 ______. Handelshögskolans i Göteborg geografiska institutionen.
Meddelanden. 1- (1941-) Irregular reprints and monographs.
One of two successor series to Göteborg. Göteborgs högskola-handelshögskola in Göteborg. Geografiska institutet. Meddelanden [2180].
In Swedish with scattered abstracts or papers or monographs in English or German.

2182 ______. Universitet. Geografiska institutionen. Göteborg.
Meddelanden. 26-90 (1941-1967) Irregular reprints and monographs.
26-41 (1941-1954) as Göteborg. Högskola. Geografiska institutionen.
In Swedish with some English or German abstracts or papers or monographs.
Superseded by 2 series: Göteborg. Universitet. Naturgeografiska institutionen. Meddelande. sA. [2186], and Göteborg. Universitet. Kulturgeografiska institutionen. Meddelande. sB. [2184].

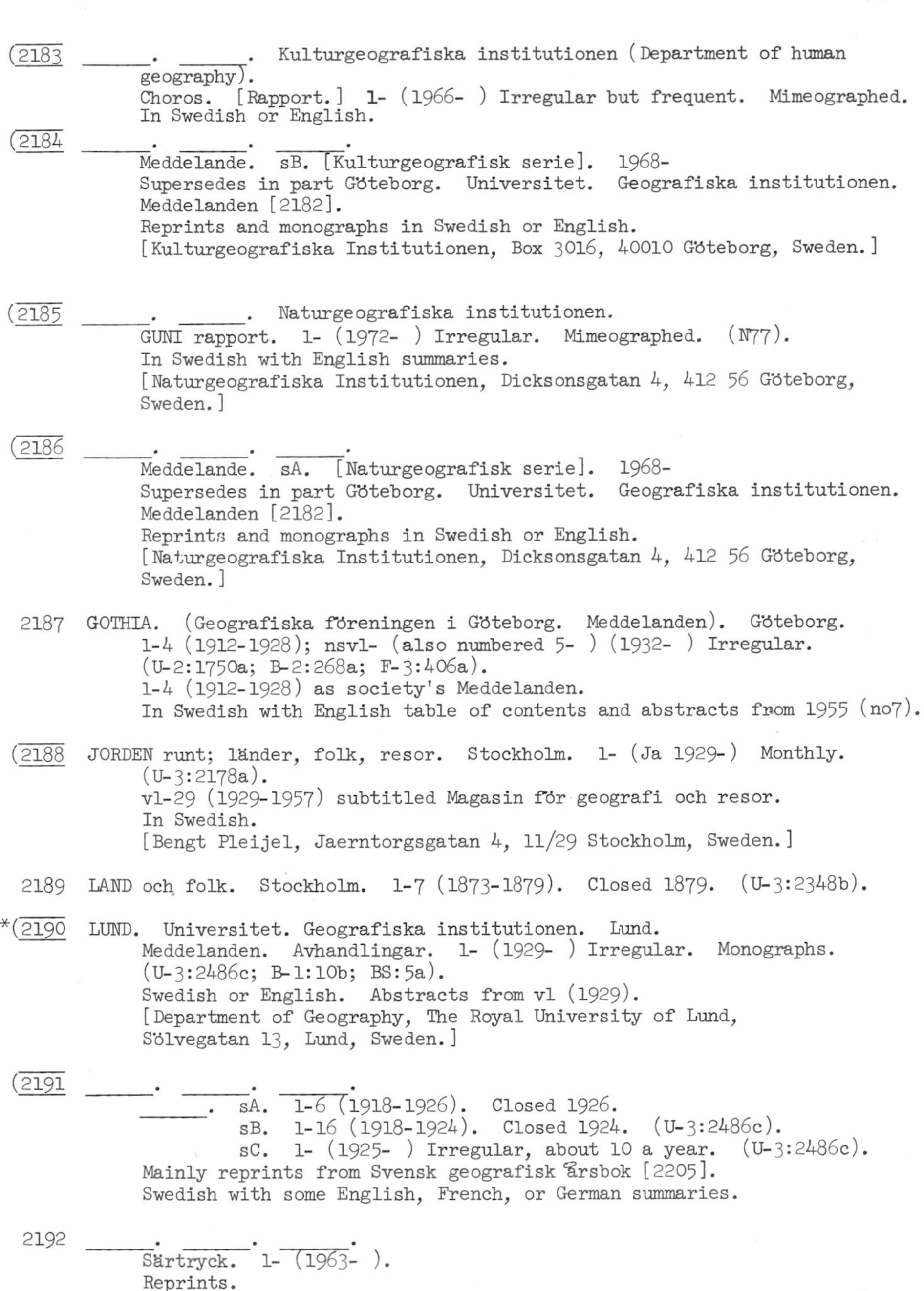

(2183 ______. ______. Kulturgeografiska institutionen (Department of human geography).
Choros. [Rapport.] 1- (1966-) Irregular but frequent. Mimeographed.
In Swedish or English.

(2184 ______. ______. ______.
Meddelande. sB. [Kulturgeografisk serie]. 1968-
Supersedes in part Göteborg. Universitet. Geografiska institutionen. Meddelanden [2182].
Reprints and monographs in Swedish or English.
[Kulturgeografiska Institutionen, Box 3016, 40010 Göteborg, Sweden.]

(2185 ______. ______. Naturgeografiska institutionen.
GUNI rapport. 1- (1972-) Irregular. Mimeographed. (N77).
In Swedish with English summaries.
[Naturgeografiska Institutionen, Dicksonsgatan 4, 412 56 Göteborg, Sweden.]

(2186 ______. ______. ______.
Meddelande. sA. [Naturgeografisk serie]. 1968-
Supersedes in part Göteborg. Universitet. Geografiska institutionen. Meddelanden [2182].
Reprints and monographs in Swedish or English.
[Naturgeografiska Institutionen, Dicksonsgatan 4, 412 56 Göteborg, Sweden.]

2187 GOTHIA. (Geografiska föreningen i Göteborg. Meddelanden). Göteborg.
1-4 (1912-1928); nsv1- (also numbered 5-) (1932-) Irregular.
(U-2:1750a; B-2:268a; F-3:406a).
1-4 (1912-1928) as society's Meddelanden.
In Swedish with English table of contents and abstracts from 1955 (no7).

(2188 JORDEN runt; länder, folk, resor. Stockholm. 1- (Ja 1929-) Monthly.
(U-3:2178a).
v1-29 (1929-1957) subtitled Magasin för geografi och resor.
In Swedish.
[Bengt Pleijel, Jaerntorgsgatan 4, 11/29 Stockholm, Sweden.]

2189 LAND och folk. Stockholm. 1-7 (1873-1879). Closed 1879. (U-3:2348b).

*(2190 LUND. Universitet. Geografiska institutionen. Lund.
Meddelanden. Avhandlingar. 1- (1929-) Irregular. Monographs.
(U-3:2486c; B-1:10b; BS:5a).
Swedish or English. Abstracts from v1 (1929).
[Department of Geography, The Royal University of Lund, Sölvegatan 13, Lund, Sweden.]

(2191 ______. ______. ______.
______. sA. 1-6 (1918-1926). Closed 1926.
sB. 1-16 (1918-1924). Closed 1924. (U-3:2486c).
sC. 1- (1925-) Irregular, about 10 a year. (U-3:2486c).
Mainly reprints from Svensk geografisk årsbok [2205].
Swedish with some English, French, or German summaries.

2192 ______. ______. ______.
Särtryck. 1- (1963-).
Reprints.
In Swedish. Some in English or with English abstracts.

(2193 ______. ______. Institutionen för kulturgeografi och ekonomisk geografi.
Rapporter och notiser. 1- (1967-). Mimeographed.
In Swedish or English. Former name: Kulturgeografiska institutionen.

(2194 ______. ______. Naturgeografiska institutionen.
Rapporter och notiser. 1- (1967-). Mimeographed. (N75-2:1369a).
In Swedish or English.

(2195 LUND studies in geography. (Lund. Universitet. Geografiska institutionen).
Lund.
sA. Physical geography. 1- (1950-) Irregular monographs.
(N70-3:3547b; B-3:106b).
Mainly in English.
[Geografiska Institutionen, Sölvegatan 13, Lund, Sweden.]

x(2196 ______.
sB. Human geography. 1- (1949-) Irregular. (U-3:2487b; B-3:106b).
Original monographs.
Mainly in English; some in French or German.
Indexes: 1949-1953; 1955-1958; 1960-1963; 1965.
[Geografiska Institutionen, Sölvegatan 13, Lund, Sweden.]

(2197 ______.
sC. General, mathematical, and regional geography. 1- (1962-)
Irregular. (N70-3:3547ab; N75-2:1369a; B68:338a).
1-9 as General and mathematical geography.
In English.

2198 STOCKHOLM. Universitet. Geografiska institutet. Stockholm.
Meddelanden. no1-83 (1929-1966). (U-5:4081c; F-3:406a).
Reprints and monographs.
Prior to 1960 as Stockholm. Högskolan. Geografiska institutet.
1966- divided into sA. (Physical geography) [2204] and sB. (Human
geography) [2202].
In Swedish, English, French, or German.

2200 ______. ______. Kulturgeografiska institutionen.
Forskningsprojekt administrativa-rumsliga system. 1-30 (1967-1976).
Irregular. Mimeographed.

(2201 ______. ______. ______.
Kulturgeografiska seminarium. 1971- Irregular number each year.
Mimeographed.

(2202 ______. ______. ______.
Meddelanden. sB. [Human geography]. 1- (1966-) Irregular.
(N70-4:5556b).

(2203 ______. ______. Naturgeografiska institutionen.
Forskningsrapport. 1- (1968-) Irregular. Mimeographed. (B73:123a).
In Swedish.

(2204 ______. ______. ______.
Meddelande. sA. [Physical geography]. (N75-2:2053b).
In Swedish, Norwegian, or English.

*(2205 SVENSK geografisk årsbok. Swedish geographical yearbook. (Sydsvenska
geografiska sällskapet i Lund). Lund. 1- (1925-) Annual.
(U-5:4127a; B-4:274a; F-4:636b).
Indexes: 1925-1934, 1935-1950.
Swedish. Titles in table of contents and abstracts in English.
[Liber Laeromedel, Box 1205, S221 01, Lund, Sweden.]

2206 SVENSKA sällskapet för antropologi och geografi.
Skrifter. Geografiska sektionen.
Tidskrift. v1 no1-13 (1878-1880). (U-5:4130b; F-4:638b).
Succeeded by Ymer [2216].
Swedish with French summaries.

(2207 SVENSKA växtgeografiska sällskapet. Uppsala.
Acta phytogeographica suecica. 1- (1929-) Irregular. (U-5:4131a; B-1:46b; F-1:60a).
In English, German, or Swedish.
[Växbiologiska Institutionen, Box 559, 751 22 Uppsala, Sweden.]

(2208 UMEÅ. Universitet. Geografiska institutionen. Umeå.
GERUM. Geografiska rapporter. A. 1- (1974-) Irregular.
In Swedish or English.

(2209 ______. ______. ______.
GERUM. Geografiska rapporter. B. 1- (1974-) Irregular.
In Swedish.

(2209a ______. ______. ______.
GERUM. Geografiska rapporter. C. 1- (1976-) Irregular but frequent.
In Swedish. Occasionally in English.

(2210 ______. ______. ______.
Geographical reports. 1- (1965-) Irregular.
In English.

(2211 ______. ______. ______.
Meddelanden. 1- (1965-) Irregular. Reprints and monographs.
In Swedish.

2212 UPPSALA. Universitet. Geografiska institutionen. Uppsala.
Avhandlingar. Naturgeografi. (Publications in physical geography). 1-3 (1958-1963). (N70-4:6108a).
nol issued in two parts.

(2213 ______. ______. ______.
Meddelanden. sA. 1- (1929-) Irregular reprints. (U-5:4341a; B-1:29b).
In English, French, German, or Swedish (partly with abstracts in English) since 1929.

(2214 ______. ______. Kulturgeografiska institutionen.
Forkningsrapporter. 1- (1965-) Irregular. Mimeographed.
In Swedish. Occasional abstracts in English. Some issues in English.

*(2215 ______. ______. Naturgeografiska institutionen.
UNGI. Rapport. 1- (1969-) Irregular. (B73:284a).
In Swedish with English abstracts or in English.

(2216 YMER. Årsbok. (Svenska sällskapet för antropologi och geografi). Stockholm. 1- (1881-) Annual. (U-5:4578a; N70-4:6428c; B-4:584a; F-4:995a).
Supersedes Svenska sällskapet för antropologi och geografi. Skrifter. Geografiska sektionen. Tidskrift [2206].
1-85 (1881-1965) without subtitle and issued quarterly. 86- (1966-) issued annually with each issue concentrated on a special topic.
Indexes: 1-30 (1881-1910), 31-45 (1911-1925), 46-70 (1926-1950).
In Swedish; 1881-1885 Resumés des séances in French; 1895-1937 French tables of contents; 1949-1965 English (occasionally German or French) summaries of articles. 1966- some English abstracts of articles.
[Generalstabens Litografiska Anstalts Förlag, Stockholm, Sweden.]

SWITZERLAND (German, French, Italian)

Schweizerische Eidgenossenschaft,

Confédération Suisse, Confederazione Svizzera

2217 ASSOCIATION des sociétés suisses de géographie. Genève.
Travaux. 1-2 (1882). Closed 1882. (U-1:520a).

2218 ATLANTIS. Länder; Völker; Reisen. Berlin, Wien, Zürich. 1- (1929-)
Monthly. (U-1:551c; B-1:254b; BS:936b; F-1:492a).
Index: v1-32 (1929-1960).
["Du Atlantis," Verlag Conzett and Huber, Morgartenstrasse 29, 8021 Zürich, Switzerland.]

2219 BASEL. Universität. Geographische Institut. Basel.
Mitteilungen und Arbeiten. 1-267 (1912-1954).
Specially numbered reprints and theses for exchange purposes.

*(2220 BASLER Beiträge zur Geographie. (Geographisch-ethnologische Gesellschaft).
Basel. 1- (1960-) Irregular. (N70-1:650b; B68:69b; G).
1-6 (1960-1965) as Basler Beiträge zur Geographie und Ethnologie.
5-6 (1963-1965) with subtitle Geographische Reihe. 1-6 (1960-1965) as Ergänzungsheft to Regio Basiliensis [2254].
In German with some French or English summaries.

(2221 BEITRÄGE zur geobotanischen Landesaufnahme der Schweiz. (Schweizerische naturforschende Gesellschaft. Pflanzengeographische Kommission).
Bern, Zürich. 1- (1916-). (U-1:622b; B-1:302a; F-1:554b; G).
Title in French: Matériaux pour le levé géobotanique de la Suisse.
Title in Italian: Contributi allo studio geobotanica della Svizzera.
1-15 (1916-1927) as Beiträge zur geobotanischen Landesaufnahme.

2222 BEITRÄGE zur Kartographie. Thun. 1- (1930-) Irregular.
[W. Kreisel, Pestalozzistr. 85, Thun, Switzerland.]

2223 BERN. Universität. Geographisches Institut.
Arbeiten. 1-4 (1888-1898). Closed 1898. (U-1:650b).
Specially numbered reprints and theses for exchange purposes.

(2224 BIBLIOGRAPHIA scientiae naturalis helvetica. (Schweizerische Landesbibliothek).
Bern. 1- (1925-) Annual. (U-1:665c; N60-1:258b; N65-1:348a; B-1:335a; F-1:611b).
Title varies: 1925-1939 as Bibliographie der schweizerischen naturwissenschaftlichen Literatur; 1940-1947 as Bibliographie der schweizerischen naturwissenschaftlichen und geographischen Literatur.
1945-1947 have added title page: Bibliographie scientifique suisse.
Bibliographia helvetica.
Indexes: v1-13 (1902-1916), 1919-1925.
[Bibliothèque Nationale Suisse, Hallwylstrasse 15, Bern, Switzerland.]

2225 BIBLIOGRAPHIE der schweizerischen Landeskunde. (Switzerland. Zentralkommission für schweizerische Landeskunde). Bern.
1-89 (1892-1945). Closed 1945. (U-1:669c; B-1:335a).

(2226 ENVIRONMENTAL conservation. (Foundation for environmental conservation).
Lausanne. 1- (1974-) Quarterly. (N75-1:779a).
In English.
[Elsevier Sequoia S.A., P.O. Box 851, 1001 Lausanne 1, Switzerland.]

2227 FERNSCHAU. (Mittelschweizerische geographisch-commercielle Gesellschaft in Aarau). Aarau. 1-6 (1886-1894). Closed 1894.
(U-2:1554a; B-2:180a; F-2:585b).

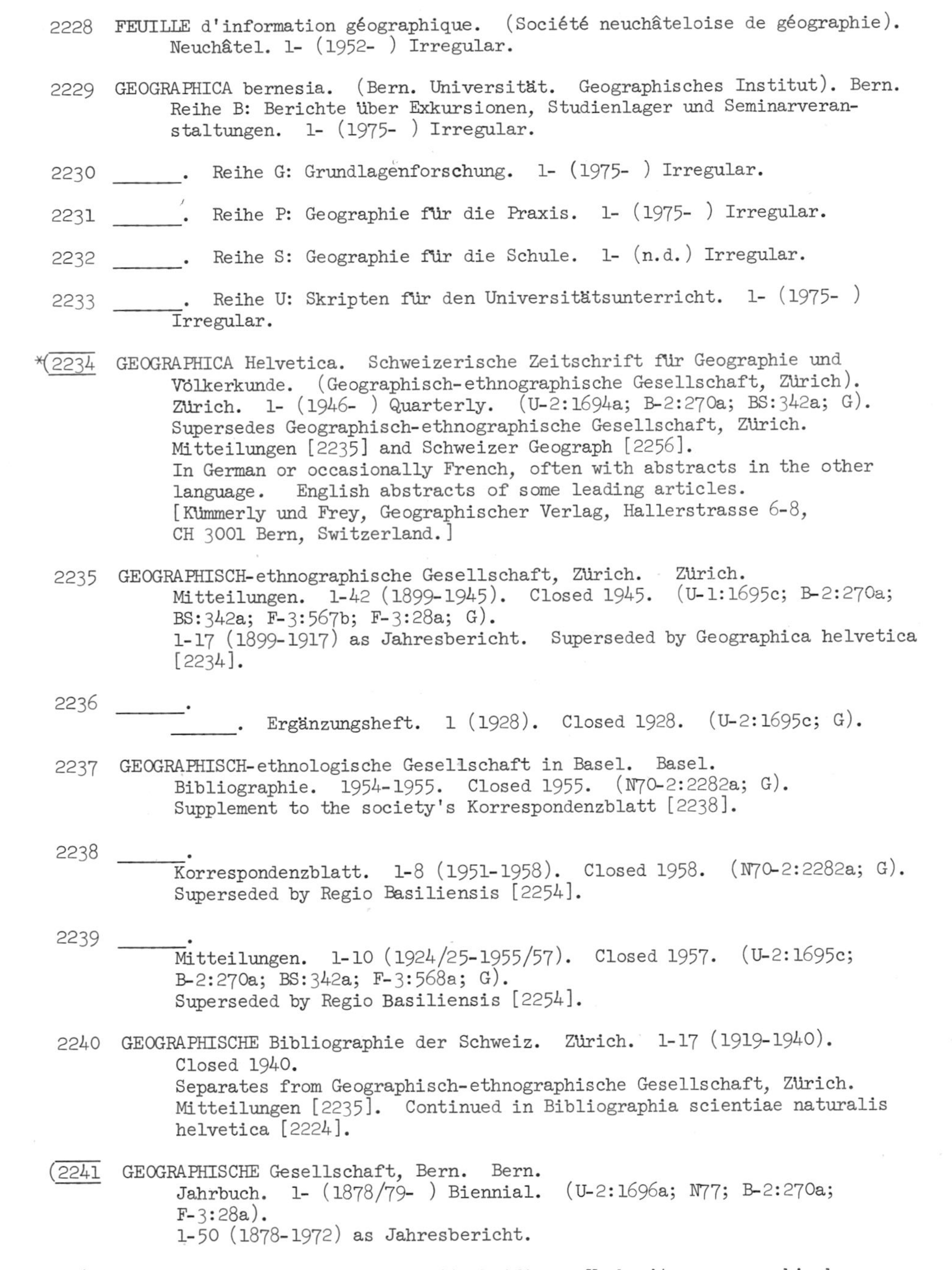

2228 FEUILLE d'information géographique. (Société neuchâteloise de géographie). Neuchâtel. 1- (1952-) Irregular.

2229 GEOGRAPHICA bernesia. (Bern. Universität. Geographisches Institut). Bern. Reihe B: Berichte über Exkursionen, Studienlager und Seminarveranstaltungen. 1- (1975-) Irregular.

2230 ______. Reihe G: Grundlagenforschung. 1- (1975-) Irregular.

2231 ______. Reihe P: Geographie für die Praxis. 1- (1975-) Irregular.

2232 ______. Reihe S: Geographie für die Schule. 1- (n.d.) Irregular.

2233 ______. Reihe U: Skripten für den Universitätsunterricht. 1- (1975-) Irregular.

*(2234 GEOGRAPHICA Helvetica. Schweizerische Zeitschrift für Geographie und Völkerkunde. (Geographisch-ethnographische Gesellschaft, Zürich). Zürich. 1- (1946-) Quarterly. (U-2:1694a; B-2:270a; BS:342a; G).
Supersedes Geographisch-ethnographische Gesellschaft, Zürich. Mitteilungen [2235] and Schweizer Geograph [2256].
In German or occasionally French, often with abstracts in the other language. English abstracts of some leading articles.
[Kümmerly und Frey, Geographischer Verlag, Hallerstrasse 6-8, CH 3001 Bern, Switzerland.]

2235 GEOGRAPHISCH-ethnographische Gesellschaft, Zürich. Zürich. Mitteilungen. 1-42 (1899-1945). Closed 1945. (U-1:1695c; B-2:270a; BS:342a; F-3:567b; F-3:28a; G).
1-17 (1899-1917) as Jahresbericht. Superseded by Geographica helvetica [2234].

2236 ______.
______. Ergänzungsheft. 1 (1928). Closed 1928. (U-2:1695c; G).

2237 GEOGRAPHISCH-ethnologische Gesellschaft in Basel. Basel. Bibliographie. 1954-1955. Closed 1955. (N70-2:2282a; G).
Supplement to the society's Korrespondenzblatt [2238].

2238 ______.
Korrespondenzblatt. 1-8 (1951-1958). Closed 1958. (N70-2:2282a; G).
Superseded by Regio Basiliensis [2254].

2239 ______.
Mitteilungen. 1-10 (1924/25-1955/57). Closed 1957. (U-2:1695c; B-2:270a; BS:342a; F-3:568a; G).
Superseded by Regio Basiliensis [2254].

2240 GEOGRAPHISCHE Bibliographie der Schweiz. Zürich. 1-17 (1919-1940). Closed 1940.
Separates from Geographisch-ethnographische Gesellschaft, Zürich. Mitteilungen [2235]. Continued in Bibliographia scientiae naturalis helvetica [2224].

(2241 GEOGRAPHISCHE Gesellschaft, Bern. Bern. Jahrbuch. 1- (1878/79-) Biennial. (U-2:1696a; N77; B-2:270a; F-3:28a).
1-50 (1878-1972) as Jahresbericht.

2242 GEOGRAPHISCHE Nachrichten. Zeitschrift zur Verbreitung geographischer Kenntnisse. Basel. 1-12 (1885-1896). Closed 1896. (F-2:728b).
1892-1894 as Zeitschrift zur Verbreitung geographischer Kenntnisse und officielles Organ der ostschweizerischen geographisch-commerciellen Gesellschaft.

2243 GEOGRAPHISCHE Rundschau. Bern; Zürich. 2 (1897). (U-2:1696c).

(2244 GLOBE. (Société de géographie de Genève). Genève. 1- (1860-) Annual. (U-2:1736b; B-4:135b; F-2:754b; F-4:534a).
1-4 (1860-1864) as Société de géographie de Genève. Mémoires et bulletin. Subtitle varies: v1-50 (1860-1911) as Journal géographique; 51- (1912-) as Bulletin (et mémoires). In some vols. no Mémoires appear.
Indexes: v1-50 (1860-1911) published separately; v51-60 (1912-1921), v61-70 (1922-1931), and v71-90 (1932-1951) published as annexes to v60 (1921), v70 (1931), and v90 (1951), respectively.

2245 HISTORISCH-geographisch-statistisches Gemälde der Schweiz. St. Gallen. v1-15 no19 (1834-1849). (B-2:404a).

2246 INSTITUT géographique international. Bern.
Bulletin. 1-4 (1881). (B-2:486b).

2247 LAUSANNE. Université. Faculté des lettres. Lausanne.
Collection géographique vaudoise. 1-10 (1932?-1937).
Specially numbered reprints for exchange.

2248 ______. ______. Laboratoires de géologie, géographie physique, minéralogie et paléontologie.
Bulletin. 1- (1901-). (U-3:2364c; B-3:20b).

2249 LEBEN und Umwelt. (Vereinigung schweizerischer Naturwissenschafts-Lehrer). Aarau. 1-? (1944-1965) Monthly. Closed 1965. (U-3:2377b).

2250 MATÉRIAUX pour l'étude des calamités. (Société de géographie de Genève). Genève. (v1-12) no1-40 (Ap/Je 1924-1937). Closed 1937. (U-3:2561c; B-3:156a).
Superseded by Revue pour l'etude des calamités (Bulletin de l'Union internationale de secours), 1- (1938-) (not in this list).

2251 MITTELSCHWEIZERISCHE geographisch-commercielle Gesellschaft in Aarau. Aarau.
Kleine Mitteilungen. v1-2 no2 (N 1892-N 1895). Closed 1895. (U-3:2708b; B-3:231a; F-3:244a).

2252 OSTSCHWEIZERISCHE geographisch-commercielle Gesellschaft, St. Gallen. St. Gallen.
Jahresbericht. 1-4 (1878-1881/1882). Closed 1882. (U-4:3213b; B-3:472a; F-3:34b).
Continued as Ostschweizerische geographische Gesellschaft. Mitteilungen [2253].

2253 ______.
Mitteilungen. 1883-1892; 1895-1944. Closed 1944. (U-4:3213b; B-3:472a; F-3:572a).
Supersedes the society's Jahresbericht [2252]. For the years 1893-1894, see Geographische Nachrichten [2242].
Index: 1878-1928.

*(2254 REGIO Basiliensis; Basler Zeitschrift für Geographie; Revue de géographie de Bâle. (Geographisch-ethnologische Gesellschaft in Basel). Basel. 1- (1959-) 2 nos. a year. (N70-3:4954c; G).
Subtitle varies: Hefte für jurassische und oberrheinische Landeskunde. Cahiers de géographie jurassienne et rhénane.
Supersedes the society's Korrespondenzblatt [2238] and its Mitteilungen [2239].
Indexes: 1-10 (1959-1969), 11-20 (1970-1979).
In German with resumé in French.
[Verlag Wepf und Co., Eisengasse 5, 4001 Basel, Switzerland.]

2255 SANKT GALLEN (Saint Gall). Hochschule für Wirtschafts- und Sozialwissenschaften. Sankt Gallen.
Veröffentlichungen: Volkswirtschaftlich-wirtschaftsgeographische Reihe. 1- (1964-). (N70-4:5151a; G).

2256 SCHWEIZER Geograph. Géographe suisse. (Verein schweizerischer Geographie-Lehrer; Geographische Gesellschaft, Bern; Geographisch-ethnographische Gesellschaft, Zürich; Ostschweizerische geographisch-commercielle Gesellschaft, St. Gall). Bern.
1-22 (1923-1945). Closed 1945. (U-5:3808b; B-4:41b; G).
Superseded by Geographica Helvetica [2234].

2257 ______.
Ergänzungsheft. 1 (1930). Only one published. (G).

(2258 SCHWEIZERISCHE naturforschende Gesellschaft. Place of publication varies.
Verhandlungen. 1- (1815-) Annual. (U-5:3812a; B-1:85b; F-2:508b-509a; G).
Contains a section for geography and cartography.
[Birkhäuser Verlag, Elisabethenstrasse 19, CH-4010 Basel, Switzerland.]

2259 SOCIÉTÉ fribourgeoise des sciences naturelles, Fribourg. Fribourg.
Mémoires: Géologie et géographie. v1-13 (1900-1947). Closed 1947. (U-5:3960c; B-4:134b; F-3:461b).

(2260 SOCIÉTÉ neuchâteloise de géographie, Neuchâtel.
Bulletin. v1- (1885-) Irregular. (U-5:3968b; B-4:145a; F-1:872a).
v50-55 (1944/47-) also as ns no1- .
[Bibliothèque de la Ville, 3, Place Numa-Droz, Neuchâtel, Switzerland.]

2261 ______.
Mémoires. 1 (1948). Only one published? (U-5:3968b; BS:805a).

2262 WELT auf Reisen. Völker, Länder, Meere. Zürich. 1-20 (1949-1968/69).
1-10 (1949-1958) as Reise; illustrierte Monatsschrift für Reisen und Länderkunde.

(2263 ZÜRICH. Eidgenössische technische Hochschule. Geographisches Institut. Zürich.
Arbeiten. 1- (1944-) Irregular. (U-5:4646a).
Reprints and new monographs.
In German or French.
[Sonneggstrasse 5, Zürich, Switzerland.]

2264 ______. ______. Institut für Landesplanung. Zürich.
Arbeiten. no1- (1946-) Irregular.
[Sonneggstrasse 5, Zürich, Switzerland.]

2265 ______. ______.
Institut für Ortes-, Regional- und Landesplanung (ORL). Zürich.
Berichte zur Orts- , Regional- und Landesplanung. 1- (1969-) Irregular.

2266 ______. ______. ______.
DISP [Dokumentations- und Informationsstelle für Planungsfragen].
Planning quarterly. 1968- Quarterly.

2267 ______. ______. ______.
Schriftenreihe zur Orts- , Regional- und Landesplanung. 1- (1969-) Irregular.

2268 ______. ______. ______.
Studienunterlagen zur Orts- , Regional- und Landesplanung. 1- (1969-) Irregular.

(2269 ______. Universität. Geographisches Institut. Zürich.
Arbeiten. nol- (1944-) Irregular. (N70-4:6483c).
Series composed of dissertations and reprints. Earliest reprint dates from 1941.
In German.
[Blümlisalpstrasse 10, 8006 Zürich, Switzerland.]

SYRIA (Arabic)

al-Jumhūriyah al-ʿArabīyah as-Sūrīyah

الجمهورية العربية السورية

(2270 المجلة الجغرافية

*(2270 al-MAJALLAH al-jughrafīyah. (al-Jam'īyah al-jughrāfīyah al-Sūrīyah).
Damascus. 1- (1976-) Annual. Supplementary title page: al-Majalla al-Joughraphia. Geographical magazine. (Syrian geographical society).
In Arabic. Supplementary title page and table of contents in English.
[Geographical Magazine, The Syrian Geographical Society, Department of Geography, Faculty of Letters, University of Damascus, Damascus, Syria.]

TANZANIA (Swahili, English)

Jamhuri ya Mwungano wa Tanzania, United Republic of Tanzania

(2271 DAR ES SALAAM. University. Bureau of resource assessment and land use planning. Dar es Salaam.
Annual reports. 1967/68- Annual. (N75-1:654a).
1967/68-1969/70 issued under earlier name: University College.
In English.
[Bureau of Resource Assessment and Land Use Planning, University of Dar es Salaam, P.O. Box 35097, Dar es Salaam, Tanzania.]

2272 ______. ______. ______.
Research notes. 1-11 (1967-1971) Irregular. Mimeographed. (N70-2:1648c; N75-1:654a).
1-9 (1967-1970) issued under earlier name: University college.
In English.
[for address: See under Bureau's Annual reports [2271.]

*(2273 ______. ______. ______.
Research papers. 1- (1968-) Irregular. Mimeographed. (70-2:1648c; N75-1:654a).
In English.
1-11 (1968-1970) issued under earlier name: University College.
[for address: See under Bureau's Annual reports [2271].]

*(2274 ______. ______. ______.
Research reports. 1-49 (1969-1971); nsl- (1973?-) Irregular.
In English.
[for address: See under Bureau's Annual reports [2271].]

2275 ______. ______. Department of geography.
Annual report.
1975/76- (N78).

*(2276 GEOGRAPHICAL association of Tanzania. Dar es Salaam.
Journal. 1- (1967-) 2 nos. a year. Mimeographed. (N75-1:906a; B73:173a).
In English.
[Geographical Association of Tanzania, University of Dar es Salaam, Box 35049, Dar es Salaam, Tanzania.]

THAILAND (Thai)

(Prathet Thai; Muang Thai)
ประเทศไทย

*(2277 GEOGRAPHICAL association of Thailand.
Geographical journal. Bangkok. 3 nos. a year.
In Thai or English.

TUNISIA (Arabic)

al-Jumhūriyah al-Tūnisīyah
الجمهورية التونسية

*(2278 المجلة الجغرافية التونسية

*(2278 REVUE tunisienne de géographie. (Faculté des lettres et des sciences humaines de Tunis). Tunis. 1- (1978-) 2 nos. a year.
In French or Arabic. Summaries in English, French, and Arabic.
[Boulevard du 9 avril 1938, B. P. 1128, Tunis, Tunisia.]

2279 SOCIÉTÉ de géographie commerciale [et d'études coloniales] de Paris. Section tunisienne. Tunis.
Bulletin. 1-4 (1908-1914); 14-15 (1931).
Others issued? (U-5:3945a).
In French.

2280 ______. ______.
______.
Revue. no18-20 (1935-1937). Closed 1937. (U-5:3945a; B-4:135a; F-4:210a).
In French.

2281 TUNIS. al-Jamiʿah al-Tūnisīyah. Centre d'études et de recherches économiques et sociales. Tunis.
Cahiers du C.E.R.E.S. Série géographique. 1- (1968-) Irregular. (N75-2:2152b).
In French.

TURKEY (Turkish)

Türkiye Cumhuriyeti

(2282 ANKARA. Üniversite. Dil ve tarih-coğrafya fakültesi. Ankara.
Dergisi. 1- (1942-). (U-1:368b; B-1:146b).
Contributions in English, French, German, and Turkish.

2283 ______. ______. ______.
Doğu anadolu araştirma istasyonu Yayimlari. 1- (1951-) Irregular.
(N70-1:353a).
Summaries in French.

(2283a ______. ______. ______. Coğrafya araştirmalari enstitüsü.
Yayinlar. 1- (1942-). Subseries of Faculty's Yayinlar (not in
this list).

(2284 COGRAFYA araştirmalari dergisi. Ankara. 1- (1966-).
In Turkish, English, French, or German.

2285 COĞRAFYA haberleri. (Türk coğrafya kurumu. Turkish geographical society).
Ankara. no1- (1959-) Irregular.

(2286 İSTANBUL. Üniversite. Coğrafya enstitüsü.
Dergisi. 1- (1951-). (N70-2:3025c; BS:250a).
In Turkish. 1-2 (1951) have supplementary titles and abstracts in
English or French.
[İstanbul Üniversitesi Coğrafya Enstitüsü, İstanbul, Turkey.]

2287 ______. ______. ______.
Doktora tezleri serisi. no1 (1942).
In Turkish and German.

2288 ______. ______. ______.
Monografiler. Monograph series. 2 (1959). (N70-2:3025c).
A subseries of the university's Yayinlar.
In Turkish. English summaries.

*(2289 ______. ______. ______.
Neşriyat. Yayinlar. (Publication). 1- (1930-) Irregular.
Monographs. (U-3:2124c; B-2:10b).
A subseries of the university's Yayinlar. no104 (1979) is no2540 of
whole series].
In Turkish. Some English or French abstracts.

*(2290 ______. ______. ______.
Review. International edition. 1- (1954-) Irregular. (N70-2:3025c;
BS:250a).
Articles in English, French, or German.
[İstanbul Üniversitesi Coğrafya Enstitusu, İstanbul, Turkey.]

(2291 JEOMORFOLOJI dergisi. (Türkiye jeomorfologlar derneği Yayini). Ankara.
v1 no1- (1969-). (N75-1:1230b).
In Turkish. Summaries in English, French, or German.
[Türkiye Jeomorfologlar Derneği, P. K. 653,Kizilay-Ankara, Turkey.]

(2292 TÜRK coğrafya dergisi. Turkish geographical review. (Türk coğrafya kurumu.
Turkish geographical society). Ankara. 1-26 (1943-1973). Annual.
(U-5:4279b; B-4:358b).
In Turkish. Abstracts in English, French, or German listed in table
of contents.
[Türk Coğrafya Kurumu, P. K. 43, Küçükesat-Ankara, Turkey.]

2293 TÜRK coğrafya kurumu. Ankara.
Yayinlar. nol- (1944-) Irregular. (U-5:4279b).
In Turkish. French summary.

UGANDA (English)

Republic of Uganda

*(2294 EAST African geographical review. (Uganda geographical association). Kampala. 1- (1963-) Annual. (N70-2:1817b; B68:179b).
[Uganda Geographical Association, Makerere University, P.O. Box 7062, Kampala, Uganda.]

2295 GEOGRAPHER. (Makerere university. Geographical society). Kampala. 6- (1970-). (N75-1:905b).

2296 MAKERERE institute of social research. Kampala.
Geography papers. 1967- Annual. (N75-2:1385c).
University of East Africa. Social sciences council conference.

*(2297 MAKERERE university. Department of geography. Kampala.
Occasional paper. 1- (1967-) Irregular, but several a year. (N70-2:3240c; N75-2:1386a).
1-16 (1967-1970) issued by Kampala. Makerere university college. Department of geography.
[P.O. Box 7062, Kampala, Uganda.]

2298 UGANDA geographical association. Kampala.
Newsletter. 1- (1967-) Annual. (N75-2:2162a). Mimeographed. Articles.
[P.O. Box 7062, Kampala, Uganda.]

UNION OF SOVIET SOCIALIST REPUBLICS

Titles in Latin letters begin on p. 300

СОЮЗ СОВЕТСКИХ СОЦИАЛИСТИЧЕСКИХ РЕСПУБЛИК

(2299 АКАДЕМИЯ наук Армянской ССР. Ереван.
Известия. Науки о земле.

(2300 АКАДЕМИЯ наук Азербайджанской ССР. Баку.
Известия. Серия наук о земле.

2301 ______. Институт географии.
Материалы научной конференции.

(2302 ______. ______.
Труды.

2303 АКАДЕМИЯ наук Грузинской ССР. Институт географии имени Вахушти. Тбилиси.
Труды.

2304 АКАДЕМИЯ наук Казахской ССР. Алма-Ата.
Известия. Серия географическая.

2305 ______. Сектор физической географии. Алма-Ата.
Гляциологические исследования в Казахстане.

2306 АКАДЕМИЯ наук Киргизской ССР. Отдел географии. Фрунзе.
Труды.

2307 ______. Тяньшанская физико-географическая станция. Фрунзе.
Работы.

*(2308 АКАДЕМИЯ наук С С С Р. Москва.
Известия. Серия географическая.

2309 ______. Ленинград, позже Москва.
Известия.Серия 7.Серия географическая и геофизическая.

2310 ______. Институт географии. Москва.
Географические сообщения.

2311 ______. ______.
Исследования ледниковых районов.

2312 ______. ______.
Материалы гляциологических исследований. Москва.

2313 ______. ______.
Труды.

2314 ______. Комиссия по естественно-историческому районированию СССР.
Труды.

(2315 АКАДЕМИЯ наук С С С Р. Москва.
Комиссия по изучению четвертичного периода.
Бюллетень.

2316 ______. Комиссия по изучению естественных производительных сил СССР. Географический отдел.
Труды.

2317 ______. Отделение геолого-географических наук. Москва.
Рефераты научно-исследовательских работ.

2318 ______. Сибирское отделение. Институт географии Сибири и Дальнего Востока. Иркутск.
Доклады.

2319 ______. ______. ______.
Материалы Обь-Иртышской экспедиции.

2320 ______. ______. Комиссия по комплексному картографированию природы, хозяйства и населения.
Труды.

2321 АКАДЕМИЯ наук Украинской ССР. Киев. Комиссия для составления историко-географического словаря Украины.
Историко-географический сборник.

2322 ______. Природно-технический отдел.
Журнал геолого-географического цикла.

2323 АЛМА-АТА. Казахский государственный университет. Алма-Ата.
Ученые записки. География.

2324 АМУРСКИЙ сборник. (Дальневосточный филиал СО АН СССР и Приамурский филиал ГО СССР). Хабаровск.

(2325 АЗЕРБАЙДЖАНСКОЕ географическое общество. Баку.
Труды.

2326 БАКУ. Азербайджанский государственный университет. Баку.
Труды. Серия геолого-географическая.

(2327 ______. ______.
Ученые записки. Серия геолого-географических наук.

2331 БИБЛИОГРАФИЯ картографической литературы и карт. (Всесоюзная книжная палата). Москва.

(2332 БИОЛОГИЯ и география. Сборник научных статей аспирантов и соискателей. (Казахский университет). Алма-Ата.

(2333 БРЯНСКИЙ краевед. (Географическое общество СССР. Брянское отделение. Областной краеведческий музей). Брянск.

2334 БУРЯТОВЕДЧЕСКИЙ сборник. (Географическое общество СССР. Восточно-Сибирский отдел. Бурят-Монгольская секция). Иркутск.

2335 ЧАРДЖОУ. Туркменский государственный педагогический институт. Чарджоу.
Серия биологических и географических наук.

2336 ЧЕЛЯБИНСК. Государственный педагогический институт. Челябинск.
Ученые записки. Серия естественно-географическая.

2337 ЧЕРНОВЦЫ. Университет. Черновцы.
Ученые записки. Серия географических наук.

2338 ______. ______.
Ученые записки. Серия геолого-географических наук.

2339 ЧИТА. Забайкальский комплексный научно-исследовательский институт. Чита.
Труды. Серия экономика и география.

2340 ДАГЕСТАНСКИЙ государственный педагогический институт. Естественно-географический факультет. Махач-Кала.
Труды.

2341 ДНЕПРОПЕТРОВСК. Университет. Геолого-географический факультет.
Сборник работ.

(2342 ДОКЛАДЫ на ежегодных чтениях памяти Л.С.Берга. (Географическое общество СССР). Москва; Ленинград.

(2343 ДОКЛАДЫ на ежегодных чтениях памяти В.А.Обручева. (Географическое общество СССР). Москва; Ленинград.

2344 ДОКЛАДЫ по экономической географии СССР. (Географическое общество СССР. Отделение экономической географии). Ленинград.

2345 ДОКЛАДЫ по геоморфологии и палеогеографии Дальнего Востока. (Географическое общество СССР). Ленинград.

2346 ДУШАНБЕ. Государственный педагогический институт. Душанбе.
Серия географическая.

*(2347 ЭСТОНСКОЕ географическое общество. Таллин.
Ежегодник.

2348 ______.
Публикации.

*(2349 ЭКОНОМИЧЕСКАЯ география; межведомственный научный сборник. Киев.

2350 ЭКОНОМИЧЕСКАЯ география; республиканский межведомственный научный сборник. (МВиССО УССР и университет). Харьков.

2351 ЭКОНОМИЧЕСКАЯ география западного Урала. (Географическое общество СССР. Пермский отдел). Пермь.

2352 ЭКОНОМИКО-ГЕОГРАФИЧЕСКИЕ проблемы формирования территориально-производственных комплексов Сибири. (Географическое общество СССР. Новосибирский отдел). Новосибирск.

2353 ЕСТЕСТВОЗНАНИЕ и география. Научно-популярный и педагогический журнал. Москва.

*(2354 ФИЗИЧЕСКАЯ география и геоморфология. Республиканский межведомственный научный сборник. Киев.

(2355 ФРУНЗЕ. Университет. Географический факультет. Фрунзе. Труды.

(2356 ГЕОБОТАНИЧЕСКОЕ картографирование. (Академия наук СССР. Ботанический институт). Ленинград.

(2357 ГЕОДЕЗИЯ и картография. (Главное управление геодезии и картографии). Москва.

(2358 ГЕОДЕЗИЯ, картография и аэрофотос'емка: республиканский межведомственный научно-технический сборник. Львов.

2359 СБОРНИК географических исследований. (Тартусский государственный университет). Тарту.

2360 ГЕОГРАФИЧЕСКИЕ науки: тематический сборник научных трудов профессорско-преподавательского состава Министерства высшего и среднего специального образования Казахской ССР.(Алма-Ата. Казахский государственный педагогический институт). Алма-Ата.

2360а ГЕОГРАФИЧЕСКИЙ месяцеслов. Петербург (Ленинград).

2361 ГЕОГРАФИЧЕСКИЙ сборник. (Географическое общество СССР. Пензенский отдел. Общество охраны природы РСФСР). Пенза.

(2361а ГЕОГРАФИЧЕСКИЙ сборник. (Казахский государственный университет). Алма-Ата.

2362 ГЕОГРАФИЧЕСКИЙ сборник. (Казань. Университет). Казань.

2363 ГЕОГРАФИЧЕСКИЙ сборник. (Всесоюзный институт научной и технической информации). Москва.

2364 ГЕОГРАФИЧЕСКИЙ вестник. (Ленинград. Географический институт). Ленинград.

2365 ГЕОГРАФИЧЕСКОЕ обозрение. Петербург (Ленинград).

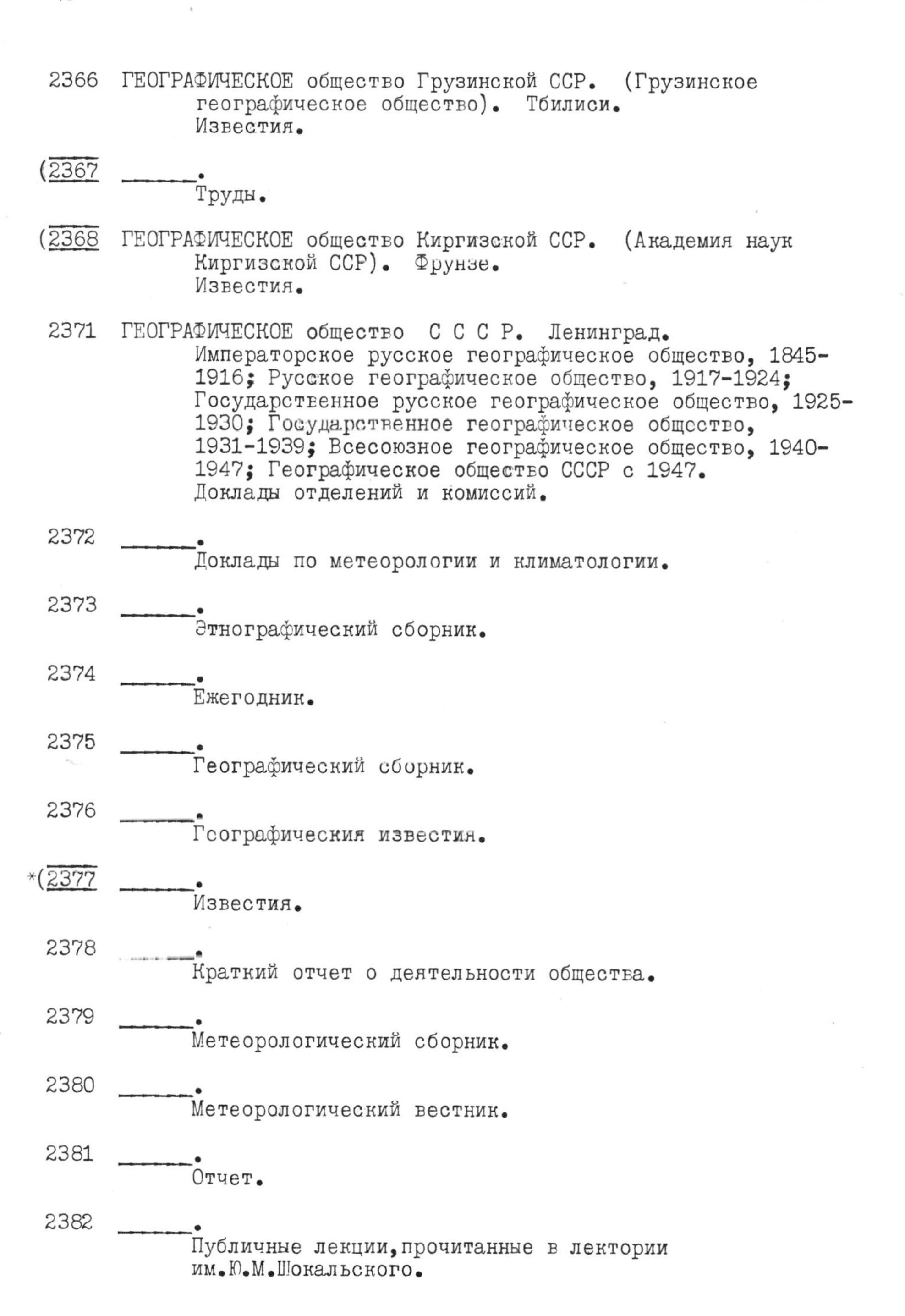

2366 ГЕОГРАФИЧЕСКОЕ общество Грузинской ССР. (Грузинское географическое общество). Тбилиси.
Известия.

(2367 ______.
Труды.

(2368 ГЕОГРАФИЧЕСКОЕ общество Киргизской ССР. (Академия наук Киргизской ССР). Фрунзе.
Известия.

2371 ГЕОГРАФИЧЕСКОЕ общество С С С Р. Ленинград.
Императорское русское географическое общество, 1845-1916; Русское географическое общество, 1917-1924; Государственное русское географическое общество, 1925-1930; Государственное географическое общество, 1931-1939; Всесоюзное географическое общество, 1940-1947; Географическое общество СССР с 1947.
Доклады отделений и комиссий.

2372 ______.
Доклады по метеорологии и климатологии.

2373 ______.
Этнографический сборник.

2374 ______.
Ежегодник.

2375 ______.
Географический сборник.

2376 ______.
Географические известия.

*(2377 ______.
Известия.

2378 ______.
Краткий отчет о деятельности общества.

2379 ______.
Метеорологический сборник.

2380 ______.
Метеорологический вестник.

2381 ______.
Отчет.

2382 ______.
Публичные лекции, прочитанные в лектории им. Ю. М. Шокальского.

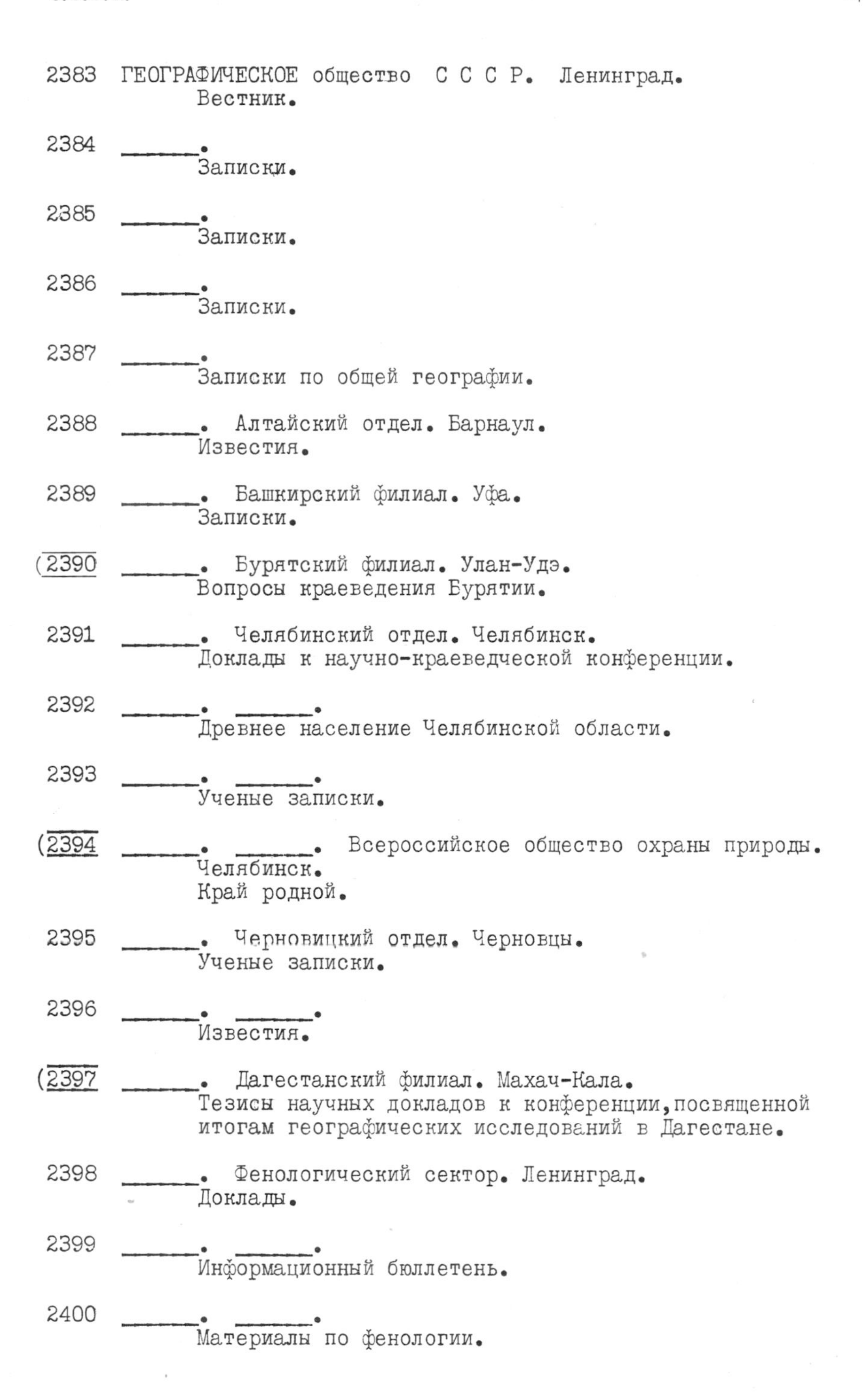

2383 ГЕОГРАФИЧЕСКОЕ общество С С С Р. Ленинград.
Вестник.

2384 ______.
Записки.

2385 ______.
Записки.

2386 ______.
Записки.

2387 ______.
Записки по общей географии.

2388 ______. Алтайский отдел. Барнаул.
Известия.

2389 ______. Башкирский филиал. Уфа.
Записки.

(2390 ______. Бурятский филиал. Улан-Удэ.
Вопросы краеведения Бурятии.

2391 ______. Челябинский отдел. Челябинск.
Доклады к научно-краеведческой конференции.

2392 ______. ______.
Древнее население Челябинской области.

2393 ______. ______.
Ученые записки.

(2394 ______. ______. Всероссийское общество охраны природы.
Челябинск.
Край родной.

2395 ______. Черновицкий отдел. Черновцы.
Ученые записки.

2396 ______. ______.
Известия.

(2397 ______. Дагестанский филиал. Махач-Кала.
Тезисы научных докладов к конференции,посвященной итогам географических исследований в Дагестане.

2398 ______. Фенологический сектор. Ленинград.
Доклады.

2399 ______. ______.
Информационный бюллетень.

2400 ______. ______.
Материалы по фенологии.

2401 ГЕОГРАФИЧЕСКОЕ общество С С С Р. Ленинград.
Якутский отдел. Якутск.
Известия.

2402 ______. ______.
Очерки по изучению Якутского края.

2403 ______. Юго-Западный отдел. Киев.
Записки.

2404 ______. Южно-Уссурийский отдел.
Известия.

2405 ______. ______. Владивосток.
Записки.

2406 ______. Калининградский отдел. (Калининград. Университет).
Калининград.
Записки.

2407 ______. Картографическая комиссия.
Журналы.

2408 ______. Кавказский отдел. Тбилиси.
Известия.

2409 ______. ______.
______. Приложение. Материалы для географии
Азиатской Турции.

2410 ______. ______.
Отчет.

2411 ______. ______.
Сборник статистических сведений о Кавказе.

2412 ______. ______.
Записки.

2413 ______. Коми филиал. Сыктывкар.
Известия.

2414 ______. Комиссия аэрос'емки и фотограмметрии. Ленинград.
Доклады.

2415 ______. Комиссия географии населения и городов. Ленинград.
Доклады по географии населения.

2416 ______. Комиссия геоморфологии и палеогеографии. Ленинград.
Доклады по геоморфологии и палеогеографии
Северо-Запада Европейской части СССР.

2417 ______. Комиссия медицинской географии. Ленинград.
Материалы.

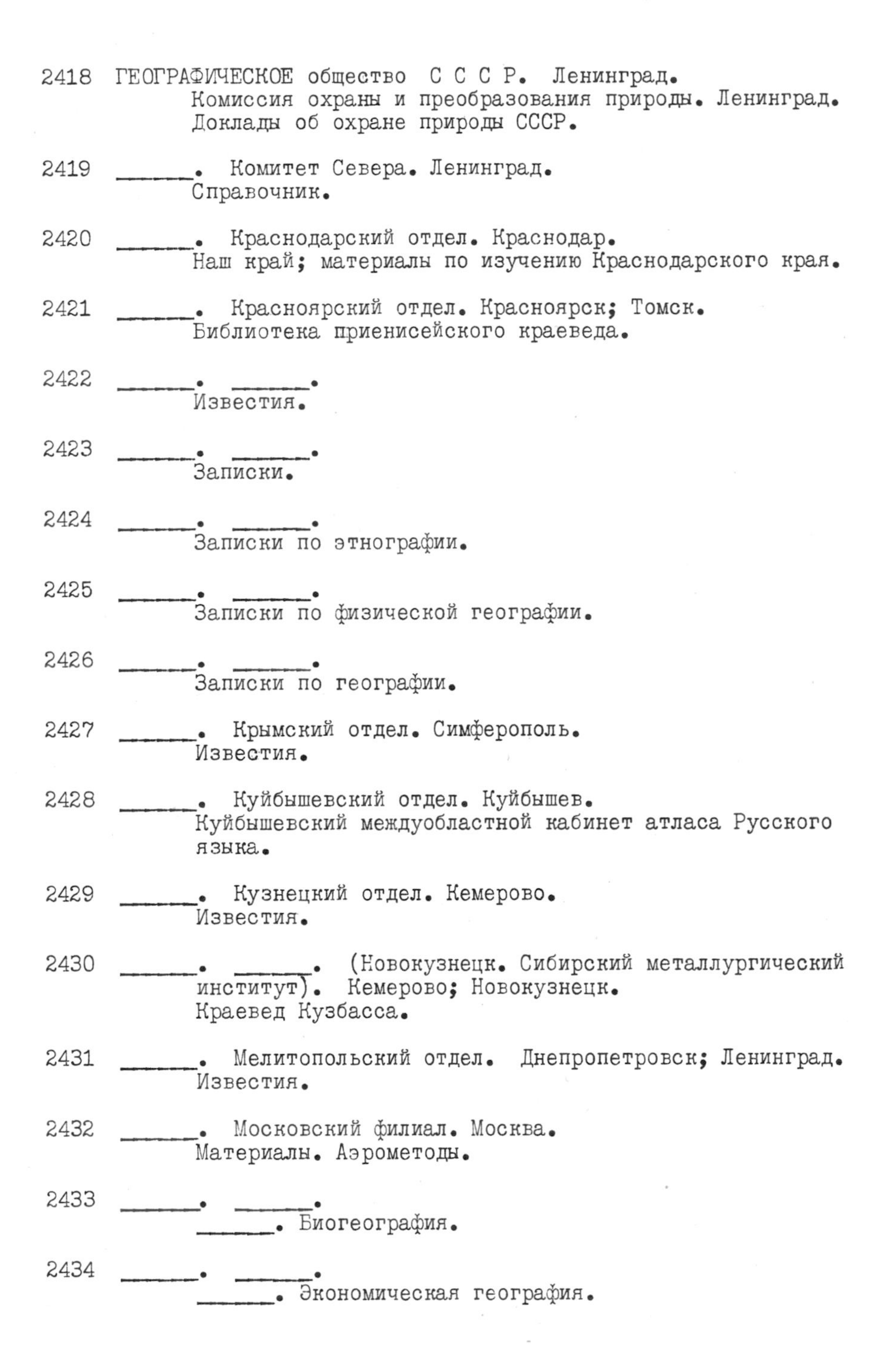

2418 ГЕОГРАФИЧЕСКОЕ общество С С С Р. Ленинград.
Комиссия охраны и преобразования природы. Ленинград.
Доклады об охране природы СССР.

2419 ______. Комитет Севера. Ленинград.
Справочник.

2420 ______. Краснодарский отдел. Краснодар.
Наш край; материалы по изучению Краснодарского края.

2421 ______. Красноярский отдел. Красноярск; Томск.
Библиотека приенисейского краеведа.

2422 ______. ______.
Известия.

2423 ______. ______.
Записки.

2424 ______. ______.
Записки по этнографии.

2425 ______. ______.
Записки по физической географии.

2426 ______. ______.
Записки по географии.

2427 ______. Крымский отдел. Симферополь.
Известия.

2428 ______. Куйбышевский отдел. Куйбышев.
Куйбышевский междуобластной кабинет атласа Русского языка.

2429 ______. Кузнецкий отдел. Кемерово.
Известия.

2430 ______. ______. (Новокузнецк. Сибирский металлургический институт). Кемерово; Новокузнецк.
Краевед Кузбасса.

2431 ______. Мелитопольский отдел. Днепропетровск; Ленинград.
Известия.

2432 ______. Московский филиал. Москва.
Материалы. Аэрометоды.

2433 ______. ______.
______. Биогеография.

2434 ______. ______.
______. Экономическая география.

2435 ГЕОГРАФИЧЕСКОЕ общество С С С Р. Ленинград.
Московский филиал. Москва.
Материалы. Фенология.

2436 ______. ______.
______. Физическая география.

2437 ______. ______.
______. География Москвы и Подмосковья. Краеведение.

2438 ______. ______.
______. География населения.

2439 ______. ______.
______. География Центра.

2440 ______. ______.
______. Геохимия ландшафта.

2441 ______. ______.
______. Геоморфология.

2442 ______. ______.
______. Гидрология.

2443 ______. ______.
______. История географических знаний и историческая география. Этнография.

(2443a ______. ______.
______. Краеведение в Центральном районе.

2444 ______. ______.
______. Медицинская география.

2445 ______. ______.
______. Метеорология и климатология.

2446 ______. ______.
______. Новые методы экономико-географических исследований. География промышленности.

2447 ______. ______.
______. Топонимика.

2448 ______. ______.
______. Учебная география.

2449 ______. Новосибирский отдел. Новосибирск.
Ежегодник.

2450 ______. ______.
Изучай свой край.

2451 ГЕОГРАФИЧЕСКОЕ общество С С С Р. Ленинград.
Новосибирский отдел. Новосибирск.
Известия.

2452 ______. Новозыбковский отдел. Новозыбков.
Известия.

2453 ______. Обнинский отдел. Обнинск.
Доклады комиссий.

2454 ______. ______.
Научные труды.

2455 ______. Омский отдел. Омск.
Известия.

2456 ______. ______.
Медицинская география.

2457 ______. ______.
Записки.

2458 ______. Оренбургский отдел. Оренбург.
Известия.

2459 ______. ______. Чкалов (Оренбург).
Известия.

2460 ______. ______.
Труды.

2461 ______. ______.
Записки.

2462 ______. ______.
Журналы.

2463 ______. Отделение экономической географии. Ленинград.
Материалы по экономической географии зарубежных стран.

2464 ______. Отделение этнографии. Ленинград.
Доклады по этнографии.

2465 ______. ______.
Материалы.

2466 ______. ______.
Записки.

2467 ______. ______. Сказочная комиссия.
Обзор работ.

2468 ______. Отделение физической географии. Ленинград.
Материалы.

2469 ГЕОГРАФИЧЕСКОЕ общество С С С Р. Ленинград.
Отделение физической географии. Ленинград.
Материалы по Арктике и Антарктике.

2470 ______. ______. Комиссия по ландшафтным исследованиям и картографированию.
Доклады.

2471 ______. ______. ______.
Материалы.

2472 ______. Отделение истории географических знаний. Ленинград.
Материалы.

2473 ______. Отделение математической географии и картографии. Ленинград.
Доклады по картографии.

2474 ______. Отделение медицинской географии. Ленинград.
Доклады по медицинской географии.

2475 ______. Отделение учебной географии. Ленинград.
Доклады.

2476 ______. ______.
Материалы.

2477 ______. Пермский отдел. Пермь.
Биогеография и краеведение.

2478 ______. ______.
Доклады.

2479 ______. ______.
Записки.

2480 ______. Постоянная природоохранительная комиссия.
Труды.

2481 ______. Приамурский филиал. Хабаровск.
Отчет.

2482 ______. ______.
Протоколы.

2483 ______. ______.
Труды.

2484 ______. ______.
Записки.

2485 ______. ______.
Журнал.

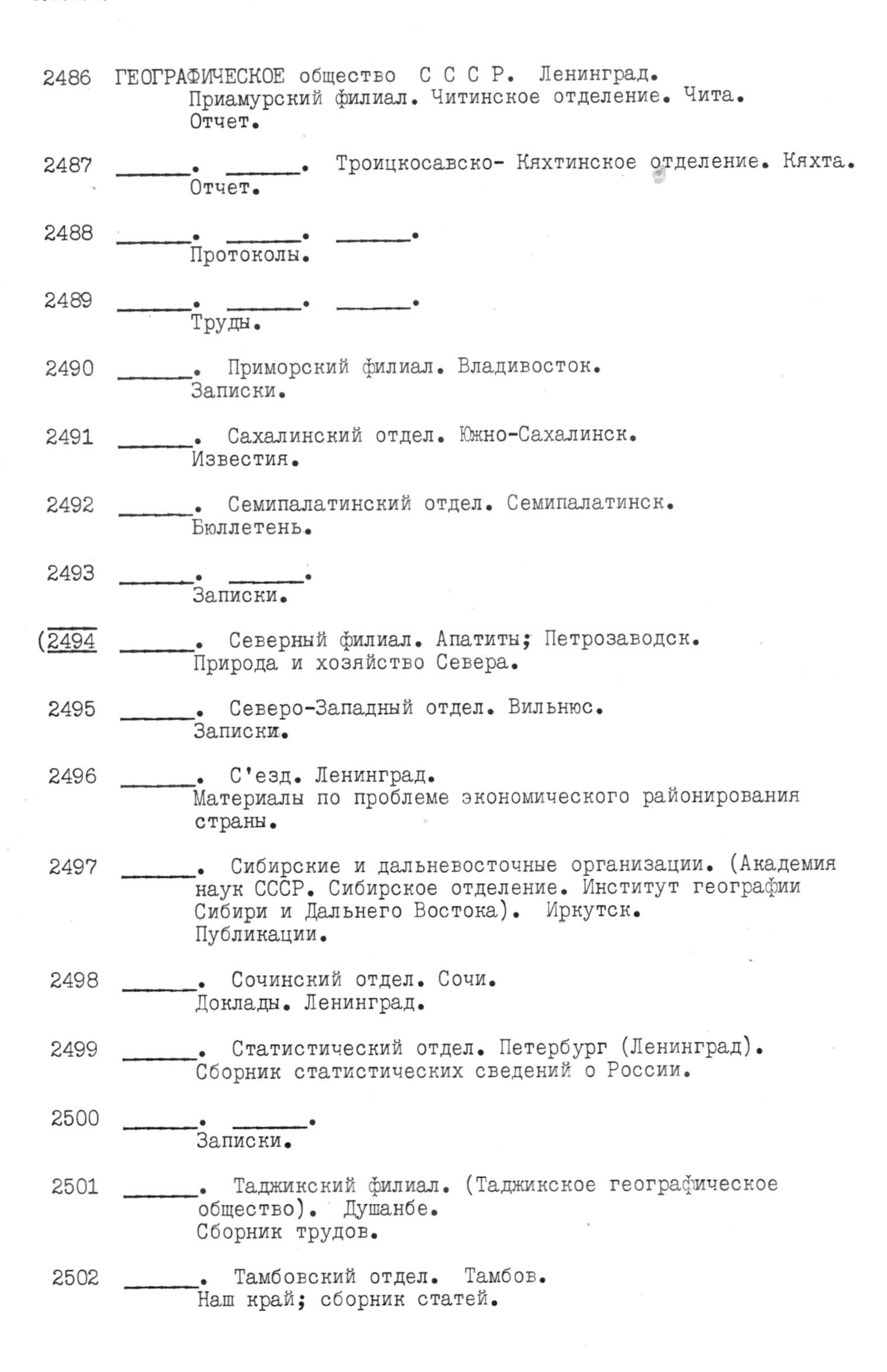

2486 ГЕОГРАФИЧЕСКОЕ общество С С С Р. Ленинград.
Приамурский филиал. Читинское отделение. Чита.
Отчет.

2487 ______. ______. Троицкосавско- Кяхтинское отделение. Кяхта.
Отчет.

2488 ______. ______. ______.
Протоколы.

2489 ______. ______. ______.
Труды.

2490 ______. Приморский филиал. Владивосток.
Записки.

2491 ______. Сахалинский отдел. Южно-Сахалинск.
Известия.

2492 ______. Семипалатинский отдел. Семипалатинск.
Бюллетень.

2493 ______. ______.
Записки.

(2494 ______. Северный филиал. Апатиты; Петрозаводск.
Природа и хозяйство Севера.

2495 ______. Северо-Западный отдел. Вильнюс.
Записки.

2496 ______. С'езд. Ленинград.
Материалы по проблеме экономического районирования страны.

2497 ______. Сибирские и дальневосточные организации. (Академия наук СССР. Сибирское отделение. Институт географии Сибири и Дальнего Востока). Иркутск.
Публикации.

2498 ______. Сочинский отдел. Сочи.
Доклады. Ленинград.

2499 ______. Статистический отдел. Петербург (Ленинград).
Сборник статистических сведений о России.

2500 ______. ______.
Записки.

2501 ______. Таджикский филиал. (Таджикское географическое общество). Душанбе.
Сборник трудов.

2502 ______. Тамбовский отдел. Тамбов.
Наш край; сборник статей.

2503 ГЕОГРАФИЧЕСКОЕ общество С С С Р. Ленинград.
Тюменский отдел. Тюмень.
Известия.

2504 ______. Томский отдел. Томск.
Доклады.

2505 ______. Центрографическая лаборатория им.Д.И.Менделеева.
Информационный бюллетень.

2506 ______. ______.
Центрография.

2507 ______. Ученый архив. Ленинград.
Описание коллекций рукописей научного архива
Географического общества СССР.

2508 ______. Уральский филиал. Свердловск.
Записки.

2509 ______. Владивостокский отдел (прежде Владивостокский
подотдел; ныне - Приморский филиал). Владивосток.
Отчет.

2510 ______. ______. Кружок юных краеведов.
Работы.

2511 ______. Воронежский отдел. Воронеж.
Известия.

2512 ______. ______.
Научные записки.

2513 ______. Восточная комиссия. Ленинград.
Доклады.

2514 ______. Восточно-Сибирская фенологическая комиссия.Иркутск.
Бюллетень.

2515 ______. Восточно-Сибирский отдел. Иркутск.
Бюллетень.

(2516 ______. ______.
Известия.

2517 ______. ______.
Отчет.

2518 ______. ______.
Труды.

2519 ______. ______.
Записки.

2520 ГЕОГРАФИЧЕСКОЕ общество С С С Р. Ленинград.
Восточно-Сибирский отдел. Иркутск.
Отделение этнографии.
Записки.

2521 ______. ______. Отделение статистики.
Записки.

2522 ______. ______. Секция землеведения.
Сборник.

2523 ______. Забайкальский филиал. Чита.
Географические аспекты горного лесоведения и лесоводства.

2524 ______. ______.
Известия.

2525 ______. ______.
Краткий отчет о деятельности.

2526 ______. ______.
Материалы по исследованию Агинской степи.

2527 ______. ______.
Проблемы строительства в условиях Забайкалья.

2528 ______. ______.
Проблемы зимоведения.

2529 ______. ______.
Снежные лавины хребта Удокан. Труды Центра научно-прикладных исследований.

2530 ______. ______.
Вестник научной информации.

2531 ______. ______.
Вопросы медицинской географии и курортологии.

(2532 ______. ______.
Записки.

2533 ______. ______. Агинский отдел. Чита.
Новости Агинских краеведов.

2534 ______. ______. Балейский отдел. Чита.
Труды.

2535 ______. ______. Комиссия охраны природы. Чита.
Охрана природы и воспроизводство естественных ресурсов.

2536 ______. ______. Отделение экономической географии. Чита.
География и хозяйство.

2537 ГЕОГРАФИЧЕСКОЕ общество С С С Р. Ленинград.
Забайкальский филиал. Чита.
Отделение физической географии. Чита.
Климат и гидрография Забайкалья.

2538 ______. ______. ______.
Труды.

2539 ______. ______. Отделение инженерной географии. Чита.
Труды.

2540 ______. ______. Отделение краеведения. Чита.
Проблемы краеведения.

2541 ______. ______.
Забайкальский краеведческий ежегодник.

(2542 ______. Западно-Казахстанский отдел.
Материалы по флоре и растительности Северного Прикаспия. Ленинград.

2543 ______. ______. Уральск.
Научные записки.

2544 ______. Западно-Сибирский отдел. Омск.
Известия.

2545 ______. ______.
Отчет.

2546 ______. ______.
Труды

2547 ______. ______.
Записки.

2548 ______. ______. Алтайский подотдел. Барнаул.
Алтайский сборник.

2549 ______. ______. ______.
Отчет.

2550 ГЕОГРАФИЯ и хозяйство. (Москва. Университет. Географический факультет). Москва.

2551 ГЕОГРАФИЯ и хозяйство Молдавии. (Академия наук Молдавской ССР. Отдел географии). Кишинев.

*(2551a ГЕОГРАФИЯ и природные ресурсы.

2552 ГЕОГРАФИЯ Пермской области. (Географическое общество СССР. Пермский отдел и Пермский университет). Пермь.

2553 ГЕОГРАФИЯ украинских и смежных земель. Львов.

*(2554 ГЕОГРАФИЯ в школе: научно-методический журнал. (Министерство просвещения СССР). Москва.

*(2555 ГЕОГРАФИЧЕСКИЙ ежегодник. (Географическое общество Литовской ССР). Вильнюс.

*(2556 ГЕОМОРФОЛОГИЯ. (Академия наук СССР). Москва.

(2557 ГИДРОГЕОЛОГИЯ и карстоведение. (Географическое общество СССР. Пермский отдел). Пермь.

(2558 ГЛЯЦИОЛОГИЧЕСКИЕ исследования. (Академия наук СССР. Междуведомственный геофизический комитет). Москва.

(2559 Гляциология Алтая. (Географическое общество СССР. Томский отдел. Томский университет). Томск.

(2560 ГЛОБУС. Географический ежегодник для детей. Ленинград.

2561 ГРОЗНЫЙ. Чечено-Ингушский государственный педагогический институт. Грозный.
Ученые записки. Серия естественно-географическая.

2562 ГЕОГРАФИЧЕСКИЕ исследования на Украине. (Украинское географическое общество). Киев.

2563 ГЕОГРАФИЧЕСКОЕ общество Украинской ССР. Киев.
Доклады и сообщения секций и комиссий.

2564 ______. Днепропетровский отдел. Днепропетровск.
Известия.

2565 ______. Харьковский отдел. Харьков.
Известия.

2566 ______. ______.
Материалы.

2567 ______. ______.
Природные и трудовые ресурсы Левобережья Украины и их использование.

(2568 ______. Львовский отдел. Львов.
Доклады и сообщения.

2569 ГЕОГРАФИЧЕСКИЙ сборник. (Академия наук Украинской ССР. Географическое общество УССР). Киев.

2570 ГЕОГРАФИЧЕСКИЙ сборник. (Львовский университет и Львовский отдел Географического общества УССР). Львов.

2571 ГЕОГРАФИЯ в школе: методический сборник. (Украинский научно-исследовательский институт педагогики). Киев.

2572 ЯКУТСКОЕ краевое географическое общество. Якутск.
Записки.

2573 ИРКУТСК. Университет. Иркутск.
Труды. Серия географическая.

2574 ______. ______. Биолого-географический научно-исследовательский институт. Иркутск.
Известия.

2575 ИСТОРИЧЕСКИЙ, статистический и географический журнал.Москва.

2576 ИТОГИ науки. (Академия наук СССР. Институт научной информации). Москва.
Экономгеографическая изученность районов капиталистического мира.

2577 ______.
Гидрология суши. Гляциология.

2578 ______.
Теоретические вопросы географии.

(2579 ИТОГИ науки и техники. (Всесоюзный институт научной и технической информации). Москва.
Биогеография.

(2580 ______.
Геодезия и аэрос'емка.

*(2581 ______.
География СССР.

(2582 ______.
География зарубежных стран.

(2583 ______.
Геоморфология.

(2584 ______.
Картография.

(2585 ______.
Медицинская география.

(2586 ______.
Океанология.

(2587 ______.
Охрана природы и воспроизводство природных ресурсов.

(2588 ______.
Теоретические и общие вопросы географии.

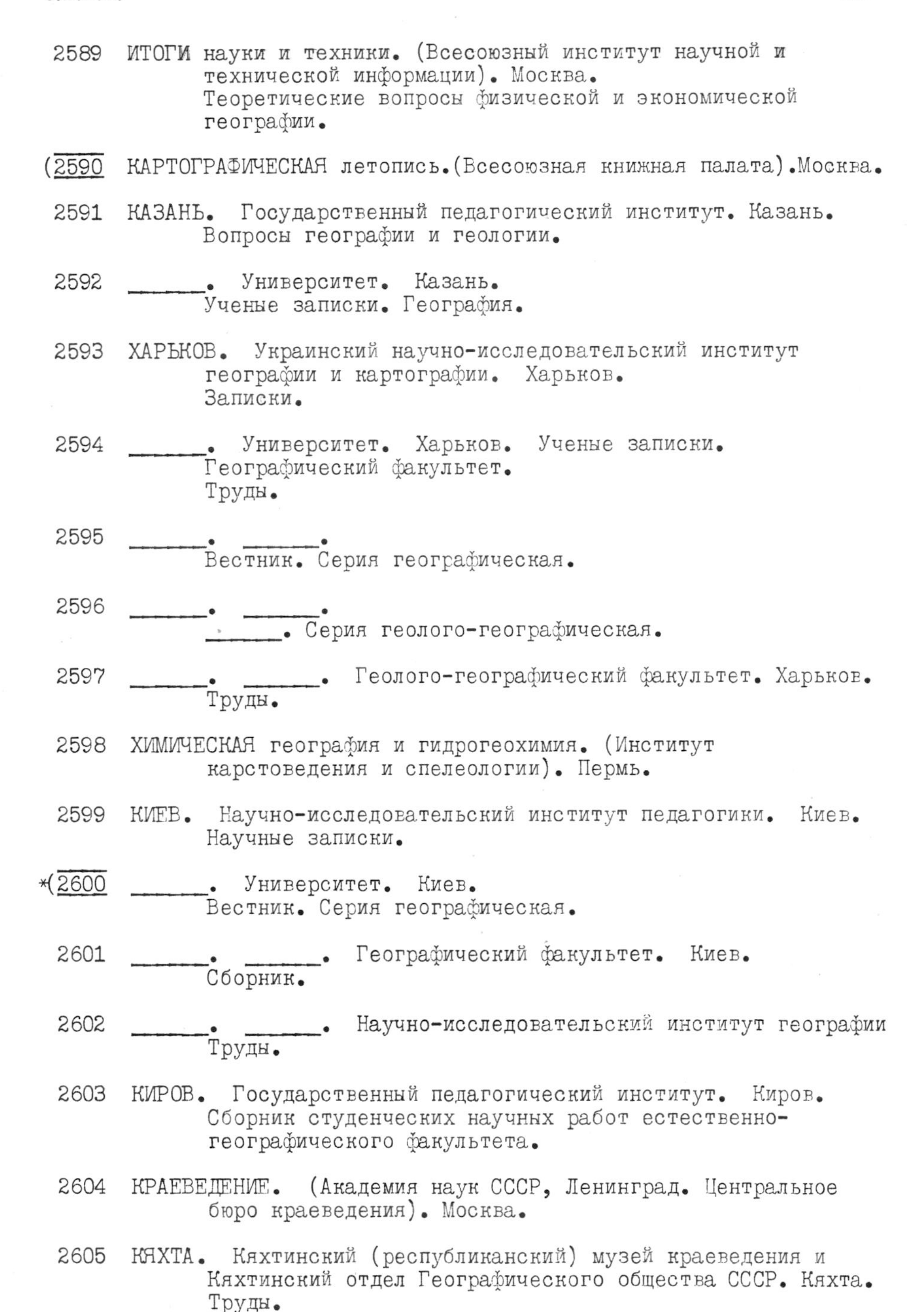

2589 ИТОГИ науки и техники. (Всесоюзный институт научной и технической информации). Москва.
Теоретические вопросы физической и экономической географии.

(2590 КАРТОГРАФИЧЕСКАЯ летопись.(Всесоюзная книжная палата).Москва.

2591 КАЗАНЬ. Государственный педагогический институт. Казань.
Вопросы географии и геологии.

2592 ______. Университет. Казань.
Ученые записки. География.

2593 ХАРЬКОВ. Украинский научно-исследовательский институт географии и картографии. Харьков.
Записки.

2594 ______. Университет. Харьков. Ученые записки.
Географический факультет.
Труды.

2595 ______. ______.
Вестник. Серия географическая.

2596 ______. ______.
______. Серия геолого-географическая.

2597 ______. ______. Геолого-географический факультет. Харьков.
Труды.

2598 ХИМИЧЕСКАЯ география и гидрогеохимия. (Институт карстоведения и спелеологии). Пермь.

2599 КИЕВ. Научно-исследовательский институт педагогики. Киев.
Научные записки.

*(2600 ______. Университет. Киев.
Вестник. Серия географическая.

2601 ______. ______. Географический факультет. Киев.
Сборник.

2602 ______. ______. Научно-исследовательский институт географии
Труды.

2603 КИРОВ. Государственный педагогический институт. Киров.
Сборник студенческих научных работ естественно-географического факультета.

2604 КРАЕВЕДЕНИЕ. (Академия наук СССР, Ленинград. Центральное бюро краеведения). Москва.

2605 КЯХТА. Кяхтинский (республиканский) музей краеведения и Кяхтинский отдел Географического общества СССР. Кяхта.
Труды.

2606 ЛАТВИЙСКОЕ географическое общество. Рига.
Географические записки.

2607 АКАДЕМИЯ наук Латвийской ССР. Геологический и географический институт. Рига.
Записки.

2608 ЛЕНИНГРАД. Географический институт. Ленинград.
Бюллетень.

2609 ______. ______.
Известия.

2610 ______. ______.
Труды.

2611 ______. Ленинградский государственный педагогический институт имени А.И.Герцена. Ленинград.
Герценовские чтения: география и геология.

2612 ______. Ленинградский государственный педагогический институт имени М.Н.Покровского. Ленинград.
Ученые записки. Естественно-географический факультет. Серия географическая.

(2613 ______. Университет. Ленинград.
Ученые записки. Серия географических наук.

2614 ______. ______.
______. Серия геолого-почвенно-географическая.

2615 ______. ______.
Вестник. Биология, география и геология.

*(2616 ______. ______.
______. Геология, география.

2617 ______. ______. Географо-экономический научно-исследовательский институт.
Сборник.

2618 ______. ______. ______.
Труды.

2619 ______. ______. Кабинет физической географии.
Труды.

2620 ______. ______. Научная сессия.
Тезисы докладов по секции географических наук.

2621 ______. ______. Общество землеведения. Петербург.
Труды.

(2622 ЛЕТОПИСЬ Севера. Сборник по вопросам исторической географии, истории географического открытия, исследования и экономического развития Севера. (СОПС). Москва.

(2623 НАУЧНЫЕ труды высших учебных заведений Литовской ССР: география и геология. Вильнюс.

2624 ФИЗИЧЕСКАЯ география Литовской ССР. (Академия наук Литовской ССР. Геологический и географический институт).Вильнюс.

(2625 АКАДЕМИЯ наук Литовской ССР. Вильнюс. Труды. Серия Б. Химия,техника,физическая география.

2626 ______. Отдел географии. Научные сообщения.

2627 ______. Геологический и географический институт. Монографическая серия.

2628 ЛИТЕРАТУРА по русской географии, статистике и этнографии. Петербург (Ленинград).

2629 ЛЬВОВ. Университет. Львов. Вестник. Серия биологическая, географическая и геологическая.

(2630 ______. ______. ______. Серия географическая.

2631 МАГАЗИН для русской истории, природных и этнографических сведений. Елгава (Митава).

2632 МАГАЗИН землеведения и путешествий, географический сборник. Москва.

2633 МАТЕРИАЛЫ для географии и статистики России. (Собранные офицерами Генерального штаба). Петербург (Ленинград).

(2634 МАТЕРИАЛЫ экспериментальных исследований на стационарах. (Академия наук СССР. Сибирское отделение. Институт географии Сибири и Дальнего Востока). Иркутск.

(2635 МАТЕРИАЛЫ гляциологических исследований: хроника,обсуждения. (Академия наук СССР. Междуведомственный геофизический комитет. Секция гляциологии. Институт географии). Москва.

(2636 МАТЕРИАЛЫ к биобиблиографии ученых СССР. Серия географических наук. (Академия наук СССР). Москва.

2637 МАТЕРИАЛЫ к истории географических исследований Сибири и Дальнего Востока. Иркутск.

2638 МАТЕРИАЛЫ по четвертичной геологии и геоморфологии СССР. Ленинград.

(2639 МАТЕРИАЛЫ по изучению Ставропольского края. (Ставропольский краеведческий музей и с 1960 Ставропольский отдел Географического общества СССР). Ставрополь.

2640 МАТЕРИАЛЫ по палеогеографии. (Москва. Университет. Географический факультет).

2641 ШКОЛЬНАЯ география. Журнал преподавателей начальной и средней школы. Казань.

2642 МЕЖДУВЕДОМСТВЕННОЕ совещание по географии населения. Москва; Ленинград.
Материалы.

2643 МИНСК. Университет. Географический факультет. Минск.
Труды.

(2644 ______. ______.
Вестник. Серия 2. Химия, биология, геология, география.

2646 МОСКОВСКОЕ общество испытателей природы. Москва.
Труды. Секция географии.

2647 МОСКВА. Институт международных отношений. Кафедра экономической географии и экономики стран Запада.
Ученые записки.

2648 ______. ______. Кафедра географии и экономики стран Востока.
Ученые записки.

2649 ______. Московский городской педагогический институт имени В.П.Потемкина. Москва.
Ученые записки. Географический факультет.

2650 ______. Московский государственный педагогический институт имени В.И.Ленина. Москва.
Ученые записки. Географический факультет.

(2651 ______. Московский институт инженеров геодезии, аэрофотос'емки и картографии. Москва.
Труды.

2652 ______. Московский областной педагогический институт имени Н.К.Крупской. Москва.
Ученые записки. Экономическая география.

2653 ______. ______.
______. Общая физическая география и геология.

(2654 ______. Центральный научно-исследовательский институт геодезии, аэрос'емки и картографии. Москва.
Труды.

2655 ______. Университет. Москва.
Ученые записки. География.

2656 ______. ______.
Вестник. Серия биологии, почвоведения, геологии, географии.

*(2657 МОСКВА. Университет. Москва.
Вестник. Серия 5. География.

2658 ______. ______. Географическая станция "Красновидово".
Труды.

2659 ______. ______. Географический факультет. Москва.
Комплексные географические исследования территории Северного Казахстана.

2660 ______. ______. Хибинская географическая станция. Москва.
Труды.

2661 ______. ______. Научно-исследовательский институт географии.
Труды.

2662 НА суше и на море. (Государственное издательство географической литературы). Москва.

2663 НАША страна. Москва.

2664 НАУЧНОЕ совещание географов Сибири и Дальнего Востока. (Академия наук СССР. Сибирское отделение. Институт географии Сибири и Дальнего Востока; Географическое общество СССР. Бюро сибирских и дальневосточных организаций. Новосибирский отдел). Иркутск.
Материалы.

2665 НАУЧНОЕ совещание по проблемам медицинской географии. (Академия наук СССР и Географическое общество СССР). Ленинград.
Доклады.

2666 НАУЧНЫЕ доклады высшей школы: геолого-географические науки. (Министерство высшего образования СССР). Москва.

2667 НАУЧНОЕ общество имени Шевченко. Географическая комиссия. Львов.
Работы. Сообщения.

2668 НОВОЕ в жизни, науке, технике: Серия наук о земле. Общество "Знание" (Всесоюзное общество по распространению политических и научных знаний). Москва.

(2669 НОВОСИБИРСК. Новосибирский институт инженеров геодезии, аэрофотос'емки и картографии. Новосибирск.
Труды.

2670 НОВОСТИ карстоведения и спелеологии. (Академия наук СССР. Междуведомственная комиссия по изучению геологии и географии карста). Москва.

2671 ОБЩЕСТВО изучения Казахстана. Оренбург.
Труды. Алма-Ата.

2672 ОБЩЕСТВО изучения Казахстана. Оренбург.
Семипалатинский отдел. Семипалатинск.
Записки.

2673 ОБЩЕСТВО любителей естествознания, антропологии и этнографии.
Географическое отделение. Москва.
Мемуары.

2674 ______. ______.
Труды.

2675 ОДЕССА. Университет. Одесса.
Научный ежегодник. Географический факультет.

2676 ______. ______.
Труды. Серия геолого-географических наук.

2677 ОХРАНА природы Молдавии. (Академия наук Молдавской ССР.
Отдел географии. Комиссия по охране природы). Кишинев.

2678 ПЕЩЕРЫ. (Географическое общество СССР. Пермский отдел.
Пермский университет. Институт карстоведения и
спелеологии).

2679 ПРИРОДА Сахалина и здоровье человека: Известия Сахалинского
отдела Географического общества СССР. Южно-Сахалинск.

2680 ПРОБЛЕМЫ физической географии. (Академия наук СССР.
Институт географии). Москва; Ленинград.

(2681 ПРОБЛЕМЫ географии Молдавии. (Академия наук Молдавской ССР.
Отдел географии. Географическое общество Молдавии).
Кишинев.

(2682 ПРОБЛЕМЫ географической науки в Украинской ССР. (Академия
наук Украинской ССР. Сектор географии). Киев.

2683 ПРОБЛЕМЫ криолитологии. (Москва. Университет.
Географический факультет. Кафедра криолитологии и
гляциологии). Москва.

2683a ПРОБЛЕМЫ медицинской географии Казахстана. (Казахское
географическое общество). Алма-Ата.

*(2684 ПРОБЛЕМЫ освоения пустынь. (Академия наук Туркменской ССР.
Научный совет по проблеме пустынь. Институт пустынь).
Ашхабад.

2685 ГЛАВНОЕ управление школ РСФСР. Москва.
Указания об использовании учебников. История,география.

*(2686 РЕФЕРАТИВНЫЙ журнал. География. (Всесоюзный институт
научной и технической информации). Москва.

2687 ______. Геология и география. (Академия наук СССР. Институт
научной информации). Москва.

2688 РЕФЕРАТИВНЫЙ журнал. Отдельный выпуск 35. Краеведение. (Академия наук СССР. Институт научной информации). Москва.

2689 РЕГИОНАЛЬНЫЙ географический прогноз. (Москва. Университет. Географический факультет). Москва.

2690 РИГА. Университет. Рига.
Географические науки.

2691 ______. ______.
Ученые записки аспирантов. Сборник работ аспирантов Географического факультета.

2692 ______. ______. Кафедра физической географии.
Вопросы физической географии Латвийской ССР.

2693 РИТМЫ природы Сибири и Дальнего Востока. (Академия наук СССР. Сибирское отделение. Институт географии Сибири и Дальнего Востока и Географическое общество СССР. Бюро сибирских и дальневосточных организаций). Иркутск.

2694 РОСТОВ-НА-ДОНУ. Университет. Ростов-на-Дону.
Ученые записки.

2695 СБОРНИК географических, топографических и статистических материалов по Азии. Петербург (Ленинград).

2696 СБОРНИК по исторической географии Грузии. (Академия наук Грузинской ССР. Институт истории, археологии и этнографии). Тбилиси.

2697 СБОРНИК статей по картографии. (Главное управление геодезии и картографии). Москва.

2698 СЕВЕРНЫЙ Кавказ. (Ставропольский государственный педагогический институт и Ставропольский отдел Географического общества СССР). Ставрополь.

(2699 СЕВЕРО-ЗАПАД Европейской части СССР. (Ленинград.Университет). Ленинград.

2700 СИБИРСКАЯ живая старина. (Географическое общество СССР. Восточно-Сибирский отдел). Иркутск.

*(2701 СИБИРСКИЙ географический сборник. (Академия наук СССР. Сибирское отделение. Институт географии Сибири и Дальнего Востока и Географическое общество СССР. Бюро сибирских и дальневосточных организаций). Москва; Новосибирск.

2702 СИСТЕМАТИЧЕСКИЙ указатель статей в иностранных журналах. Геология, геофизика и география. (Всесоюзная государственная библиотека иностранной литературы). Москва.

2703 СОВЕТСКОЕ краеведение. (Центральное бюро краеведения и Коммунистическая академия. Краеведческая секция). Москва.

2704 СРЕДНЕ-АЗИАТСКОЕ географическое общество. Ташкент. Известия.

2705 ______. Отчет.

2706 СТАВРОПОЛЬСКОЕ общество для изучения Северо-Кавказского края в естественно-историческом, географическом и антропологическом отношениях. Петербург, Киев. Труды.

2707 СТРАНЫ Азии. Географические справки. Москва.

*(2708 СТРАНЫ и народы Востока. (Географическое общество СССР. Восточная комиссия). Москва.

2709 СВЕРДЛОВСКИЙ государственный педагогический институт и Географическое общество СССР. Свердловский отдел. Свердловск. Вопросы физической и экономической географии.

2710 ТАМБОВ. Государственный педагогический институт. Тамбов. Научные работы студентов. География, биология.

(2711 ТАРТУ. Университет. Тарту. Труды по географии.

2712 ______. ______. Ученые записки. Геология и география.

2713 ______. ______. Институт географии. Тарту. Публикации.

2714 ______. ______. Географический семинар. Тарту. Публикации.

2715 ТАШКЕНТ. Средне-Азиатский университет. Ташкент. Научные труды. Географические науки.

2716 ______. ______. Труды.

2717 ______. ______. Географический факультет. Научно-исследовательский отдел. Труды.

2718 ТБИЛИСИ. Университет. Тбилиси. Труды. Серия геолого-географических наук.

2719 ______. ______. Географический институт. Тбилиси. Труды.

2720 ТЕМАТИЧЕСКИЙ сборник научных работ по биологии и географии. (Министерство высшего и среднего специального образования Казахской ССР). Алма-Ата.

2721 ТОМСК. Институт исследования Сибири. Географический отдел. Томск.
Труды.

2722 ТУРКМЕНСКОЕ географическое общество. Ашхабад.
Труды.

2723 УФА. Башкирский государственный педагогический институт.
Ученые записки. Серия геолого-географическая.

2724 ______. Башкирский государственный университет. Уфа.
Серия географических наук.

2725 УКРАИНСКИЙ историко-географический сборник. (Институт истории). Киев.

2726 УЗБЕКИСТАНСКОЕ географическое общество. Ташкент.
Известия.

2727 ______.
Труды.

2728 В ПОМОЩЬ учителю географии. (Калмыцкий республиканский институт усовершенствования учителей). Элиста.

2729 В ПОМОЩЬ учителю географии; методическое пособие. (Новосибирский государственный педагогический институт). Новосибирск.

2730 В ПОМОЩЬ учителю географии средней школы и неполной средней школы. (Омский областной институт усовершенстования учителей. Кабинет географии). Омск.

2731 ВИЛЬНЮС. Университет. Вильнюс.
Биология, география, геология.

2732 ВЛАДИВОСТОК. Дальневосточный государственный университет.
Труды. Серия 9. География.

(2733 ВОКРУГ света. Москва.

2734 ВОЛОГОДСКИЙ край. (Географическое общество СССР. Вологодский отдел). Вологда.

2735 ВОПРОСЫ экономической географии Урала и Западной Сибири. (Пермский университет и Пермский отдел Географического общества СССР). Пермь.

2736 ВОПРОСЫ физической географии. (Географическое общество СССР. Саратовский отдел). Саратов.

2736a ВОПРОСЫ физической географии и геоморфологии нижнего Поволжья. (Саратовский университет). Саратов.

2737 ВОПРОСЫ физической географии Урала. Пермь.

2738 ВОПРОСЫ географической патологии; сборник статей. (Якутский университет. Медицинский факультет). Якутск.

*(2739 ВОПРОСЫ географии. (Географическое общество СССР. Московский филиал. Научные сборники). Москва.

2740 ВОПРОСЫ географии Белоруссии. (Географическое общество Белорусской ССР). Минск.

(2741 ВОПРОСЫ географии Дальнего Востока. (Географическое общество СССР. Приамурский филиал). Хабаровск.

2742 ВОПРОСЫ географии и геологии. (Казанский педагогический институт). Казань.

2743 ВОПРОСЫ географии и картографии; сборник. (Москва. Научно-издательский институт Большого советского атласа мира). Москва.

2744 ВОПРОСЫ географии Якутии. (Академия наук СССР. Сибирское отделение. Якутский филиал и Географическое общество СССР. Якутский филиал). Якутск.

2745 ВОПРОСЫ географии Южного Урала. (Челябинский государственный педагогический институт). Челябинск.

(2746 ВОПРОСЫ географии Камчатки. (Географическое общество СССР. Камчатский отдел). Петропавловск-Камчатский.

2747 ВОПРОСЫ географии Казахстана. (Академия наук Казахской ССР. Сектор физической географии и Географическое общество Казахской ССР). Алма-Ата.

2748 ВОПРОСЫ географии Кузбасса и Горного Алтая. (Кемеровский государственный педагогический институт).Междуреченск; Кемерово; Новокузнецк.

2749 ВОПРОСЫ географии Северного Прикаспия. (Географическое общество СССР. Западно-Казахстанский отдел). Ленинград.

(2750 ВОПРОСЫ географии Сибири. (Географическое общество СССР. Томский отдел и Томский университет). Томск.

2751 ВОПРОСЫ географии Туркменистана. (Географическое общество Туркменской ССР. Академия наук Туркменской ССР). Ашхабад.

2752 ВОПРОСЫ геоморфологии и геологии Башкирии. (Башкирский филиал АН СССР.Горно-геологический институт). Уфа.

2753 ВОПРОСЫ истории социально-экономической и культурной жизни Сибири и Дальнего Востока. (Академия наук СССР. Сибирское отделение. Институт истории, филологии и философии и Географическое общество СССР.Новосибирское отделение.Секция исторической географии). Новосибирск.

2754 ВОПРОСЫ медицинской географии Западной Сибири.(Географическое общество СССР.Новосибирский отдел). Новосибирск.

2755 ВСЕМИРНЫЙ путешественник. Иллюстрированный журнал путешествий и географических открытий. Петербург.

2756 ВСЕСОЮЗНОЕ общество по распространению политических и научных знаний. Москва.
(Издания). Серия 12. Геология и география.

2757 ВСЕСОЮЗНОЕ совещание по вопросам ландшафтоведения. Москва.
Тексты докладов; материалы к совещанию.

2758 ВСЕСОЮЗНЫЙ географический с'езд. Ленинград.
Труды.

2759 ЯКУТСК. Университет. Биолого-географический факультет. Якутск.
Труды.

2760 ЕРЕВАН. Армянский педагогический институт. Ереван.
Научные труды. Серия географических наук.

2761 ______. Университет. Ереван.
Серия географических наук.

2762 ______. ______.
Серия геолого-географических наук.

*(2763 ЗЕМЛЕВЕДЕНИЕ. (Московское общество испытателей природы). Москва.

2764 ЗЕМЛЕВЕДЕНИЕ; географический журнал. (Общество любителей естествознания,антропологии и этнографии. Географическое отделение). Москва.

(2765 ЗЕМЛЯ и люди. Географический календарь. Москва.

2766 ЗЕМЛЯ и люди. Научно-популярный географический сборник. (Академия наук Узбекской ССР). Ташкент.

2767 ЖИВАЯ старина.(Императорское русское географическое общество. Отделение этнографии). Петербург (Ленинград).

(2768 ЖИЗНЬ земли: сборник.(Москва.Университет.Музей землеведения). Москва.

2769 ЗОНАЛЬНОЕ совещание представителей кафедр географии педагогических институтов Сибири. Новосибирск.
Труды.

УКРАЇНСЬКА РАДЯНСЬКА СОЦІАЛІСТИЧНА РЕСПУБЛІКА

2321 АКАДЕМІЯ наук УРСР. Київ. Комісія для складання історично-географічного словника України. Історично-геграфічний збірник.

2322 ______. Природничо-технічний відділ. Журнал геолого-географічного циклу.

2337 ЧЕРНІВЦІ. Університет. Чернівці. Наукові записки. Серія географічних наук.

2338 ______. ______. Наукові записки. Серія геолого-географічних наук.

2341 ДНІПРОПЕТРОВСЬК. Університет. Геолого-географічний факультет. Збірник праць.

*(2349 ЕКОНОМІЧНА географія; Міжвідомчий науковий збірник. Київ.

*(2354 ФІЗИЧНА географія; Міжвідомчий науковий збірник. Київ.

2396 ГЕОГРАФІЧНЕ Товариство СРСР. Чернівецький відділ. Чернівці. Вісті.

2562 ГЕОГРАФІЧНІ дослідження на Україні. (Академія наук УРСР. Географічне товариство УРСР). Київ.

2563 ГЕОГРАФІЧНЕ товариство Української РСР. Київ. Доповіді секцій та комісій.

2569 ГЕОГРАФІЧНИЙ збірник. (Академія наук УРСР та Географічне товариство УРСР). Київ.

2570 ГЕОГРАФІЧНИЙ збірник. (Львів. Університет та Географічне товариство УРСР. Львівський відділ). Львів.

2593 ХАРКІВ. Український науково-дослідний інститут географії та картографії. Харків. Записки.

2594 ______. Університет. Наукові записки. Географічний факультет.

2597 ______. ______. Праці.

2599 КИЇВ. Науково-дослідний інститут педагогіки. Київ. Наукові записки. Серія хімічна та географічна.

*(2600 КИЇВ. Університет. Київ.
Вісник. Серія географії.

2601 _____. _____. Географічний факультет. Київ.
Збірник.

2602 _____. _____. Науково-дослідний інститут географії. Київ.
Праці.

2629 ЛЬВІВ. Університет. Львів.
Вісник. Серія біологічна, географічна та геологічна.

(2630 _____. _____.
_____. Серія географічна.

2667 НАУКОВЕ товариство імені Шевченка. Географічна комісія.
Львів.
Праці.

2676 ОДЕСА. Університет. Одеса.
Праці. Серія геолого-географічних наук.

(2682 ПРОБЛЕМИ географічної науки в Українській РСР. Київ.

2725 Український історико-географічний збірник. (Інститут
історії). Київ.

Азәрбајҹан ССР

(2300 АЗӘРБАЈҸАН ССР ЕЛМЛӘР АКАДЕМИЈАСЫНЫН ХӘБӘРЛӘРИ
ЈЕР ЕЛМЛӘРИ СЕРИЈАСЫ

(2325 АЗӘРБАЈҸАН ҸОҒРАФИЈА ҸӘМИЈЈӘТИНИН ӘСӘРЛӘРИ

საქართველოს სსრ

2303 საქართველოს სსრ მეცნ. აკადემია. — ვახუშტის სახ. გეოგრაფიის ინსტ. შრ.,

(2367 საქ. სსრ გეოგრაფიული საზოგადოების შრომები,

ՀԱՅԿԱԿԱՆ ՍՍՀ

(2299 ՀԱՅԿԱԿԱՆ ՍՍՀ ԳԻՏՈՒԹՅՈՒՆՆԵՐԻ ԱԿԱԴԵՄԻԱՅԻ ՏԵՂԵԿԱԳԻՐ
ԳԻՏՈՒԹՅՈՒՆՆԵՐ ԵՐԿՐԻ ՄԱՍԻՆ

U.S.S.R.

Union of Soviet Socialist Republics (Russian and 14 other languages of union republics)

(Soyuz Sovetskikh Sotsialisticheskikh Respublik)

Note: Entries for irregular serials in the U.S.S.R. have been checked against official Soviet bibliographies of serials published (see Sources 103-121, p. 24-25) and numbers published recorded in this list; such serials are not to be regarded as closed unless so stated.

(2299 AKADEMIIA nauk Armianskoi S.S.R. Armenian: Haykakam S.S.H. gitowt'yownneri akademia. Yerevan.
Izvestiia. Nauki o zemle. v10- (1957-) 6 nos. a year. (N70-1:161ab; BS:36a; B68:283a; S200; FC-S:175; GR-1:313-314).
Supersedes in part academy's Izvestiia. Fiziko-matematicheskie, estestvennye i tekhnicheskie nauki (not in this list) and continues its numbering. v10 (1957) as Izvestiia. Geologicheskie i geograficheskie nauki; v11-16 (1958-1963) as Izvestiia. Seriia geologicheskikh i geograficheskikh nauk.
In Russian with Armenian summaries.

(2300 AKADEMIIA nauk Azerbaidzhanskoi S.S.R. Azerbaijani: Azerbajdzhan S.S.R. elmler akademijasyny. Baku.
Izvestiia. Seriia nauk o zemle. Kheberleri. Jer elmleri. 1958- Bimonthly. (N70-1:162b; BS:36a; B73:163a; S225, 226; FC-S:174; GR-1:314).
Supersedes in part the academy's Izvestiia (not in this list). 1958-1965 as Izvestiia, Kheberleri. Seriia geologo-geograficheskikh nauk [1961-1963, ... i nefti]. Keolokiia-dzhografiia elmleri seriiasy.
In Russian. Summaries in Azerbaijani. Summaries also in English from 1973, no2.

2301 ______. Institut geografii.
Materialy nauchnoi konferentsi. 1- (1962-) Annual. (N75-1:59b).
1971- with added title page in Azerbaijani.
In Azerbaijani or Russian.

(2302 ______. ______.
Trudy. 1- (1948-) Irregular. (U-1:110a; S235; FC-S:395).
Not issued in order. Title varies. Added title page in Azerbaijani: Dzhografija institutunun eserleri.
In Russian or Azerbaijani with summaries in other language and in English.

2303 AKADEMIIA nauk Gruzinskoi S.S.R. Institut geografii imeni Vakhushti. Tbilisi.
Trudy. 1-20 (1947-1964). Closed 1964. (S276; FC-S:395; GR-3:1830).
In Georgian. Russian summaries.

2304 AKADEMIIA nauk Kazakhskoi S.S.R. Alma-Ata.
Izvestiia. Seriia geograficheskaia. 1-3 (1948-1951). (S326).
v1 (1948) is v57 of parent series; 2 (1948), 58; 3 (1961), 112.

2305 ______.
Sektor fizicheskoi geografii. Alma-Ata.
Gliatsiologicheskie issledovaniia v Kazakhstane. 1-9 (1961-1971). (B73:131a).
1-3 (1961-1963) issued by Academy's Otdel geografii.
v1-4 (1961-1964) as Gliatsiologicheskie issledovaniia v period MGG.
In Russian. Table of contents and abstracts in English.

2306 AKADEMIIA nauk Kirgizskoi S.S.R. Otdel geografii. Frunze.
Trudy. 1 (1958). Only one issued. (N70-1:166c; S414).
v1 (1958) issued jointly as academy's Tian'shanskaia fiziko-geograficheskaia stantsiia. Raboty [2307]. After v1 absorbed by academy's Izvestiia (not in this list).

2307 ______. Tian'shanskaia fiziko-geograficheskaia stantsiia. Frunze.
Raboty. 1-12 (1958-1966). Frunze. (N70-1:166c; S418).
Title varies. 1 (1958) as Trudy (Akademiia nauk Kirgizskoi S.S.R. Otdel geografii [and] Tian'-Shanskaia fiziko-geograficheskaia stantsiia). v2-4 (1960-1961) and 8-10 (1964) as Materialy gliatsiologicheskikh issledovanii. Variant name of station: Tian'shanskaia vysokogornaia fiziko-geograficheskaia stantsiia.
Predecessor: Tian'-Shanskaia fiziko-geograficheskaia stantsiia. Raboty. 1-7 (1950-1962), a subseries of Akademiia nauk SSSR. Institut geografii. Trudy [2313].

*(2308 AKADEMIIA nauk S.S.S.R. Moskva. 1725-1914 as Petersburgskaia akademiia nauk; 1915-1925 Rossiiskaia akademiia nauk; 1925- Akademiia nauk S.S.S.R.
Izvestiia. Seriia geograficheskaia. 1951- 6 nos. a year. (N60-1:57b; N68-1:74c [not in N70]; B-1:18b; BS:8a; S443; FC-S:184, 185; GR-1:322).
Supersedes in part Seriia geograficheskaia i geofizicheskaia [2309].
Index: 1951-1966.
Supplementary English table of contents from 1957 no1.

2309 ______. Leningrad; later Moskva.
Izvestiia. s7. Seriia geograficheskaia i geofizicheskaia. 1-15 no1 (1937-F 1951). Closed 1951. (U-1:112b; B-1:18a; S444; FC-1:281; FC-S:184; GR-1:322).
Superseded in part by Seriia geograficheskaia [2308].
Index: 1937-1951 for geographical articles.

2310 ______. Institut geografii. Moskva.
Geograficheskie soobshcheniia. 1-3 (1959-1966) Irregular. (N70-2:2278a; S4264).
v1 unnumbered.

2311 ______. ______.
Issledovaniia lednikovykh raionov. 1-3 (1961-1963).

AKADEMIIA nauk S.S.S.R. Institut geografii.

2312 ______. ______.
Materialy gliatsiologicheskikh issledovanii. Moskva.
1960-1964 (N70-1:172a; B73; 205a; S 613-619).
Consists of many series:
______. Khronika obsuzhdenii. 1- (1961-). See separate entry under title [2635].
______. Novaia Zemlia: meteorologiia. 1-4 (1961).
______. Novaia Zemlia: snezhnyi pokrov. 1-3 (1962).
______. Novaia Zemlia: temperatura snega, firna i l'da. 1-2 (1963-1964).
______. Piataia sovetskaia antarkticheskaia ekspeditsiia. Gliatsiologicheskie issledovaniia na amerikanskoi antarkticheskoi stantsii Mak-Mérdo. 1-3 (1963).
______. Poliarnyi Ural: meteorologiia. 1-8 (1961-1963).
______. Tret'ia kontinental'naia antarkticheskaia ekspeditsiia: snezhnyi pokrov. (1) (1960).
______. Vtoraia kontinental'naia antarkticheskaia ekspeditsiia: snezhnyi pokrov. 1-4 (1960).
______. Vtoraia kontinental'naia antarkticheskaia ekspeditsiia: strukturnye issledovaniia. 1-2 (1963).
______. Zagorsk: teplovoi balans. 1-4 (1960-1961).
______. Zemlia Frantsa-Iosifa: gliatsiogeomorfologiia. 1-2 (1962).
______. Zemlia Frantsa-Iosifa: meteorologiia. 1-5 (1960-1964).
______. Zemlia Frantsa-Iosifa: snezhnyi pokrov. 1 (1960).
______. Zemlia Frantsa-Iosifa: temperatura snega i l'da. 1-2 (1960-1963).

2313 ______. ______.
Trudy. 1-81 (1931-1962). Closed 1962. (U-1:116b; B-1:25a; B68:568a; S620; FC-1:16, 18; GR-2:1245).
Supersedes the academy's Komissiia po izucheniiu estestvennykh proizvoditel'nykh sil S.S.S.R. Geograficheskii otdel [2316].
v1-12 (1931-1934) as Geomorfologicheskii institut. Trudy;
v14-24 (1935-1937) as Institut fizicheskoi geografii. Trudy.
1-11 (1931-1934) published in Leningrad.
Includes subseries:
______. Materialy po biogeografii S.S.S.R. 1-2 (1953-1955). (54 and 66 of parent series). (GR-1:523).
______. Materialy po fizicheskoi geografii S.S.S.R. 1-2 (1955-1957). (64, 71). (N70-3:3680b; GR-1:524).
______. Materialy po geomorfologii i paleogeografii SSSR. 1-24 (1948-1960). (42, 43, 46, 47, 50, 51, 52, 53, 55, 58, 61, 62, 63, 65, 68, 69, 72, 73, 74, 76, 77, 78, 79, 80). (GR-1:526).
______. Raboty Tian'-Shanskoi fiziko-geograficheskoi stantsii. 1-7 (1950-1962). (45, 49, 56, 60, 67, 75, 81). (GR-2:834).
______. Voprosy geografii kapitalisticheskikh stran. 1-2. (1953-1956). (57, 70). (N70-4:6225c; GR-3:1553).
______. Voprosy geografii stran narodnoi demokratii. 1 (1954). (59). (N70-4:6225c; GR-3:1553).
Index: 1-2, 14-74, 76 (1931-1958).

2314 ______. Komissiia po estestvenno-istoricheskomu raionirovaniiu S.S.S.R. Moskva.
Trudy. v1-2 (1947-1948). Closed 1948. (U-1:118a; S787).

(2315 ______. Komissiia po izucheniiu chetvertichnogo perioda.
Biulleten'. 1- (1929-). (U-1:118a; B-1:24b; S794; FC-S:14).

2316 ______. Komissiia po izucheniiu estestvennykh proizvoditel'nykh sil S.S.S.R. Geograficheskii otdel.
Trudy. no1-2 (1928-1930). Closed 1930. (U-1:118b; B-1:25a; S801; FC-1:10; GR-2:1205).
Superseded by the academy's Institut geografii. Trudy [2313].

AKADEMIIA nauk S.S.S.R.

2317 ______. Otdelenie geologo-geograficheskikh nauk. Moskva.
Referaty nauchno-issledovatel'skikh rabot. 1940-1945.
(U-1:119b; B-1:26b; S885).

2318 ______. Sibirskoe otdelenie. Institut geografii Sibiri i Dal'nego Vostoka. Irkutsk.
Doklady. 1-51 (1962-1977). Closed 1977. (N70-1:180c; B68:176b; S944).
Indexes nos. 1-10 (1962-1965) in no10, pp. 80-85; 11-20 (1966-1968) in no20 (1968), p. 66-76; 21-30 (1969-1971) in no30 (1971), p. 72-75; 31-40 (1971-1973) in no40 (1973), p. 70-73; 41-50 (1973-1976) in no50 (1976), p. 72-81.

2319 ______. ______. ______.
Materialy Ob'-Irtyshskoi ekspeditsii. 1-2 (1969-1971). Closed 1971.

2320 ______. ______. Komissiia po kompleksnomu kartografirovaniiu prirody, khoziaistva i naseleniia.
Trudy. 1-4 (1965-1970).
(Institut geografii Sibiri i Dal'nego Vostoka).

2321 AKADEMIIA nauk Ukrainskoi S.S.R. (Ukrainska akademiia nauk), Kiev. Komisiia dlia skladannia istorychno-heohrafichnoho slovnyka Ukrainy.
Istorychno-heohrafichnyi zbirnyk. 1-4 (1927-1931). Closed 1931.
(U-1:123a; B-4:370b; S1218; FC-1:338).
1-3 (1927-1929) also as no46, 46b, and 46v of the academy's Istorychno-filolohichnyi viddil. Zbirnyk (not in this list).
In Ukrainian.

2322 ______. Pryrodnychno-tekhnichnyi viddil.
Zhurnal heoloho-heohrafichnogo tsiklu. no1-8 (1932-1934).
Closed 1934. (U-1:123b; B-4:371a; S1117; FC-1:215; GR-3:1671).
Russian title: Zhurnal geologo-geograficheskogo tsikla.
Superseded by Heolohichnyi zhurnal (not in this list).
In Ukrainian.

2323 ALMA-ATA. Kazakhskii gosudarstvennyi universitet. Alma-Ata.
Uchenye zapiski. Geografiia. 1-5 (1954-1960) Irregular. (S1472).
1-4 (1954-1958) as Geologiia i geografiia.
v2 (1954) is v18 of parent series; 3(1957), 27; 4 (1958), 37; 5 (1960), 46.
In Kazakh and Russian.

2324 AMURSKII sbornik. (Akademiia nauk S.S.S.R. Sibirskoe otdelenie. Dal'nevostochnyi filial [and] Geograficheskoe obshchestvo S.S.S.R. Priamurskii filial). Khabarovsk. v1-4 (1959-1963). (N70-1:334c-335a; S1518).
Absorbed by Voprosy geografii Dal'nego Vostoka [2741].

2325 AZERBAIDZHANSKOE geograficheskoe obshchestvo. Baku.
Trudy. 1-5 (1960-1977) Irregular. (N70-1:599c; S1885).
Title in Azerbaijani: Azerbajdzhan dzhografija dzhemijjetinin eserleri.
Alternate listings: Geograficheskoe obshchestvo S.S.S.R. Azerbaidzhanskii filial. Trudy; Geograficheskoe obshchestvo Azerbaidzhanskoi S.S.R. [Akademiia nauk Azerbaidzhanskoi S.S.R.]. Trudy; Azerbajdzhan Dzhografija dzhemejjeti. Eserleri. v1 unnumbered.
In Russian or Azerbaijani with summary in other language.

2326 BAKU. Azerbaidzhanskii gosudarstvennyi universitet. Baku.
Trudy. Seriia geologo-geograficheskaia. 1-2 (1952-1955). Closed 1955.
(S2001).
In Azerbaijani and Russian.

BAKU. Azerbaidzhanskii gosudarstvennyi universitet.

(2327 ______. ______.
Uchenye zapiski: Seriia geologo-geograficheskikh nauk. 1958-
Bimonthly. (N70-1:615b; S2010).
Title varies: Uchenye zapiski. Geologo-geograficheskaia seriia;
Geologiia i geografiia.
Title in Azerbaijani: Elmi eserler. Keolokiia-dzhografiia seriiasy.
Superseded in part university's Uchenye zapiski (not in this list).
Text in Azerbaijani or Russian. No. 2, 4, 6 in Russian.

*(2328 BALTICA (Baltistika). (Lietuvos T.S.R. mokslų akademija. Geografijos
skyrius. INQUA tarbinė sekcija; Akademiia nauk Litvoskoi S.S.R.
Otdel geografii. Sovetskaia sektsiia INKVA; Academy of sciences of
the Lithuanian S.S.R. Department of geography. INQUA Soviet section).
Vil'nyus. 1- (1963-) Irregular. (N70-1:619a; B68:69a; S2088c).
International yearbook for the quaternary geology and palaeogeography,
coastal morphology and shore processes, marine geology and recent
tectonics of the Baltic Sea area.
Articles in Russian, English, or German.
Abstracts in English, Russian, or German.

2329 BEITRÄGE zur Kenntnis des russischen Reiches und der angränzenden Länder
Asiens. (Akademiia nauk). (St. Petersburg) Leningrad. 1-26 (1839-
1871); s2 v1-9 (1879-1886); s3 v1-8 (1886-1900); s4 v1-2 (1893-1896).
Closed 1896. (U-1:624c; B-1:303b; F-1:557b; GR-1:50).
Indexes: 1-15, 17-25 (1839-1868); 16, 26. s2 v3 (1871-1880); in
index of publications of the Academy. v16 was published in 1872.

2330 BERICHT über die Arbeiten zur Landeskunde der Bukowina. 1891- .

2331 BIBLIOGRAFIIA kartograficheskoi literatury i kart. (Vsesoiuznaia
knizhnaia palata). Moskva. 1931-1940 Quarterly. (U-3:2273b;
B-2:640a; BS:478b; S2244; FC-1:348; GR-1:411).
1-8 (1931-1938) as Kartograficheskaia letopis'
Superseded by Kartograficheskaia letopis' [2590].

(2332 BIOLOGIIA i geografiia. Sbornik nauchnykh statei aspirantov i soiskatelei.
(Alma-Ata. Kazakhskii gosudarstvennyi universitet). Alma-Ata.
5- (1968-) Irregular.

(2333 BRIANSKII kraeved. (Geograficheskoe obshchestvo S.S.S.R. Brianskoe otdelenie.
Oblastnoi kraevedcheskii muzei). Bryansk. 1- (1957-) Irregular.
(N75-1:336c).

2334 BURIATOVEDCHESKII sbornik. (Geograficheskoe obshchestvo S.S.S.R.
Vostochno-Sibirskii otdel. Buriat-Mongol'skaia sekstiia). Irkutsk.
1-6 (1926-1930). Closed 1930. (U-1:849c; S2823).

2335 CHARDZHOU. Turkmenskii gosudarstvennyi pedagogicheskii institut. Chardzhou.
Seriia biologicheskikh i geograficheskikh nauk. 1-2 (1964-1965).
Subseries of institute's Uchenye zapiski (not in this list).
Title of 1: Seriia biologo-geograficheskikh nauk.

2336 CHELYABINSK. Gosudarstvennyi pedagogicheskii institut. Chelyabinsk.
Uchenye zapiski. Seriia estestvenno-geograficheskaia. 1 (1957).
v1 (1957) is v 3 of parent series.

2337 CHERNOVTSY [Czernowitz]. Universytet. Chernovtsy.
Naukovi zapysky; Uchenye zapiski. Seriia geograficheskikh nauk.
1-4 (1955-1961). (S2965).
Subseries of university's Naukovi zapysky (not in this list).
nol (1955) is no 13 of whole series; 2 (1956), 22; 3 (1957), 25;
4 (1961), 52.
In Ukrainian or Russian.

2338 ______. ______.
Naukovi zapysky. Uchenye zapiski. Seriia geologo-geograficheskikh nauk. 1-3 (1949-1953).
v1 (1949) as Seriia pochvenno-geograficheskikh nauk.
v1 (1949) as v 3 of parent series; 2 (1950), 8; 3 (1953), 10.
2-3 published by Izdatel'stvo Kievskogo Universiteta.

2339 CHITA. Zabaikal'skii kompleksnyi nauchno-issledovatel'skii institut. Chita.
Trudy. Seriia ekonomika i geografiia. 1 (1963). (N70-1:1267c; S2999).
After nol series changed to unnumbered monographs.

2340 DAGESTANSKII gosudarstvennyi pedagogicheskii institut. Estestvenno-geograficheskii fakul'tet. Makhachkala.
Trudy. 3-7 (1968-1972). (N75-1:648b).

2341 DNEPROPETROVSK. Universytet. Heoloho-heohrafichnyi fakul'tet.
Sbornik rabot. 1-[5] (1940-1958). (S3297).
Subseries of university's Naukovi zapysky. v1 (1940) is v17 of parent series; v2 (1941) is v27; 3 volumes unnumbered in subseries are v31 (1948), v39 (1954), and v67 (1958) of parent series.
v1 (1940) issued by Heolohichnyi fakul'tet.
Title in Russian: Universitet. Geologo-geograficheskii fakul'tet.
Sbornik rabot. (Nauchnye zapiski).
In Ukrainian or Russian.

2342 DOKLADY na exhegodnykh chteniiakh pamiati L. S. Berga. (Geograficheskoe obshchestvo S.S.S.R.). Moskva; Leningrad. 1- (1952-) Annual.
(N70-2:1772b; S3333; GR-1:181).
1972-1975 not published.
Cover title: Chteniia pamiati L'va Semenovicha Berga.

2343 DOKLADY na ezhegodnykh chteniiakh pamiati V. A. Obrucheva.
(Geograficheskoe obshchestvo S.S.S.R.). Moskva; Leningrad.
1/5, 1956/60; 8/14, 1960/66 (1968); 15/19, 1967/71 (1973); 20, 1976 (1978). (N70-2:1772b; S3335; GR-3:1736).
Cover title: Chteniia pamiati Vladimira Afanas'evicha Obrucheva.

2344 DOKLADY po ekonomicheskoi geografii S.S.S.R. (Géograficheskoe obshchestvo S.S.S.R. Otdelenie ekonomicheskoi geografii).
Leningrad. 1 (1965). (N70-2:1772b; S4267a).

2345 DOKLADY po geomorfologii i paleogeografii Dal'nego Vostoka.
(Geograficheskoe obshchestvo S.S.S.R.). Leningrad. 1 (1964).
(N70-2:1772b; S3335a).

2346 DUSHANBE [Stalinabad]. Gosudarstvennyi pedagogicheskii institut. Dushanbe.
Seriia geograficheskaia. [1]-2 (1959-1962). Irregular. (N70-2:1801a; S3513).
Subseries of Institute's Uchenye zapiski (not in this list).
v1 is unnumbered and is v21 of main series. v2 (1962) is v35 of main series.

*(2347 EESTI geograafia selts (Eesti NSV teaduste akadeemia). Tallin. Aastaraamat. 1957- Annual. (N70-2:1876b; BS:282a; S3586).
Russian title: Estonskoe geograficheskoe obshchestvo (Akademiia nauk Estonskoi S.S.R.). Ezhegodnik.
In Estonian or Russian with summaries in the other language and in English or German.

2348 ______.
Publikatsioonid (Pulikatsii; Publications). 1-5 (1960-1962). (N70-2:1876c; S3587, 3588, 3589).
In Estonian, English, German, or Russian.

*(2349 EKONOMICHESKAIA geografiia; mezhvedomstvennyi nauchnyi sbornik. Ekonomichna heohrafiia; mizhvidomchyi naukovyi zbirnyk (Kiev. Universytet. Heohrafichnyi fakul'tet. Kafedra ekonomichnoi heohrafii). Kiev. 1- (1966-). (N70-2:1888a; S3690a).
In Russian. v1-21 (1966-1976) in Ukrainian.

2350 EKONOMICHESKAIA geografiia; respublikanskii mezhvedomstvennyi nauchnyi sbornik. (Ministerstvo vysshego i srednego spetsial'nogo obrazovaniia Ukrainskoi S.S.R.) (Khar'kov. Universitet). Khar'kov. 1-5 (1964-1965). (N70-2:1887b; S3657a).

2351 EKONOMICHESKAIA geografiia zapadnogo Urala. (Geograficheskoe obshchestvo S.S.S.R. Permskii otdel). Perm'. 1-5 (1963-1970). (N70-2:1887b).
Issued in Perm'. Universitet. Uchenye zapiski.
n1 (1963) as no 101 of whole series; 2 (1964), 123; 3 (1966), 144; 4 (1967), 168; 5 (1970), 242.

2352 EKONOMIKO-GEOGRAFICHESKIE problemy formirovaniia territorial'no-proizvodstvennykh komplesksov Sibiri. (Geograficheskoe obshchestvo S.S.S.R. Novosibirskii otdel). Novosibirsk. 1-6 (1969-1974). (N75-1:755c).

2353 ESTESTVOZNANIE i geografiia. Nauchno-populiarnyi i pedagogicheskii zhurnal. Moskva. 1-22 (1896-1917) Monthly. Closed 1917. (S3925).
Index: 1896-1905 in 1905 no10, p. 1-21.

*(2354 FIZICHESKAIA geografiia i geomorfologiia. Respublikanskii mezhvedomstvennyi nauchnyi sbornik. Fizychna heohrafiia ta heomorfolohiia. Mizhvidomchyi naukovyi zbirnyk. Kiev. 1- (1970-). (N75-1:846a; N78).
In Russian. Supplementary table of contents in English. Abstracts in English and Russian. v1-16 (1970-1976) in Ukrainian.

(2355 FRUNZE. Universitet. Geograficheskii fakul'tet. Frunze. Trudy. 1- (1955-) Irregular. (N70-2:2228b; N76; S4223).
Title varies. v1-2 (1955-1956) as Uchenye zapiski.
v6 (1975) as Universitet. Trudy. Seriia geograficheskikh nauk.

(2356 GEOBOTANICHESKOE kartografirovanie. (Akademiia nauk SSSR. Botanicheskii institut). Leningrad. 1963- Annual. (N70-2:2275b).
In Russian with supplementary English table of contents and summary of articles.
Index: 1963-1972 in 1972.

(2357 GEODEZIIA i kartografiia. (Glavnoe upravlenie geodezii i kartografii). Moskva. 1956- Monthly. (N70-2:2277a; BS:341a; S4253; FC-S:113; GR-1:246).

(2358 GEODEZIIA, kartografiia i aerofotos"emka: respublikanskii mezhvedomstvennyi nauchno-tekhnicheskii sbornik. L'vov. 1- (1964-) Irregular. (N70-2:2277a; S4253a).

2359 GEOGRAAFILISTE tööde kogumik. (Tartu. Ülikool). Tartu. 1-2 (1962-1964). (S4263).
Russian title: Sbornik geograficheskikh issledovanii. (Tartuskii gosudarstvennyi universitet).
In Estonian. Summaries in Russian.

2360 GEOGRAFICHESKIE nauki: tematicheskii sbornik nauchnykh trudov professorsko-prepodavatel'skogo sostava Ministerstva vysshego i srednego spetsial'nogo obrazovaniia Kazakhskoi S.S.R. (Alma-Ata. Kazakhskii gosudarstvennyi pedagogicheskii institut). Alma-Ata. 1-3 (1969-1972) Irregular.
v. 2 (1970) as Geografiia. Subtitle varies.

2360a GEOGRAFICHESKII mesiatseslov. St. Petersburg [Leningrad]. 1773-1775? (GR-1:247).
Continued as Mesiatseslov istoricheskii i geograficheskii.
German edition as: Geographischer Calender.

2361 GEOGRAFICHESKII sbornik. (Geograficheskoe obshchestvo S.S.S.R. Penzenskii otdel. Obshchestvo okhrany prirody R.S.F.S.R.). Penza. 1-3 (1968-1971).

(2361a GEOGRAFICHESKII sbornik. (Kazakhskii gosudarstvennyi universitet). Alma Ata. 1- (1974-). (GZS).

2362 GEOGRAFICHESKII sbornik. (Kazan'. Universitet). Kazan'. 1-5 (1966-1970). (N70-2:2278a; S5904a).

2363 GEOGRAFICHESKII sbornik. (Vsesoiuznyi institut nauchnoi i tekhnicheskoi informatsii). Moskva. 1-5 (1963-1975) Irregular.
v1 (1963) not numbered.
v1-2 (1963-1966) issued by Akademiia nauk S.S.S.R. Institut nauchnoi informatsii. v3-4 (1969-1970) have supplementary English tables of contents.

2364 GEOGRAFICHESKII vestnik. (Leningrad. Geograficheskii institut). Leningrad. v1-2 (1922-1925). Closed 1925. (U-2:1692a; S4265; FC-1:153; GR-1:247).
v1 (1922) has 3 nos. v2 (1923-1925), 4 nos.
Supersedes the institute's Biulleten' [2608].

2365 GEOGRAFICHESKOE obozrenie. St. Petersburg [Leningrad]. -1881.

2366 GEOGRAFICHESKOE obshchestvo Gruzinskoi S.S.R. (Gruzinskoe geograficheskoe obshchestvo). Tbilisi.
Izvestiia. 1-2 (1924-1946). Only two numbers issued. (S4266).
Title in Georgian: Sakartvelos S.S.R. geografiuli sazogadoebis moambe.
In Georgian and Russian.

(2367 ______.
Trudy. 1 (1939); 1/2- (1949-) Irregular. (N70-2:2278a; S4267).
Title in Georgian: Sakartvelos S.S.R. geografiuli sazogadoebis shromebi.
In Georgian with Russian summaries, or in Russian.
[Ul. Ketskhoveli 11, Tbilisi, U.S.S.R.]

(2368 GEOGRAFICHESKOE obshchestvo Kirgizskoi S.S.R. (Akademiia nauk Kirgizskoi S.S.R.). Frunze.
Izvestiia. 1-13 (1959-) Irregular. (N70-2:3304b; S6872a; FC-S:193).
1-5 (1959-1965) issued under former name: Geograficheskoe obshchestvo S.S.S.R. Kirgizskii filial.
Also as: Kirgizskoe geograficheskoe obshchestvo.
In Russian.

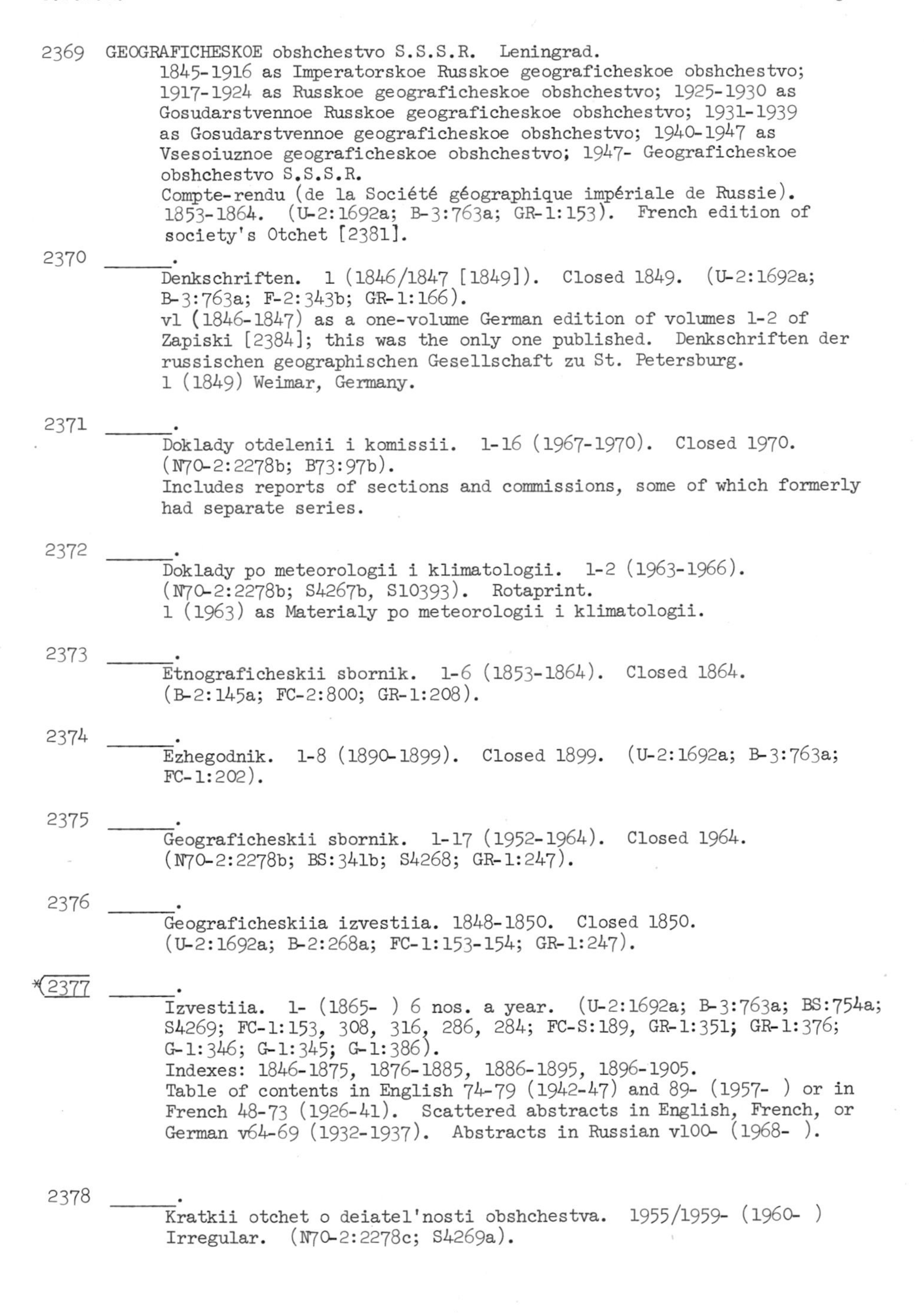

2369 GEOGRAFICHESKOE obshchestvo S.S.S.R. Leningrad.
1845-1916 as Imperatorskoe Russkoe geograficheskoe obshchestvo; 1917-1924 as Russkoe geograficheskoe obshchestvo; 1925-1930 as Gosudarstvennoe Russkoe geograficheskoe obshchestvo; 1931-1939 as Gosudarstvennoe geograficheskoe obshchestvo; 1940-1947 as Vsesoiuznoe geograficheskoe obshchestvo; 1947- Geograficheskoe obshchestvo S.S.S.R.
Compte-rendu (de la Société géographique impériale de Russie). 1853-1864. (U-2:1692a; B-3:763a; GR-1:153). French edition of society's Otchet [2381].

2370 ______.
Denkschriften. 1 (1846/1847 [1849]). Closed 1849. (U-2:1692a; B-3:763a; F-2:343b; GR-1:166).
v1 (1846-1847) as a one-volume German edition of volumes 1-2 of Zapiski [2384]; this was the only one published. Denkschriften der russischen geographischen Gesellschaft zu St. Petersburg.
1 (1849) Weimar, Germany.

2371 ______.
Doklady otdelenii i komissii. 1-16 (1967-1970). Closed 1970. (N70-2:2278b; B73:97b).
Includes reports of sections and commissions, some of which formerly had separate series.

2372 ______.
Doklady po meteorologii i klimatologii. 1-2 (1963-1966). (N70-2:2278b; S4267b, S10393). Rotaprint.
1 (1963) as Materialy po meteorologii i klimatologii.

2373 ______.
Etnograficheskii sbornik. 1-6 (1853-1864). Closed 1864. (B-2:145a; FC-2:800; GR-1:208).

2374 ______.
Ezhegodnik. 1-8 (1890-1899). Closed 1899. (U-2:1692a; B-3:763a; FC-1:202).

2375 ______.
Geograficheskii sbornik. 1-17 (1952-1964). Closed 1964. (N70-2:2278b; BS:341b; S4268; GR-1:247).

2376 ______.
Geograficheskiia izvestiia. 1848-1850. Closed 1850. (U-2:1692a; B-2:268a; FC-1:153-154; GR-1:247).

*2377 ______.
Izvestiia. 1- (1865-) 6 nos. a year. (U-2:1692a; B-3:763a; BS:754a; S4269; FC-1:153, 308, 316, 286, 284; FC-S:189, GR-1:351; GR-1:376; G-1:346; G-1:345; G-1:386).
Indexes: 1846-1875, 1876-1885, 1886-1895, 1896-1905.
Table of contents in English 74-79 (1942-47) and 89- (1957-) or in French 48-73 (1926-41). Scattered abstracts in English, French, or German v64-69 (1932-1937). Abstracts in Russian v100- (1968-).

2378 ______.
Kratkii otchet o deiatel'nosti obshchestva. 1955/1959- (1960-) Irregular. (N70-2:2278c; S4269a).

GEOGRAFICHESKOE obshchestvo S.S.S.R.

2379 ______.
Meteorologicheskii sbornik; Repertorium für Meteorologie. 1-3 (1860-1864); s2 v1-17 (1870-1894). Closed 1894. (U-1:113b; B-3:683a; B-3:192b; GR-1:564).
s2 v1-17 issued by Akademiia nauk.
In French and German.

2380 ______.
Meteorologicheskii vestnik. 1891-1935. Closed 1935. (U-3:2621a; B-3:192b; BS:555ab; S10622; FC-1:411; GR-1:564).
United with Vestnik edinoi gidrometeorologicheskoi sluzhby to form Meteorologiia i gidrologiia (not in this list).

2381 ______.
Otchet (Comptes rendus). 1845-1929. Closed 1929. (U-2:1692a; B-3:763a; S4270; FC-2:490; GR-2:701; GR-2:705).
1845-1911, 1915, 1918-1929 as part of or supplement to society's Izvestiia (1739). 1912-1914, and 1917-1926 never published.
Continued intermittently 1936- as part of society's Izvestiia [2377].

2382 ______.
Publichnye lektsii, prochitannye v lektorii im. Iu. M. Shokal'skogo. 1-14 (1967-1970). (N70-3:4837c).

2383 ______.
Vestnik. 1-30 (1851-1860). Closed 1860. (U-2:1692b; B-3:763b; BS:754a).
Superseded by society's Zapiski [2385].
Contents indexed in index to society's publications 1846-1875.

2384 ______.
Zapiski. 1-13 (1846-1859). Closed 1859. (U-2:1692b; B-3:763b; FC-1:252; GR-3:1610).

2385 ______.
Zapiski. 1-4 (1861-1864). Closed 1864. (U-2:1692b; B-3:763b; BS:754a; FC-1:252; GR-1:1610).
Supersedes society's Vestnik [2383]. After 1865 divided into three distinct journals, of which society's Zapiski...po obshchei geografii [2387] is included in this list.

2386 ______.
Zapiski. nsv 1-24 (1940-1965). (U-2:1692b; BS:754b; S4271; FC-1:237; GR-3:1632).

2387 ______.
Zapiski...po obshchei geografii. 1-51 (1867-1916). Closed 1916. (U-2:1692b; B-3:763b; FC-1:240; GR-3:1610).
v43-46 never published.
Index for 1867-1875 in index to society's publications 1846-1875.

2388 ______. Altaiskii otdel. Barnaul.
Izvestiia. 1-14 (1961-1970) Irregular. (N70-2:2278a; S4273).
Place of publication varies. Barnaul 1 (1961), Gorno-Altaysk 2-4 (1963), Barnaul 5-12 (1965-1970). Biisk 13 (1970), and Novosibirsk 14 (1970).

2389 ______. Bashkirskii filial. Ufa.
Zapiski. 1-8 (1957-1973) Irregular. (N70; S4274).
no8 published in 1971 in no7 in 1973.

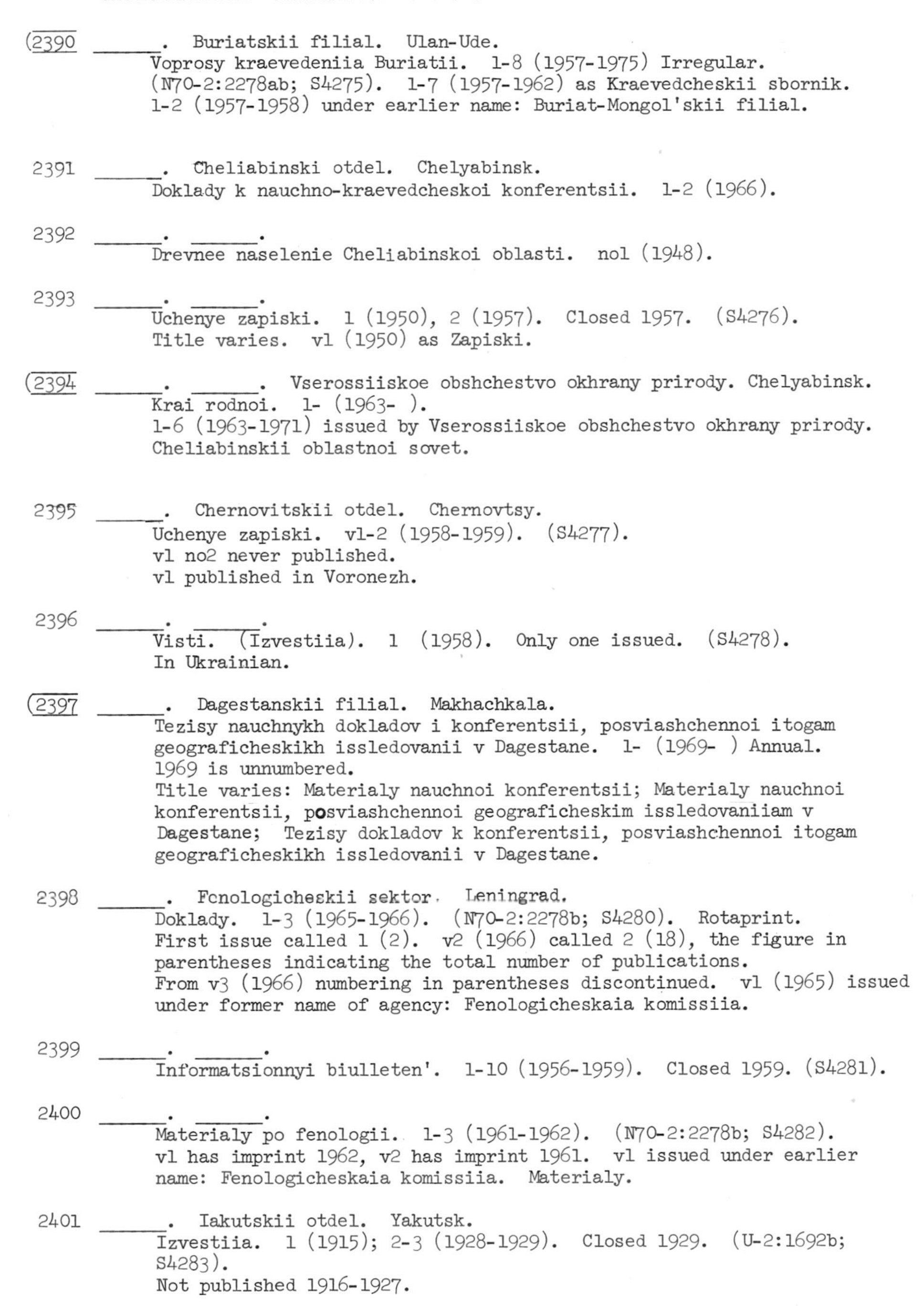

GEOGRAFICHESKOE obshchestvo S.S.S.R.

(2390 ______. Buriatskii filial. Ulan-Ude.
Voprosy kraevedeniia Buriatii. 1-8 (1957-1975) Irregular.
(N70-2:2278ab; S4275). 1-7 (1957-1962) as Kraevedcheskii sbornik.
1-2 (1957-1958) under earlier name: Buriat-Mongol'skii filial.

2391 ______. Cheliabinski otdel. Chelyabinsk.
Doklady k nauchno-kraevedcheskoi konferentsii. 1-2 (1966).

2392 ______. ______.
Drevnee naselenie Cheliabinskoi oblasti. nol (1948).

2393 ______. ______.
Uchenye zapiski. 1 (1950), 2 (1957). Closed 1957. (S4276).
Title varies. vl (1950) as Zapiski.

(2394 ______. ______. Vserossiiskoe obshchestvo okhrany prirody. Chelyabinsk.
Krai rodnoi. 1- (1963-).
1-6 (1963-1971) issued by Vserossiiskoe obshchestvo okhrany prirody.
Cheliabinskii oblastnoi sovet.

2395 ______. Chernovitskii otdel. Chernovtsy.
Uchenye zapiski. vl-2 (1958-1959). (S4277).
vl no2 never published.
vl published in Voronezh.

2396 ______. ______.
Visti. (Izvestiia). 1 (1958). Only one issued. (S4278).
In Ukrainian.

(2397 ______. Dagestanskii filial. Makhachkala.
Tezisy nauchnykh dokladov i konferentsii, posviashchennoi itogam
geograficheskikh issledovanii v Dagestane. 1- (1969-) Annual.
1969 is unnumbered.
Title varies: Materialy nauchnoi konferentsii; Materialy nauchnoi
konferentsii, posviashchennoi geograficheskim issledovaniiam v
Dagestane; Tezisy dokladov k konferentsii, posviashchennoi itogam
geograficheskikh issledovanii v Dagestane.

2398 ______. Fenologicheskii sektor. Leningrad.
Doklady. 1-3 (1965-1966). (N70-2:2278b; S4280). Rotaprint.
First issue called 1 (2). v2 (1966) called 2 (18), the figure in
parentheses indicating the total number of publications.
From v3 (1966) numbering in parentheses discontinued. vl (1965) issued
under former name of agency: Fenologicheskaia komissiia.

2399 ______. ______.
Informatsionnyi biulleten'. 1-10 (1956-1959). Closed 1959. (S4281).

2400 ______. ______.
Materialy po fenologii. 1-3 (1961-1962). (N70-2:2278b; S4282).
vl has imprint 1962, v2 has imprint 1961. vl issued under earlier
name: Fenologicheskaia komissiia. Materialy.

2401 ______. Iakutskii otdel. Yakutsk.
Izvestiia. 1 (1915); 2-3 (1928-1929). Closed 1929. (U-2:1692b;
S4283).
Not published 1916-1927.

GEOGRAFICHESKOE obshchestvo S.S.S.R. Iakutskii otdel.

2402 ______. ______.
Ocherki po izucheniiu Iakutskogo kraia. 1-2 (1927-1928). (S4319; GR-2:680-681).
Alternate listing: Gosudarstvennoe Russkoe geograficheskoe obshchestvo. Vostochno-Sibirskii otdel. Iakutskaia sektsiia.

2403 ______. Iugo-Zapadnyi otdel. Kiev.
Zapiski. 1-2 (1873-1874). Closed 1874. (U-2:1692b; FC-1:256).

2404 ______. Iuzhno-Ussurskii otdel.
Izvestiia. Ussuriysk (Nikol'sk-Ussuriyski). no1-16 (1922-1928). Closed 1928. (U-2:1692b; S4284; GR-1:356).
no1-4 (1922) as society's branch: Priamurskii otdel. Iuzhno-Ussuriiskoe otdelenie. Izvestiia. 5-11 (1924-1925) as Russkoe geograficheskoe obshchestvo. Iuzhno-Ussurskoe otdelenie. Izvestiia.

2405 ______. ______. Vladivostok.
Zapiski. 1-3 (1922-1929). Closed 1929. (U-2:1692b; S4285; GR-3:1614).
v1 (1922) by Priamurskii otdel. Iuzhno-Ussuriiskoe otdelenie.

2406 ______. Kaliningradskii otdel. (Kaliningrad. Universitet). Kaliningrad.
Zapiski. 1 (1972).

2407 ______. Kartograficheskaia komissiia.
Zhurnaly. no1-7 (1905-1907). Closed 1907.

2408 ______. Kavkazskii otdel. Tbilisi.
Izvestiia. 1-25 (1872-1917). Closed 1917. (U-2:1692b; B-3:763b; FC-1:311; GR-1:356).
Index: 1872-1906 as supplement to 19 no3 (1907).

2409 ______. ______.
______. Prilozhenie. Materialy dlia geografii Aziatskoi Turtsii. 1 (1893).

2410 ______. ______.
Otchet. 1852, 1859-1864, 1865, 1870, 1877-1878, 1879, 1880, 1881, 1882, 1883, 1884, 1885, 1886, 1887, 1888, 1889, 1909, 1915. (U-2:1692b; BS:754b; FC-2:491).

2411 ______. ______.
Sbornik statisticheskikh svedenii o Kavkaze. 1 (1869). Only one issued. (FC-2:605).

2412 ______. ______.
Zapiski. 1-30 (1852-1914 [1919]). Closed 1919. (U-2:1692b; B-3:763b; FC-1:245; GR-3:1615).
Index: 1852-1906 as supplement to Izvestiia v19 no3 (1907); 1852-1916 in v29 no2 (1916), p. 2-4.

2413 ______. Komi filial. Syktyvkar.
Izvestiia. 1-16 (1951-1973) Irregular. (N70-2:2278b; BS:754b; S4288).
11 (1967) also called v2 no1 (1967); 12 also called v2 no2 (1969); 13 also as no3 (1970); 14 also as v2 no4 (1972), 15 as v2 no5 (1973).

2414 ______. Komissiia aeros"ëmki i fotogrammetrii. Leningrad.
Doklady. 1-7 (1964-1969). Rotaprint. (N70-2:2278b; S4288a).
v1 (1964): Doklady po voprosam aerofotos"ëmki.

2415 ______. Komissiia geografii naseleniia i gorodov. Leningrad.
Doklady po geografii naseleniia. 1-5 (1962-1965). Rotaprint. (N70-2:2278b; N70-3:3680b; S4288b; S10323b).
Title varies: 1-2 (1962-1963) as Materialy po geografii naseleniia. no1 issued by society's Otdelenie ekonomicheskoi geografii. no1 (3) (1965) as Doklady.

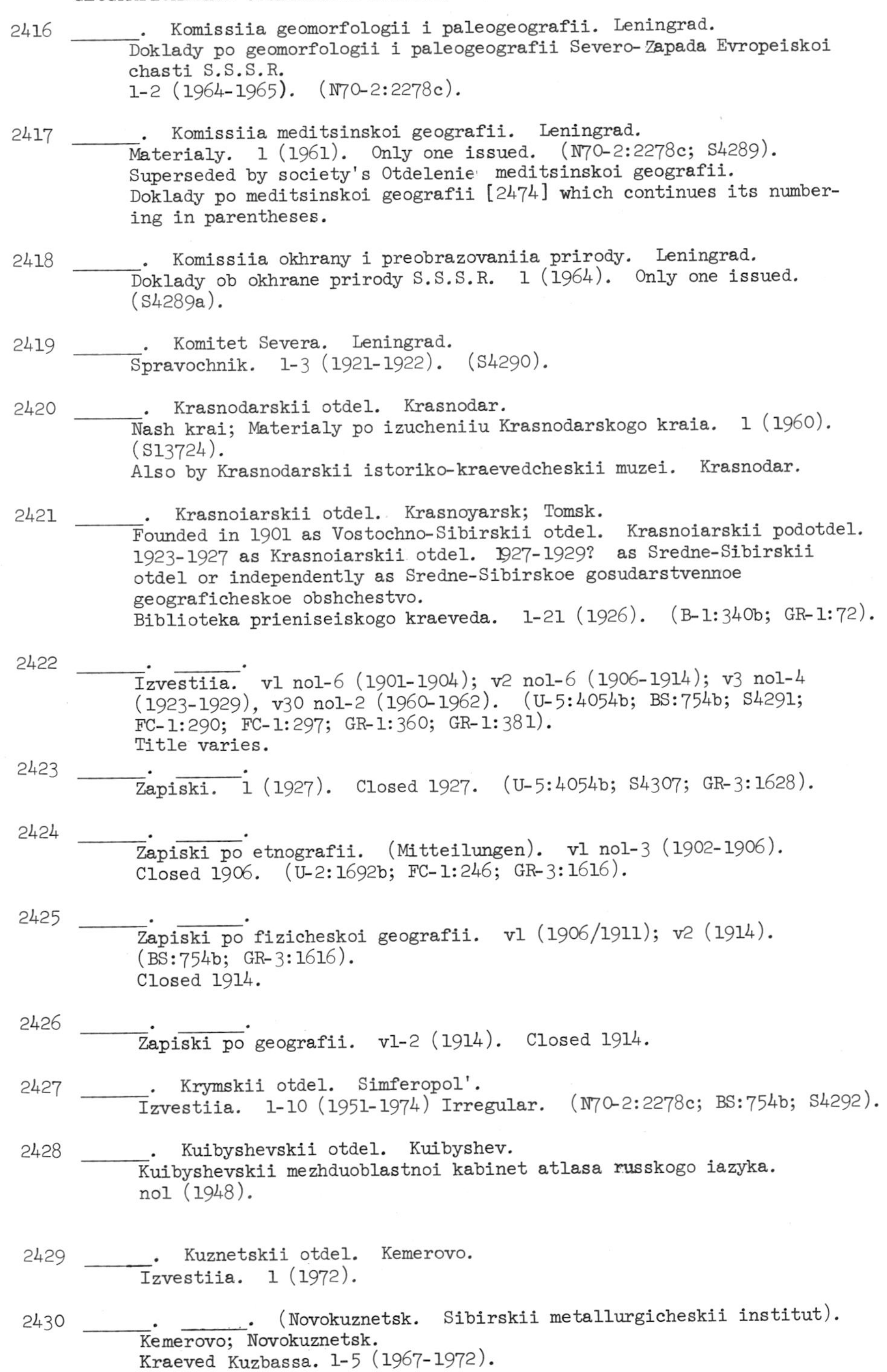

GEOGRAFICHESKOE obshchestvo S.S.S.R.

2416 ______. Komissiia geomorfologii i paleogeografii. Leningrad.
Doklady po geomorfologii i paleogeografii Severo-Zapada Evropeiskoi chasti S.S.S.R.
1-2 (1964-1965). (N70-2:2278c).

2417 ______. Komissiia meditsinskoi geografii. Leningrad.
Materialy. 1 (1961). Only one issued. (N70-2:2278c; S4289).
Superseded by society's Otdelenie meditsinskoi geografii. Doklady po meditsinskoi geografii [2474] which continues its numbering in parentheses.

2418 ______. Komissiia okhrany i preobrazovaniia prirody. Leningrad.
Doklady ob okhrane prirody S.S.S.R. 1 (1964). Only one issued. (S4289a).

2419 ______. Komitet Severa. Leningrad.
Spravochnik. 1-3 (1921-1922). (S4290).

2420 ______. Krasnodarskii otdel. Krasnodar.
Nash krai; Materialy po izucheniiu Krasnodarskogo kraia. 1 (1960). (S13724).
Also by Krasnodarskii istoriko-kraevedcheskii muzei. Krasnodar.

2421 ______. Krasnoiarskii otdel. Krasnoyarsk; Tomsk.
Founded in 1901 as Vostochno-Sibirskii otdel. Krasnoiarskii podotdel. 1923-1927 as Krasnoiarskii otdel. 1927-1929? as Sredne-Sibirskii otdel or independently as Sredne-Sibirskoe gosudarstvennoe geograficheskoe obshchestvo.
Biblioteka prieniseiskogo kraeveda. 1-21 (1926). (B-1:340b; GR-1:72).

2422 ______. ______.
Izvestiia. v1 no1-6 (1901-1904); v2 no1-6 (1906-1914); v3 no1-4 (1923-1929), v30 no1-2 (1960-1962). (U-5:4054b; BS:754b; S4291; FC-1:290; FC-1:297; GR-1:360; GR-1:381).
Title varies.

2423 ______. ______.
Zapiski. 1 (1927). Closed 1927. (U-5:4054b; S4307; GR-3:1628).

2424 ______. ______.
Zapiski po etnografii. (Mitteilungen). v1 no1-3 (1902-1906). Closed 1906. (U-2:1692b; FC-1:246; GR-3:1616).

2425 ______. ______.
Zapiski po fizicheskoi geografii. v1 (1906/1911); v2 (1914). (BS:754b; GR-3:1616).
Closed 1914.

2426 ______. ______.
Zapiski po geografii. v1-2 (1914). Closed 1914.

2427 ______. Krymskii otdel. Simferopol'.
Izvestiia. 1-10 (1951-1974) Irregular. (N70-2:2278c; BS:754b; S4292).

2428 ______. Kuibyshevskii otdel. Kuibyshev.
Kuibyshevskii mezhduoblastnoi kabinet atlasa russkogo iazyka. no1 (1948).

2429 ______. Kuznetskii otdel. Kemerovo.
Izvestiia. 1 (1972).

2430 ______. ______. (Novokuznetsk. Sibirskii metallurgicheskii institut). Kemerovo; Novokuznetsk.
Kraeved Kuzbassa. 1-5 (1967-1972).

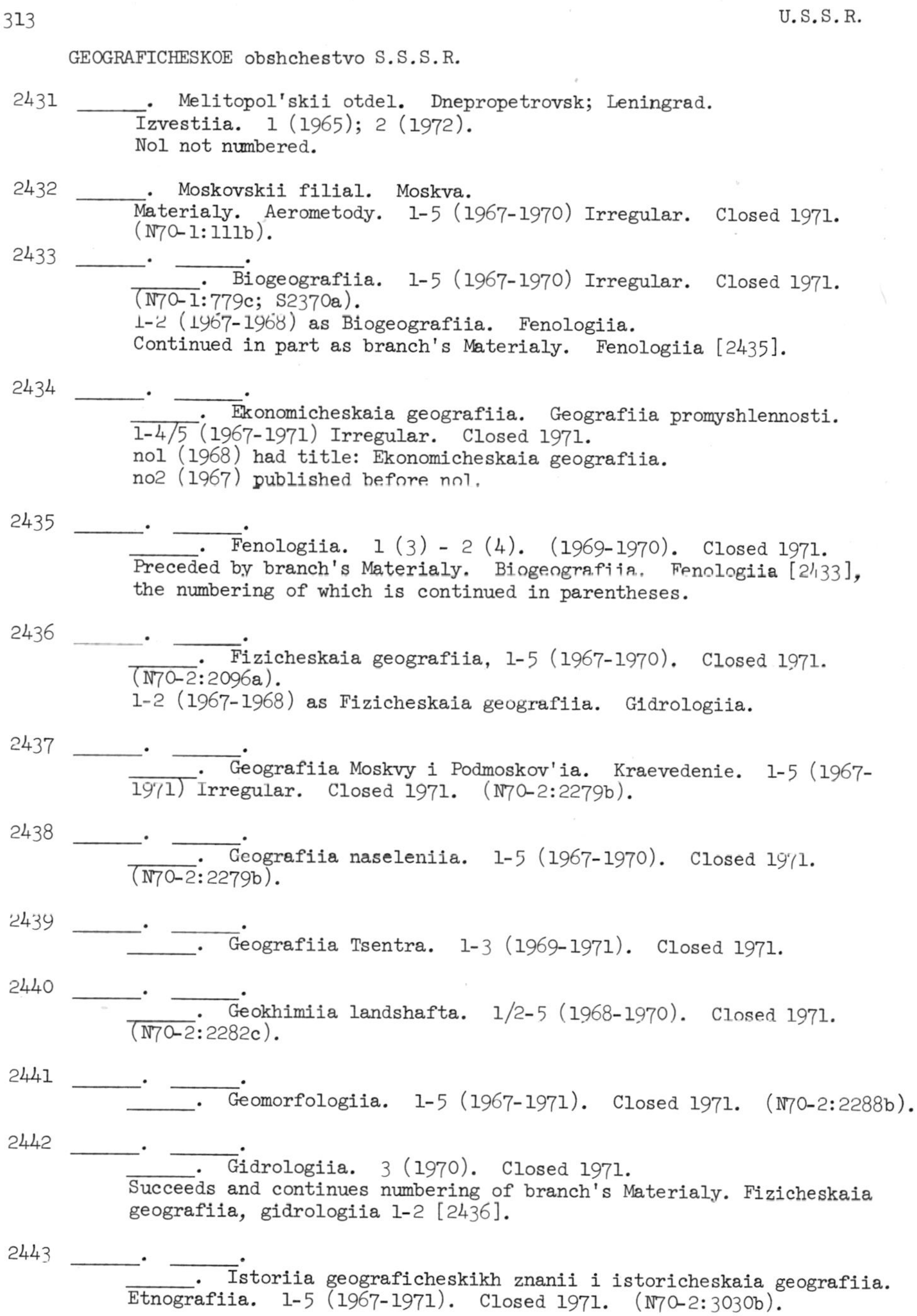
GEOGRAFICHESKOE obshchestvo S.S.S.R.

2431 ______. Melitopol'skii otdel. Dnepropetrovsk; Leningrad.
Izvestiia. 1 (1965); 2 (1972).
No1 not numbered.

2432 ______. Moskovskii filial. Moskva.
Materialy. Aerometody. 1-5 (1967-1970) Irregular. Closed 1971.
(N70-1:111b).

2433 ______. ______.
______. Biogeografiia. 1-5 (1967-1970) Irregular. Closed 1971.
(N70-1:779c; S2370a).
1-2 (1967-1968) as Biogeografiia. Fenologiia.
Continued in part as branch's Materialy. Fenologiia [2435].

2434 ______. ______.
______. Ekonomicheskaia geografiia. Geografiia promyshlennosti.
1-4/5 (1967-1971) Irregular. Closed 1971.
no1 (1968) had title: Ekonomicheskaia geografiia.
no2 (1967) published before no1.

2435 ______. ______.
______. Fenologiia. 1 (3) - 2 (4). (1969-1970). Closed 1971.
Preceded by branch's Materialy. Biogeografiia. Fenologiia [2433],
the numbering of which is continued in parentheses.

2436 ______. ______.
______. Fizicheskaia geografiia, 1-5 (1967-1970). Closed 1971.
(N70-2:2096a).
1-2 (1967-1968) as Fizicheskaia geografiia. Gidrologiia.

2437 ______. ______.
______. Geografiia Moskvy i Podmoskov'ia. Kraevedenie. 1-5 (1967-1971) Irregular. Closed 1971. (N70-2:2279b).

2438 ______. ______.
______. Geografiia naseleniia. 1-5 (1967-1970). Closed 1971.
(N70-2:2279b).

2439 ______. ______.
______. Geografiia Tsentra. 1-3 (1969-1971). Closed 1971.

2440 ______. ______.
______. Geokhimiia landshafta. 1/2-5 (1968-1970). Closed 1971.
(N70-2:2282c).

2441 ______. ______.
______. Geomorfologiia. 1-5 (1967-1971). Closed 1971. (N70-2:2288b).

2442 ______. ______.
______. Gidrologiia. 3 (1970). Closed 1971.
Succeeds and continues numbering of branch's Materialy. Fizicheskaia geografiia, gidrologiia 1-2 [2436].

2443 ______. ______.
______. Istoriia geograficheskikh znanii i istoricheskaia geografiia.
Etnografiia. 1-5 (1967-1971). Closed 1971. (N70-2:3030b).

2443a ______. ______.
Kraevedenie v Tsentral'nom raione. 1-2 (1974-1976). Irregular.

2444 ______. ______.
______. Meditsinskaia geografiia. 1-5 (1967-1970). Closed 1971.
(N70-3:3714b; S10484a).

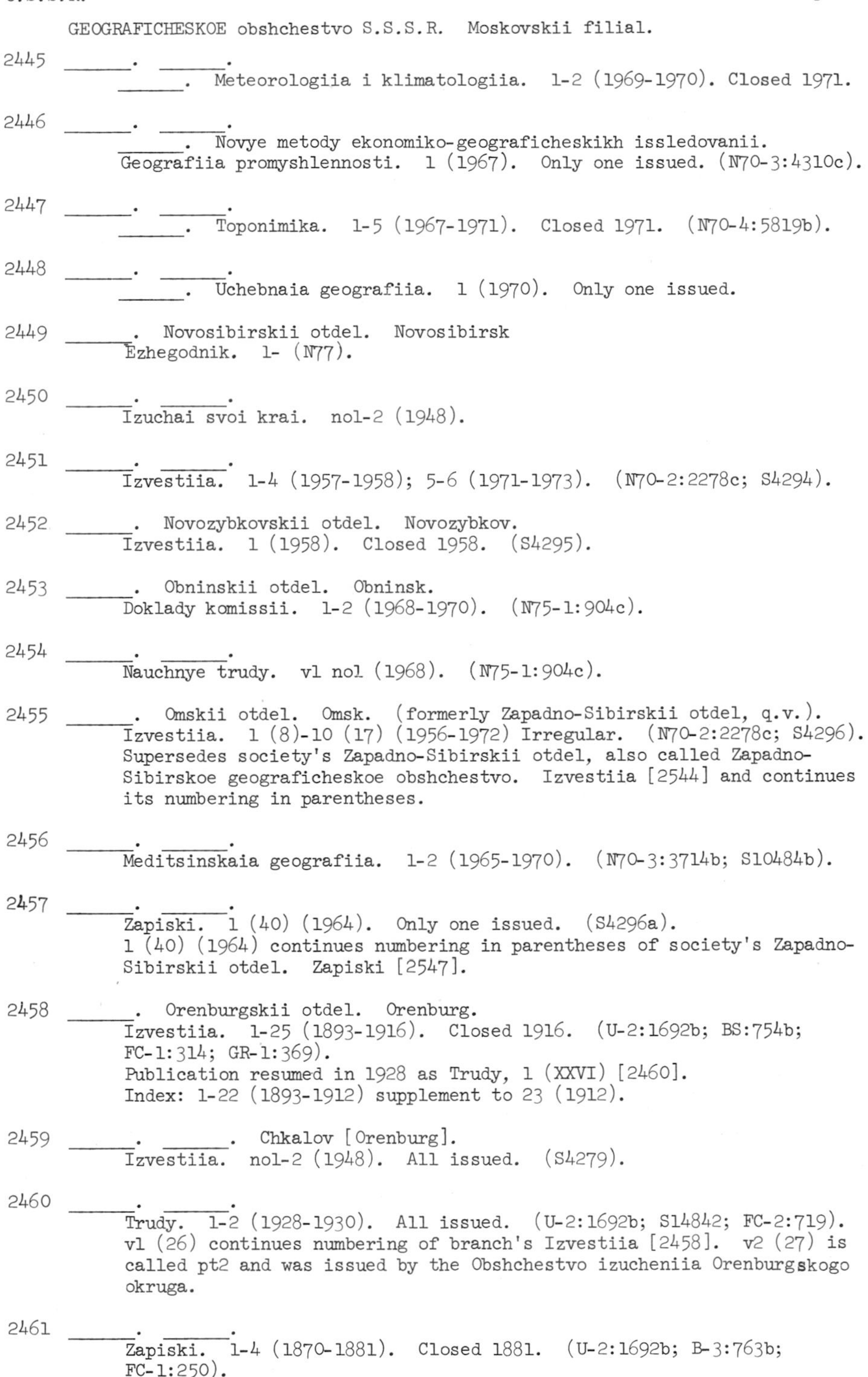

GEOGRAFICHESKOE obshchestvo S.S.S.R. Moskovskii filial.

2445 ______. ______.
______. Meteorologiia i klimatologiia. 1-2 (1969-1970). Closed 1971.

2446 ______. ______.
______. Novye metody ekonomiko-geograficheskikh issledovanii. Geografiia promyshlennosti. 1 (1967). Only one issued. (N70-3:4310c).

2447 ______. ______.
______. Toponimika. 1-5 (1967-1971). Closed 1971. (N70-4:5819b).

2448 ______. ______.
______. Uchebnaia geografiia. 1 (1970). Only one issued.

2449 ______. Novosibirskii otdel. Novosibirsk
Ezhegodnik. 1- (N77).

2450 ______. ______.
Izuchai svoi krai. no1-2 (1948).

2451 ______. ______.
Izvestiia. 1-4 (1957-1958); 5-6 (1971-1973). (N70-2:2278c; S4294).

2452 ______. Novozybkovskii otdel. Novozybkov.
Izvestiia. 1 (1958). Closed 1958. (S4295).

2453 ______. Obninskii otdel. Obninsk.
Doklady komissii. 1-2 (1968-1970). (N75-1:904c).

2454 ______. ______.
Nauchnye trudy. v1 no1 (1968). (N75-1:904c).

2455 ______. Omskii otdel. Omsk. (formerly Zapadno-Sibirskii otdel, q.v.).
Izvestiia. 1 (8)-10 (17) (1956-1972) Irregular. (N70-2:2278c; S4296).
Supersedes society's Zapadno-Sibirskii otdel, also called Zapadno-Sibirskoe geograficheskoe obshchestvo. Izvestiia [2544] and continues its numbering in parentheses.

2456 ______. ______.
Meditsinskaia geografiia. 1-2 (1965-1970). (N70-3:3714b; S10484b).

2457 ______. ______.
Zapiski. 1 (40) (1964). Only one issued. (S4296a).
1 (40) (1964) continues numbering in parentheses of society's Zapadno-Sibirskii otdel. Zapiski [2547].

2458 ______. Orenburgskii otdel. Orenburg.
Izvestiia. 1-25 (1893-1916). Closed 1916. (U-2:1692b; BS:754b; FC-1:314; GR-1:369).
Publication resumed in 1928 as Trudy, 1 (XXVI) [2460].
Index: 1-22 (1893-1912) supplement to 23 (1912).

2459 ______. ______. Chkalov [Orenburg].
Izvestiia. no1-2 (1948). All issued. (S4279).

2460 ______. ______.
Trudy. 1-2 (1928-1930). All issued. (U-2:1692b; S14842; FC-2:719).
v1 (26) continues numbering of branch's Izvestiia [2458]. v2 (27) is called pt2 and was issued by the Obshchestvo izucheniia Orenburgskogo okruga.

2461 ______. ______.
Zapiski. 1-4 (1870-1881). Closed 1881. (U-2:1692b; B-3:763b; FC-1:250).

GEOGRAFICHESKOE obshchestvo S.S.S.R. Orenburgskii otdel.

2462 ______. ______.
Zhurnaly. 1874-1891. Closed 1891. (U-2:1692b).

2463 ______. Otdelenie ekonomicheskoi geografii. Leningrad.
Materialy po ekonomicheskoi geografii zarubezhnykh stran. 1 (1961). (N70-2:2278c; S10320).

2464 ______. Otdelenie etnografii. Leningrad.
Doklady po etnografii. 1-6 (1965-1968). Closed 1968. (N70-2:2278c; S4296b).
Rotaprint.
Supersedes the section's Materialy po etnografii [2405] and continues its numbering in parentheses.

2465 ______. ______.
Materialy. 1-3 (1958/61-1963). Closed 1963. (N70-2:2278c; S4297).
Cover title: Materialy po etnografii.
Superseded by the section's Doklady po etnografii [2464].

2466 ______. ______.
Zapiski. 1-44 no2 (1867-1925). Closed 1925. (U-2:1692c; B-3:763b; S4272; FC-1:241; GR-3:1610; GR-3:1607).
v43 published in 1925.

2467 ______. ______. Skazochnaia komissiia.
Obzor rabot. 1924-1927. Closed 1927. (U-2:1692c; B-3:763b; S4306; GR-2:678).

2468 ______. Otdelenie fizicheskoi geografii. Leningrad.
Materialy. 1 (1961). Only one issued. (N70-2:2278c; S4298).

2469 ______. ______.
Materialy po Arktike i Antarktike. 1 (1961). Only one issued. (N70-3:3680a).

2470 ______. ______. Komissiia po landshaftnym issledovaniiam i kartografirovaniiu. Leningrad.
Doklady. 1-3 (1961-1963); 1 (1965). (S4299).

2471 ______. ______. ______.
Materialy. 1-3 (1961-1963. (N70-2:2278c; S4299a).
Former name: Komissiia po landshaftnym kartam. Cover title on v1-2 (1961-1962): Materialy po fizicheskoi geografii.

2472 ______. Otdelenie istorii geograficheskikh znaii. Leningrad.
Materialy. 1 (1962). Only one issued. (N70-2:2278c: S4300).

2473 ______. Otdelenie matematicheskoi geografii i kartografii. Leningrad.
Doklady po kartografii. 1-3 (1961-1966). Rotaprint. (N70-2:2278c; S4301).
Title varies: v1-2 (1961-1962) as Materialy. Cover title on v1-2: Materialy po kartografii.

2474 ______. Otdelenie meditsinskoi geografii. Leningrad.
Doklady po meditsinskoi geografii. 1(2)-2(3) (1964-1965). Rotaprint. (N70-2:2279a; S4301a).
Supersedes society's Komissiia meditsinskoi geografii. Materialy [2417] and continues its numbering in parentheses.

2475 ______. Otdelenie uchebnoi geografii. Leningrad.
Doklady. 1 (3) (1964). Closed 1964. (S4301b).
Supersedes section's Materialy [2476] and continues its numbering in parentheses. Cover title: Doklady po uchebnoi geografii.

GEOGRAFICHESKOE obshchestvo S.S.S.R. Otdelenie uchebnoi geografii.

2476 ______. ______.
Materialy. 1-2 (1951-1963). Closed 1963. (N70-2:2279a; S4302).
Superseded by section's Doklady [2475] which continues its numbering in parentheses.

(2477 ______. Permskii otdel. Perm'.
Biogeografiia i kraevednie. 1-4 (1971-1976). (N77).

2478 ______. ______.
Doklady. 1 (1959).

2479 ______. ______.
Zapiski. 1 (1960); 2 (1964). (S4303).
V2 (1964) is also v2 of Geografiia Permskoi oblasti [2552] and v118 of Perm'. Universitet. Uchenye zapiski (not in this list).

2480 ______. Postoiannaia prirodookhranitel'naia komissiia. Petrograd [Leningrad].
Trudy. 1-3 (1914-1918). Closed 1918. (U-2:1692c).

2481 ______. Priamurskii filial (formerly Priamurskii otdel). (Obshchestvo izucheniia Amurskogo kraia). Khabarovsk
Otchet. 1-17 (1894-1922)? (U-2:1692c).

2482 ______. ______.
Protokoly. no1-10 (1894-1895). Closed 1895? (U-2:1692c).

2483 ______. ______.
Trudy. 1-4 (1895-1901). (U-2:1692c).

2484 ______. ______.
Zapiski. v1-10 (1894-1914). Closed 1914. (U-2:1692c; B-3:763b; FC-1:251; GR-3:1623).

2485 ______. ______.
Zhurnal. no1-9 (1894-1895). Closed 1895. (U-2:1692c).

2486 ______. ______. Chitinskoe otdelenie. Chita.
Otchet. 1895, 1896, 1902-1906, 1907-1908, 1909-1910. (U-2:1692c; FC-2:498).

2487 ______. ______. Troitskosavsko-Kiakhtinskoe otdelenie. Kyakhta.
Otchet. 1894-1897? (U-2:1692c; B-3:763b; GR-2:708).

2488 ______. ______. ______.
Protokoly. 1-3 (1894-1895). Closed 1897. (U-2:1692c; B-3:763b).
Continued in society's Trudy (1835).

2489 ______. ______. ______.
Trudy. 1-17 (1898-1922). Closed 1922. (U-2:1692c; B-3:763b; FC-2:736).

2490 ______. Primorskii filial. Vladivostok.
Zapiski. 1-27 (1888-1968) Irregular. (U-2:1692c; N70-2:2279a; S4304, S4304a; FC-1:249; GR-3:1621, 1630, 1601).
1-17 (1888-1922); 1(18)-6(23) (1928-1930); 1(24) (1965); 25 (1966).
1-17 (1888-1922) by Obshchestvo izucheniia Amurskogo kraia; 1 (18)-5 (22) (1928-1930) by Vladivostokskoe otdelenie of Gosudarstvennoe Russkoe geograficheskoe obshchestvo. 6(23) (1936) jointly with Dal'nevostochnaia kraevaia planovaia komissia. 1(24)-27 (1965-1968) by Primorskii filial.
Indexes: v1-17 (1888-1922) in v1 (18) (1928) pi-iv; 1-20 (1888-1929) in v3(20) no2 (1929), p. 379-382; v1(18)-5(22) no1 (1928-1929) in v5(22) no2 (1930), p. 214-216.

GEOGRAFICHESKOE obshchestvo S.S.S.R.

2491 ______. Sakhalinskii otdel. Iuzhno-Sakhalinsk.
Izvestiia. 1-4 (1970-1973).

2492 ______. Semipalatinskii otdel. Semipalatinsk.
Biulleten'. 1-3 (Jl-S 1924). (S4305).

2493 ______. ______.
Zapiski. nol-17 (1903-1928). Closed 1928. (BS:754b; S14837; FC:1-253; GR-3:1627).
1-14 (1903-1923) as Zapadno-Sibirski otdel. Semipalatinskii podotdel. Zapiski. Succeeded by Obshchestvo izucheniia Kazakhstana. Semipalatinskii otdel. Zapiski [2673], which partly continues numbering.
Index: 1-7 (1903-1913) in 8 (1914), p. 3-4.

2494 ______. Severnyi filial. Apatity; Petrozavodsk.
Priroda i khoziaistvo Severa. 1- (1969-) Irregular. (N75-2:1775c).

2495 ______. Severo-Zapadnyi otdel. Vilnius.
Zapiski. 1-4 (1911-1913). Closed 1913. (U-2:1692c; FC-1:254; GR-3:1627).

2496 ______. S"ezd. Leningrad.
Materialy po probleme ekonomicheskogo raionirovaniia strany. (N70-2:2279a).

2497 ______. Sibirskie i dal'nevostochnie organizatsii. (Akademiia nauk S.S.S.R. Sibirskoe otdelenie. Institut geografii Sibirii i Dal'nego Vostoka). Irkutsk.
Publikatsii. 2 nol-2 (1964/1969) [1971].

2498 ______. Sochinskii otdel. Sochi.
Doklady. Leningrad.
1-2 (1968-1971). (N75-1:904c).

2499 ______. Statisticheskii otdel.
Sbornik statisticheskikh svedenii o Rossii. 1-3 (1851-1858). Closed 1858. (U-2:1692c; GR-2:945).

2500 ______. ______.
Zapiski. 1-14 (1866-1915). Closed 1915. (U-2:1692c; B-3:763b; FC-1:241).

2501 ______. Tadzhikskii filial. (Tadzhikskoe geograficheskoe obshchestvo). Dushanbe.
Sbornik trudov. 1-2 (1958-1961); 3 (1975). (N70-2:2279a; S4303).
vl as Sbornik statei is also Akademiia nauk Tadzhikskoi S.S.R. Trudy. v99 (1958).
Also as: Geograficheskoe obshchestvo Tadzhikskoi S.S.R.

2502 ______. Tambovskii otdel. Tambov.
Nash krai; sbornik statei. 1 (1964). Only one issued.

2503 ______. Tiumenskii otdel. Tyumen'.
Izvestiia. nol (1972).

2504 ______. Tomskii otdel. Tomsk.
Doklady. Leningrad. 1 (1970). Only one issued.

2505 ______. Tsentrograficheskaia laboratoriia im. D. I. Mendeleeva.
Informatsionnyi biulleten', 1917-1927; Informatsiia, 1928. Closed 1928. (U-2:1692c; B-3:763b; GR-1:277).

GEOGRAFICHESKOE obshchestvo S.S.S.R. Tsentrograficheskaia laboratoriia.

2506 ______. ______.
Tsentrografiia. 1 (1933). Closed 1933. (U-2:1692c; GR-1:135).

2507 ______. Uchenyi arkhiv. Leningrad.
Opisanie kollektsii rukopisei nauchnogo arkhiva Geograficheskogo obshchestva S.S.S.R. 1 (1973). (N76).

2508 ______. Ural'skii filial. Sverdlovsk.
Zapiski. 1-4 (1954-1961). (N70-2:2279a; S4310; GR-3:1629).
v1-3 (1954-1960) under earlier name Ural'skii otdel. v3 (1960) called v1(3). v2 (1955) as Uchenye zapiski.

2509 ______. Vladivostokskii otdel (earlier Vladivostokskii podotdel; now Primorskii filial). Vladivostok.
Otchet. 1884-1927. (U-2:1692c; S4311).
Title varies: Godovoi otchet.

2510 ______. ______. Kruzhok iunykh kraevedov.
Raboty. 1-3 (1927-1929). Closed 1929. (U-2:1692c; S4312).

2511 ______. Voronezhskii otdel. Voronezh.
Izvestiia. 1-4 (1957-1962). Closed 1962. (N70-2:2279a; S4313).

2512 ______. ______.
Nauchnye zapiski. 1963, 1965, 1967-1968, 1970-1974. Annual. (N70-2:2279a; S4314).
None issued in 1964, 1966, and 1969?

2513 ______. Vostochnaia komissiia. Leningrad.
Doklady. 1-4 (1962-1967). (N70-2:2279a; S4315, S4316).
v1 (1962) as Materialy. v1 (2) (1965) as Doklady continuing numbering of Materialy in parenthesis. Next issue v3 (1966).

2514 ______. Vostochno-Sibirskaia fenologicheskaia komissiia. Irkutsk.
Biulleten'. 2/3 (1963). (N70-2:2279a).

2515 ______. Vostochno-Sibirskii otdel. Irkutsk.
Biulleten'. 1-6 (1923-1925). Closed 1925. (U-2:1693a; S4317).
1-4 (1923) carry subtitle Etnograficheskii biulleten'.

(2516 ______. ______.
Izvestiia. 1-55 (1870-1929), 56-57 (1936-1937), 58 (1954), 59-69 (1960-1976) Irregular. (U-2:1693c; B-3:764a; S4318; FC-1:318; GR-1:384).
v1-8 (1870-1877) as Sibirskii otdel. Izvestiia. v56-57 (1936-1937) issued by Obshchestvo izucheniia Vostochno-Sibirskogo kraia and also numbered vol. 1-2.
Indexes: v1-22 (1870-1891); v23-31 (1892-1900); v32-41 (1901-1910); v46-49 (1921-1926).

2517 ______. ______.
Otchet. 1856-1924/25. (U-2:1693a; B-3:764a; GR-2:715).
Indexes: 1863-1873; 1891-1899.

2518 ______. ______.
Trudy. no1-8 (1898-1914). Closed 1914. (U-3:1693a; BS:754b; FC-2:688; GR-2:1389).
Index: 1-4 (1897-1901).

GEOGRAFICHESKOE obshchestvo S.S.S.R. Vostochno-Sibirskii otdel.

2519 _____. _____.
Zapiski. 1-12 (1856-1886); nsv1-3 (1890-1896). Closed 1896. (U-2:1693a; B-3:764a; FC-1:237; GR-3:1631; GR-3:1627).
1-11 (1850-1877) as Sibirskii otdel. Zapiski. In 1889 divided into three series: Zapiski po obshchei geografii [2519] [nsv1- (1890-) continuation of this entry, 2519]; branch's Otdelenie etnografii. Zapiski [2520]; and branch's Otdelenie statistiki. Zapiski [2521].
Index: 1-2 (1856-1886).

2520 _____. _____. Otdelenie etnografii. Irkutsk.
Zapiski. 1-4 (1889-1892). Closed 1896. (U-2:1693a; B-3:764a).

2521 _____. _____. Otdelenie statistiki.
Zapiski. 1 (1889). Closed 1889? (B-3:764a).

2522 _____. _____. Sektsiia zemlevedeniia.
Sbornik. 2-3 (1926). (U-2:1693a).

2523 _____. Zabaikal'skii filial. Chita.
Geograficheskie aspekty gornogo lesovedeniia i lesovodstva. 1-3 (1967-1972) Irregular. (GZS).
no1 not numbered; no2 is no54 of whole series (1971); no3 is no 71 (1972).
2-3 as subseries of filial's Zapiski [2532].

2524 _____. _____.
Izvestiia. v1-v7 no6 (1965-1971). Closed 1971. (N70-2:2279a; N75-1:904c; S4319a).
v1 under earlier name: Zabaikal'skii otdel.

2525 _____. _____.
Kratkii otchet o deiatel'nosti. My 1924-O 1926. (S4320).

2526 _____. _____.
Materialy po issledovaniiu Aginskoi stepi. no1-7 (1910-1913). Closed 1913.
Subtitled Trudy Aginskoi ekspeditsii.

2527 _____. _____.
Problemy stroitel'stva v usloviiakh Zabaikal'ia. 1-2 (1967-1970) Irregular.
no2 is no44 of branch's Zapiski [2532].

2528 _____. _____.
Problemy zimovedeniia. 1-4 (1966-1972) Biennial. (N70-3:4786c; B73:269a).
1-3 (1966-1970) as Problemy regional'nogo zimovedeniia. 1-2 (1966-1968) as an independent series; 3-4 (1970-1972) as subseries of branch's Zapiski [2532]. no3 (1970) is no40 of whole series; no4 (1972) is no65.

2529 _____. _____.
Snezhnye laviny khrebta Udokan. Trudy Tsentra nauchno-prikladnykh issledovanii. 1 (1971).
no1 is no60 of branch's Zapiski [2532].

2530 _____. _____.
Vestnik nauchnoi informatsii. 1-12 (1965-1975) Irregular. (N70-2:2279a; S4320a).
1-5 (1965-1966) by Zabaikal'skii otdel.

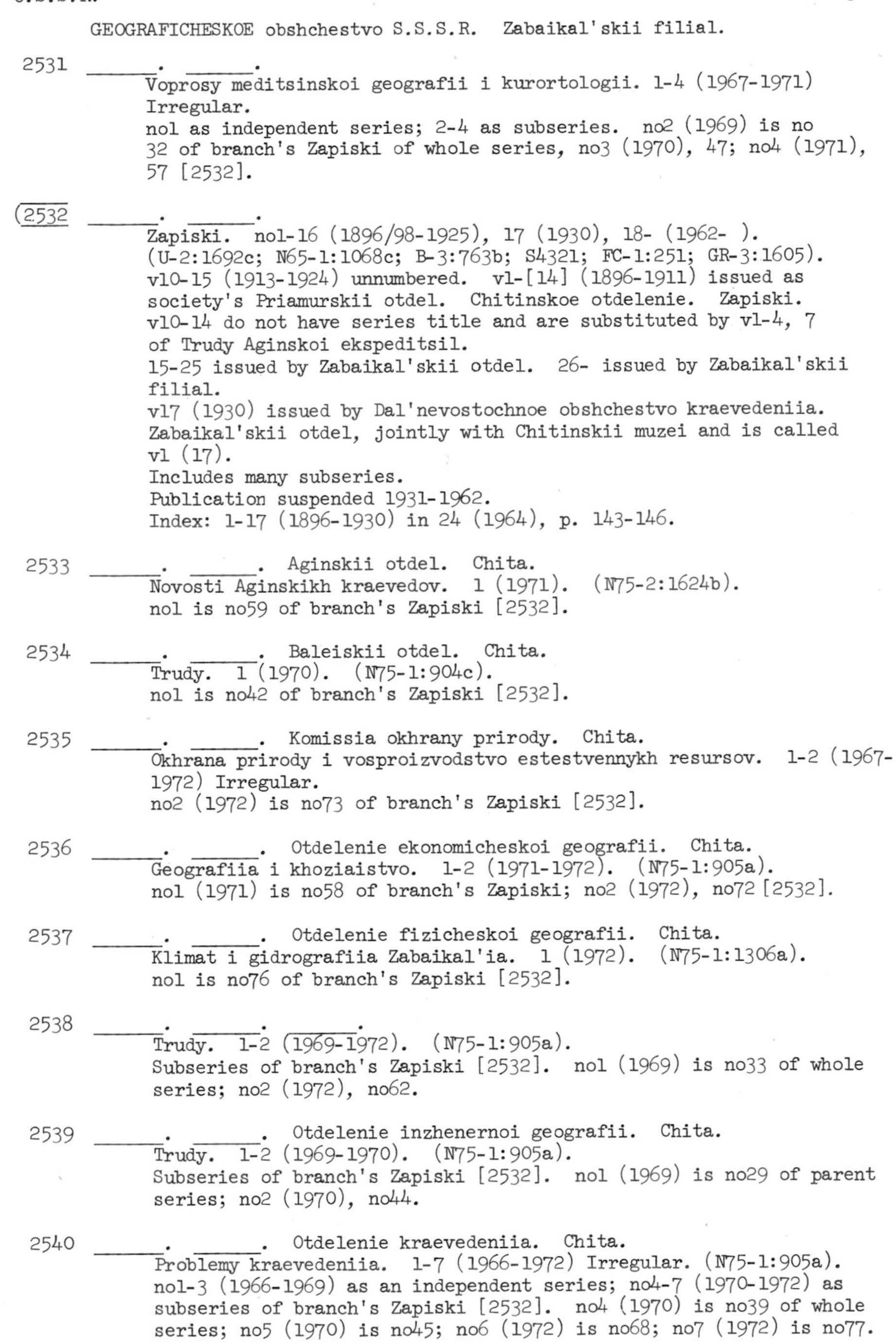

GEOGRAFICHESKOE obshchestvo S.S.S.R. Zabaikal'skii filial.

2531 ______. ______.
Voprosy meditsinskoi geografii i kurortologii. 1-4 (1967-1971)
Irregular.
no1 as independent series; 2-4 as subseries. no2 (1969) is no 32 of branch's Zapiski of whole series, no3 (1970), 47; no4 (1971), 57 [2532].

2532 ______. ______.
Zapiski. no1-16 (1896/98-1925), 17 (1930), 18- (1962-).
(U-2:1692c; N65-1:1068c; B-3:763b; S4321; FC-1:251; GR-3:1605).
v10-15 (1913-1924) unnumbered. v1-[14] (1896-1911) issued as society's Priamurskii otdel. Chitinskoe otdelenie. Zapiski.
v10-14 do not have series title and are substituted by v1-4, 7 of Trudy Aginskoi ekspeditsil.
15-25 issued by Zabaikal'skii otdel. 26- issued by Zabaikal'skii filial.
v17 (1930) issued by Dal'nevostochnoe obshchestvo kraevedeniia. Zabaikal'skii otdel, jointly with Chitinskii muzei and is called v1 (17).
Includes many subseries.
Publication suspended 1931-1962.
Index: 1-17 (1896-1930) in 24 (1964), p. 143-146.

2533 ______. ______. Aginskii otdel. Chita.
Novosti Aginskikh kraevedov. 1 (1971). (N75-2:1624b).
no1 is no59 of branch's Zapiski [2532].

2534 ______. ______. Baleiskii otdel. Chita.
Trudy. 1 (1970). (N75-1:904c).
no1 is no42 of branch's Zapiski [2532].

2535 ______. ______. Komissia okhrany prirody. Chita.
Okhrana prirody i vosproizvodstvo estestvennykh resursov. 1-2 (1967-1972) Irregular.
no2 (1972) is no73 of branch's Zapiski [2532].

2536 ______. ______. Otdelenie ekonomicheskoi geografii. Chita.
Geografiia i khoziaistvo. 1-2 (1971-1972). (N75-1:905a).
no1 (1971) is no58 of branch's Zapiski; no2 (1972), no72 [2532].

2537 ______. ______. Otdelenie fizicheskoi geografii. Chita.
Klimat i gidrografiia Zabaikal'ia. 1 (1972). (N75-1:1306a).
no1 is no76 of branch's Zapiski [2532].

2538 ______. ______. ______.
Trudy. 1-2 (1969-1972). (N75-1:905a).
Subseries of branch's Zapiski [2532]. no1 (1969) is no33 of whole series; no2 (1972), no62.

2539 ______. ______. Otdelenie inzhenernoi geografii. Chita.
Trudy. 1-2 (1969-1970). (N75-1:905a).
Subseries of branch's Zapiski [2532]. no1 (1969) is no29 of parent series; no2 (1970), no44.

2540 ______. ______. Otdelenie kraevedeniia. Chita.
Problemy kraevedeniia. 1-7 (1966-1972) Irregular. (N75-1:905a).
no1-3 (1966-1969) as an independent series; no4-7 (1970-1972) as subseries of branch's Zapiski [2532]. no4 (1970) is no39 of whole series; no5 (1970) is no45; no6 (1972) is no68; no7 (1972) is no77.

2541 ______. ______.
Zabaikal'skii kraevedcheskii ezhegodnik. 1-6 (1967-1972). (S27181).

GEOGRAFICHESKOE obshchestvo S.S.S.R.

2542 ______. Zapadno-Kazakhstanskii otdel.
Materialy po flore i rastitel'nosti Severnogo Prikaspiia. Leningrad.
v1-6 no2 (1964-1972); 7 (1974); 1975; 1977 Rotaprint.
7 (1974) as Botanicheskaia geografiia. Severnogo Prikaspiia: materialy po flore i rastitel'nosti Severnogo Prikaspiia.
1975 as Flora i rastitel'nost' Severnogo Prikaspiia.
1977 as Botanicheskaia geografiia Severnogo Prikaspiia.

2543 ______. ______. Ural'sk.
Nauchnye zapiski. 1-11 (1947-1958). (N70-2:2279b; S4322).
v6 (1949) and 10 (1957) unnumbered.

2544 ______. Zapadno-Sibirskii otdel. (Later Omskii otdel, q.v.). Omsk.
Izvestiia. 1-7 (1913-1930). Closed 1930. (U-2:1693a; B-3:764a; BS:754b; S27274; FC-1:306).
v7 (1930) as Zapadno-Sibirskoe geograficheskoe obshchestvo. Izvestiia.
Index: 1-5 (1913-1926).

2545 ______. ______.
Otchet. (Comptes-rendus). 1878-1912, 1924/25-1935-/36.
(U-2:1693a; BS:754b; FC-2:493; GR-2:716).

2546 ______. ______.
Trudy. no1-5 (1922). Closed 1922. (S27276).

2547 ______. ______.
Zapiski. 1-38 (1879-1916); 39 (1927). Closed 1927. (U-3:1693a; B-3:764a; S27277; FC-1:238; GR-3:1632).
Continued by society's Omskii otdel. Zapiski [2457].
Index: 1-35 (1879-1909); in v36 no1 (1912), p.1-4; 1-38 (1879-1926) in 39 (1927).

2548 ______. ______. Altaiskii podotdel. Barnaul.
Altaiskii sbornik. v1-11 (1894-1912); 12 (1930).
1894-1903 as Obshchestvo liubitelei issledovaniia Altaia, Tomsk. Altaiskii sbornik.
12 (1930) issued by Altaiskoe geograficheskoe obshchestvo and Barnaulskii estestvenno-istoricheskii muzei.

2549 ______. ______. ______.
Otchet. 1904.

2550 GEOGRAFIIA i khoziaistvo. (Moskva. Universitet. Geograficheskii fakul'tet). Moskva. 1-12 (1958-1963). Closed 1963. (N70-2:2279b; S4325).

2551 GEOGRAFIIA i khoziaistvo Moldavii. (Akademiia nauk Moldavskoi S.S.R. Otdel geografii). Kishinëv. 1-2 (1969-1970). (N70-2:2279b; B73:127b).

*2551a GEOGRAFIIA i prirodnye resursy. (Akademiia nauk SSSR. Sibirskoe otdelenie). Novosibirsk. 1980- . Quarterly.

2552 GEOGRAFIIA permskoi oblasti. (Geograficheskoe obshchestvo S.S.S.R. Permskii otdel. [and] Perm'. Universitet). Perm'. 1-3 (1962-1966). (S4326).
v2 (1964) and v3 (1966) issued in university's Uchenye zapiski, v118 and v138. v2 also issued as Geograficheskoe obshchestvo S.S.S.R. Permskii otdel. Zapiski, 2 [2479].

2553 GEOGRAFIIA ukrainskykh i smezhnykh zemel'. L'vov. 1 (1938).

*(2554 GEOGRAFIIA v shkole: nauchno-metodicheskii zhurnal. (Ministerstvo prosveshcheniia S.S.S.R.). Moskva. 1934- 6 nos a year. (U-3:1693a; B-2:268b; BS:341b; S4327; FC-1:154; FC-S:113; GR-1:248).
4 nos in 1934, 3 nos in 1941.
Publication suspended 1942-Dec. 1945.
Indexes in 1935 no6, p. 100-104; 1947 no1, p. 77-79; 1967 no3, p. 69-76.

*(2555 GEOGRAFINIS metraštis. (Lietuvos T.S.R. mokslų akademija. Geografijos skyrius. Lietuvos T.S.R. geografinė draugija).
Vil'nyus. 1- (1958-) Annual. (N70-3:2279b; S4329).
Russian: Geograficheskii ezhegodnik. (Geograficheskoe obshchestvo Litovskoi S.S.R.).
In Lithuanian. Supplementary table of contents and abstracts in Russian and German. or Russian and English.

*(2556 GEOMORFOLOGIIA. (Akademiia nauk S.S.S.R.). Moskva. 1970- Quarterly. (N75-1:910b; B73:129a).
Title in English: Geomorphology.
Supplementary title, table of contents, and abstracts in English.

(2557 GIDROGEOLOGIIA i karstovedenie. (Geograficheskoe obshchestvo S.S.S.R. Permskii otdel). Perm'. 1- (1962-)Irregular. (N70-2:2350b; S4412b).
v1-2 (1962-1964) issued by Perm'. Universitet. Institut karstovedeniia i speleologii [and] Laboratoriia geologii, and published as part of the university's Uchenye zapiski: v1 (v24, pt3) and v2 (v119). v3-8 (1966-1977) issued independently.

(2558 GLIATSIOLOGICHESKIE issledovaniia. (Akademiia nauk S.S.S.R. Mezhduvedomstvennyi geofizicheskii komitet). Moskva. 1- (1959-) Irregular. (N70-2:2359c).
Title in English: Glaciological researches.
Title page, table of contents, and abstracts also in English.

(2559 GLIATSIOLOGIIA Altaia. (Geograficheskoe obshchestvo S.S.S.R. Tomskii otdel; Tomsk. Universitet). Tomsk. 1- (1962-) Irregular. (B73:131a).

(2560 GLOBUS. Geograficheskii ezhegodnik dlia detei. Leningrad. 1-3 (1938-1949); 1957- Annual. (S4446).

2561 GROZNYY. Chechen-Ingushskii gosudarstvennyi pedagogicheskii institut. Groznyy.
Uchenye zapiski. Seriia estestvenno-geograficheskaia. 1-2 (1964-1968). (S4772).
no1 (1964) is no23 of the whole series; no2 (1968), no30.

2562 HEOHRAFICHNI doslidzhenniia na Ukraini (Akademiia nauk Ukrainskoi R.S.R. Heohrafichne tovarystvo U.R.S.R.). Kiev. 1-3 (1969-1971) Annual. (N70-2:2279b; BL).
Title in Russian: Geograficheskie issledovaniia na Ukraine (Ukrainskoe geograficheskoe obshchestvo).
In Ukrainian. Summary in Russian.

2563 HEOHRAFICHNE tovarystvo Ukrainskoi R.S.R. (Geograficheskoe obshchestvo Ukrainskoi S.S.R.). Kiev.
Doklady i soobshcheniia sektsii i komissi. 1-2 (1974-1976).

2564 ______. Dnepropetrovskii otdel. Dnepropetrovsk.
Izvestiia. 1-3 (1970-1973) Irregular.

HEOHRAFICHNE tovarystvo Ukrainskoi R.S.R.

2565 ______. Kharkivs'kyi viddil. (Geograficheskoe obshchestvo SSSR. Khar'kovskii otdel). Kharkov.
Izvestiia. 1961. (S4286).

2566 ______. ______.
Materialy. 1-12 (1964-1973). (N70-2:2523b).
Title in Russian: Geograficheskoe obshchestvo Ukrainskoi S.S.R. Khar'kovskii otdel. Materialy.

2567 ______. ______.
Prirodnye i trudovye resursy Levoberezh'ia Ukrainy i ikh ispol'zovanie. 1-15 (1965-1972).
v1-2 (1965) also as Materialy Khar'kovskogo otdela Geograficheskogo obshchestva Soiuza S.S.R.
Variously numbered: v1-7 (1965-1966); v1-5 (1967); 15 (1970); 11 (1971); 13 (1972).

2568 ______. L'vivs'kyi viddil. (Geograficheskoe obshchestvo Ukrainskoi S.S.R. L'vovskii otdel). L'vov.
Doklady i soobshcheniia. 1964 (1965), 1965 (1967), 1966 (1969), 5 (1975), 6 (1977). (N70-2:2523b; S4322a).

2569 HEOHRAFICHNYI zbirnyk. (Akademiia nauk Ukrainskoi R.S.R. [and] Heohrafichne tovarystvo U.R.S.R.). Kiev. 1-9 (1956-1969). (N70-2:2279b; S4324; GR-1:270).
Russian: Geograficheskoe obshchestvo Ukrainskoi S.S.R. [or] Geograficheskoe obshchestvo S.S.S.R. Ukrainskii filial. Geograficheskii sbornik.
In Ukrainian.

2570 HEOHRAFICHNYI zbirnyk. (L'vov. Universytet [and] Heohrafichne tovarystvo U.R.S.R. L'vivskyi viddil). L'vov. 1-9 (1951-1969) Irregular. (N70-2:2523b; N70-3:3552a; BS:525b; S4293; FC-S:113; GR-3:1746).
v1-4 (1951-1957) in L'vov. Universytet. Naukovi zapysky as v18, 28, 39, and 40. v5- (1959-) as independent publication. Also sponsored by geographical society.
Russian: Geograficheskii sbornik. (L'vov. Universitet; Geograficheskoe obshchestvo S.S.S.R. L'vovskii otdel).
In Ukrainian or Russian.

2571 HEOHRAFIIA v shkoli: metodychnyi zbirnyk. (Ukrainskii naukovo-doslidnyi instytut pedahohy). Kiev. 1-11 (1948-1964). (S4328).
Russian: Geografiia v shkole: metodicheskii sbornik. (Ukrainskii nauchno-issledovatel'skii institut pedagogiki).
In Ukrainian.

2572 IAKUTSKOE kraevoe geograficheskoe obshchestvo. Yakutsk.
Zapiski. 1 (1923). Only one issued. (S4846).

2573 IRKUTSK. Universitet. Irkutsk.
Trudy. Seriia geograficheskaia. 1 (1958). Only one issued.
v1 (1958) is v24 of parent series.

2574 ______. ______. Biologo-geograficheskii nauchno-issledovatel'skii institut. Irkutsk.
Izvestiia. v1-9 no4 (1924-1942); v10-24 (1947-1971). (U-3:2116c; B-2:556b; S5084; FC-1:283; GR-1:334; GR-1:333).

2575 ISTORICHESKII, statisticheskii i geograficheskii zhurnal. Moskva. 1-41 (1790-1830). Closed 1830. (U-3:2130b; B-3:573b; FC-2:515; GR-1:295).
1790-1809 as Politicheskii zhurnal.
Index: 1790-1802 (1874).

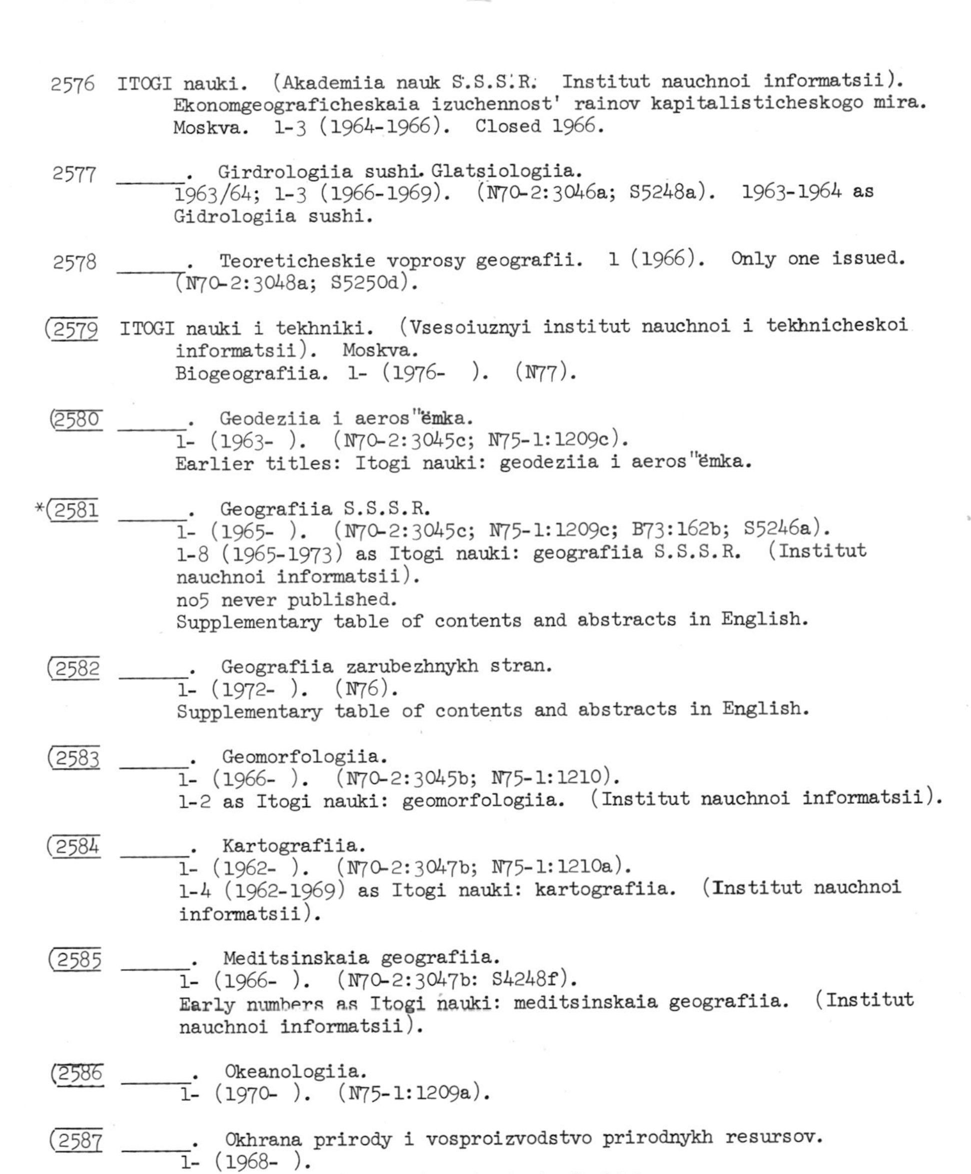

2576 ITOGI nauki. (Akademiia nauk S.S.S.R. Institut nauchnoi informatsii). Ekonomgeograficheskaia izuchennost' rainov kapitalisticheskogo mira. Moskva. 1-3 (1964-1966). Closed 1966.

2577 ______. Girdrologiia sushi. Glatsiologiia. 1963/64; 1-3 (1966-1969). (N70-2:3046a; S5248a). 1963-1964 as Gidrologiia sushi.

2578 ______. Teoreticheskie voprosy geografii. 1 (1966). Only one issued. (N70-2:3048a; S5250d).

(2579 ITOGI nauki i tekhniki. (Vsesoiuznyi institut nauchnoi i tekhnicheskoi informatsii). Moskva. Biogeografiia. 1- (1976-). (N77).

(2580 ______. Geodeziia i aeros"ëmka. 1- (1963-). (N70-2:3045c; N75-1:1209c). Earlier titles: Itogi nauki: geodeziia i aeros"ëmka.

*(2581 ______. Geografiia S.S.S.R. 1- (1965-). (N70-2:3045c; N75-1:1209c; B73:162b; S5246a). 1-8 (1965-1973) as Itogi nauki: geografiia S.S.S.R. (Institut nauchnoi informatsii). no5 never published. Supplementary table of contents and abstracts in English.

(2582 ______. Geografiia zarubezhnykh stran. 1- (1972-). (N76). Supplementary table of contents and abstracts in English.

(2583 ______. Geomorfologiia. 1- (1966-). (N70-2:3045b; N75-1:1210). 1-2 as Itogi nauki: geomorfologiia. (Institut nauchnoi informatsii).

(2584 ______. Kartografiia. 1- (1962-). (N70-2:3047b; N75-1:1210a). 1-4 (1962-1969) as Itogi nauki: kartografiia. (Institut nauchnoi informatsii).

(2585 ______. Meditsinskaia geografiia. 1- (1966-). (N70-2:3047b: S4248f). Early numbers as Itogi nauki: meditsinskaia geografiia. (Institut nauchnoi informatsii).

(2586 ______. Okeanologiia. 1- (1970-). (N75-1:1209a).

(2587 ______. Okhrana prirody i vosproizvodstvo prirodnykh resursov. 1- (1968-). Supplementary table of contents in English.

(2588 ______. Teoreticheskie i obshchie voprosy geografii. 1- (1974-).

2589 ITOGI nauki i tekhniki.
Teoreticheskie voprosy fizicheskoi i ekonomicheskoi geografii. 1 (1972). (N75-1:1211a).

(2590 KARTOGRAFICHESKAIA letopis'; Organ gosudarstvennoi bibliografii. (Vsesoiuznaia knizhnaia palata). Moskva. 1- (1941-) Annual since 1954. (U-3:2273b; B-2:640a; BS:478b; S5720; FC-1:348; GR-1:411).
First issue covers period 1941-1950; second, 1951-1953.
Supersedes Bibliografiia kartograficheskoi literatury i kart [2331].

2591 KAZAN'. Gosudarstvennyi pedagogicheskii institut. Kazan'.
Voprosy geografii i geologii. 4-6 (1967-1971).
Subseries of institute's Uchenye zapisi. (S5874).
no4 (1967);no5 (1970) is no81 of whole series; no6 (1971), no93.

2592 ______. Universitet. Kazan'.
Uchenye zapiski. Geografiia. 1-2 (1935-1940). (S5905; GR-3:1435, 1436).
v1 (1935) is v95 no9 of whole series; v2 (1940), v100 no3.
v1 (1935) as Ekonomicheskaia geografiia.

2593 KHAR'KOV. Ukrains'kyi naukovo-doslidnyi instytut heohrafii ta kartohrafii. Khar'kov.
Zapysky. 1-2 (1927/28-1928/29). Closed 1930. (U-3:2285b; B-4:373a; S6206; FC-1:255).
In Ukrainian.
Russian: Ukrainskii nauchno-issledovatel'skii institut geografii i kartografii. Zapiski.
German title: Ukrainisches Forschungsinstitut für Geographie und Kartographie. Mitteilungen.

2594 ______. Universytet. Khar'kov. Ucheni zapysky. Heohrafichnyi fakultet.
Trudy. 1-5 (1952-1963). (N70-2:3291b; GR-2:1205).
5 (1963) as Geologo-geograficheskii fakul'tet. Geograficheskoe otdelenie. Trudy.
no1 (1952), no42 of whole series; 2 (1955), 56; 3 (1957), 81; 4 (1958), 97; 5 (1963), 136.
In Ukrainian or Russian.

2595 ______. ______.
Vestnik. Seriia geograficheskaia. 1 (1964). Only one issued. (N70-2:3291b).
No1 (1964) is no2 of parent series.
Superseded by Seriia geologo-geograficheskaia [2596].
Subseries of university's Vestnik.

2596 ______. ______.
______. Seriia geologo-geograficheskaia. 2 (1967). Only one issued. (N70-2:3291b).
Subseries of university's Vestnik. Visnyk. no2 is no25 of whole series.
Supersedes and continues numbering of Seriia geograficheskaia [2595].
In Russian.

2597 ______. ______. Heoloho-heohrafichnyi fakul'tet. Khar'kov.
Trudy. 1 (1940). (S6225).
Also as v18 of university's Ucheni Zapysky.
In Ukrainian and Russian.

2598 KHIMICHESKAIA geografiia i gidrogeokhimiia. (Geograficheskoe obshchestvo S.S.S.R. Permskii otdel. [Perm'. Universitet. Institut karstovedeniia i speleologii]). Perm'. 1-5 (1961-1967). (N70-2:3292a; S6316, S6317).
1 (1961) as Khimicheskaia geografiia.
Title in English: Chemical geography and hydrogeochemistry.

2599 KIEV. Naukovo-doslidnyi instytut pedahohiki. Kiev.
Naukovy zapysky. Seriia khimichnyi i heohrafichnyi.
1 (1962). Also as no19 of whole series. (S6597).
Russian: Nauchno-issledovatel'skii institut pedagogiki. Nauchnye zapiski.
Seriia khimicheskaia i geograficheskaia.
In Ukrainian.

*(2600 ______. Universytet. Kiev.
Visnyk. Seriia heohrafii. no1- (1958-) Irregular but usually annual.
(N70-2:3298b; N75-1:1303c; S67616; FC-S:92). 1-8 (1958-1967) as Seriia
heolohii ta heohrafii; 9- (1967-) as Seriia heohrafii.
In Ukrainian with summaries in English and Russian.

2601 ______. ______. Heohrafichnyi fakul'tet. Kiev.
Zbirnyk. 1-5 (1950-1958). 1-3 (1950-1954) as Trudy.
Subseries of university's Naukovi zapysky. v1 (1950) as v9 no3;
v2 (1953) as v12 no2; 3 (1954) as v13 no3; v4 (1956) as v15 no10;
v5 (1958) as v17 no1 of whole series.
Russian: Geograficheskii fakul'tet. Sbornik. Subseries of Nauchnye
zapiski.
In Ukrainian or Russian.

2602 ______. ______. Naukovo-doslidnii instytut heohrafii. Kiev.
Trudy. 1-5 (1948-1952). (S6706).
In university's Naukovi zapysky. v1 (1948) as v7 no1; 2 (1948) as
v7 no8; 3 (1949) as v8 no8; 4 (1950) as v9 no1; 5 (1952) as v11 no6.
Russian: Nauchno-issledovatel'skii institut. Trudy. Nauchnye zapiski.

2603 KIROV. Gosudarstvennyi pedagogicheskii institut. Kirov.
Sbornik studencheskikh nauchnykh rabot estestvenno-geograficheskogo
fakul'teta. 1-2 (1956-1961). (S6878a).
1 (1956) entitled Sbornik studencheskikh nauchnykh rabot.

2604 KRAEVEDENIE. (Akademiia nauk S.S.S.R., Leningrad. Tsentral'noe biuro
kraevedeniia). Moskva. v1-6 no10 (1923-1929). Closed 1929.
(U-3:2316a; S7800; FC-1:364; GR-1:445).
v1-4 (1923-1927) 4 nos. per vol; v5-6 (1928-1929) 10 nos. per vol.
Merged with bureau's Izvestiia (not in this list) to form Sovetskoe
kraevedenie [2703].

2605 KYAKHTA. Kiakhtinskii (respublikanskii) muzei kraevedeniia [and] Geograficheskoe obshchestvo S.S.S.R. Kiakhtinskii otdel. Kyakhta.
Trudy. 1-18 (1899-1961). (S6487).

2606 LATVIJAS geografijas biedrība, Riga. (Société de géographie de Latvie).
Rīga.
Geōgrafiski raksti. Folia geographica. 1-6 (1928-1938). Closed
1938. (U-3:2363b; B-2:268b; F-4:516a).
In Latvian with abstracts, occasional articles and supplementary
table of contents in English, French, or German.

2607 LATVIJAS P.S.R. Zinātņu akadēmija. Ģeoloģijas un geografijas institūts. Riga.
Raksti. 1 (1947). (S8489; GR2:1249).
Superseded by Ģeoloģijas un derīgo izraktenu institūts and later by
Ģeoloģijas institūts. Trudy. 2- (1958-) (not in this list).
Russian: Akademiia nauk Latviiskoi S.S.R. Institut geologii i
geografii. Trudy.
In Latvian.

2608 LENINGRAD. Geograficheskii institut. Leningrad.
Biulleten'. no1-8 (Mr-D 1921). Closed 1921. (U-3:2389b; S9141; GR-1:89).
In 1926 name of issuing agency changed to Leningrad. Universitet. Geograficheskii fakul'tet. Superseded by Geograficheskii vestnik [2364].

2609 ______. ______.
Izvestiia. 1-6 (1919-1926). Closed 1926. (U-3:2389b; B-3:41b; S9142; FC-1:285; GR-1:341).

2610 ______. ______.
Trudy. 1 (1922). Only one issued. (B-3:41b; S9143).

2611 ______. Leningradskii gosudarstvennyi pedagogicheskii institut imeni. A. I. Gertsena. Leningrad.
Gertsenovskie chteniia: geografiia i geologiia. 18- (1965-). (N70-3:3443b; S8746d).

2612 ______. Leningradskii gosudarstvennyi pedagogicheskii institut imeni M. N. Pokrovskogo. Leningrad.
Uchenye zapiski. Estestvenno-geograficheskii fakul'tet. Seriia geograficheskaia. 1-4 (1940-1957). Closed 1958. (S8851).
1-3 as Geograficheskii fakul'tet.
v1 (1940) is v3 of parent series; 2 (1948), 6; v3 (1955) 10; 4 (1957), 35.
In 1958 the Institute was absorbed by Leningradskii gosudarstvenyi pedagogicheskii institut imeni A. I. Gertsena.

(2613 ______. Universitet.
Uchenye zapiski. Seriia geograficheskikh nauk. 1- (1939-) Irregular. (U-3:2393a; B-3:41a; FC-1:375; GR-3:1444).
v1 (1939) as Trudy Sablinskoi nauchno-uchebnoi stantsii.
v2 as v25 of parent series; 3, 50; 4, 56; 5, 104; 6 (1949), 124; 7 (1950), 125; 8 (1952), 152; 9 (1954), 166; 10 (1955), 199; 11 (1956), 213; 12 (1958), 226; 13 (1959), 269; 14 (1960), 292; 15 (1961), 298; 16 (1961), 309; 17 (1962), 315; [18] (1962), 317; 19 (1967), 327; 20 (1970), 342; 21 (1971), 358; 22 (1971), 360; 25 (1977), 388. Some numbers have supplementary title and table of contents in English.

2614 ______. ______.
______. Seriia geologo-pochvenno-geograficheskaia. no1-4 (1935-1937). Closed 1937. (U-3:2392a; B-3:41a; S9121; FC-1:375).
no1-4 published as Uchenye zapiski no1, 9, 10, 16 (1935-1937).
After no4 "geograficheskaia" dropped from title and therefore this serial from this list. Seriia geograficheskikh nauk [2613] began a new series.

2615 ______. ______.
Vestnik. Seriia biologii, geografii i geologii. 1954-1955 Quarterly. Closed 1955. (S9124; GR-3:1512).
Numbered also as Vestnik, no1, 4, 7, and 10 each year.
Superseded for geography by Vestnik Geologii i geografii [2616].

*(2616 ______. ______.
______. Geologiia, geografiia. 1956- Quarterly. (N70-3:3446b; N70-3:3446c; BS:506b; S9127; FC-S:79; GR-3:1511-1512).
Title varies: 1956-66: Vestnik. Seriia geologiia, geografiia.
Numbered also as Vestnik, no6, 12, 18, 24 each year.
In Russian with supplementary English table of contents from 1956 no3 and English abstracts from 1957 no1.

LENINGRAD. Universitet.

2617 ______. ______. Geografo-ekonomicheskii nauchno-issledovatel'skii institut. Sbornik. 1926-1928. Closed 1930. (U-3:2393b; B-3:41b; S8625; FC-2:594; GR-2:922).

2618 ______. ______. ______. Trudy. 1-5 (1922-1934); v3, 5 (1934) do not bear series title. (U-3:2393b; B-3:41b; S8626; FC-2:691).

2619 ______. ______. Kabinet fizicheskoi geografii. Trudy. (Sbornik trudov). 1-3 (1899-1906). (U-3:2393c; B-3:41b; FC-2:606).

2620 ______. ______. Nauchnaia sessiia. Tezisy dokladov po sektsii geograficheskikh nauk. 1950- (N70-3:3446a).

2621 ______. ______. Obshchestvo zemlevedeniia. St. Petersburg (Leningrad). Trudy. v1-3 (1902/1904-1909/1913).
Title: Trudy Obshchestva zemlevedeniia pri S.-Peterburgskom universitete.

(2622 LETOPIS' Severa. Sbornik po voprosam istoricheskoi geografii, istorii geograficheskogo otkrytia, issledovaniia i ekonomicheskogo razvitiia Severa. (Sovet po izucheniiu proizvoditel'nykh sil pri Gosplane S.S.S.R. Mezhduvedomstvennaia komissiia po problemam Severa). Moskva. 1- (1949-) Irregular. (U-3:2397a; S9606; GR-1:474).
1 (1949), 2 (1957), 3 (1962), 4 (1964), 5 (1971), 6 (1972), 7 (1975), 8 (1977).
Subtitle and issuing agency vary. v8 (1977) issued by Tsentral'nyi nauchno-issledovatel'skii ekonomicheskii institut pri Gosplane R.S.F.S.R. and Geograficheskoe obshchestvo S.S.S.R. Moskovskii filial, and bears subtitle: Sbornik po voprosam istorii ekonomicheskogo razvitiia i istoricheskoi geografii Severa.

(2623 LIETUVOS T.S.R. aukštųjų mokyklų mokslo darbai: Geografija ir geologija. Vil'nyus. 1- (1962-) Irregular but typically annual. (N70-3:3476c; N70-2:2279b; B73:195a; S9647).
Added title page in Russian: Nauchnye trudy vysshikh uchebnykh zavedenii Litovoskoi S.S.R.: geografiia i geologiia.
In Lithuanian or Russian.
Abstracts in other language and in German.

2624 LIETUVOS T.S.R. fizinė geografija. (Lietuvos T.S.R. mokslų akademija, Vil'nyus. Geologijos ir geografijos institutas). Vil'nyus. 1- (1958-). (BS:511b).

(2625 LIETUVOS T.S.R. mokslų akademija. (Akademiia nauk Litovskoi S.S.R.). Vilnius Darbai. Trudy. sB. Chemija, technika, fizinė geografija. Khimiia, tekhnika, fizicheskaia geografiia. 1955- Bimonthly. (N70-3:3477c; S9665; BS:511b).
Series B dates from 1955. Title of series B as chemistry, technology, and physical geography [1967-1970 as geography] dates from 1967.
In Russian or Lithuanian, with abstracts in other language. Abstracts also in English.

2626 ______. Geografijos skyrius. Vil'nyus. Moksliniai pranešimai. 1-15 (1955-1963). Closed 1963. (N70-3:3478a; BS:512a; S9680; GR-3:1783).
v1-14, no2 (1955-1962) issued by academy's Geologijos ir geografijos institutas.
Russian: Akademiia nauk Litovskoi S.S.R. Otdel geografii (earlier: Institut geologii i geografii). Nauchnye soobshcheniia.
In Lithuanian with Russian and English or German summaries.

LIETUVOS T.S.R. mokslų akademija.

2627 ______. Geologijos ir geografijos institutas.
Monografinė serija. Monograficheskaia seriia. 1- (1959-) Irregular. (N70-3:3478a; S9782, S9681).
In Lithuanian. Added title page in English and Russian.

2628 LITERATURA po russkoi geografii, statistike i etnografii. St.Petersburg. 1-9 (1864-1883). Closed 1883. (B-3:67b; FC-1:383; GR-1:485).
Title varies: Literatura po russkoi geografii, etnografii i statistike.
Supplement to Imperatorskoe russkoe geograficheskoe obshchestvo. Izvestiia [2377].

2629 L'VOV. Universytet. L'vov.
Visnyk. Seriia biolohichna, heohrafichna i heolohichna. 4 (1968).
Only one issued. (N70-3:3552b).
Supersedes and is in turn superseded by Visnyk in three separate series: Seriia biolohichna, Seriia heohrafichna [2630], and Seriia heolohichna.

(2630 ______. ______.
______. Seriia heohrafichna. 1-4 (1962-1966); 5- (1970-) Irregular but usually annual. (N70-3:3552b; S10034).
Russian: L'vov. Universitet. Vestnik. Seriia geograficheskaia.
In Ukrainian with Russian summaries.

2631 MAGAZIN für Russlands Geschichte, Länder- und Völkerkunde. Jelgava (Mitau). 1-2 (1825-1826). Closed 1826. (U-3:2510c; B-3:119b; GR-1:497).

2632 MAGAZIN zemlevedeniia i puteshestvii, geograficheskii sbornik. Moskva. 1-6 (1852-1860). Closed 1860? (U-3:2510c; B-3:119b; GR-1:497).

2633 MATERIALY dlia geografii i statistiki Rossi. (Sobrannye ofitserami general'nago shtaba). St. Petersburg [Leningrad]. 1-29 (1860-1868). (B-3:155a).

(2634 MATERIALY eksperimental'nykh issledovanii na statsionarakh. (Akademiia nauk S.S.S.R. Sibirskoe otdelenie. Institut geografii Sibiri i Dal'nego Vostoka). Irkutsk. 1- (1970-).
Abstracts in English.

(2635 MATERIALY gliatsiologicheskikh issledovanii: khronika, obsuzhdeniia. (Akademiia nauk S.S.S.R. Mezhduvedomstvennyi geofizicheskii komitet. Sektsiia gliatsiologii [and] Institut geografii). Moskva. 1- (1961-) Irregular.
Title in English: Academy of Sciences of the U.S.S.R. Soviet geophysical committee and Institute of geography. Data of glaciological studies: chronicles, discussion.
See Akademiia nauk S.S.S.R. Institut geografii. Materialy gliatsiologicheskikh issledovanii [2312] for other subseries.
Supplementary table of contents and abstracts in English.

(2636 MATERIALY k biobibliografii uchenykh S.S.S.R. Seriia geograficheskikh nauk. (Akademiia nauk S.S.S.R.). Moskva. 1- (1947-) Irregular. 1 (1947); 2 (1952); 3 (1961); 4 (1971); 6 (1976).
(U-1:112c; B-3:154b; S10276; GR-1:513, 515).

2637 MATERIALY k istorii geograficheskikh issledovanii Sibiri i Dal'nego Vostoka. Irkutsk. (N70-3:3679b).

2638 MATERIALY po chetvertichnoi geologii i geomorfologii S.S.S.R. Leningrad. 1-6 (1956-1967). (N70-3:3680a).
1-4 (1956-1961) subseries of Vsesoiuznyi nauchno-issledovatel'skii geologicheskii institut (VSEGEI): Materialy (not in this list). no1 (1956) is no17 of main series; no2 (1959), no27; no3 (1961), no42; no4 (1961), no34; Parent series closed 1961.
5-6 (1963-1967) subseries of institute's Trudy (not in this list). 5 (1963) no90 of this parent series; 6 (1967), no145.

(2639 MATERIALY po izucheniiu Stavropol'skogo kraia. (Stavropol'skii kraevedcheskii muzei [and, since 1960] Geograficheskoe obshchestvo S.S.S.R. Stravropol'skii otdel). Stavropol'. 1-14 (1949-1976) Irregular. (S10386).

2640 MATERIALY po paleogeografii. (Moskva. Universitet. Geograficheskii fakul'tet). 1-2 (1954-1961). (N70-3:3681b; S10413).

2641 MEKTITA geografiia. Geografiia v shkole: zhurnal dlia prepodavatelei nachal'noi i srednei shkoly. Kazan'. 1 (1935). Only one issued. In Tatar.

2642 MEZHDUVEDOMSTVENNOE soveshchanie po geografii naseleniia. Moskva; Leningrad. Materialy. 1-7 (1961-1965). (N70-3:3765a; S10755). Rotaprint except no6, printed.

2643 MINSK. Universitet. Geograficheskii fakul'tet. Minsk. Trudy. 1-2 (1958). (N70-3:3830c; S10920; GR-2:1205).

(2644 ______. ______.
Vestnik. s2. Khimiia, biologiia, geologiia, geografiia. 1969- 3 nos. a year.
1969 as Biologiia, khimiia, geologiia, geografiia.
In Russian.

2645 MONATSSCHRIFT zur Kenntniss der Geschichte und Geographie des russischen Reichs. Rīga. no1-12 (1790-1791). Closed 1791. (U-3:2728a). In German.

2646 MOSKOVSKOE obshchestvo ispytatelei prirody. Moskva.
Trudy. Sektsiia geografii.
Subseries not separately numbered. no8 (1964); no 36 (1970) of main series.
In Russian. Titles and abstracts in English.

2647 MOSKVA (Moscow). Institut mezhdunarodnykh otnoshenii. Kafedra ekonomicheskoi geografii i ekonomiki stran Zapada. Moskva.
Uchenye zapiski. 1 (1958). (N70-3:3902b; S11534).
Also as Voprosy ekonomicheskoi i politicheskoi geografii.

2648 ______. ______. Kafedra geografii i ekonomiki stran Vostoka. Moskva.
Uchenye zapiski. 1-3 (1956-1960). (N70-3:3902b; S11535).
v1 unnumbered and entitled: Voprosy ekonomiki stran Vostoka.

2649 ______. Moskovskii gorodskoi pedagogicheskii institut imeni V. P. Potemkina. Moskva.
Uchenye zapiski. Geograficheskii fakul'tet. 1-6 (1948-1959). Closed 1960. (S11229).
1 (1948) as Kafedra ekonomicheskoi geografii i kafedra fizicheskoi geografii.
3 (1953) as Kafedra fizicheskogo stranovedeniia i obshchei fizicheskoi geografii. 4-5 (1955-1957) as Trudy Geograficheskogo fakul'teta.
v1 (1949) as v9 of parent series; 2 (1950), 11; 3 (1953), 21; 4 (1955), 39; 5 (1957), 66, 6 (1959), 101.

MOSKVA

In 1960 the institute was united with Moskovskii gosudarstvennyi pedagogicheskii institut imeni V. I. Lenina.
Index: 1-6 (1948-1959) in 6 (1959), p. 281-282.

2650 ______. Moskovskii gosudarstvennyi pedagogicheskii institut imeni V. I. Lenina. Moskva.
Uchenye zapiski. Geograficheskii fakul'tet. 1938; 1-6 (1949-1960).
Unnumbered volume in 1938; v1 (1949) is v54 of parent series; 2 (1957), 106; 3 (1958), 120; 4 (1959), 121; 5 (1960), [117]; 6 (1960), 159.
Unnumbered volume as Uchenye zapiski Geograficheskogo fakul'teta; v3 (1958) as Geografiia; 4-5 (1959-1960) as Kafedra ekonomicheskoi geografii; 6 (1960) as Kafedra fizicheskoi geografii.

(2651 ______. Moskovskii institut inzhenerov geodezii, aerofotos"emki i kartografii. Moskva.
Trudy. 1- (1940-) Irregular. (S11504; GR-2:1301).

2652 ______. Moskovskii oblastnoi pedagogicheskii institut imeni N. K. Krupskoi. Moskva.
Uchenye zapiski. Ekonomicheskaia geografiia. 1-2 (1968-1969). (N70-3:3904c).
no1 (1968) as 208 of whole series; no2 (1969), as 243.

2653 ______. ______.
______. Obshchaia fizicheskaia geografiia i geologiia. 1-14 (1937-1971) Irregular. (N70-3:3905b; N70-3904c-3905a; S11697).
[1] (1937) and 2 (1939) unnumbered in main series Uchenye zapiski; 1 (3) (1940) v3 of parent series; 4 (1947), 9; 5 (1951), 17; 6 (1956), 47; 7 (1961), 93; 8 (1961), 94; 9 (1961), 97; 10 (1963), 124; unnumbered volume (1964), 153; 11 (1968), 180; 12 (1968), 181; 13 (1970), 267; 14 (1971), 303.
[1] (1937) entitled Miachkovskaia geograficheskaia stantsiia; 2 (1939) as Zapiski Geograficheskogo fakul'teta; 1 [3] (1940) as Uchenye zapiski Geograficheskogo fakul'teta; 4-6 (1949-1956) as Trudy kafedr Geograficheskogo fakul'teta; 7-9 (1961) as Trudy Kafedr geografii; 10 (1963) Fizicheskaia geografiia; unnumbered v (1964) Estestvenno-geograficheskii [vypusk]; 11 (1968) Obshchaia fizicheskaia geografiia i geologiia; 12 (1968) Obshchaia fizicheskaia geografiia; 13-14 (1970-1971) Obshchaia fizicheskaia geografiia i geologiia.

(2654 ______. Tsentral'nyi nauchno-issledovatel'skii institut geodezii, aeros"emki i kartografii. Moskva.
Trudy. 1- (1931-) Irregular. (U-3:2766b; S12263; GR-2:1190).
1-5 (1931-1932) issued by Gosudarstvennyi institut geodezii i kartografii.

2655 ______. Universitet. Moskva.
Uchenye zapiski. Geografiia. no5 (1936); 14 (1938); 21 (1938); 25 (1939); 29 (1939); 31 (1940); 37 (1939); 35 (1940); 38 (1940); 119 (1946); 160 (1952); 170 (1954); 189 (1959), Biogeografiia.
Parent series closed 1961. (U-3:2766c; B-3:261b; BS:573a; S12383; FC-2:765; GR-3:1429).

2656 ______. ______.
Vestnik. Seriia biologii, pochvovedeniia, geologii, geografii. 11-14 (Jl 1956-1959). Closed 1959. (N70-3:3911c; BS:573a; S12386; FC-S:85; GR-3:1516).
Supersedes in part university's Vestnik 1-11 (S 1946-Je 1956) (not in this list) and continues its numbering. Superseded in part by university's Vestnik, seriia 5: Geografiia [2657].

MOSKVA. Universitet. Vestnik.

*(2657 ______. ______.
______. Seriia 5. Geografiia. v15- (1960-) 6 nos. a year. (N70-3:3911c; BS:573b; B68:580a; S12395; FC-S:81; GR-3:1517). Supersedes in part university's Vestnik. Seriia biologii, pochvovedeniia, geologii, geografii [2656] and continues its numbering. Supplementary table of contents in English from v15 (1960) and English abstracts from v23 (1968).

2658 ______. ______. Geograficheskaia stantsiia "Krasnovidovo." Moskva. Trudy. no1-2 (1948). Closed 1948. (S12413).

2659 ______. ______. Geograficheskii fakul'tet. Moskva. Kompleksnye geograficheskie issledovaniia territorii Severnogo Kazakhstana. (N70-2:3333c).

2660 ______. ______. Khibinskaia geograficheskaia stantsiia. Apatity. Trudy. 1. Moskva. (1960). (N70-3:3910c; S12423).

2661 ______. ______. Nauchno-issledovatel'skii institut geografii. Moskva. Trudy. 1-4 (1925-1929). Closed 1929. (U-3:2767a; B-3:262a; S12436; FC-2:715; GR-2:1311).

2662 NA sushe i na more. (Gosudarstvennoe izdatel'stvo geograficheskoi literatury; 1964- Mysl'). Moskva. 1960- Annual. (N70-3:3975; S13535).

2663 NASHA strana. Moskva. 1-5 (Ap 1937-Jl 1941) Monthly. (U-4:2823c; S13774; GR-1:615).

2664 NAUCHNOE soveshchanie geografov Sibiri i Dal'nego Vostoka. (Akademiia nauk S.S.S.R. Sibirskoe otdelenie. Institut geografii Sibiri i Dal'nego Vostoka; Geograficheskoe obshchestvo S.S.S.R. Biuro sibirskikh i dal'nevostochnykh organizatsii. Novosibirskii otdel). Irkutsk. Materialy. 1- (1959-). (N68-2:1544b; S13998a).
Conferences held every few years: 1 (1959), 2 (1962), 3 (1966); 4 (1969).

2665 NAUCHNOE soveshchanie po problemam meditsinskoi geografii. (Akademiia nauk S.S.S.R. [and] Geograficheskoe obshchestvo S.S.S.R.). Leningrad. Doklady. 1- (1962-). (N70-3:4080b; S14001).
Conferences held every few years: 1 (1962), 2 (1965), 3 (1968), 4 (1972).
Each volume has also a distinctive title. v1: Problemy meditsinskoi geografii: doklady.

2666 NAUCHNYE doklady vysshei shkoly: geologo-geograficheskie nauki. (Ministerstvo vysshego obrazovaniia S.S.S.R.). Moskva. 1958-1959 no2 Quarterly. Closed 1959. (N70-3:4081b; BS:602a; S18972; FC-S:277; GR-1:624).
In Russian with supplementary English table of contents.

2667 NAUKOVE tovarystvo imeny Shevchenka, Lwów (L'vov). Heohrafichna komisiia. L'vov.
Pratsi. Abhandlungen. 1 (1935). All issued. (U-4:2939b; FC-2:526; GR-2:762).
German: Abhandlungen der geographischen kommission. Ukrainische Ševčenko Gesellschaft der Wissenschaften in Lemberg.
In Ukrainian and German.

2668 NOVOE v zhizni, nauke, tekhnike: Seriia nauk o zemle. (Vsesoiuznoe obshchestvo po rasprostraneniiu politicheskikh i nauchnykh znanii; Znanie). Moskva. 1966- 12 issues a year. (N70-3:4308b; S14398).
Subtitle varies: 1966-1967: Seriia XIII: Nauk o zemle.
Supersedes society's (Izdaniia). seriia 12. Geologiia i geografiia [2756].

(2669 NOVOSIBIRSK. Novosibirskii institut inzhenerov geodezii, aerofotos"emki i kartografii. Novosibirsk.
Trudy. 1- (1947-) Irregular. (S14433; GR-2:1323).

2670 NOVOSTI karstovedeniia i speleologii. (Akademiia nauk S.S.S.R. Mezhduvedomstvennaia komissiia po izucheniiu geologii i geografii karsta). Moskva. 1-3 (1960-1963). (N70-3:4309c; S14489).
Title varies: v1 (1960) as Informatsionnyi sbornik; materialy komissii po izucheniiu geologii i geografii karsta.

2671 OBSHCHESTVO izucheniia Kazakhstana. Orenburg.
Trudy. Alma-Ata. 1-10 (1921-1929). Closed 1929. (U-4:3130c; S14836; FC-2:718; GR-2:1325).
1 (1921) as Orenburgskoe obshchestvo izucheniia Kirgizskogo kraia; 2-6 (1922-1925) Obshchestvo izucheniia Kirgizskogo kraia.

2672 _____. Semipalatinskii otdel. Semipalatinsk.
Zapiski. 18-20 (1929-1932). Closed 1932. (U-4:3130c; S14837; FC-1:254).
v18 also as Geograficheskoe obshchestvo S.S.S.R. Semipalatinskii otdel Zapiski [2493].

2673 OBSHCHESTVO liubitelei estestvoznaniia, antropologii i etnografii. Geograficheskoe otdelenie. Moskva.
Memuary. 1-2 (1925). Closed 1925. (U-4:3131a; S14861; FC-1:409; GR-1:555).

2674 _____. _____.
Trudy. 1-2 (1894-1910). (FC-1:308; GR-2:1205).
v2 (1910) also as v119 of society's Izvestiia (not in this list).

2675 ODESSA. Universytet. Odessa.
Nauchnyi ezhegodnik. Geograficheskii fakul'tet.
1-2 (1957-1960). (S15060).
1 (1957) joint with Biologicheskii fakul'tet.

2676 _____. _____.
Pratsi-Seriia heoloho-heohrafichnykh nauk. 1-11 (1949-1962). (N70-3:4347a; S15067; GR-2:1327).
Subseries of university's Trudy, pratsi (not in this list).
Title varies: 1 (1949), Sbornik geologicheskogo fakul'teta; 2 (1954), Sbornik geologo-geograficheskogo fakul'teta; 3 (10) (1955-1962) as Seriia heolohichnykh i heohrafichnykh nauk.
In Ukrainian or Russian.

2677 OKHRANA prirody Moldavii. (Akademiia nauk Moldavskoi S.S.R. Otdel geografii. Komissiia po okhrane prirody). Kishinëv. 1-11 (1960-1972). (N70-3:4376a; S15183).
no1-6 (1960-1968) issued by Akademiia nauk Moldavskoi S.S.R. Komissiia po okhrane prirody.

2678 PESHCHËRY. Caves. (Geograficheskoe obshchestvo S.S.S.R. Permskii otdel [and] Perm'. Universitet. Institut karstovedeniia i speleologii). Perm'. 1 [2]-15 (1961-1974) Irregular. (B73:254b; S15841).
Also continues numbering of Speleologicheskii biulleten' 1 (1947).

2679 PRIRODA Sakhalina i zdorov'e cheloveka: Izvestiia Sakhalinskogo otdela Geograficheskogo obshchestva S.S.S.R. (Geograficheskoe obshchestvo S.S.S.R. Sakhalinskii otdel [and] Geograficheskoe obshchestvo S.S.S.R. Komissiia meditsinskoi geografii). Iuzhno-Sakhalinsk. 1-2 (1962-1971) Irregular.

2680 PROBLEMY fizicheskoi geografii. (Akademiia nauk S.S.S.R. Institut geografii). Leningrad, Moskva. 1-17 (1934-1951). Closed 1951. (U-4:3446b; B-3:608a; S16345; FC-2:537; GR-2:788).
Name of issuing agency: 1 (1934) Geomorfologicheskii institut; 2-3 (1935-1936) Institut fizicheskoi geografii.
Index: 1-16 (1934-1951).

(2681 PROBLEMY geografii Moldavii. (Akademiia nauk Moldavskoi S.S.R. Otdel geografii [and] Geograficheskoe obshchestvo Moldavii). Kishinëv. 1-10 (1966-1975) Annual. (N70-3:4785b; B73:268b; S16352a).
In Russian.

(2682 PROBLEMY heohrafichnoi nauky v Ukrains'kii R.S.R. (Akademiia nauk Ukrains'koi R.S.R. Sektor heohrafii). Kiev. 1- (1972-) Irregular. (N76).
In Ukrainian. Abstracts in Russian. v1 (1972) also has supplementary table of contents and abstracts in English.

2683 PROBLEMY kriolitologii. (Moskva. Universitet. Geograficheskii fakul'tet. Kafedra kriolitologii i gliatsiologii). Moskva. 1-4 (1969-1974). (N70-3:4785c).
Title in English: Problems of cryolithology.
Table of contents and preface also in English.

2683a PROBLEMY meditsinskoi geografii Kazakhstana. (Kazakhskoe geograficheskoe obshchestvo). Alma-Ata. 4 (1971). Rotaprint.

*(2684 PROBLEMY osvoeniia pustyn'. (Akademiia nauk Turkmenskoi S.S.R. Nauchnyi sovet po probleme pustyn'. Institut pustyn'). Ashkhabad. 1967- 6 nos. a year. (N70-3:4786a; B73:268b).
Title in English: Problems of desert development.
Title in Turkmen: Chelleri ezleshdirmegin problemalary.
In Russian. Supplementary English table of contents and abstracts.

2685 R.S.F.S.R. Glavnoe upravlenie shkol. Moskva.
Ukazaniia ob ispol'zovanii uchebnikov. Istoriia, geografiia. 1954/55- Annual. (N70-3:5116a).

*(2686 REFERATIVNYI zhurnal. Geografiia. (Vsesoiuznyi institut nauchnoi i tekhnicheskoi informatsii). Moskva. 1956- 12 issues a year. (N70-3:4948b; BS:724a; S17271; FC-S:323; GR-2:851).
Annual indexes: Predmetnyi i geograficheskii ukazateli [subject and geographical indexes]; avtorskii ukazatel' [author index], as separate volumes.
Author index in each issue from 1979, no1.
Supersedes in part: Referativnyi zhurnal. Geologiia i geografiia [2687].

2687 ______. Geologiia i geografiia. (Akademiia nauk S.S.S.R. Institut nauchnoi informatsii). Moskva. 1954-1955. Closed 1955. (N60-2:1719b; BS:7242; S17273; FC-S:323, 324; GR-2:851).
Superseded in part by Referativnyi zhurnal. Geografiia [2686].
Author index: 1954-1955.
Subject and geographical index: 1954-1955.

2688 ______. Otdel'nyi vypusk 35. Kraevedenie. (Akademiia nauk S.S.S.R. Institut nauchnoi informatsii). Moskva. 1962, no1-12. (N70-3:4949a; S17290).
Closed with 1962, no12.

2689 REGIONAL'NYI geograficheskii prognoz. (Moskva. Universitet. Geograficheskii fakul'tet). Moskva. 1- (1977-) Irregular.

2690 RĪGA. Universitāte. Rīga.
Geogrāfijas zinātes. Geograficheskie nauki. 1-4 and unnumbered issue. (1956-1965). (N70-3:5061a; S17473).
Issued in university's Zinātniskie raksti; Uchenye zapiski (not in this list).
no1 (1956) is no7 of whole series, 2 (1957), 27; 3 (1959), 31; no4 (1961) no37 of whole series; unnumbered issue (1965), no65.
In Latvian.

2691 ______. ______.
Uchenye zapiski aspirantov. Aspirantu zinātinskie raksti.
Sbornik rabot aspirantov Geograficheskogo fakul'teta. 1-2 (1963-1965). (N70-3:5060c; S17482).
no1 (1963) is no1 of main series; no2 (1965) is no3 of main series.
Abstracts in English.

2692 ______. ______. Fizikalas geogrāfijas katedra.
Voprosy fizicheskoi geografii Latviiskoi S.S.R. 1-2 (1972-1973). (N75-2:1885c).
Subseries of university's Zinātniskie raksti. Uchenyi zapiski.
no1 (1972) as no162 of parent series; no2 (1973) as no186.
Russian: Latviiskii gosudarstvennyi universitet imeni Petra Stuchki. Kafedra fizicheskoi geografii.
Also listed as: Geograficheskoe obshchestvo Latviiskoi S.S.R. (Akademiia nauk Latviiskoi S.S.R.).
In Russian and Latvian.

2693 RITMY prirody Sibiri i Dal'nego Vostoka. (Akademiia nauk S.S.S.R. Sibirskoe otdelenie. Institut geografii Sibiri i Dal'nego Vostoka [and] Geograficheskoe obshchestvo S.S.S.R. Biuro sibirskikh i dal'nevostochnykh organizatsii). Irkutsk.
1-2 (1967-1975) Irregular. (N70-3:5071a).

2694 ROSTOV-NA-DONU. (Rostov-on-the-Don). Universitet. Rostov-na-Donu.
Uchenye zapiski. (S17607).
Trudy geologo-geograficheskogo fakul'teta. 1-10 (1948-1958) Irregular.
1 (1948) as Trudy Geograficheskogo fakul'teta.
v1 (1948) as v13 of parent series; 4 (1952), 17; 5 (1954), 23; 6 (1955). 33; 7 (1956), 34; 8 (1959), 44; 9 (1958), 53; 10 (1958), 55.

2695 SBORNIK geograficheskikh, topograficheskikh i statisticheskikh materialov po Azii. (Russia. Voenno-uchenyi komitet). St. Petersburg [Leningrad]. no1-87 (1883-1914). Closed 1914. (U-5:3788c; B-4:31b; FC-2:597; GR-2:921-922).
Index: 1-86 (1883-1913) in 87 (1914), p. iii-xii.

2696 SBORNIK po istoricheskoi geografii Gruzii. (Akademiia nauk Gruzinskoi S.S.R. Institut istorii, arkheologii i etnografii). Tbilisi. 1-2 (1960-1964). (N70-4:5205a; S19780a).
Georgian: Sakartvelos istoriuli geografmis krebuli.
In Georgian or Russian.

2697 SBORNIK statei po kartografii. (Glavnoe upravlenie geodezii i kartografii). Moskva. 1-13 (1952-1961). Closed 1961. (N70-4:5206a; S19905; GR-2:944).

2698 SEVERNYI Kavkaz. (Stavropol'skii gosudarstvennyi pedagogicheskii institut [and] Geograficheskoe obshchestvo S.S.S.R. Stavropol'skii otdel). Stavropol'. 1-2 (1970-1973).

(2699 SEVERO-ZAPAD Evropeiskoi chasti S.S.S.R. (Leningrad. Universitet). Leningrad. 1- (1959-) Irregular. (N70-4:5298b; B73:314b). 1-2 (1959) as Severo-Zapad.

2700 SIBIRSKAIA zhivaia starina. (Geograficheskoe obshchestvo S.S.S.R. Vostochno-Sibirskii otdel). Irkutsk. 1-8/9 (1923-1929). Closed 1929. (U-5:3868b; S20400; GR-2:977).

*(2701 SIBIRSKII geograficheskii sbornik. (Akademiia nauk S.S.S.R. Sibirskoe otdelenie. Institut geografii Sibiri i Dal'nego Vostoka [and] Geograficheskoe obshchestvo S.S.S.R. Biuro sibirskikh i dal'nevostochnykh organizatsii). Novosibirsk, Moskva. 1- (1962-) Irregular. (N70-4:5322b; B68:509b; S20407; GR-2:978). Index: 1-5 (1962-1967).

2702 SISTEMATICHESKII ukazatel' statei v innostrannykh zhurnalakh. Geologiia, geofizika i geografiia. (Vsesoiuznaia gosudarstvennaia biblioteka innostrannoi literatury). Moskva. 1-12 (1948-1953). Closed 1953. (S12524). Subtitle varies. no1-2 (1948) Geologiia i geografiia.

2703 SOVETSKOE kraevedenie. (Tsentral'noe biuro kraevedeniia; Kommunisticheskaia akademiia. Kraevedcheskaia sektsiia). Moskva. 1930-1936. Closed 1936. (U-5:4032b; S21146; FC-2:634; GR-2:1029-1030). Succeeds Kraevedenie [2604]. 1930-1931 called years 7, 8 in continuation of Kraevedenie.

2704 SREDNE-AZIATSKOE geograficheskoe obshchestvo. Tashkent. Izvestiia. Zhurnal (Title also in English). 1-20 (1898-1932). Closed 1932. (U-5:4054b; B-3:764a; S21441; FC-1:319; FC-S:201; GR-1:383). v1-17(1898-1924) as (Imperatorskoe) Russkoe geograficheskoe obshchestvo. Turkestanskii otdel. Izvestiia. v18 (1928) as Gosudarstvennoe Russkoe geograficheskoe obshchestvo. Sredne-Aziatskii otdel. Izvestiia. Continued as Uzbekistanskoe geograficheskoe obshchestvo, Tashkent. Trudy [2727]. Index: v12-13 (1915-1917) in v14 no1 (1918), p. 2. v1-10 (1898-1914) in v10 (1914), no1, p. 2-4.

2705 ______. Otchet. 1897-1927 (1927). (BS:754b).

2706 STAVROPOL'SKOE obshchestvo dlia izucheniia Severo-Kavkazskogo kraia v estestvenno-istoricheskom, geograficheskom i antropologicheskom otnosheniiakh. Trudy. 1-3. St. Petersburg, Kiev, Stavropol? (1911-1913 [1914]). (GR-2:1372).

2707 STRANY Azii. Geograficheskie spravki. Moskva. 1955- (BS:839b).

*(2708 STRANY i narody Vostoka. Countries and peoples of the East. (Geograficheskoe obshchestvo S.S.S.R. Vostochnaia komissia). Moskva. 1- (1959-) Irregular, about one a year. (N70-4:5561b; S21660).

2709 SVERDLOVSK. Gosudarstvennyi pedagogicheskii institut [and] Geograficheskoe obshchestvo S.S.S.R. Sverdlovskii otdel. Sverdlovsk. Voprosy fizicheskoi i ekonomicheskoi geografii. 3-8 (1967- 1971). Irregular. (N75-2:2258b). Subseries of institute's Uchenye zapiski. no5 (1968), 80 of whole series as Geograficheskii sbornik; no8 (1971), 174 of whole series.

2710 TAMBOV. Gosudarstvennyi pedagogicheskii institut. Tambov. Nauchnye raboty studentov. Geografiia, biologiia. 1 (1963). Only one issued. (S22038).

*2711 TARTU. Ülikool (Universitet). Tartu.
Geograafia-alaseid töid. (Trudy po geografii). 1- (1960-)
Irregular. (N70-4:5700b; S22079).
Subseries of the university's Toimetised (Uchenye zapiski) (not in this list). no1 (1960), no88 of whole series; no2 (1962), no 128; no3 (1963), no144; no4 (1964), no156 in English; no5 (1969), no227; no6 (1969), no237; no7 (1969), no242; no8 (1971), no282, in English; no9 (1971), no288; no10 (1972), no296; no11 (1974), no317; no12 (1974), no341; 13 (1976), no393 in English; 14 (1976), no388; no15 (1978), no440; 16 (1977), no432.
In Estonian or Russian with summaried in other language. Supplementary table of contents and abstracts in English or German. Special volumes in English.

2712 ______. ______.
Toimetised. 1-4 (1946-1948). Closed 1948. (S22079).
Russian: Tartu. Universitet. Uchenye zapiski. Geologiia i geografiia.
In Estonian or Russian.

2713 ______. ______. Institutum geographicum. Tartu.
Publicationes. 1-24 (1925-1940). Closed 1940. (U-5:4154b; B-4:285b; F-3:1082b).
In English, Estonian, French, or German.

2714 ______. ______. Majandusgeograafia seminar. Tartu.
Üllitised...Publicationes. 1-27 (1931-1939). Closed 1939. (U-5:4154c; F-3:1082b).
In Estonian with English, French, or German summaries.

2715 TASHKENT. Sredne-Aziatskii universitet. Tashkent.
Nauchnye trudy. Geograficheskie nauki. 1-32 (1950-1965). (N70-4:5703c; S22204).
Subseries of university's Nauchnye trudy. 1-4 (1949-1953) as Geologo-geograficheskie nauki. Continues university's Trudy. s12a. Geografiia [2716].
1 (1949), is no12 of whole series; 2 (1950), 21; 3 (1952), 34?; 4 (1953), 38; 5 (1954), 50; 6 (1955), 64; 6a (1955), 70; 8 (1956), 80; 9 (1957), 95; 10 (1957), 99; 11 (1957), 107; 12 (1958), 120; 13 (1958), 129; 14 (1959), 155; 15 (1959), 156; 16 (1960), 157; 17 (1960) 167; 18 (1960), 175; 19 (1960), 182; 20 (1960), 183; 21 (1961), 185; 22 (1961), 186; [23] (1962), 193; 24 (1963), 213; 25 (1964), 217; 26 (1964), 226; 27 (1964), 231; 28 (1964), 236; [29] (1964), 237; 30 (1964), 248; unnumbered (1964), 251; 31 (1965), 266; and 32 (1965), 269. Later geographerical volumes not separately numbered.
In Uzbek or Russian.

2716 ______. ______.
Trudy. s12a. Geografiia. no1-17 (1928-1938). (U-5:4156b; B-4:213a; S22235; FC-2:733; GR-2:1371).

2717 ______. ______. Geograficheskii fakul'tet. Nauchno-issledovatel'skii otdel.
Trudy. 1-3 (1964). (N70-4:5703c; S22242).
Subseries of university's Nauchnye trudy. Geograficheskie nauki [2715]. no1 is no26 (226 of whole series); no2 is no30 (248 of whole series); and no3 is unnumbered (251 of whole series).

2718 TBILISI. Universitet. Tbilisi.
Trudy. Seriia geografo-geologicheskikh nauk. 1-4 (1959-1967). (S22740).
Georgian: Shromberi.
no1 (1959) is no72 of whole series; 2 (1963), 90; 3 (1965), 111; and 4 (1967), 122.
In Georgian and Russian.

TBILISI. Universitet.

2719 ______. ______.
Geograficheskii institut. Tbilisi.
Trudy. 1 (1936). Only one issued?
In Georgian with Russian and French or German summaries.

2720 TEMATICHESKII sbornik nauchnykh rabot po biologii i geografii. (Ministerstvo vysshego i srednego spetsial'nogo obrazovaniia Kazakhskoi S.S.R.).
Alma-Ata. 1 (1964). (N70-4:5733b).
General title: Sbornik statei aspirantov i soiskatelei.

2721 TOMSK. Institut issledovaniia Sibiri. Geograficheskii otdel. Tomsk.
Trudy 1-2 (1920-1921).
1-2 also as 2, 5 of Institute's Izvestiia (S22808), (not in this list).

2722 TURKMENSKOE geograficheskoe obshchestvo. Ashkhabad.
Trudy. 1-2 (1958-1961). Closed 1961. (N70-4:5888b; S23342).
(Geograficheskoe obshchestvo S.S.S.R. Turkmenskii filial).
Superseded by Voprosy geografii Turkmenistana [2751].

2723 UFA. Bashkirskii gosudarstvennyi pedagogicheskii institut. Ufa.
Uchenye zapiski. Seriia geologo-geograficheskaia. 1 (1955).
(N70-4:5907a; S23511).
v1 (1955) is v6 of parent series.

2724 ______. Bashkirskii gosudarstvennyi universitet. Ufa.
Uchenye zapiski. Seriia geograficheskikh nauk. 1-4 (1964-1972). (S23514).
Subseries of university's Uchenye zapiski. no1 (1964), no16 of whole series; no2 (1967), no30; no3 (1969), no37; no4 (1972), no58.
Title varies: Seriia geograficheskaia.

2725 UKRAINS'KYI istoryko-heohrafichnyi zbirnyk. (Instytut istorii). Kiev.
1-2 (1971-1972).
Russian: Ukrainskii istoriko-geograficheskii sbornik. (Institut istorii).
In Ukrainian.

2726 UZBEKISTANSKOE geograficheskoe obshchestvo. Tashkent.
Izvestiia. 1-13 (1955-1971) Irregular. (N70-4:6126c; B73:164b; S42191; FC-S:201).
Supersedes Uzbekistanskoe geograficheskoe obshchestvo. Trudy [2727] and Sredne-Aziatskoe geograficheskoe obshchestvo [formerly Russkoe geograficheskoe obshchestvo. Turkestanskii otdel]. Izvestiia [2704].
v1-2 also carry in parentheses numbers (22-23) as continuation of Trudy. v1-6 (1955-1962) issued by society under earlier name: Geograficheskoe obshchestvo S.S.S.R. Uzbekistanskii filial.

2727 ______.
Trudy. nsv1-2, no2 (21-22) (1937-1948). (S24192; FC-S:201).
Supersedes Sredne-Aziatskoe geograficheskoe obshchestvo. Izvestiia [2704] and continues its numbering in parentheses.
Superseded by Uzbekistanskoe geograficheskoe obshchestvo. Izvestiia [2726].
Index: 1897-1947 includes predecessors and other publications of the society in v2 (21) (1948), p. 171-201.

2728 V POMOSHCH' uchiteliu geografii. (Kalmytskii respublikanskii institut usovershchenstvovaniia uchitelei). Elista. 1-2 (1964-1968).

2729 V POMOSHCH' uchiteliu geografii; metodicheskoe posobie. (Novosibirsk. Gosudarstvennyi pedagogicheskii institut). Novosibirsk. 1-2 (1960-1962). (S24468).
Title varies: v1 (1960) has title: V pomoshch' uchiteliu geografii i khimii.

2730 V POMOSHCH' uchiteliu geografii SSh [sredeni shkoly] i NSSh [nepol'noi srednei shkoly]. (Omskii oblastnoi institut usovershenstvovaniia uchitelei. Kabinet geografii). Omsk. [1]-2 (1939). (S24469).
1 unnumbered and entitled: Metodicheskii sbornik v pomoshch' prepodavateliu geografii.

2731 VILNIUS. (Vil'nyus. Vilna). Universitetas. Vil'nyus.
Mokslo darbai; uchenye zapiski. Biologija, geografija, geologija; Biologiia, geografiia, geologiia.
1-7 (1949-1960). Closed 1960. (N70-4:6192c; S24823). 1-2 (1949-1954) are unnumbered and have title: Seriia estestvenno-matematicheskikh nauk. v3-4 (1955-1957) have title: Seriia biologicheskikh, geograficheskikh i geologicheskikh nauk; biologijos, geografijos ir geologijos mokslų serija.
no3 (1955), is no 7 of parent series; 4 (1957), 12; 5 (1958), 19; 6 (1959), 23; 7 (1960), 36.
In Lithuanian and Russian.

2732 VLADIVISTOK. Dal'nevostochnyi gosudarstvennyi universitet. Vladivostok. Trudy. Seriia 9. Geografiia. no1 (1927). Only one issued. (U-5:4415a; S24963; FC-2:694; GR-2:1215-1217).

2733 VOKRUG sveta. Moskva. 1861- Monthly. (U-5:4422a; B-4:485b; BS:944a; S25086; FC-1:110; GR-3:1545).
Suspended 1869-1884, 1917-1926, 1941-1945.
1861-1868 published in St. Petersburg.

2734 VOLOGODSKII krai. (Geograficheskoe obshchestvo S.S.S.R. Vologodskii otdel). Vologda. 1-4 (1959-1963). (N70-4:6222b; S25156).

2735 VOPROSY ekonomicheskoi geografii Urala i Zapadnoi Sibiri. (Perm'. Universitet [and] Geograficheskoe obshchestvo S.S.S.R. Permskii otdel). Perm'. 1-2 (1969-1970). (N75-2:2258a).
Subseries of university's Uchenye zapiski.
v1 (1969) as v202 of parent series; v2 (1970), 211.

2736 VOPROSY fizicheskoi geografii. (Geograficheskoe obshchestvo S.S.S.R. Saratovskii otdel). Saratov. 1-4 (1962-1971) Irregular. (N75-2:2258b; S25279).
Izdatel'stvo Saratovskogo universiteta.

2736a VOPROSY fizicheskoi geografii i geomorfologii nizhnego Povolzh'ia. (Saratov. Universitet). Saratov. 2-3 (1974-1975).

2737 VOPROSY fizicheskoi geografii Urala. Perm'. 1 (1973).
no1 is no308 of Perm'. Universitet. Uchenye zapiski (not in this list).

2738 VOPROSY geograficheskoi patologii; sbornik statei. (Yakutsk. Universitet. Meditsinskii fakul'tet). Yakutsk. 1-2 (1966). (S25295).

*2739 VOPROSY geografii. (Geograficheskoe obshchestvo S.S.S.R. Moskovskii filial. Nauchnye sborniki). Moskva. 1- (1946-) Irregular. Several each year. (U-5:4428b; BS:945a; S25296; FC-1:447; GR-3:1553).
In Russian; supplementary titles in table of contents in English 2-3 (1946-47), 46-55 (1959-61) and 70- (1966-) and in Esperanto 45-63 (1959-63). Abstracts in English 64- (1964-) and in Russian 67- (1965-).
Indexes: v1-90 (1946-1972), 91-100 (1972-1976).

2740 VOPROSY geografii Belorussii. (Geograficheskoe obshchestvo Belorusskoi S.S.R.). Minsk. 1-3 (1960-1972) Irregular. (N77; S25297).

(2741 VOPROSY geografii Dal'nego Vostoka. (Geograficheskoe obshchestvo S.S.S.R. Priamurskii [Khabarovskii] filial). Khabarovsk.
1- (1949-) Irregular. (S25298).
Index: 1-3 (1949-1957) in 4 (1960), p. 410-411.

2742 VOPROSY geografii i geologii. (Kazan'. Pedagogicheskii institut). Kazan'.
2-4 (1965-1967).
4 (1967) is v45 of institute's Uchenye zapiski.
Title varies: 2 (1965) as Voprosy geografii, geologii i metodiki prepodavaniia geologo-geograficheskikh distsiplin v pedagogicheskikh vuzakh.
3 (1966) as Voprosy geologii, geografii i kraevedeniia. (N70-4:6225c).

2743 VOPROSY geografii i kartografii; sbornik. (Moskva. Nauchno-izdatel'skii institut bol'shogo sovetskogo atlasa mira). Moskva.
nol (1935). Closed 1935. (U-5:4428b; S25299).

2744 VOPROSY geografii Iakutii. (Akademiia nauk S.S.S.R. Sibirskoe otdelenie. Iakutskii filial [and] Geograficheskoe obshchestvo S.S.S.R. Iakutskii filial). Yakutsk. 1-6 (1961-1973) Irregular. (N70-4:6225b; S25300).
vl unnumbered.

2745 VOPROSY geografii Iuzhnogo Urala.(Cheliabinskii gosudarstvennyi pedagogicheskii institut). Chelyabinsk. 1-5/6 (1966-1972) Irregular. (N70-4:6225c).
nol (1966) entitled Voprosy fizicheskoi geografii Iuzhnogo Urala.

(2746 VOPROSY geografii Kamchatki. (Geograficheskoe obshchestvo S.S.S.R. Kamachatskii otdel). Petropavlovsk-Kamchatskiy. 1- (1963-) Irregular. (N70-4:6225c; B73:361a; S25301).

2747 VOPROSY geografii Kazakhstana. (Akademiia nauk Kazakhskoi S.S.R., Alma Ata. Sektor fizicheskoi geografii [and] Geograficheskoe obshchestvo Kazakhskoi S.S.R.). Alma Ata. 1-17 (1956-1975) Irregular. (N70-4:6225c; BS:945a; S390; FC-S:98; GR-3:1553). Section's earlier names: Sektor geografii; Otdel geografii. Society earlier as Geograficheskoe obshchestvo S.S.S.R. Kazakhskii filial.
In Russian. 10 (1963) in Kazakh and Russian.

2748 VOPROSY geografii Kuzbassa i Gornogo Altaia. (Kemerovskii gosudarstvennyi pedagogicheskii institut). Mezhdurechensk; Kemerovo; Novokuznetsk.
1-6 (1968-1972). Rotaprint. (N75-2:2258c).
vl-4 no2 (1968-1971) issued by Geograficheskoe obshchestvo S.S.S.R. Kuznetskii otdel.

2749 VOPROSY geografii Severnogo Prikaspiia. (Geograficheskoe obshchestvo S.S.S.R. Zapadno-Kazakhstanskii otdel). Leningrad.
1 (1971). Rotaprint.

(2750 VOPROSY geografii Sibiri. (Geograficheskoe obshchestvo S.S.S.R. Tomskii otdel [and] Tomsk. Universitet). Tomsk. 1- (1949-) Irregular. (N70-4:6225c; S25302).

2751 VOPROSY geografii Turkmenistana. (Geograficheskoe obshchestvo Turkmenskoi S.S.R. Akademiia nauk Turkmenskoi S.S.R.). Ashkhabad. 3 (1976). Supersedes and continues numbering of Turkmenskoe geograficheskoe obshchestvo. Trudy [2722].

2752 VOPROSY geomorfologii i geologii Bashkirii. (Akademiia nauk S.S.S.R. Bashkirskii filial. Gorno-geologicheskii institut [and] Geograficheskoe obshchestvo S.S.S.R. Bashkirskii filial). Ufa. 1-2 (1957-1959). Closed 1959. (N70-4:6226a; S25318).

2753 VOPROSY istorii sotsial'no-ekonomicheskoi i kul'turnoi zhizni Sibiri i Dal'nego Vostoka. (Akademiia nauk S.S.S.R. Sibirskoe otdelenie. Institut istorii, filologii i filosofii [and] Geograficheskoe obshchestvo S.S.S.R. Novosibirskoe otdelenie. Sektsiia istoricheskoi geografii). Novosibirsk. 1-2 (1968). (N70-4:6227b; B73:361a).

2754 VOPROSY meditsinskoi geografii Zapadnoi Sibiri. (Geograficheskoe obshchestvo S.S.S.R. Novosibirskii otdel). Novosibirsk. 1-2 (1969-1970).

2755 VSEMIRNYI puteshestvennik. Illiustriovannyi zhurnal puteshestvii i geograficheskikh otkrytii. St. Petersburg. 1-12 (1867-1878). Closed 1878. (U-5:4433b; FC-1:122).

2756 VSESOIUZNOE obshchestvo po rasprostraneniiu politicheskikh i nauchnykh znanii. Moskva.
(Izdaniia). seriia 12. Geologiia i geografiia. 1959-1965? (N70-4:6240b; S26318).
Superseded by Novoe v zhizni, nauke, tekhnike: Seriia nauk o zemle [2668].

2757 VSESOIUZNOE soveshchanie po voprosam landshaftovedeniia. Moskva.
Teksty dokladov; materialy k soveshchaniiu. (N70-4:6242a; S26386).

2758 VSESOIUZNYI geograficheskii s"zed. Leningrad.
Trudy. 1st. Leningrad, 1934. 4v; 2nd. Leningrad, 1947. 3v. (S26402; FC-2:689).

2759 YAKUTSK. Universitet. Biologo-geograficheskii fakul'tet. Yakutsk.
Trudy. 1-2 (1970).

2760 YEREVAN (Erivan). Armianskii pedagogicheskii institut. Yerevan.
Nauchnye trudy. Seriia geograficheskikh nauk. 1-3 (1963-1968). Irregular. (N70-2:1944a; S3856).
1-2 (1963-1967) as Sbornik nauchnykh trudov. Gitakan ashkhatutiunneri zhokhovatsu.
Added title page in Armenian: Gitakan ashkhatutiunner.
In Armenian and Russian.

2761 ______. Universitet. Yerevan.
Nauchnye trudy. Seriia geograficheskikh nauk. 1-4 (1954-1958). (N70-2:1944b; S3911).
v1 (1954), v43 of whole series; v2 (1955), 51; v3 (1956), 58; v4 (1958), 63.
Later as seriia geologo-geograficheskikh nauk [2762], then as Estestvennye nauki (not in this list).
Armenian: Gitakan ashkhatutiunner.
In Armenian or Russian.

2762 ______. ______.
Seriia geologo-geograficheskikh nauk. 1965 (not separately numbered).
v99 of university's Uchenye zapiski. Gitakan tekhekagir.
Superseded by Estestvennye nauki (not in this list).
In Armenian or Russian.

*(2763 ZEMLEVEDENIE. (Moskovskoe obshchestvo ispytatelei prirody). Moskva.
nsv1(41)- (1940-) Irregular. (U-5:4625a; B-4:615a; S27394; GR-3:1656).
Continues numbering of old series [2764] in parentheses: 1(41)- .
Suspended 1941-1947.

2764 ZEMLEVEDENIE; geograficheskii zhurnal. (Obshchestvo liubitelei estestvoznaniia, antropologii i etnografii. Geograficheskoe otdelenie). Moskva. v1-40 no2 (1892-1938). Closed 1938.
(U-5:4625a; B-4:615a; S27394; FC-1:271; GR-3:1656).
Suspended 1918-1921, 1923. Superseded by Zemlevedenie [2763].
Indexes: 1894-1905, 1906-1917.

(2765 ZEMLIA i liudi. Geograficheskii kalendar'. Moskva. 1958- Annual.
(N70-4:6469b; FC-S:170; GR-3:1656).

2766 ZEMLIA i liudi. Nauchno-populiarnyi geograficheskii sbornik. (Akademiia nauk Uzbekskoi S.S.R.). Tashkent. 1962-1963 no1 Quarterly.
Closed 1963.
Subtitle of 1963 no1: Sbornik po geografii i geologii.
Uzbek: Er va el.
In Uzbek.

2767 ZHIVAIA starina. (Imperatorskoe Russkoe geograficheskoe obshchestvo. Otdelenie etnografii). St. Petersburg-Petrograd. 1-27 (1890-1916).
Closed 1916. (U-5:4634c; B-4:620a; BS:986b; FC-1:211; GR-3:1659).

(2768 ZHIZN' zemli: sbornik. (Moskva. Universitet. Muzei zemlevedeniia).
Moskva. 1-9 (1961-1973). (N70-4:6475b; B68:602a; S27530; GR-3:1663).

2769 ZONAL'NOE soveshchanie predstavitelei kafedr geografii pedagogicheskikh institutov Sibiri. Novosibirsk.
Trudy. 1- (1960-). (N70-4:6480a; S27641).

UNITED KINGDOM (English)

United Kingdom of Great Britain and Northern Ireland

(2770 ABERDEEN. University. Department of geography. Aberdeen.
O'Dell memorial monographs. 1- (1968-) Irregular. (B73:243b).
1 (1968) as O'Dell memorial essays in geography.

2771 ADVANCEMENT of science. (British association for the advancement of science). London. 1-27, no34 (1939-Jn 1971). Closed 1971.
(U-1:67c; N75-2:2343c; B-1:401b-402a; B73:6b).
Supersedes British association for the advancement of science. Report of the. . . meetings [2780]. Section E covers Geography.

*(2772 AREA. (Institute of British geographers). London. 1- (1969-) Quarterly.
(N70-1:433a; B73:29a).
[Institute of British Geographers, 1 Kensington Gore, London SW7 2AR, England.]

2773 BEDE transects. (Durham. Bede college. Department of geography). Durham.
1- (1974-).

(2774 BIBLIOGRAPHY series. (Geo abstracts). Norwich. 1- (1973-) Irregular.
[Geo Abstracts, University of East Anglia, Norwich NR4 7TJ, England.]

2775 BIRMINGHAM. University. Department of geography. Birmingham.
Occasional publications. (DP).
[Department of Geography, University of Birmingham, P.O. Box 363, Birmingham B15, 2TT, England.]

(2776 BLOOMSBURY geographer. (London. University college. Geographical society).
London. 1- (1968-) Annual. (N70-1:796a; B73:46a).
[The Bloomsbury Geographer, University College, London Union, 25 Gordon Street, London WCIH ORH, England.]

2777 BRATHAY exploration group, Ambleside, Westmorland.
Annual report and account of expeditions. 1959- Annual.

(2778 BRIGHTON polytechnic geographical society magazine. (Brighton polytechnic. Department of humanities). Falmer, Brighton, Sussex.
1- (1977-) 2 nos. a year.

2779 BRISTOL. University. Department of Geography. Bristol.
Seminar paper series.
Series A. no5-25 (1967-1974).
Series B. no1- (1968-).

2780 BRITISH association for the advancement of science.
Report of the. . . meetings. 1-108 (1831-1938). Closed 1938.
(U-1:776b; B-1:401b-402a; F-4:90a).
Superseded by Advancement of Science [2771]. See Section E. Geography, presidential addresses and sectional transactions.
In the early years geography often was grouped with geology or ethnology.
Index: 1831-1860; 1861-1890.

(2781 BRITISH cartographic society. Boreham Wood, Hertfordshire.
Special publications. 1- (1974-). (N75-1:339a; B75:61b).
[J. K. Wilcox, British Cartographic Society, 9 Kenilworth Close, Boreham Wood, Hertfordshire, England.]

(2782 BRITISH geomorphological research group. Norwich.
Current research in geomorphology. 1963/65- Irregular. (N70-1:875b; B68:163a).
[Geo Abstracts, University of East Anglia, Norwich NO4 7TJ, England.]

(2783 ______.
Occasional paper. 1- (1964-). (N70-1:875b; B68:398a; B73:240a).
(1-2) as Occasional publication.
[Geo Abstracts, University of East Anglia, Norwich NR4 7TJ, England.]

(2784 ______.
Technical bulletin. Norfolk. (N75-1:342c; B73:338a).
[Geo Abstracts, University of East Anglia, University Village, Norwich NR5 7TJ, England.]

2785 BRITISH landscapes through maps. (Geographical association). Sheffield.
1- (1960-) Irregular. (N70-1:882c).
[The Geographical Association, 343 Fulwood Road, Sheffield S10 3BP, England.]

(2786 BRITISH schools exploring society, London.
Report. 1937/39, 1947- Annual. (N70-1:886c; BS:711a).
1937-1947 as Public schools exploring society.
[British Schools Exploring Society, 175 Temple Chamber, Temple Avenue, London E.C. 44, England.]

2787 BRYCGSTOWE. (Bristol. University. Geographical society). Bristol.
1- (Spring 1965-). (N70-1:903c; B68:96a).
Mimeographed student publication.
[Department of Geography, University of Bristol, University Road, Bristol 8, England.]

*(2788 CAMBRIA. A Welsh geographical review. (Cylchgrawn daearyddol cymreig). Lampeter. 1- (1974-) 2 nos. a year. (N75-1:406c; B74:13b).
[D. A. Davidson, Department of Geography, Saint David's University College, Lampeter, Dyfed, Wales, United Kingdom.]

2789 CAMBRIDGE. University. Department of geography. Cambridge.
Geographical articles. 1- (1963-) Irregular. (DP).
[Department of Geography, Downing Place, Cambridge CB2 3EN, England.]

2790 CAMBRIDGE expeditions journal. (Cambridge. University. Explorers' and travellers' club. Journal). Cambridge. 1- (1965-) Annual. (N70-1:1048b; B68;121a).

*(2791 CAMBRIDGE geographical studies. (Cambridge university press). Cambridge. 1- (1969-) Irregular monographs.
[Cambridge University Press, Bentley House, 200 Euston Road, London N.W.1, England.]

*(2792 CARTOGRAPHIC journal. (British cartographic society). London.
1- (1964-) Semiannual. (N70-1:1151c; B68:127b).
Index: v1-8 (1964-1971).
[Department of Geography, King's College, Strand, London WC2R 2LS, England.]

2793 CARTOGRAPHICAL progress. (Royal geographical society). London.
1- (1960-) Irregular. (N70-1:1151c).
Reprints from the Geographical journal [2847].

*(2794 CATMOG. Concepts and techniques in modern geography. (Institute of British geographers. Quantitative methods study group). Norwich. 1- (1975-) Irregular. (B75:10b).
[Geo Abstracts, University of East Anglia, Norwich NR4 7TJ, England.]

2795 CITY of Leeds training college geographer. Leeds. 5- (1964-). (B68:136b).

*(2796 CLASSROOM geographer. Luton, Bedfordshire. D 1972- 8 nos. a year. (B75:12a).
[Classroom Geographer, Humanities Department, Brighton Polytechnic, Falmer, Brighton BN1 9PH, England.]

2797 COMPASS. (Cambridge. University. Geographical society). Cambridge.
v1 no1-3 (1948-1949). Closed 1949. (U-2:1150c; B-1:621a).

2798 COMPUTERS in the environmental sciences. (Geo abstracts). Norwich.
1970- Annual. (N75-1:578c; B73:80a).
1970 as Computers in geography.

2799 CONCEPTS in geography. London. 1- (1969-) Irregular. (N75-1:579c).
Introductory theoretical syntheses in human geography.
[Longmans Green and Co., Ltd., London, England.]

2800 CONFLUENCE. (Leicester university geographical society). Leicester.
1- (1964-) Irregular.
Supersedes Leicester geographical journal [2892].
[Leicester University Geographical Society, University Road, Leicester, England.]

2801 COSMOS. (London. University. Queen Mary college. Geographical society magazine). London. 1-12 (1946-1957). Closed 1957. (U-2:1213b; B-1:660b).
1946-1948: Geographical and geological society.
Superseded by the issuing agency's Geographical society magazine [2898].

2802 COSMOS. A geographical, philosophical and educational review. London.
1-2 (1883-1885). Closed 1885. (U-2:1213b; B-1:660b).

2803 DON. (Sheffield. University. Geographical society. Journal).
Sheffield. 1- (1957-) Annual. (N70-2:1778b; BS:266a).

(2804 DRUMLIN. (Glasgow. University. Geographical society. Magazine). Glasgow.
2- (1956-) Annual. (N78; BS:267b).

2805 DUNDEE. University. Department of geography. Dundee.
Occasional papers series. 1- (1971-). (N76; DP).
[Department of Geography, University of Dundee, Tower Block, Perth Road, Dundee DD1 4HN, Scotland, United Kingdom.]

(2806 DURHAM. University. Centre for Middle Eastern and Islamic studies. Durham.
Occasional papers series. 1- (1972-).
(N75-1:492c; B74:48a). [Bibliographies].

2807 ______. ______. ______.
Publications. 1- (1969-). (N70-1:1184a; B73:276a).
Subseries of Durham. University. Department of geography. Research paper series [2811].
[The Librarian, Department of Geography, Science Laboratories, South Road, Durham, England.]

2808 ______. ______. Department of geography. Durham.
Census research unit working papers. 1-5 (1975). (DP).

2809 ______. ______. ______.
Occasional papers series. 1-13 (1957-1962). Closed 1972.
(N70-2:1800b; DP).
Superseded by Department's Occasional publication [2810].

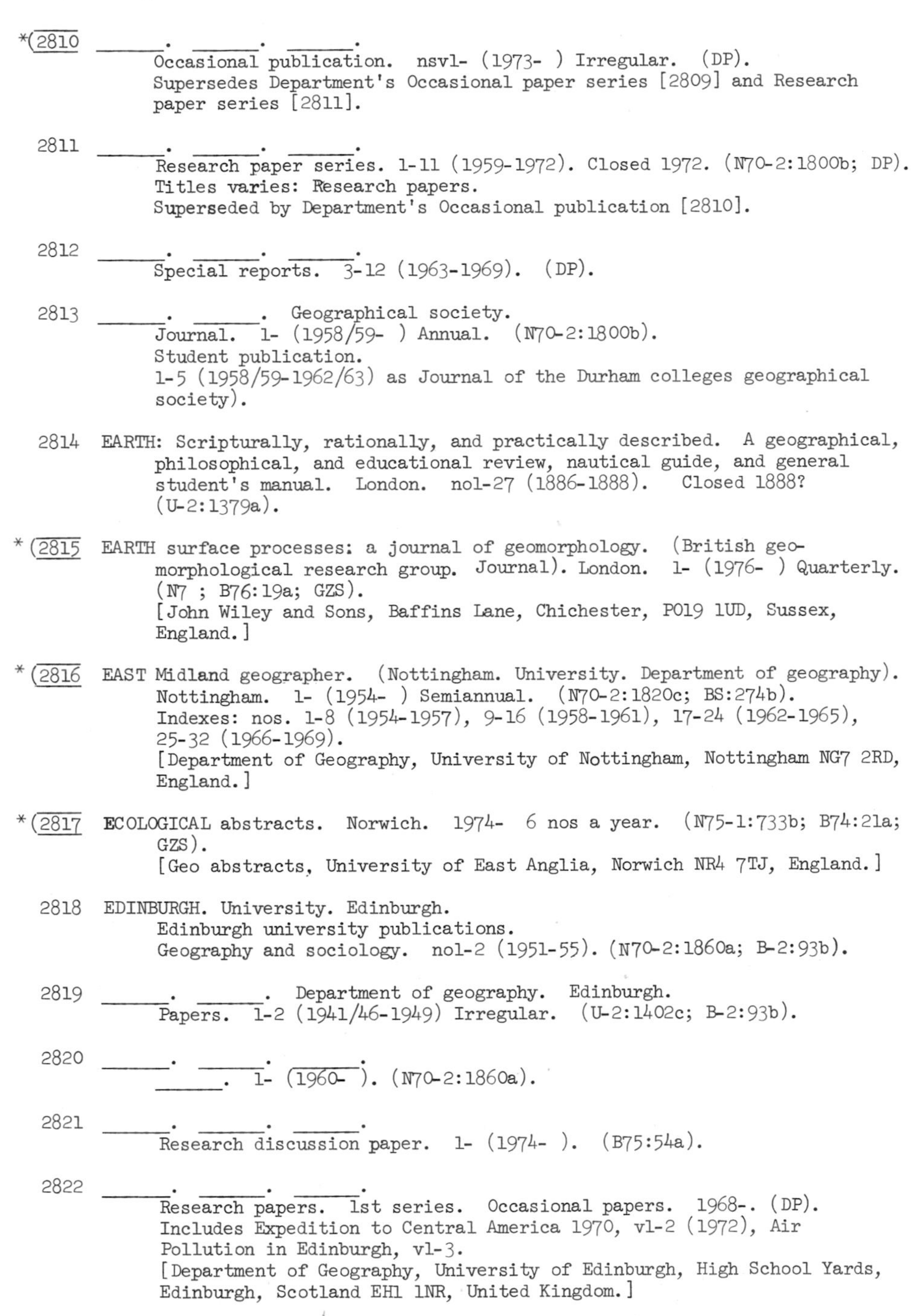

*(2810 ______. ______. ______.
Occasional publication. nsv1- (1973-) Irregular. (DP).
Supersedes Department's Occasional paper series [2809] and Research paper series [2811].

2811 ______. ______. ______.
Research paper series. 1-11 (1959-1972). Closed 1972. (N70-2:1800b; DP).
Titles varies: Research papers.
Superseded by Department's Occasional publication [2810].

2812 ______. ______. ______.
Special reports. 3-12 (1963-1969). (DP).

2813 ______. ______. Geographical society.
Journal. 1- (1958/59-) Annual. (N70-2:1800b).
Student publication.
1-5 (1958/59-1962/63) as Journal of the Durham colleges geographical society).

2814 EARTH: Scripturally, rationally, and practically described. A geographical, philosophical, and educational review, nautical guide, and general student's manual. London. no1-27 (1886-1888). Closed 1888? (U-2:1379a).

*(2815 EARTH surface processes: a journal of geomorphology. (British geomorphological research group. Journal). London. 1- (1976-) Quarterly. (N7 ; B76:19a; GZS).
[John Wiley and Sons, Baffins Lane, Chichester, PO19 1UD, Sussex, England.]

*(2816 EAST Midland geographer. (Nottingham. University. Department of geography). Nottingham. 1- (1954-) Semiannual. (N70-2:1820c; BS:274b).
Indexes: nos. 1-8 (1954-1957), 9-16 (1958-1961), 17-24 (1962-1965), 25-32 (1966-1969).
[Department of Geography, University of Nottingham, Nottingham NG7 2RD, England.]

*(2817 ECOLOGICAL abstracts. Norwich. 1974- 6 nos a year. (N75-1:733b; B74:21a; GZS).
[Geo abstracts, University of East Anglia, Norwich NR4 7TJ, England.]

2818 EDINBURGH. University. Edinburgh.
Edinburgh university publications.
Geography and sociology. no1-2 (1951-55). (N70-2:1860a; B-2:93b).

2819 ______. ______. Department of geography. Edinburgh.
Papers. 1-2 (1941/46-1949) Irregular. (U-2:1402c; B-2:93b).

2820 ______. ______. ______.
______. 1- (1960-). (N70-2:1860a).

2821 ______. ______. ______.
Research discussion paper. 1- (1974-). (B75:54a).

2822 ______. ______. ______.
Research papers. 1st series. Occasional papers. 1968-. (DP).
Includes Expedition to Central America 1970, v1-2 (1972), Air Pollution in Edinburgh, v1-3.
[Department of Geography, University of Edinburgh, High School Yards, Edinburgh, Scotland EH1 1NR, United Kingdom.]

2823 ______. ______. ______.
______. 2nd series. TRRU research reports. (Tourism and recreation unit). 1- (1970-). (DP).
[Department of Geography, University of Edinburgh, High School Yards, Edinburgh, Scotland, EH1 1NR, United Kingdom.]

2824 EDINBURGH journal of natural and geographical science. Edinburgh. 1-3 (O 1829-1831). Closed 1831. (U-2:1403b; B-2:95b; F-2:456b).

*(2825 ENVIRONMENT and planning: Series A. International journal of urban and regional research (Pion). London. 1- (1969-) Monthly. (N70-2:1935a; B73:107b).
[Pion Ltd., 207 Brondesbury Park, London NW2 5JN, England.]

2826 EXPEDITION. (World expeditionary association). London. 4- (1974-) 6 nos. a year.
[World Expeditionary Association, 45 Brompton Street, London SW3, England.]

2827 EXPLORATION. London. 1 (My 1961). Closed 1961. (N70-2:2015b; B68:204b).

2828 EXPLORATION review. (London. Imperial college of science and technology. Imperial college exploration society. Journal). London. 1- (1960-) Annual. (N70-2:2015c; B68:204b).

(2829 FIELD studies. (Field studies council). London. no1- (My 1959-) Annual. (N70-2:2072b; BS:311b).
Indexes: every 5 years.
[Field Studies Council, 9 Devereux Court, Strand, London, WC2R 3JR, England.]

2830 G.E. (Edinburgh. University. Geographical society. Magazine). Edinburgh. 1- (1950-). (N70-2:2243c; B-2:258b; BS:339a).
Student publication.

*(2831 GEO ABSTRACTS. Annual index. Norwich. 1966- Annual. (N75-1:903c; GZS).
1966-1971 in single volume for each year, 1972- in two separate volumes: v1 covers series A, B, and E (1972, 1973) and series A, B, E, and G (1974-); v2 covers series C, D, and F (1972-).
1966-1971 as Geographical abstracts. Annual index.
[Geo Abstracts, University of East Anglia, Norwich, NR4 7TJ, England.]

*(2832 ______. A. Landforms and the quaternary. Norwich. 1966- Bimonthly. (N70-2:2280bc; N75-1:902b; B68:224b; B73:127a; GZS).
1966-1971 as Geographical abstracts. Series A. Geomorphology.
Supersedes Geomorphological abstracts [2859].
Indexes: 1966-1970, 1971-1975.
[Geo Abstracts, University of East Anglia, Norwich NR4 7TJ, England.]

*(2833 ______. B: Climatology and hydrology. Norwich. 1966- Bimonthly. (N70-2:2280c; N75-1:902c; B68:225a; B73:127a; GZS).
1966-1971 as Geographical abstracts. Series B. Biogeography and climatology. 1972-1973 as Geo abstracts. Series B. Biogeography and climatology.
Index: 1966-1970.
[Geo Abstracts, University of East Anglia, Norwich, NR4 7TJ, England.]

*(2834 ______. C: Economic geography. Norwich. 1966- Bimonthly. (N70-2:2280c; N75-1:902c; B68:127b; B73:127b; GZS).
1966-1971 as Geographical abstracts. Series C. Economic geography.
Index: 1966-1970.
[Geo Abstracts, University of East Anglia, Norwich, NR4 7TJ, England.]

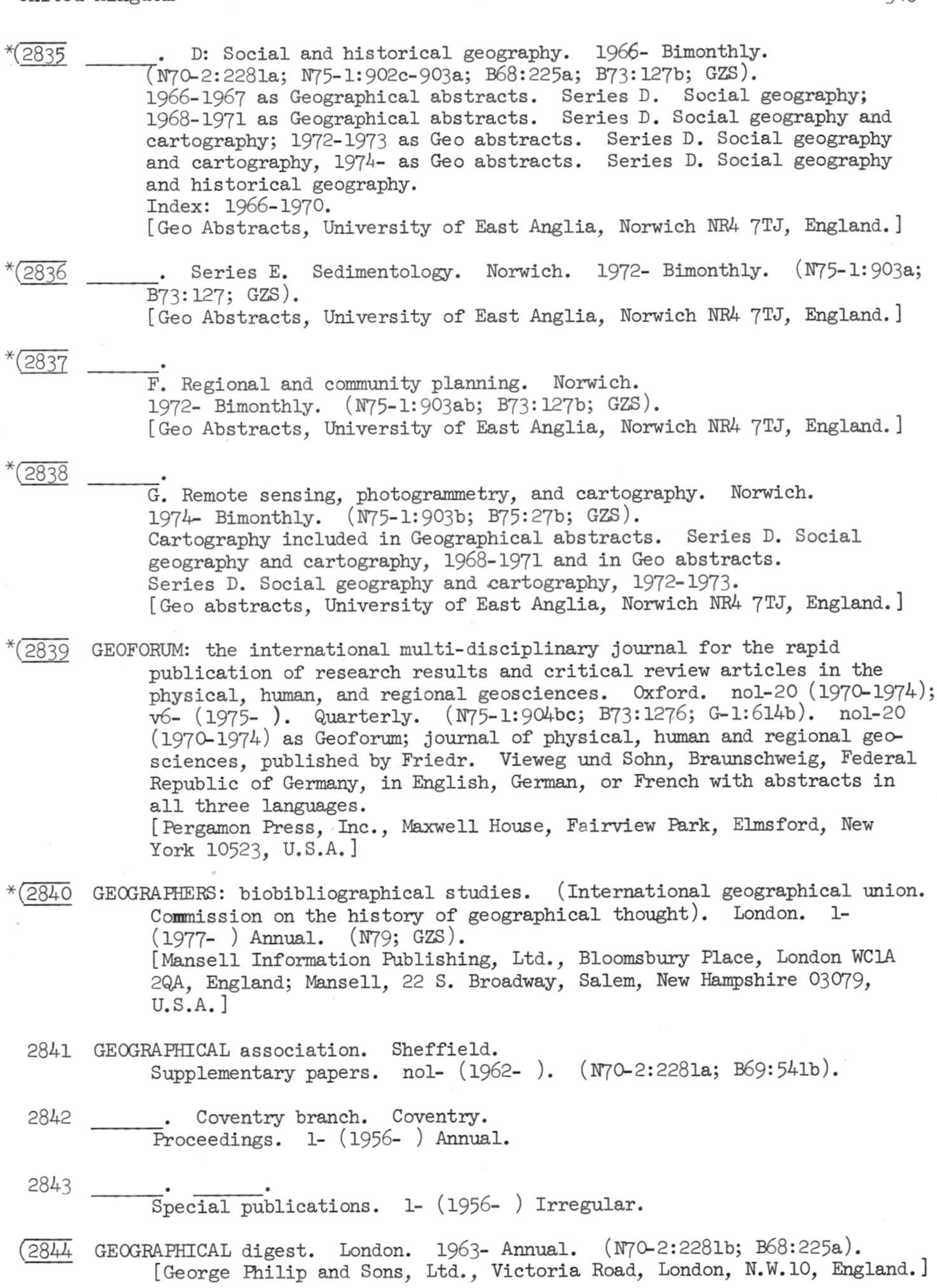

*(2835 ______. D: Social and historical geography. 1966- Bimonthly. (N70-2:2281a; N75-1:902c-903a; B68:225a; B73:127b; GZS).
1966-1967 as Geographical abstracts. Series D. Social geography; 1968-1971 as Geographical abstracts. Series D. Social geography and cartography; 1972-1973 as Geo abstracts. Series D. Social geography and cartography, 1974- as Geo abstracts. Series D. Social geography and historical geography.
Index: 1966-1970.
[Geo Abstracts, University of East Anglia, Norwich NR4 7TJ, England.]

*(2836 ______. Series E. Sedimentology. Norwich. 1972- Bimonthly. (N75-1:903a; B73:127; GZS).
[Geo Abstracts, University of East Anglia, Norwich NR4 7TJ, England.]

*(2837 ______.
F. Regional and community planning. Norwich.
1972- Bimonthly. (N75-1:903ab; B73:127b; GZS).
[Geo Abstracts, University of East Anglia, Norwich NR4 7TJ, England.]

*(2838 ______.
G. Remote sensing, photogrammetry, and cartography. Norwich.
1974- Bimonthly. (N75-1:903b; B75:27b; GZS).
Cartography included in Geographical abstracts. Series D. Social geography and cartography, 1968-1971 and in Geo abstracts. Series D. Social geography and cartography, 1972-1973.
[Geo abstracts, University of East Anglia, Norwich NR4 7TJ, England.]

*(2839 GEOFORUM: the international multi-disciplinary journal for the rapid publication of research results and critical review articles in the physical, human, and regional geosciences. Oxford. no1-20 (1970-1974); v6- (1975-). Quarterly. (N75-1:904bc; B73:1276; G-1:614b). no1-20 (1970-1974) as Geoforum; journal of physical, human and regional geosciences, published by Friedr. Vieweg und Sohn, Braunschweig, Federal Republic of Germany, in English, German, or French with abstracts in all three languages.
[Pergamon Press, Inc., Maxwell House, Fairview Park, Elmsford, New York 10523, U.S.A.]

*(2840 GEOGRAPHERS: biobibliographical studies. (International geographical union. Commission on the history of geographical thought). London. 1- (1977-) Annual. (N79; GZS).
[Mansell Information Publishing, Ltd., Bloomsbury Place, London WC1A 2QA, England; Mansell, 22 S. Broadway, Salem, New Hampshire 03079, U.S.A.]

2841 GEOGRAPHICAL association. Sheffield.
Supplementary papers. no1- (1962-). (N70-2:2281a; B69:541b).

2842 ______. Coventry branch. Coventry.
Proceedings. 1- (1956-) Annual.

2843 ______. ______.
Special publications. 1- (1956-) Irregular.

(2844 GEOGRAPHICAL digest. London. 1963- Annual. (N70-2:2281b; B68:225a).
[George Philip and Sons, Ltd., Victoria Road, London, N.W.10, England.]

2845 GEOGRAPHICAL field group. Nottingham.
Regional studies. 1- (1957-) Irregular. (N70-2:2281b).
[Geographical Field Group, c/o Department of Geography, University of Nottingham, Nottingham, NG7 2RD, England.]

2846 GEOGRAPHICAL intelligence, for the better understanding of foreign news, The. London. 1 (19 Je, 1689). No more published? (B-2:268b).

*(2847 GEOGRAPHICAL journal. (Royal geographical society, London). London. 1- (1893-) Quarterly. (U-2:1694a; B-2:269a; F-2:727b).
Includes the society's proceedings. Supersedes Royal geographical society. Proceedings. ns [2954].
Indexes: no1 v1-20 (1893-1902), no2 v21-40 (1903-1912), no3 v41-60 (1913-1922), no4 v61-80 (1923-1932), no5 v81-100 (1933-1942), no6 v101-120 (1943-1954), no7 v121-130 (1955-1964), no8 v131-140 (1965-1974).
[Royal Geographical Society, 1 Kensington Gore, London, S.W.7 2 AR, England.]

2848 GEOGRAPHICAL magazine. London. 1-5 (Ap 1874-D 1878). Closed 1878. (U-2:1694b; B-3:476b; F-2:727b).
Supersedes Ocean highways [2925].

*(2849 GEOGRAPHICAL magazine. London. 1- (My 1935-) Monthly. (U-2:1694c; B-2:268b; BS:341b).
Index: v23-45 (1950-1972).
[1 Kensington Gore, London, S.W.7 2AR, England.]

2850 GEOGRAPHICAL magazine, or the universe display'd. London. 1-3 (1790-1792). Closed 1792? (B-2:268b).

2851 GEOGRAPHICAL studies. London. 1-5 (1954-1958). Closed 1958. (N70-2:2281c; B-2:269b; BS:342a).

*(2852 GEOGRAPHY. (Geographical association). London; Sheffield. 1- (O 1901-) Quarterly. (U-2:1697a; B-2:269b; BS:342a; F-2:728a).
1-13 (1901-1926) as Geographical teacher.
Index: v1-54 (1901-1969).
[G. Philip and Son, Ltd., 12-14 Long Acre, London WC2E 9LP, England.]

2853 ______.
Supplement. 1-2 (1925-1926). Closed 1926. (U-2:1697a; B-2:269b).

2854 GEOGRAPHY. Notices and notes. (Manchester geographical society). Manchester. no1-93/94 (S 1895-1904). Closed 1904. (U-2:1697a; B-2:270b; F-2:729b).

(2855 GEOGRAPHY departments of universities and polytechnics in the British Isles. Theses in preparation 1967/68- . Theses completed 1967- Annual.
Issued by Heads of geography departments conference and Human geography subcommittee, Social science research council. Title varies slightly. Place of publication varies.
[Professor James H. Johnson, Department of Geography, University of Lancaster, Lancaster LA1 4YR, England.]

2856 GEOGRAPHY forum. Belfast.
[Ernest Brown, Editorial committee, Down County Education Office, Windsor Avenue, Belfast 9, Northern Ireland.]

2857 GEOGRAPHY in colour series. (Ward Lock educational company). London 1- (1967-). (N70-2:2282b).

2858 GEOGRAPHY of railways. London. 1 (1886). (B-2:270b).

2859 GEOMORPHOLOGICAL abstracts. London. no1-27 (Je 1960-D 1965). Closed 1965. (N70-2:2288b; BS:343b; B68:227a).
Superseded by Geo abstracts. Series A: Landforms and the quaternary [2832].
Index: 1960-1965.

2860 GEOMORPHOLOGICAL REPORT. (Nottingham. University. Department of geography). Nottingham.
nol- (1970-). (N76; B73:129a).

2861 GEOPHILE. (Cambridgeshire college of arts and technology, Cambridge. Geography department). Cambridge.
1- (1967/68-). (B73:129a).

2862 GEOPHYSICAL abstracts. Norwich. 1977- 6 nos. a year. (N77; B78).
Earlier publication of the same title issued jointly by the U. S. Bureau of Mines and U. S. Geological Survey to 1971.
[Geo abstracts, University of East Anglia, Norwich NR4 7TJ, England.]

2863 GEOSCIENCE documentation: journal for geoscience information. (Geosystems), London. 1- (Jl 1969-) Bimonthly. (N70-2:2302a; B73:129b).
[Geosystems, P.O. Box 1024, Westminster, London SW1P 2JL, England.]

2864 GLASGOW. University. Department of geography. Geographical field group Glasgow.
Regional studies. 1- (1957-). (N70-2:2357a).
Some numbers lack series title and number.

2865 GLOBE. (Manchester. University. Geographical society). Manchester.
nol-17 (1924-1937). Closed 1937? (U-2:1736a; B-2:304b).

2866 GREAT BRITAIN. Permanent committee on geographical names for British official use. London.
Gazetteers. 1942; 1946. No series numbers.

2867 ______. ______.
General lists and lists. 1921-1941. Closed 1941.
Each continent, region and country a separate series.

2868 ______. ______.
Glossaries. 1-8 (1942-1954). Closed 1954. (B-2:269b).

2869 ______. ______.
List of names, new series. 1- (1954-) Irregular. (N70-3:4594b).
Name of issuing office varies.

2870 ______. ______.
Mimeographed circulars. 1-21 (1932-1949). Closed 1949.

2871 GRIFFIN. (London. North-western polytechnic. Geography department. Journal). London.
1-2 (1970-1971). Closed 1971. (N76; B73:133a).
Superseded by North London geographer [2917].
Student publication.

2872 HORIZON. (London. University. King's college [and] London school of economics geographical association). London. 1951- (N70-2:2580a; B-2:418b).
Student publication.

2873 HUDDERSFIELD polytechnic. Department of geography and geology. Huddersfield, West Yorkshire.
Local information paper. 1973-. (N77; B74:41a).

2874 ______. ______.
Occasional paper. 1- (1974-). (B77:44b).

2875 HULL. University. Department of geography. Hull.
Departmental monographs.

2876 ______. ______. ______.
Miscellaneous series in geography. no1- (1956-) Irregular. Processed. (N70-2:2603a; DP).
Title varies: Miscellaneous series.
[Secretary, Department of Geography, University of Hull, Hull, Yorkshire, England.]

*(2877 ______. ______. ______.
Occasional papers in geography. no1- (1963-) Irregular, about two a year. (N70-2:2603a; B68:397a; DP).
Abstracts in English, French, and German, no17, 19-22 (1970-1975) and in English and German, 23-24 (1978).
[The Secretary, The Publications Committee, The University, Hull HU6 7RX, England.]

2878 IDEAS in geography. (Nottingham. University. Department of geography). Nottingham. 22-52 (1969-1975). Closed 1975. (N75-1:1043b).
Continues numbering of Nottingham. University. Department of geography. Bulletin of quantitative data for geographers [2922].

2879 ILLUSTRATED travels: record of geography, discovery and adventure. London. 1-6 (1869-1874). Closed 1874. (U-3:1937a; B-2:446a).

*(2879a IMAGO mundi. Journal of the International society for the history of cartography. Berlin, Amsterdam, Stockholm; London. 1- (1935-) Annual. (U-3:1941c; BS:402a; F-2:857b).
Suspended 1940-1946, 1957-1958. Subtitle varies.
Index: 1-10 (1935-1953); 11-20 (1954-1966).
In English.
[Imago Mundi, c/o Lympne Castle, Kent, England.]

(2879b ______.
Supplement. 1- (1958-) Irregular. (N70-2:2684a; BS:402a).

*(2880 INSTITUTE of British geographers. London.
Special publications. 1- (1968-) Irregular. (N70-2:2816a; B73:324a).
[Institute of British Geographers, 1 Kensington Gore, London, SW7 2AR, England.]

*(2881 ______.
Transactions. 1-66 (1935-1975). nsv1- (1976-) Quarterly. (U-3:2007b; N76; B-2:496a; BS:422a).
1-35 (1935-1964) as Transactions and Papers.
Index: 1-42 (1935-1967) in no42 (1967).
[1, Kensington Gore, London, SW7 2AR, England.]

2882 ______.
Urban study group. London.
Occasional publication. 1- (1972-). (N75-1:1117b; B73:242a).
[1, Kensington Gore, London, SW7 2AR, England.]

2883 INTERNATIONAL catalogue of scientific literature (published for the International council by the Royal society of London.) London.
Section J. Geography. Mathematical and physical. v1-14 (1902-1916). Closed 1916. (B-2:520a; F-2:912b).

*(2883a INTERNATIONAL journal of urban and regional research. London. 1- (1977-) Quarterly.
[Edward Arnold (Publishers) Ltd., 41 Bedford Square, London WC1B 3DQ, England.]

*(2884 JOURNAL of biogeography. (Blackwell scientific publications). Oxford. 1- (1974-) Quarterly. (N75-1:1242c-1243a; B74:35b; GZS).
[Blackwell Scientific Publications, Ltd., Osney Mead, Oxford, OX2 0EL, England.]

*(2885 JOURNAL of geography in higher education. (Oxford. Polytechnic. Faculty of modern studies). Oxford. 1- (1977-) 2 nos. a year. (N77; B77:28a).
[Geography Publications, Faculty of Modern Studies, Oxford Polytechnic, Headington, Oxford OX3 OBP, England.]

(2886 JOURNAL of glaciology. (International glaciological society). London; later Cambridge.
v1- (1947-) Semiannual. (U-3:2218b; BS:91a).
1-3 (1947-1961) published by British glaciological society; 4-10 (1962-1971) by the Glaciological society.
In English. Abstracts in English, French, and German.
[Scott Polar Research Institute, Cambridge CB2 1ER, England.]

*(2887 JOURNAL of historical geography. (Academic press). London; New York.
1- (1975-) Quarterly. (N75-1:1257c-1258a; B75:30a; GZS).
[Academic Press, 24-28 Oval Road, London NW1, England, or
111 Fifth Avenue, New York, New York 10003, U.S.A.]

2888 KINGSTON geographer. (Kingston-upon-Thames. Kingston college of technology. Geography club [and] Geographical association. Kingston branch. Journal). Kingston-upon-Thames. v1 no1- (O 1968-). Mimeographed. (B73:186b).

2889 KINGSTON polytechnic. Kingston-upon-Thames. School of geography. Kingston-upon-Thames.
Occasional papers. 1- (1973-).

2890 LAND utilisation survey of Britain. London.
Report. (The land of Britain). 1-92 (1936-1950). (B-3:12a).

*(2891 LEEDS. University. Department of geography. Leeds.
Working paper. 1- (1970-) Irregular. (N75-2:1337c; DP).
Most of the papers in this series are intended for later publication in journals and books. A list of working papers, available from the department, records availability of the working papers, and place of publication of revised papers; 241 working papers were included in the January 1979 list.
[Department of Geography, University of Leeds, Leeds LS2 9JT, England.]

2892 LEICESTER geographical journal. (University of Leicester students' union geographical society). Leicester. 1960-1964. Annual. Closed 1964. (N70-3:3438a).
Superseded by Confluence [2800].

2893 LIVERPOOL. University. Department of geography. Liverpool.
Research paper. 1- (1962-) Irregular. (N70-3:3504c; B68:472; DP).
[P.O. Box 147, Liverpool, L69 3BX, England.]

(2893a ______. ______. ______.
Discussion papers. 1- (Jl 1978-) Irregular.
[Secretary, Discussion Paper Series, Department of Geography, University of Liverpool, Roxby Building, P.O. Box 147, Liverpool L69 3BX, England.]

2894 ______. ______.
Geographical society. Liverpool.
Journal. 1-3 (1972-1974). (B75:67a).

2895 LIVERPOOL geographical society. Liverpool.
Transactions and annual report. 1-41 (1892-1932). Closed 1932. (U-3:2447c; B-3:73b).
1-4 (1892-1895) as Annual reports [of the council].

2896 LONDON. University. King's college. Department of geography. London.
Occasional papers. 1- (1973-).

(2897 ______. ______. Queen Mary college. Department of geography. London.
Occasional papers in geography. 1- (1974-). (N76; B75:43b; DP).
[Department of Geography, Queen Mary College, Mill End Road, London E1 6NS, England.]

2898 ______. ______. ______. Geographical society. London.
Geographical society magazine. 1- (1959-) Annual. (N70-2:2281c).
Supersedes Cosmos [2801].
[Geographical Society, Queen Mary College, London, E.1, England.]

2899 ______. ______. School of Oriental and African studies. Department of geography. London.
Occasional papers. 1- (1976-).
[Publications Officer, Room 205, School of Oriental and African Studies, Malet Street, London WC1E 7HP, England.]

2900 ______. ______. University College. Department of Geography. London.
Occasional paper. 1- (1969-) Irregular. (B73:240a; DP).
[Department of Geography, University College London, Gower Street, London WC1E 6DT, England.]

2901 LONDON school of economics and political science. London.
Studies in economics and political science.
Geographical studies. no1-3 (1910-1924). Closed 1924.
(U-3:2466b; B-4:241a).

2902 ______. Department of geography.
Geographical papers. 1- (1964-). (N70-3:3521a; B68:225a; DP).

(2903 ______. Graduate school of geography.
Discussion paper. 1- (1964-) Irregular but frequent. Mimeographed.
[London School of Economics, Houghton Street, Aldwych, London, WC2, England.]

2904 LONDONER. (City of London polytechnic. Geography section. Journal).
London. 1- (Jn 1972-). (B73:199a).

2905 MAKING of the English landscape. (Hodder and Stoughton). London.
v1-5 (1955-1957). (N70-3:3606a).
Superseded by an unnumbered series, 1972-

2906 MANCHESTER. University. School of geography. Manchester.
Research paper series. 1

2907 MANCHESTER geographical society. Manchester.
Journal. 1-57 (1885-1952/1954). (U-3:2530c; B-3:136b; F-3:206a).

(2908 MAP collector. (Map collector publications). Tring, Hertfordshire.
1- (D 1977-) Quarterly. (N78).
[Map Collector Publications, 48 High Street, Tring, Herts. HP23 5AE, England.]

2909 MAP collectors' circle. London.
Map collectors' series. 1-110 (1963-1975). Closed 1975. (N70-3:3639a; B68:344a).

2910 MERIDIAN. (London. University. Goldsmiths' college. Department of geography. Bulletin). London. 1-6 (1960-1964/65) Annual. (N70-3:3733c).
Mimeographed student publication.
[Department of Geography, Goldsmith's College, London, SE4, England.]

2911 MIDDLESEX polytechnic. London.
Monographs in geography.

2912 NEW geography. (Abelard-Schuman). London. 1966/67- Every two years. (N70-3:4134c-4135a; B73:228a).
[Abelard-Schuman, London, England, or 6 West 57th Street, New York, New York, U.S.A.]

2913 NEWCASTLE-UPON-TYNE. University. Department of geography. Newcastle-upon-Tyne.
Migration and mobility series. nol- (1965-). (N70-3:4204c; B73:251b).
Title varies: 1-2 as Papers on migration and mobility in northeast England.

2914 ______. ______. ______.
Research series. 1- (1954-) Irregular. (N70-3:4204c).
1-3 (1954-1964) as Durham. University. King's college, Newcastle-upon-Tyne. Department of geography. Research series.
[1/2 Sydenham Terrace, Newcastle-upon-Tyne, England.]

(2915 ______. ______. ______.
Seminar papers. 1- (1968-). Mimeographed. (DP).
[c/o Seminar Committee, Department of Geography, University of Newcastle-upon-Tyne, Newcastle-upon-Tyne, NEL 7RU, England.]

2916 ______. ______. Geographical society.
Geographical journal. 1- (1948-) Annual student publication. (U-2:1374c; N65-3:38c; B-2:75b; BS:271b; B68:225a).
1-14 (1948-1963) as Durham. University. King's college, Newcastle-upon-Tyne. Geographical society. Geographical journal (magazine).

2917 NORTH LONDON geographer. (London. Polytechnic of North London. Department of Geography). London. 1- (1973-). (B75:42b).
Supersedes Griffin [2871].
[Department of Geography, The Polytechnic of North London, Prince of Wales Road, London, N.W.S., England.]

2918 NORTH Staffordshire journal of field studies. (University of North Staffordshire). Keele. 1- (1961-). (N70-3:4272b; B68:388a).
Incorporates North Staffordshire field club. Transactions (not in this list).

2919 NORTHERN universities' geographical journal. 1- (1960-) Annual student publication. (N70-3:4280b; BS:630b; B68:389a).
Sponsorship of publication rotates among the geographical societies of the universities of Birmingham, Durham, Hull, Leeds, Leicester, Liverpool, Manchester, Newcastle, Nottingham, and Sheffield.

2920 NORWICH. University of East Anglia. School of development studies. Norwich.
Development studies discussion papers. 1- (1973-). (DP).
[Publications Secretary, School of Development Studies, University of East Anglia, Norwich NR4 7TJ, England.]

2921 NOTTINGHAM. University. Department of geography. Nottingham.
Bulletin of material on the geography of the U.S.S.R. 1-12 (1958-1961). Closed 1961. (N70-3:4295b; BS:162b; B68:108b).

2922 ______. ______. ______.
Bulletin of quantitative data for geographers. 1-21 (1965-1969). Closed 1969. Mimeographed. (N70-3:4295b).
Continued as Ideas in geography [2878].

(2923 ______. ______. ______.
Computer applications. s1 no1-15 (1967-1973) Irregular; s2 v1- (1973-) 4 a year. Not published in 1975.
[Computer Applications, Department of Geography, The University, Nottingham NG7 2RD, England.]

2924 ______. ______. Geographical society.
Magazine. 1- (1963-). (N70-3:4295b; B68:390b).

2925 OCEAN highways: the geographical review. London. v1-2 no12 (1870-1873); nsv1 (Ap 1873-Mr 1874). Closed 1874. (U-4:3135a; B-3:476b; F-3:859a; F-3:816a).
v1-2 no3 (1870-1872) as Our ocean highways. The geographical record and travellers register. Superseded by Geographical magazine [2848].

2926 ORB. (Aberdeen. University. Geographical society). Aberdeen. 1- (1949-) Annual student publication. (B-3:457b).

2927 ORBIT. (Leeds university geographical society. Journal). Leeds. 1- (1950-). (BS:647b).
Title varies: v1-v3 no1 (1950--Mr 1953) as Leeds university geographical society. Journal [or] magazine.

2928 OUTPOST. (Belfast. Queen's university. Geographical society). Belfast. 1952- Irregular. (BS:656a).

2929 OXFORD. Polytechnic. Oxford.
Discussion papers in geography. 1- (1976-).

*(2930 ______. University. School of geography. Oxford.
Research papers. 1- (1972-) Irregular. (N77; B73:291b; GZS; DP).
[Editor, Research papers, School of Geography, Mansfield Road, Oxford OX1 3TB, England.]

2931 OXFORD geographical studies. Oxford. 1 (1922). (B-3:485a).

2932 OXFORD university exploration club. Oxford.
Annual report. 1-10 (1928/29-1937/38). Closed 1938. (U-4:3226c; B-3:483a).
Superseded by club's Bulletin [2933].

2933 ______.
Bulletin. 1- (1948-). (U-4:3226a; B-3:483a).
Supersedes club's Annual report [2932].

2934 PANORAMA. (Isle of Thanet geographical association). Ramsgate, Kent. 1- (1956-) Annual.

(2934a POLAR record. (Scott polar research institute). Cambridge. 1- (1931-) 3 nos. a year. (U-4:3381b; B-3:570b; F-3:979b).
Index: 1-8. Cumulative index every 2 years.
[Scott Polar Research Institute, Cambridge CB2 1ER, England.]

2935 PORTSMOUTH. Polytechnic. Department of geography. Portsmouth.
Climatological series. 1- (1972-). (DP).
[Department of Geography, Portsmouth Polytechnic, Lion Terrace, Portsmouth PO1 3HE, England.]

2936 ______. ______. ______.
Discussion paper series. 1- (1975-). (DP).

2937 ______. ______. ______.
Occasional paper series. 1- (1972-). (DP).

PORTSMOUTH. Polytechnic. Department of geography. Portsmouth.

2938 ______. ______. ______.
Other departmental papers. 1- (1972-). (DP).

(2939 PROFILE. (St. Paul's and St. Mary's colleges of education geographical society). Cheltenham. 1- (1969-) 3 nos. a year. (N77).
[St. Paul's College, Swindon Road, Cheltenham, Gloucestershire GL30 4AZ, England.]

2940 PROGRESS in geography; international reviews of current research. (Edward Arnold). London. 1-9 (1969-1976) Annual. Closed 1976. (N70-3:4800c; B73:272a).
Superseded by Progress in human geography [2941], and by Progress in physical geography [2942].

*(2941 PROGRESS in human geography: an international review of geographical work in the social sciences and humanities. London.
1- (1977-) Quarterly. (N78; B77:51a; GZS).
v1-2 (1977-1978) had 3 nos. each year.
Supersedes in part Progress in geography [2940].
[Edward Arnold, 41 Bedford Square, London WC1B 3DP, England.]

*(2942 PROGRESS in physical geography: an international review of geographical work in the natural and environmental sciences. London.
1- (1977-) Quarterly. (N78; B77:51a; GZS).
v1-2 (1977-1978) had 3 nos. each year.
Supersedes in part Progress in geography [2940].
[Edward Arnold, 41 Bedford Square, London WC1B 3DP, England.]

*(2943 READING. University. Department of geography. Reading.
Geographical papers.
GP 1- (1970-) Irregular, about 8 a year. (N75-2:1835a; B75:21b; DP).
[Department of Geography, University of Reading, Whiteknights, Reading, RG6 2AB, England.]

2944 ______. ______. ______. Urban systems research unit.
Working papers. URSU WP. 1- (1970-). (B73:366b; DP).
[Urban Systems Research Unit, Department of Geography, University of Reading, Whiteknights, Reading RG6 2AU, England.]

(2945 READING geographer. (Reading. University. Department of Geography). Whiteknights Park, Reading. 1- (1970-) Annual. (N75-2:1835a; B73:285a).
[Publications Officer, Department of Geography, University of Reading, Whiteknights, Reading RG6 2AF, England.]

(2946 RECENT polar literature. (Cambridge. Scott polar research institute). Cambridge. 1- (1973-). (N75-2:1837c; B73:285b; GZS).
Supplement to Polar record [2934a].
Part of Polar record, 1931-1972.
[Scott Polar Research Institute, Lensfield Road, Cambridge CB2 1ER, England.]

*(2947 REGIONAL studies. (Regional studies association. Journal). Oxford.
1- (1967-) 6 nos. a year. (N70-3:4957a; B68:462b).
Abstracts in English, French, and German.
[Pergamon Press, Ltd., Headington Hill Hall, Oxford OX3 OBW, England or Fairview Park, Elmsford, New York 10523, U.S.A.]

2948 REGIONAL studies association. Oxford.
Discussion papers. 1- (1972?). (GZS).

2949 ROYAL geographical society. London.
Address at the anniversary meeting. 1836-1878. Closed 1878.
After 1878 printed in society's Journal [2951] and its successors.

2950 ______.
Extra publication.

2951 ______.
Journal. 1-50 (1830-1880). Closed 1880. (U-4:3720b; B-2:269a; F-3:209a).
Succeeded by the society's Proceedings ns [2954].
Indexes: 1-10 (1831-1841); 11-20 (1841-1850); 21-30 (1851-1860); 31-40 (1861-1870); 41-50 (1871-1880).

2952 ______.
Library series. 1-4 (1953-1958) Mimeographed; 5-7 (1961-1964) Printed.
No more issued. (N70-3:5106a).

*2953 ______.
New geographical literature and maps. nsl- (1951-) 2 nos. a year. (N70-3:5106ab; B-2:269b; BS:342a).
Succeeds society's Recent geographical literature, maps, and photographs added to the society's collections [2955].
[Royal Geographical Society, 1 Kensington Gore, London SW7 2AR, England.]

2954 ______.
Proceedings. 1-22 (1855-1878); nsv1-14 (1879-1892). Closed 1892. (U-4:3720b; B-2:269a; F-3:1038a).
Succeeded by Geographical Journal [2847]. Supersedes the society's Journal [2951].
Indexes: v1-22 (1855-1878); nsv1-14 (1879-1892).

2955 ______.
Recent geographical literature, maps, and photographs added to the society's collections. no1-64 (Jn 1918-Ja 1941). (U-5:3720b; B-2:269b; F-4:29b).
Supplement to Geographical journal [2847]. Publication suspended with no64 (1941) but later continued as "Additions to the library" published as an integral part of the Geographical journal v105-116 (1945-1950). Succeeded by the society's New geographical literature and maps [2953].
Index: no1-41 (1918-1932).

2956 ______.
Reproductions of early engraved maps. 1- (1927-) Irregular. (U-4:3720c; B-3:686a).

2957 ______.
Reproductions of early manuscript maps. 1-2 (1929-1934). (U4:3720c; B-3:686b).

2958 ______.
Research series. 1- (1948-) Irregular. (U-4:3720c; B-3:647a).
Supersedes society's Technical series [2960]. no4 was 1960.

2959 ______.
Supplementary papers. 1-4 (1882/86-1890). Closed 1890. (U-4:3720c; B-2:269b; F-4:388b).

2960 ______.
Technical series. 1-5 (1920-1928). Closed 1928. (U-4:3720c; B-3:647a).
Succeeded by the society's Research series [2958].

2961 ______.
Yearbook and record. 1-18 (1898-1916). Closed 1916. (U-4:3721a; B-2:269b; F-4:388b).

2962 ST. ANDREWS geographer. (St. Andrews. University. Department of geography). St. Andrews. (1971-).
[StAG Magazine, c/o Department of Geography, The University, St. Andrews, Fife, Scotland, United Kingdom.]

2963 SAINT LUKE'S college geographical society. Exeter.
[Geographical magazine]. 1966.

2964 SALFORD geographer. (Salford. University. Geographical society. Magazine). Salford. 1- (1969-). (B75:57).

2965 SCOTTISH association of geography teachers. Glasgow.
Journal. 1- (1971-). (B73:181b).

2966 ______.
Occasional papers. 1- (1975-). (B77:45a).

*2967 SCOTTISH geographical magazine. (Royal Scottish geographical society). Edinburgh. 1- (1885-) 3 nos. a year. (U-5:3830b; B-4:56b-57a; F-4:458b).
Indexes: v1-50 (1885-1934), 51-81 (1935-1965).
[Royal Scottish Geographical Society, 10 Randolph Crescent, Edinburgh EH3 7TU, Scotland, United Kingdom.]

2968 SERIAL map service. Letchworth; London. v1 no1-v9 no6 (S 1939-Mr 1948). Closed Mr 1948. (U-5:3850c; B-4:72b).
Index: 1-5.

*2968a SOCIAL science and medicine. Part D. Medical geography. v12D- (1978-) 4 per annum.
Supersedes in part Social science and medicine.
1-11 (1967-1977) not divided into subseries.
[Pergamon Press, Inc., Journals Dept., Headington Hill Hall, Oxford OX3 0BW, England, or Maxwell House, Fairview Park, Elmsford, New York 10523, U.S.A.]

2969 SOCIETY of university cartographers. Liverpool.
Bulletin. 1- (1966-) 2 nos. a year. (N70-4:5396c; B73:60b).
Mimeographed.
[Subscriptions: G. Kingdom, Department of Geography, Portsmouth College of Technology, 111 High Street, Portsmouth, Hampshire, England.]

2970 SOUTH Hampshire geographer. (Portsmouth polytechnic. Geographical society. Journal). Portsmouth. 1- (1968-) Annual. (N70-4:5442b; B73:322a).
[Department of Geography, Portsmouth Polytechnic, Lion Terrace, Portsmouth PO1 3HE, England.]

2971 SOUTHAMPTON. University. Department of geography.
Southampton research series in geography. 1- (1965-) Irregular. (N70-4:5443c; B68:519b; DP).

2972 SOUTHAMPTON geographical society. Southampton.
Report. 1898-1901. Closed 1901?

2973 STUDENT geographer. (Cambridge. University. N.U.S. Geography committee). no2 (1943).

2974 STUDIES in regional science. (Regional science association. British section). London. 1- (1969-) Annual. (B73:33a).
[Pion Ltd., 207 Brondesbury Park, London, NW2, England.]

*(2975 SWANSEA geographer. (Swansea. University college. Geographical society). Swansea. 1- (1959-) Annual. (N70-4:5641b).
[Department of Geography, University College of Swansea, Singleton Park, Swansea, Glamorganshire SA2 8PP, Wales, United Kingdom.]

2976 TEACHING geography. (Geographical association). Sheffield. 1-23 (1967-1974). (N70-4:5712c; B73:337b).
[Geographical Association, 343 Fulwood Road, Sheffield, S10 3BP, England.]

*(2977 TEACHING geography. (Geographical association). Harlow, Essex. 1- (Ap 1975-). (N76).
[Longman Group Ltd., Journals Division, 43/45 Annandale Street, Edinburgh EH7 4AT, Scotland, United Kingdom.]

2978 TYNESIDE geographical society. Newcastle-upon-Tyne.
Journal. v1-6 no5 (1889-Ja 1915); nsv1-3 (1936-1939). (U-5:4284a; B-4:362a; F-3:212a).

2979 ______.
Lecture. 1961- Irregular. (N70-4:5892b).

(2980 URBAN studies. (Glasgow. University). Edinburgh. 1- (1964-) 3 nos. a year. (N70-4:6112b; B68:576a).
[Longmans Group Ltd., 43-45 Annandale Street, Edinburgh EH7 4AT, Scotland, United Kingdom.]

2981 WALES. University college, Aberystwyth. Department of geography. Aberstwyth.
Memorandum. 1- (1958-). (N70-4:6257c).

2982 ______. ______.
Exploration society.
Journal. 1966- Annual.

(2983 WESSEX geographer. (Southampton. University. Geographical society). Southampton. 1- (1960-) Annual. (N70-4:630a; B68:587b).
1-3 mimeographed, 4- printed. Student publication.
[H. G. Walters, Publishers, Ltd., Market Square, Narbeth, Pembrokeshire, England, for the Southampton Geographical Society.]

2984 WESSEX geographical year. (Southampton. University. Geographical society). Southampton. 1/3- (1959/62-). (N70-4:6302a).
[H. G. Walters, Publishers, Ltd., Market Square, Narbeth, Pembrokeshire, England, for Southampton University Geographical Society.]

2985 WORLD land use survey. Bude-Stratton.
Monograph. 1-5 (1958-1968). Closed. (N70-4:6390c).

2986 ______.
Occasional papers. 1-11 (1956-1971). Closed. (N70-4:6390c; BS:970a).

2987 WORLD'S landscapes. (Longmans). London. 1- (1970-) Irregular monographs. (N75-2:2313a).

2988 WYE college. Wye, England. Department of agricultural economics. Ashford, Kent.
Studies in rural land use. Report. 1- (1954-). (N70-4:6405c; BS:844b).
First issue unnumbered and does not bear series title.

UNITED STATES (English)

United States of America

(2989 ALABAMA geographer; a newsletter for Alabama's geographers (Society of Alabama geographers). University, Alabama. 1- (1969-) Quarterly. (N75-1:65a).
[Dr. Walter F. Koch, P. O. Box 1945, University, Alabama 35486, U.S.A.]

2990 ALASKA. State geographic board. Juneau.
Report. 1 (1966/67). All published?

2991 ALASKA and northwest quarterly. (Alaska geographical society). Seattle, Washington. v1 no1-4 (Ja-D 1899). Closed 1899. (U-1:129b). Superseded by the society's Bulletin [2993].

*(2992 ALASKA geographic. (Alaska geographic society). Anchorage. 1- (1972-) Quarterly. (N75-1:68c; B73:13a).
[The Alaska Geographic, Box 4-EEE, Anchorage, Alaska 99509, U.S.A.]

2993 ALASKA geographical society. Seattle, Washington.
Bulletin. v1 no1 no2 (1900). Closed 1900. (U-1:129c).
Supersedes Alaska and northwest quarterly [2991].

2994 AMERICAN bureau of geography. Winona, Minnesota.
Bulletin. v1-2 (Mr 1900-D 1901). Closed 1901. (U-1:192b; B-1:106b).
United with Journal of school geography [3157] to form Journal of geography [3154].
Index: 1-2 with index to Journal of geography 1897-1921.

*(2995 AMERICAN cartographer. (American congress on surveying and mapping). Falls Church, Virginia.
1- (1974-) Semiannual. (N75-1:91a).
[American Congress on Surveying and Mapping. 210 Little Falls Street, Falls Church, Virginia 22046, U.S.A.]

(2996 AMERICAN congress on surveying and mapping. Falls Church, Virginia.
Bulletin. 1- (1971-) Quarterly. (N77).
[American Congress on Surveying and Mapping, 210 Little Falls Street, Falls Church, Virginia 22046, U.S.A.]

(2996a ______. Cartography division. Falls Church, Virginia.
Technical monograph series. CA-1-5 (1962-1971).
(N70-1:260b).

2997 AMERICAN geographical society of New York. New York, New York.
Antarctic map folio series. 1-19 (1964-1975). Closed. (N70-1:272a; B68:47b).
[American Geographical Society, Broadway at 156th Street, New York, New York 10032, U.S.A.]

2998 ______.
Around the world program. Ag 1955-S 1973 Monthly. Closed.

2999 ______.
Atlas of the Americas. Text series. 1-2 (1942-1944). (U-1:228a).

3000 ______.
Bowman memorial lectures. 1-4 (1951-1958). (N70-1:837b; BS:136b).

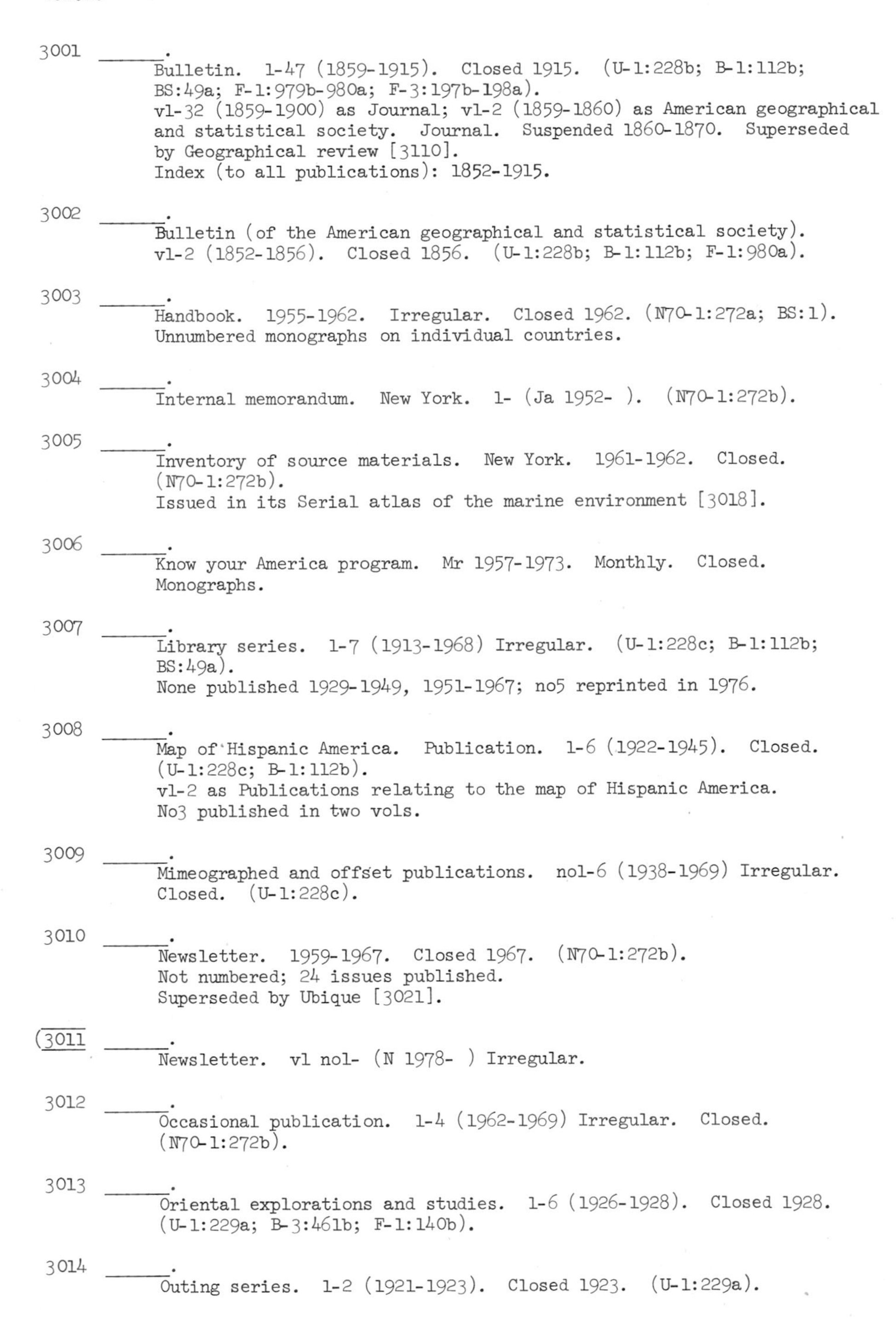

3001 ______.
Bulletin. 1-47 (1859-1915). Closed 1915. (U-1:228b; B-1:112b; BS:49a; F-1:979b-980a; F-3:197b-198a).
v1-32 (1859-1900) as Journal; v1-2 (1859-1860) as American geographical and statistical society. Journal. Suspended 1860-1870. Superseded by Geographical review [3110].
Index (to all publications): 1852-1915.

3002 ______.
Bulletin (of the American geographical and statistical society).
v1-2 (1852-1856). Closed 1856. (U-1:228b; B-1:112b; F-1:980a).

3003 ______.
Handbook. 1955-1962. Irregular. Closed 1962. (N70-1:272a; BS:1).
Unnumbered monographs on individual countries.

3004 ______.
Internal memorandum. New York. 1- (Ja 1952-). (N70-1:272b).

3005 ______.
Inventory of source materials. New York. 1961-1962. Closed. (N70-1:272b).
Issued in its Serial atlas of the marine environment [3018].

3006 ______.
Know your America program. Mr 1957-1973. Monthly. Closed.
Monographs.

3007 ______.
Library series. 1-7 (1913-1968) Irregular. (U-1:228c; B-1:112b; BS:49a).
None published 1929-1949, 1951-1967; no5 reprinted in 1976.

3008 ______.
Map of Hispanic America. Publication. 1-6 (1922-1945). Closed. (U-1:228c; B-1:112b).
v1-2 as Publications relating to the map of Hispanic America.
No3 published in two vols.

3009 ______.
Mimeographed and offset publications. no1-6 (1938-1969) Irregular. Closed. (U-1:228c).

3010 ______.
Newsletter. 1959-1967. Closed 1967. (N70-1:272b).
Not numbered; 24 issues published.
Superseded by Ubique [3021].

(3011 ______.
Newsletter. v1 no1- (N 1978-) Irregular.

3012 ______.
Occasional publication. 1-4 (1962-1969) Irregular. Closed. (N70-1:272b).

3013 ______.
Oriental explorations and studies. 1-6 (1926-1928). Closed 1928. (U-1:229a; B-3:461b; F-1:140b).

3014 ______.
Outing series. 1-2 (1921-1923). Closed 1923. (U-1:229a).

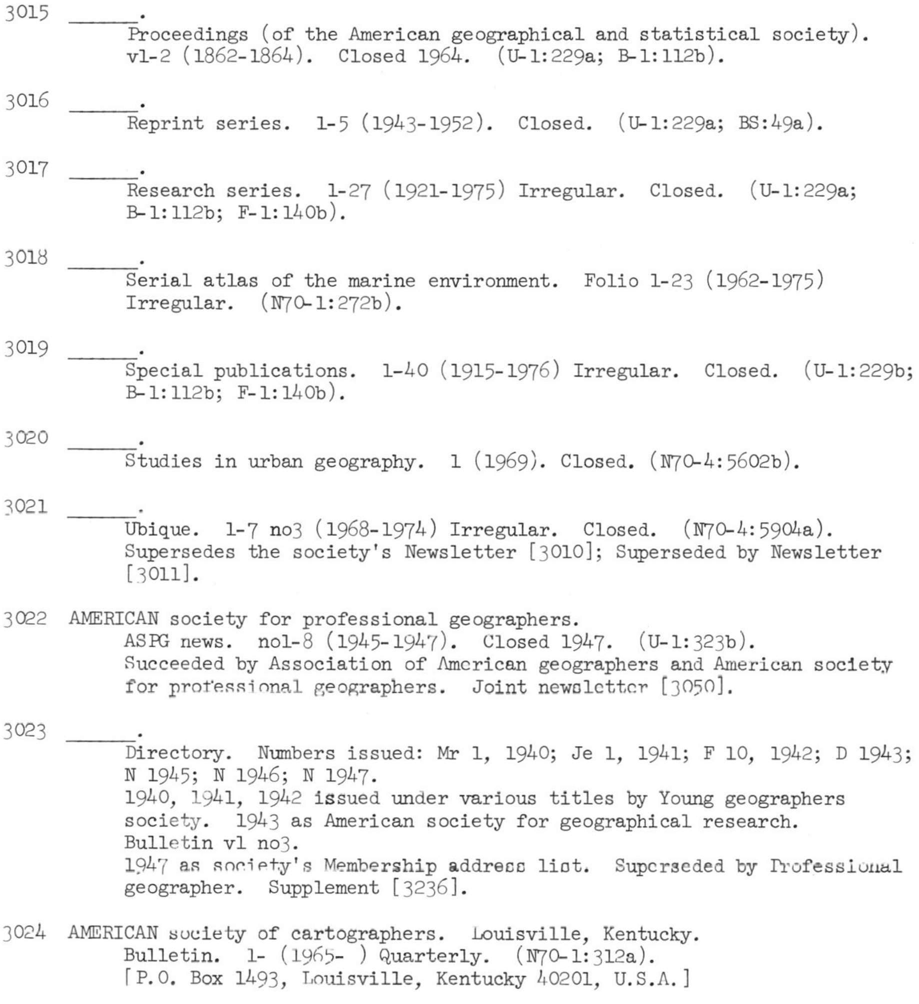

3015 ______.
Proceedings (of the American geographical and statistical society). v1-2 (1862-1864). Closed 1964. (U-1:229a; B-1:112b).

3016 ______.
Reprint series. 1-5 (1943-1952). Closed. (U-1:229a; BS:49a).

3017 ______.
Research series. 1-27 (1921-1975) Irregular. Closed. (U-1:229a; B-1:112b; F-1:140b).

3018 ______.
Serial atlas of the marine environment. Folio 1-23 (1962-1975) Irregular. (N70-1:272b).

3019 ______.
Special publications. 1-40 (1915-1976) Irregular. Closed. (U-1:229b; B-1:112b; F-1:140b).

3020 ______.
Studies in urban geography. 1 (1969). Closed. (N70-4:5602b).

3021 ______.
Ubique. 1-7 no3 (1968-1974) Irregular. Closed. (N70-4:5904a).
Supersedes the society's Newsletter [3010]; Superseded by Newsletter [3011].

3022 AMERICAN society for professional geographers.
ASPG news. no1-8 (1945-1947). Closed 1947. (U-1:323b).
Succeeded by Association of American geographers and American society for professional geographers. Joint newsletter [3050].

3023 ______.
Directory. Numbers issued: Mr 1, 1940; Je 1, 1941; F 10, 1942; D 1943; N 1945; N 1946; N 1947.
1940, 1941, 1942 issued under various titles by Young geographers society. 1943 as American society for geographical research. Bulletin v1 no3.
1947 as society's Membership address list. Superseded by Professional geographer. Supplement [3236].

3024 AMERICAN society of cartographers. Louisville, Kentucky.
Bulletin. 1- (1965-) Quarterly. (N70-1:312a).
[P.O. Box 1493, Louisville, Kentucky 40201, U.S.A.]

(3024a ANNALS of regional science: a journal of urban, regional, and environmental research and policy. (Western regional science association and Western Washington university. College of business and economics). Bellingham, Washington. 1- (1967-) 3 per annum. (N70-1:361c; B73:20a).
[c/o Department of Economics, Western Washington University, Bellingham, Washington 98225, U.S.A.]

(3025 ANTARCTIC bibliography. Washington, D.C. (U. S. National science foundation. Office of polar programs and U. S. Library of congress. Science and technology division). 1951/61- ; v1- (1962-) Every 18 months. (N70-1:378b).
[Superintendent of documents, Government printing office, Washington, D.C. 20402, U.S.A.]

*(3026 ANTIPODE: a radical journal of geography. Worcester, Massachusetts.
1- (Ag 1969-) Irregular. (N75-1:145c; B73:24a).
Index: 1969-1978 in v10 no3/v11 no1 (1978/1979), p. 143-171.
[P.O. Box 225, West Side Station, Worcester, Massachusetts 01602, U.S.A.]

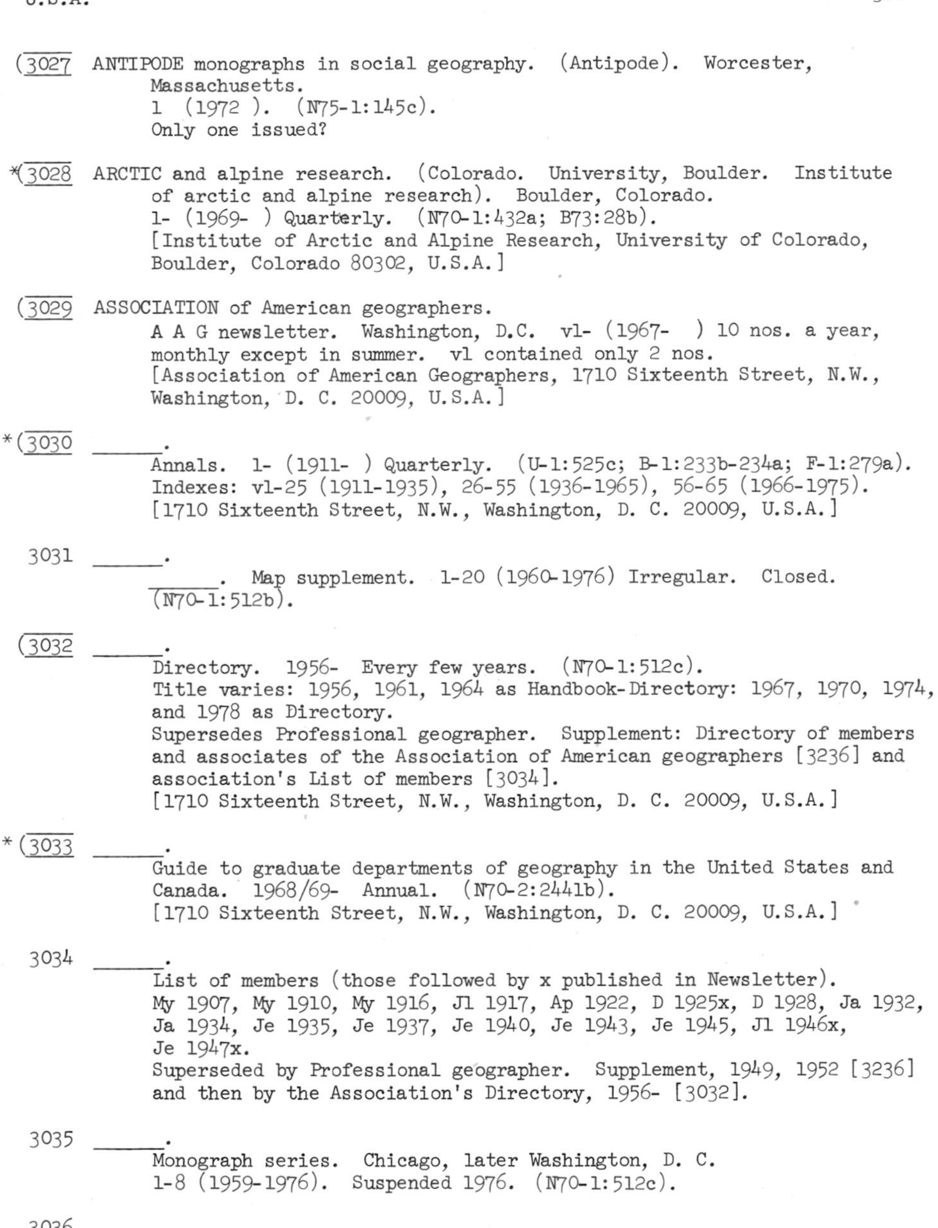

(3027 ANTIPODE monographs in social geography. (Antipode). Worcester, Massachusetts.
1 (1972). (N75-1:145c).
Only one issued?

*(3028 ARCTIC and alpine research. (Colorado. University, Boulder. Institute of arctic and alpine research). Boulder, Colorado.
1- (1969-) Quarterly. (N70-1:432a; B73:28b).
[Institute of Arctic and Alpine Research, University of Colorado, Boulder, Colorado 80302, U.S.A.]

(3029 ASSOCIATION of American geographers.
A A G newsletter. Washington, D.C. v1- (1967-) 10 nos. a year, monthly except in summer. v1 contained only 2 nos.
[Association of American Geographers, 1710 Sixteenth Street, N.W., Washington, D. C. 20009, U.S.A.]

*(3030 ______.
Annals. 1- (1911-) Quarterly. (U-1:525c; B-1:233b-234a; F-1:279a).
Indexes: v1-25 (1911-1935), 26-55 (1936-1965), 56-65 (1966-1975).
[1710 Sixteenth Street, N.W., Washington, D. C. 20009, U.S.A.]

3031 ______.
______. Map supplement. 1-20 (1960-1976) Irregular. Closed. (N70-1:512b).

(3032 ______.
Directory. 1956- Every few years. (N70-1:512c).
Title varies: 1956, 1961, 1964 as Handbook-Directory: 1967, 1970, 1974, and 1978 as Directory.
Supersedes Professional geographer. Supplement: Directory of members and associates of the Association of American geographers [3236] and association's List of members [3034].
[1710 Sixteenth Street, N.W., Washington, D. C. 20009, U.S.A.]

*(3033 ______.
Guide to graduate departments of geography in the United States and Canada. 1968/69- Annual. (N70-2:2441b).
[1710 Sixteenth Street, N.W., Washington, D. C. 20009, U.S.A.]

3034 ______.
List of members (those followed by x published in Newsletter).
My 1907, My 1910, My 1916, Jl 1917, Ap 1922, D 1925x, D 1928, Ja 1932, Ja 1934, Je 1935, Je 1937, Je 1940, Je 1943, Je 1945, Jl 1946x, Je 1947x.
Superseded by Professional geographer. Supplement, 1949, 1952 [3236] and then by the Association's Directory, 1956- [3032].

3035 ______.
Monograph series. Chicago, later Washington, D. C.
1-8 (1959-1976). Suspended 1976. (N70-1:512c).

3036 ______.
Newsletter.
Copies issued: Mr, Je 1923; D 1925; 1935 n.d.; My 1936, Ap 1937; My 1938; My 1939; Ap 1940; Ap 1941; My 1942; Je, N 1943; Jl 1944; Je 1945; Jl, O 1946; Je, O 1947. Closed 1947.
The first 3 as Supplements to the Association of American geographers. Annals [3030]. Succeeded by Joint newsletter of the Association of American geographers and the American society for professional geographers [3050].

3037 ______.
Newsletter. Recreation/Tourism and Sport. College Station, Texas. vl nol-3 (Ja 1975-M 1977).
[Department of Recreation and Parks, Texas A and M University, College Station, Texas 77840, U.S.A.]

3038 ______.
Physical environmental newsletter. 1-2 (1971). Mimeographed.

3039 ______.
Proceedings. 1-8 (1969-1976) Annual. Closed 1976. (N70-1:512c; B73:269b).

(3040 ______.
Program of annual meetings. 1905- Annual.

*(3040a ______.
Resource papers for college geography. 75/1- (1975-). 4 per annum. Numbered individually by years 75/1-4, 76/1-4, 77/1-4, 78/1-4. Not published in 1979-1980. To be resumed in 1981.
Supersedes Association of American geographers. Commission on college geography. Resource papers [3043).

3041 ______. Commission on college geography.
Newsletter. nol- (D 1965-). Closed. (N70-1:512b).

3042 ______. ______.
Publication. General series. 1-13 (1965-1974). Closed 1974. Irregular, about two a year. (N70-1:512c; B73:276b).

3043 ______. ______.
Resource paper. 1-28 (1968-1975). Closed 1974. Superseded by Association of American geographers. Resource papers for college geography [3040a). (N70-1:512c; B73:293a).

3044 ______. ______.
Technical paper. 1-10 (1968-1972). Closed. (N70-1:512c; B73:339a).

3045 ______. Committee on Asian studies.
Bulletin. nol-7 (1948-1950). Closed 1950. (U-1:526a). Mimeographed.

3046 ______. High school geography project.
Newsletter. 1-20 (1962-1971). Closed 1971. (N70-1:512c).

(3047 ______.
Middle states division. Buffalo.
Proceedings. 1- (1966-) Annual. Mimeographed.

3048 ______.
New York-New Jersey division.
Proceedings of the annual meeting. 8- (1967-). (N75-1:197c).
Proceedings of the 1st-7th conferences never published.
vl (1967) is Proceedings of the 8th Annual meeting. Mimeographed.

3049 ______. Southeastern division. Chapel Hill, North Carolina.
Memorandum folio. 1-18 (1950-1966). Closed 1966. (N70-1:512c).
Papers presented at meetings.
Superseded by the Southeastern geographer [3251].

3050 ASSOCIATION of American geographers and American society for professional geographers.
Joint newsletter. Mr, Je, S, N 1948. Closed 1948. (U-1:525c).
Supersedes American society for professional geographers. ASPG news [3022]; beginning with 1949 combined with the Professional geographer [3235].

(3051 ASSOCIATION of North Dakota geographers. Grand Forks, North Dakota.
Bulletin. 1- (1947-) Annual. (N79).
[Department of Geography, University of North Dakota, Grand Forks, North Dakota 58202, U.S.A.]

3052 ______.
Special publications in geography.
1- (1973-) Irregular.
[Association of North Dakota Geographers, University of North Dakota, Grand Forks, North Dakota 58202, U.S.A.]

*(3053 ASSOCIATION of Pacific coast geographers. Corvallis, Oregon.
Yearbook. 1-8 (1935-1942); v9- (1947-) Annual. (U-1:533c; B-1:242a).
Suspended 1943-1946.
Indexes: 1-27 (1935-1965), 28-35 (1966-1973) in v35 (1973), p. 203-220, 36-40 (1974-1978) in v40 (1978), p. 175-189.
[Oregon State University Press, 101 Waldo, Corvallis, Oregon 97331, U.S.A.]

3054 ATLAS of distribution of diseases. (American geographical society of New York). New York, New York. 1-17 (1950-1955). Closed 1955. (N70-1:536a).

(3055 BULLETIN of Asian geography. (Association of American geographers. Committee on Asian geography. Newsletter). Akron, Ohio. 1- (1976-) two nos. a year.
[P. P. Karan, Department of Geography, University of Kentucky, Lexington, Kentucky 40506, U.S.A.]

3056 CALIFORNIA. University. Berkeley, California.
Map series for students and teachers. 1-4, 6-10 (1915-1928?).
Closed. (U-2:872c).

*(3057 ______. ______. Berkeley and Los Angeles, California.
Publications in geography. 1- (1913-) Irregular. (U-2:874c; B-1:475a; BS:172b; F-4:822b).
[University of California Press, Berkeley, California 94720, or Los Angeles, California 90024, U.S.A.]

3058 CALIFORNIA council of geography teachers. Fresno, California.
Bulletin. 1-7 (1953-1960) 2 or 3 nos. during school year. Closed. (N70-1:1034a).
Mimeographed.
Becomes a newsletter after v6 no2.

*(3059 CALIFORNIA geographer. (California council for geographic education). Hayward, California. 1- (1960-) Annual. (NO-1:1035c; B68:120b).
Index: v1-10 in v12.
[The California Geographer, Department of Geography, California State University, Hayward, California 94542, U.S.A.]

3060 CALIFORNIA physical geography club, Oakland. Oakland, California.
Bulletin. v1-2 no2 (Ap 1907-Mr 1909). Closed 1909. (U-2:888c).

(3061 CAROLINA geographical symposium. Chapel Hill, North Carolina.
Papers. 1- (1974-).
Subseries of North Carolina. University at Chapel Hill. Department of geography. Studies in geography. no1 (1974) is no6 of main series; 2 is no9; 4 is no11.
[Department of Geography, University of North Carolina at Chapel Hill, Chapel Hill, North Carolina 27514, U.S.A.]

(3062 CHESAPEAKE regional research papers. (George Mason university. Department of public affairs. Fairfax, Virginia. 1- (D 1978-) Irregular monographs.
[Editor, Department of Public Affairs, George Mason University, Fairfax, Virginia 22030, U.S.A.]

*(3063 CHICAGO. University. Department of geography. Chicago, Illinois.
Research papers. 1- (1948-) Monographs; several each year. (U-2:1003c; B-1:544b; BS:200a; DP). Complete list at back of this volume.
[Department of Geography, University of Chicago, 5828 University Avenue, Chicago, Illinois 60637, U.S.A.]

*(3064 CHINA geographer. Boulder, Colorado.
no1-10 (Spring 1975-Spring 1978); no11- (1980-) Annual. (N76).
1975-1978, 3 nos. a year.
Index: no1-10 in no10, p. 53-62.
no1-10 (1975-1978) published in Los Angeles, at U.C.L.A.
[Westview Press, Inc., 5800 Central Avenue, Boulder, Colorado 80301, U.S.A.]

3065 CLARK university, Worcester, Massachusetts. Graduate school of geography.
Place perception project reports. 1-10 (1969-1971). (DP).

3066 COASTAL geography conference. (Louisiana. State university. Coastal studies institute). Baton Rouge.
Proceedings. 1 (1954); 2 (1959); 1961.
Sponsored by the Office of Naval Research, Geography Branch, and National Research Council, Committee on Geography.

(3067 COASTAL research notes (Florida. State university, Tallahassee. Department of geology). Tallahassee. v1- (F 1962-) 3 nos. a year. (N70-1:1327c; B68:140a).
Issues for F 1962-N 1965 as nos1-12.

(3068 COLORADO. University. Boulder. Institute of arctic and alpine research. Boulder, Colorado.
Occasional paper. 1- (1971-) Irregular. (N75-1:551a).

3069 COLTON'S journal of geography and collateral sciences: a record of discovery, exploration, and survey. New York, New York. no1-7 (O 1867-Jl 1871). Closed 1871. (U-2:1115a; B-1:605b).

*(3070 CONFERENCE of Latin Americanist geographers. Place of publication varies with each conference.
Proceedings. 1- (1970-) Irregular but nearly annual. (N75 1:584a).
Note: N75-1:584a lists two series, Proceedings, and Publication series, separately, but they are the same.
[Orders: Tom L. Martinson, Department of Geography and Geology, Ball State University. Muncie. Indiana 37206. USA.]

3072 COUNCIL of planning librarians. Exchange bibliography. Monticello, Illinois.
1- 1564/1565 (1958-1978). Closed July 1978. (N70-1:565a).
Indexes: 1-715 (1958-D 1974) as 716/717 (Ja 1975); 716-1193 (Ja 1975-D 1976) as 1194/1195 (Ja 1977); 1196-1563 (Ja 1977-Jl 1978) as 1564/1565 (Jl 1978).

*(3073 CURRENT geographical publications: additions to the research catalogue of the American geographical society collection of the University of Wisconsin-Milwaukee library. New York, later Milwaukee. 1- (1938-) 10 times a year. (U-2:1243b; B-1:682b; F-2:320b).
v1-41 no2 (1938-February 1978) published by the American geographical society, New York. v41 no3 (March 1978-) by the American geographical society collection, the University of Wisconsin-Milwaukee library, Milwaukee.
[The American Geographical Society Collection, The University of Wisconsin-Milwaukee Library, P.O. Box 399, Milwaukee, Wisconsin 53201, U.S.A.]

3074 ______. Photographic supplement. 1941-1943. Closed 1943.

3075 DAKOTA geographic. Little Wound, South Dakota. 1- (1973-) Annual. (N76).
[Tinpsila Press, Little Wound, South Dakota, U.S.A.]

3076 DARTMOUTH college. Department of geography. Hanover, New Hampshire. Geography publications at Dartmouth. 1- (1964-) Irregular. (N70-2:1650b).

3077 DENVER. University. Department of geography. Denver, Colorado. Technical paper. 64:1-7, 65:1, 66:1-2, 68:1, 71:1-2; 74:1. (N70-2:1684b).

3078 DESCRIPTIVE America, a geographical and industrial monthly magazine. New York, New York. v1 no1-6 (My 1884-Mr 1885). Closed 1885. (U-2:1287a).

3079 DIRECTIONS. (New Jersey council for geographic education). 1953/54- Quarterly.

(3080 DIRECTORY of college geography of the United States. Richmond, Kentucky. 1948/49- Annual. (N70-2:1739a).
1948/49 as College geography directory of the United States.
Earlier published in Lexington, Kentucky.
[Geographical Studies and Research Center, Eastern Kentucky University, Richmond, Kentucky, U.S.A.]

3081 EARTH and man. Des Moines, Iowa; Detroit, Michigan. v1-3 (1895-1897). Title varies: Reports from the geographical societies of the globe. London, Paris, etc...The Earth, a monthly magazine devoted to geography and correlative subjects, Des Moines. v1 (1895/96); Earth. Des Moines. v2 (1896); Earth and man. Detroit, v2-3 (1896-1897).

3082 EARTH journal. (Minnesota geographic society). Minneapolis. 1-8? (1970-1977). Closed 1977. (N75-2:1463a).
Early issues as Minnesota earth journal.

*(3083 EAST Lakes geographer. Bowling Green, Ohio. 1- (1964-) Annual. (N70-2:1820c; B68:180b). 1-7 (1964-1971) published in Columbus, Ohio, for members of the East Lakes Division, Association of American Geographers.
[John Hiltner, Editor, Department of geography, Bowling Green State University, Bowling Green, Ohio 43403, U.S.A.]

3084 ECOLOG. (California. University, Santa Barbara. Department of geography). Santa Barbara. Ja 15, 1970- Weekly. (N70-2:1834b).

*(3085 ECONOMIC geography. (Clark university). Worcester, Massachusetts. 1- (1925-) Quarterly. (U-2:1394c; B-2:90a; BS:277a; F-2:450b).
Indexes: v1-25 (1925-1949), 26-41 (1950-1965).
[Economic geography, 950 Main Street, Worcester, Massachusetts 01610, U.S.A.]

*(3086 ECUMENE. (East Texas state university). Commerce, Texas. 1- (0 1968-) Annual. (N70-2:1852a).
[P.O. Box 3036, E. T. Station, Commerce, Texas 75428, U.S.A.]

(3087 ENVIRONMENT and behavior. Beverly Hills, California. 1- (1969-). (N70-2:1934c-1935a; B73:107b).
[Sage Publications, Inc., 275 South Beverly Drive, Beverly Hills, California 90212, U.S.A.]

(3088 ENVIRONMENTAL MANAGEMENT. New York. 1- (1976-) Bimonthly. (N77).
[Springer-Verlag New York, Inc., 175 Fifth Avenue, New York, New York 10010, U.S.A.]

3089 FLORIDA Atlantic university, Boca Raton. Department of geography.
Occasional publication. 1 (1968).
1 issued also as v1 no2 of department's Paracas papers [3224].

3090 FLORIDA. University. Gainesville, Florida.
Publications. Geography series. v1 no1 (Je 1930). Closed 1930.
(U-2:1577c; B-2:197b).

3091 ______. ______. Department of geography. Gainesville, Florida.
Eminent geographers. 1- (1972-) Irregular.

3092 ______. ______. ______.
Ryukyu islands project.
Research and information papers. Gainesville. 1-26 (1970-1975).
(N75-1:851c).

(3093 FLORIDA geographer. (Florida society of geographers). Tallahasee.
1- (1964-) 2 nos. a year. (N75-1:853a). 1-2 (1964-1965) as Florida society of geographers Newsletter. Not published 1974-1975.
[Florida society of geographers, Department of geography, Florida Atlantic university, Boca Raton, Florida 33431, U.S.A.]

*(3094 FOCUS. (American geographical society of New York). New York, New York.
1- (O 1950-) Bimonthly except July and August. (N70-2:2120c-2121a; B-2:199b; BS:317b).
[American Geographical Society, Broadway at 156th Street, New York, New York 10032, U.S.A.]

3095 FRED K. SCHAFER memorial lecture. (Iowa. University. Department of geography). Iowa City. 1- (1972-). (N77).
[Department of Geography, University of Iowa, Iowa City, Iowa 52240, U.S.A.]

3096 GEOGRAM. (Virginia geographical society). Fredericksburg, Virginia.
1- ? (1959-1965). Closed 1965. (N70-2:2280a).
Superseded by World [3297].

3097 GEOGRAPHIC news. (U. S. Bureau of foreign and domestic commerce. Division of regional information. Geographic section). Washington, D. C. v1-2, no1-27 (1931-1933). Closed 1933. (U-2:1693c).
Issuing office varies.

3098 GEOGRAPHIC quarterly. (Austin Peay state college. Department of geography). Clarksville, Tennessee. v1-10 no3 (1943-1953) Quarterly.
Mimeographed.

3099 GEOGRAPHIC society of Chicago. Chicago, Illinois.
Bulletin. 1-9 (1899-1933). Closed 1933. (U-2:1693c; B-2:268b; F-2:727b).
Some issued in revised editions.

3100 ______.
Excursion bulletin. 1-3 (1911-1917). Closed 1917. (U-2:1694a).

3101 ______.
Papers on Chicago. 1 (1955). (N70-2:2280b; BS:663b).

3102 ______.
Program bulletin. no1-212 (1918-1947); nsno1- (1947-) Annual (since 1953).
no1-212 (1918-1947) as Monthly bulletin.

3103 ______.
Yearbook. 1912- Biennial (earlier annual).
Some as Handbook.

*(3104 GEOGRAPHICAL analysis; an international journal of theoretical geography. Columbus, Ohio. 1- (1969-) Quarterly. (N70-2:2281a; N75-1:906a; B73:128a).
Index: v1-10 (1969-1978) in v10 (1978), p. 427-453.
[The Ohio State University Press, 2070 Neil Avenue, Columbus, Ohio 43210, U.S.A.]

3105 GEOGRAPHICAL and commercial gazette. New York, New York. v1 no1-5 (Ja-My 1855). Closed 1855. (U-2:1694a; B-2:268b).

3106 GEOGRAPHICAL and military museum. Albany, New York. v1 no1-15 (F 28-Je 6 1814). Closed 1814. (U-2:1694a).

(3107 GEOGRAPHICAL bulletin. (Gamma Theta Upsilon). Monmouth, Oregon. 1- (1970-) 2 per annum. (N75-1:906ab).
[James W. Gallagher, Oregon College of Education, Monmouth, Oregon 97361, U.S.A.]

3108 GEOGRAPHICAL, historical and statistical repository. Philadelphia, Pennsylvania. v1 no1-2 (1824). Closed 1824. (U-2:1694a).
Caption title: Darby's repository.

*(3109 GEOGRAPHICAL perspectives. (University of Northern Iowa. Department of geography). Cedar Falls. no 33- (Spring, 1974-) two nos. a year. (N75-1:906b).
Continues Iowa geographer [3150].
[Editor, Department of Geography, University of Northern Iowa, Cedar Falls, Iowa 50613, U.S.A.]

*(3110 GEOGRAPHICAL review. (American geographical society of New York). New York, New York. 1- (1916-) Quarterly. (U-2:1694c; B-1:112b; BS:49a; F-2:728a).
Supersedes the society's Bulletin (formerly Journal) [3001].
Indexes: v1-15 (1916-1925), 16-25 (1926-1935), 26-35 (1936-1945), 36-45 (1946-1955), 46-55 (1956-1965).
[American Geographical Society, Broadway at 156th Street, New York, New York 10032, U.S.A.]

3111 GEOGRAPHICAL society of California. San Francisco, California. Bulletin. 1-2 (Mr 1893-My 1894). Closed 1894. (U-2:1695a; B-2:269a).

3112 ______.
Proceedings. 1892-1893. (U-2:1695a).

3113 ______.
Publication. no1-9 (1891?-1894). Closed 1894. (U-2:1695a).
1 lacks date; 8 also called v1 pt 1; 9 also called v2.

3114 GEOGRAPHICAL society of Philadelphia. Philadelphia, Pennsylvania. Bulletin. 1-36 (1893/94-1938). Closed 1938. (U-2:1695b; B-2:268b; F-1:984b).
Through 1896 as Geographical club of Philadelphia.
Index: 1-36, 1893-1938, in v36.

3115 ______.
Special publication. no1-3 (1908-1928). Closed 1928? (U-2:1695b).

3116 GEOGRAPHICAL society of the Pacific. San Francisco, California. Bulletin. s2 v3-6 (1904-1910). Closed 1910. (U-2:1695b; B-2:269b).
Takes numbering of its Transactions and proceedings [3117].

3117 ______.
Transactions and proceedings. v1-3 (1881-1892); s2 v1-6 no2 (1901-1910). Closed 1910. (U-2:1695b; B-2:269b; F-4:707a).
See also Kosmos; an eclectic monthly journal of nature, science and art [3162].

*(3118 GEOGRAPHICAL survey: a quarterly of research and commentary. (Ball state university, Muncie, Indiana. Department of geography and geology). Muncie, Indiana. 1- (1972-) Quarterly. (N75-1:906c).
vl-4 (1972-1975) published by the Blue earth county geographical society, Mankato, Minnesota.
[Department of Geography and Geology, Ball State University, Muncie, Indiana 47306, U.S.A.]

3119 GEOGRAPHY teacher. (New York society for the study of education. Geography section). New York, New York. 1939-1947.

3121 GEOMORPHOLOGY. (Consultants bureau). New York. vl nol-4 (1970). All published. Closed. (N75-1:910c; B73:129a).
English translation of Geomorfologiia, Moskva. Dating and numbering of issues follow the original.

3122 GEOPUB review of geographical literature. (Geographic and area study publications). Tualatin, Oregon. vl nol-11, v2 nol-3 (S 1974-My/Je 1976). Suspended. (N75-1:911b).

(3123 GEORGIA. University. Geography curriculum project. Athens.
Publication: Occasional paper. 1- (1969-) Irregular. (N70-2:2297b; DP).
no3 (1973) Bibliography on geographic thought, philosophy, and methodology by James O. Wheeler revised and reissued in F 1975 and Jl 1978.

3124 ______. ______. Institute of community and area development. Athens, Georgia.
Publications. 1-19 (1962-1967) Irregular. (N70-2:2297b).
nol-11 lack series title and numbering. Title varies: Monograph.
Some in co-operation with the university's Department of geography.

3125 GEORGIA geographer. (Georgia. University. Department of Geography). Atlanta. 1- (1969-). (N75-1:914b).

(3126 GEOSCIENCE and man. (Louisiana. State university. School of geoscience). Baton Rouge. 1- (1970-) Irregular. (B73:129b).
[School of Geoscience, Louisiana State University, Baton Rouge, Louisiana 70803, U.S.A.]

(3126a GLACIOLOGICAL data (World data center A: glaciology [Snow and ice].). Boulder, Colorado.
Report GD-1- (1977-) Irregular.
[World Data Center A for Glaciology (Snow and Ice), Institute of Arctic and Alpine Research, University of Colorado 80309, U.S.A.]

(3126b GLACIOLOGICAL notes (World data center A: glaciology). New York, Seattle. 1- (1960-). (N70-2:2355c: BS:352a).
[World Data Center A: Glaciology, U.S. Geological Survey, 1305 Tacoma Avenue, Tacoma, Washington 98402, U.S.A.]

3127 GOLDTHWAITE'S geographical magazine. New York. vl-7 no6 (1891-1895). Closed 1895. (U-2:1743a; B-2:308b).

*(3128 GREAT Plains-Rocky Mountain geographical journal. (Association of American geographers. Great Plains-Rocky Mountain division. Laramie, Wyoming. 1972- Annual. (N75-1:955c).
[Richard G. Reider, Editor, Department of Geography, University of Wyoming, Laramie, Wyoming 82071, U.S.A.]

*(3129 HANDBOOK of Latin American studies. Gainesville, Florida. 1- (1935-) Annual. (U-3:1793b; B-2:370ab; F-2:790b).
1-13 (1936-1947) published in Cambridge, Massachusetts, by Harvard University Press. Includes works in geography.
From v26 (1964) divided into two volumes; one for Humanities, the other for Social Sciences, in alternating years. Geography constitutes a

section in the social science volume.
Author index: v1-28 (1936-1966).
[University of Florida Press, 15 N.W. 15th Street, Gainesville, Florida 32601, U.S.A.

(3130 HARVARD papers in theoretical cartography. (Harvard university. Laboratory for computer graphics and spatial analysis). Cambridge, Massachusetts. 1- (1978-) Irregular.
[Laboratory for Computer Graphics and Spatial Analysis, Center for Environmental Design Studies, Graduate School of Design, Harvard University, 116 Memorial Hall, Cambridge, Massachusetts 02138, U.S.A.]

3131 HARVARD papers in theoretical geography. (Harvard university. Laboratory for computer graphics and spatial analysis). Cambridge, Massachusetts. Geography and the properties of surfaces series. 1-47 (1967-1975). (N70-2:2487a).
Processed technical reports on Office of naval research contracts.

3131a ______.
Geography of income series. no1- (1970-) (N75-1:907a).
Technical reports on National science foundation grants GS-2833.

*(3132 HISTORICAL geography: a newsletter for historical geographers. Northridge, California. 1- (1971-) Semiannual. (N75-1:1004c; B74:30a).
[Department of Geography, California State University, Northridge, California 91330, U.S.A.]

3133 HOME geographic monthly. (Home geographic society. Clark university). Worcester, Massachusetts. v1-2 no6 (1931-1932). Closed 1932. (U-3:1876b).

(3134 HUMAN ecology: an interdisciplinary journal. New York.
1- (1972-) Quarterly. (N75-1:1019c-1020a; B73:141a).
Abstracts.
[Plenum Publishing Corporation, 277 West 17th Street, New York, New York 10011, U.S.A.]

3135 ILLINOIS. Southern Illinois university, Carbondale. Department of geography. Carbondale, Illinois.
Discussion paper. 1 (1970). (N75-1:1049a).

(3136 ______. ______. ______.
Occasional papers in geography. 1- (1963-) Irregular. (N70-2:2667c; B68:396b; DP).
Title varies slightly: 1 as Occasional paper.

3137 ______. State university, Normal, Illinois. Department of geography, Normal, Illinois.
Monographs. 1-2 (1973). (DP).

(3138 ______. University at Urbana-Champaign. Department of geography. Urbana, Illinois.
Occasional publication. 1- (1972-). (N75-1:1050b; N75-1:1051a; DP).
Early numbers issued by Illinois. University. Geography graduate students association.
[Department of Geography, Davenport Hall, University of Illinois, Urbana, Illinois 61808, U.S.A.]

*(3139 ILLINOIS geographical society. (Affiliated with the National council for geographic education).
Bulletin. v4-14 no2 (D 1941-Ja 1951); v5-6 no1 (N 1951-D 1952); nsv1- (Ap 1955-) 2 nos. a year. (N70-2:2678a).
v4 no1-v13 no6 (D 1941-Ap 1950) as Illinois bulletin of geography. (National council of geography teachers. Illinois chapter).
[Department of geography, University of Illinois at Chicago Circle, Chicago, Illinois 60680, U.S.A.]

3140 INDIANA. State university, Terre Haute. Department of geography and geology. Terre Haute, Indiana.
Occasional papers. 1 (1968). Closed. (N70-2:2750b; B73:240a).
Superseded by department's Professional paper [3141].

*(3141 ______. ______. ______.
Professional paper. 2- (1971-) Irregular. (N75-1:1085; B73:271a; DP).
Supersedes and continues numbering of Department's Occasional paper [3140].

3142 INDIANA. University. Department of geography. Bloomington, Indiana.
Discussion paper. 1-7 (1971-1972). (N75-1:1083c; DP).

3143 ______. ______. ______.
Geographic monograph series. 1-5 (1966-1974). (N70-2:2747b; DP).
[Department of Geography, Indiana University, Bloomington, Indiana 47401, U.S.A.]

3144 ______. ______. ______.
Occasional publication. 1-6 (1965-1971) Irregular. (N70-2:2747a; DP).

3145 ______. ______. ______.
Technical paper. 1-12 (1960-1961). (N70-2:2747a; B68:550b).

3146 INTERNATIONAL geographical union. Commission on quantitative methods.
Current research notes in quantitative and theoretical geography.
Honolulu, Hawaii. 1 (Ag 1971). Only one issued. (B73:89b).

(3147 INTERNATIONAL regional science review. (Regional science association).
Philadelphia. 1- (1975-) 2 nos. a year. (N77).
[Regional science association, Department of regional science, 3718 Locust Street, University of Pennsylvania, Philadelphia, Pennsylvania 19174, U.S.A.]

(3148 IOWA. University. Department of geography. Iowa City, Iowa.
Discussion paper series. 1- (My 1966-) Irregular. (N70-2:2994a; DP).
Mimeographed.

3149 ______. ______. ______.
Iowa studies in geography. 1-6 (1956-1973) Irregular. (N70-2:2994a; DP).
nol-4 (1956-1963) as department's Publication. nol (1956) without series numbering.

3150 IOWA geographer. (Iowa council for geographic education, Inc. [and] University of Northern Iowa. Department of geography). Cedar Falls. nol-32 (1963-fall 1973). Closed 1973. (N75-1:1193a; N77).
Continued as Geographical perspectives [3109].

3151 ITOGI, summaries of scientific progress: theoretical problems in physical and economic geography. (G. K. Hall). Boston, Massachusetts.
1 (1974).
Subseries of geography series. Spine title, vl. Physical and economic geography. Translation of Itogi nauki i tekhniki: teoreticheskie voprosy fizicheskoi i ekonomicheskoi geografii.

3152 JOBS in geography. (Association of American geographers. Placement committee). Washington, D.C. 1-165 (1958-Jn/Jl 1973) Monthly.
Closed. 156-165 (Ag 1972-Jn/Jl 1973) included in AAG Newsletter.
From August 1973 no longer separately numbered and fully incorporated into AAG Newsletter [3029]. (N70-2:3120c).

3153 JOHNS Hopkins university. Isaiah Bowman department of geography. Baltimore, Maryland.
Progress reports on soils, terraces and time in the Chesapeake Bay region. 1-5 (1951-1952). Closed 1952. (N70-2:3124c-3125a).
Issued under the department's earlier name: Isaiah Bowman school of geography.

*(3154 JOURNAL of geography. (National council for geographic education). Houston. 1- (1902-) 7 nos. a year. (U-3:2217a; B-2:624a; F-3:190a).
Succeeds Journal of school geography [3157] and American bureau of geography. Bulletin [2994]. To 1957 as National council of geography teachers.
Indexes: v1-20 (1897-1921), v21-55 (1922-1956), 1-69 (1902-1970), 71-75 (1972-1976).
[The National Council for Geographic Education, Department of Geography, Western Illinois University, Macomb, Illinois 61455, U.S.A.]

3155 JOURNAL of geomorphology; Journal de géomorphologie; Geomorphologische Zeitschrift. New York, New York. v1-5 no4 (1938-1942). Closed 1942. (U-3:2217c; B-2:611b; F-3:190a).
English, with occasional articles or abstracts in French or German.

(3156 JOURNAL of regional science. (Regional science research institute in cooperation with Pennsylvania. University. Wharton school. Department of regional science). Philadelphia, Pennsylvania.
1- (1958-) 3 nos. a year from 1969. (N70-2:3198bc).
Index: v1-10 (1958-1970).
[Regional Science Research Institute, G.P.O. Box 8776, Philadelphia, Pennsylvania 19101, U.S.A.]

3157 JOURNAL of school geography, devoted to the interests of geography teachers. Lancaster, Pennsylvania. 1-5 (1897-1901). Closed 1901. (U-3:2238c; B-2:624a; F-3:195b).
Subtitle varies. United with American bureau of geography. Bulletin [2994] to form Journal of geography [3154].
Index: 1-5 in index to Journal of geography.

(3158 KANSAS. University. Department of geography. Lawrence, Kansas. Occasional paper. 1- (1977-) Irregular. (N78).
[Department of Geography, University of Kansas, Lawrence, Kansas 66045, U.S.A.]

(3159 KANSAS geographer. (Kansas council for geographic education [and] Kansas. State university. Department of geography). Manhattan, Kansas.
1-3; 4- (1968/69-). (N75-1:1290c).
Not published 1962-1967.

(3160 KENT state university. Department of geography. Kent, Ohio. Discussion papers. 1- (1976- (Irregular. (DP).

3161 KENTUCKY study series. (Kentucky. University. Department of geography). Lexington, Kentucky. 1- (1965-). (N70-2:3278c; DP).

3162 KOSMOS; an eclectic monthly journal of nature, science and art. (Geographical society of the Pacific). San Francisco, California. no1-4 (1887). Closed 1887. (U-3:2315a; B-2:667b).

*(3163 LANDSCAPE. Berkeley, California. 1-18 (1951-1968); 19- (1975-). 3 issues a year. (N70-3:3407c; B-3:13b; BS:498b).
Publication suspended 1969-1974.
Indexes: v1-6 (1951-Spring 1956); v11-15 (Fall 1961-Spring 1966).
Early issues published in Santa Fe, New Mexico.
[Landscape, P.O. Box 7107, Berkeley, California 94707.]

3164 LOS ANGELES geographical society.
Publication. Palo Alto, California. 1- (1964), 2 (1973). (N70-3:3527b).

3165 LOUISIANA. State university. Coastal studies institute. Baton Rouge, Louisiana.
Coastal studies bulletin. 1-7 (1967-1973). Closed.

3166 ______. ______. ______.
Technical report. 1- (1953-). (N70-3:3534b).
Partly reprints, partly mimeographed or processed.

3167 LOUISIANA state university studies. Coastal studies series. Baton Rouge, Louisiana. 1- (1958-) Irregular. (N70-3:3537c-3538a).
[Louisiana State University Press, Baton Rouge, Louisiana 70803, U.S.A]

3168 MAINE geographer. (Farmington state college). Farmington, Maine. 1967- Semiannual. (N70-3:3599c).

(3169 MAPLINE: a quarterly newsletter. (Newberry library. Hermon Dunlap Smith center for the history of cartography). Chicago, Illinois. no1- (1976-) Quarterly. (N77).
[Hermon Dunlap Smith Center for the History of Cartography, Newberry Library, 60 West Walton Street, Chicago, Illinois 60610, U.S.A.]

3170 MAPMAKERS. (Aero service corporation, Philadelphia). Philadelphia, Pennsylvania. 1- (1963-) Quarterly. (N70-3:3639b).
[Aero Service Corporation, 210 East Courtland Street, Philadelphia, Pennsylvania 19120, U.S.A.]

3171 MARYLAND. University. Department of geography. College Park, Maryland.
University of Maryland occasional papers in geography. 1- (1971-) Irregular. (N75-2:1409b; DP).

3172 MARYLAND geographer. Baltimore, Maryland. 3- (1970) Semiannual. (N75-2:1409c).
[Department of Geography, Morgan State College, Hillen Road and Cold Spring Lane, Baltimore, Maryland 21212, U.S.A.]

3173 MIAMI. University, Coral Gables, Florida. Miami geographical society.
Transactions. 1- (1970-). (DP).
[The Miami Geographical Society, P. O. Box 8152, University of Miami, Coral Gables, Florida 33124, U.S.A.]

3173a MICHIGAN. University. Department of geography. Ann Arbor, Michigan.
Cartographic laboratory reports. 1-12 (1974-1975).

3174 ______. ______. ______.
Technical reports. 1-3 (1964). (B68:550b).

3175 MICHIGAN academy of science, arts and letters. Ann Arbor, Michigan.
Papers. Pt3. Geography. v24-30 (1938-1944); v32-40 (1946-1954) Annual. (U-3:2641b).
Supersedes Michigan papers in geography [3178].

*(3176 MICHIGAN geographical publications. (Michigan. University. Department of geography). Ann Arbor, Michigan. 1- (1970-) Irregular. (N75-2:1447c; DP).
[Department of Geography, University of Michigan, Ann Arbor, Michigan 48104, U.S.A.]

3177 MICHIGAN inter-university community of mathematical geographers. Ann Arbor, Michigan.
Discussion papers. no1-12 (1965-1968). Closed 1968. (N70-3:3785c).
Issued by Department of geography, University of Michigan, Ann Arbor, Michigan. Incorporated into Geographical analysis: an international journal of theoretical geography [3104].

3178 MICHIGAN papers in geography. Ann Arbor, Michigan. v1-8 (1930-1937). Closed 1937. (U-3:2645a; B-3:203b).
Reprinted from Michigan academy of science, arts and letters. Papers. Succeeded by Michigan academy of science, arts and letters. Papers. Pt3, Geography [3175].

(3179 MIDDLE Atlantic: newsletter of the Middle Atlantic division of the Association of American geographers. 1- (1975-) Irregular, usually 8 nos. per year.
Some expanded issues carry title Mad-Cap, newsletter and journal (N77).
[Secretary, Elldor Pederson, Department of Geography, George Washington University, Washington, D.C. 20052, U.S.A.]

3180 MINNESOTA. University. Department of geography. Minneapolis, Minnesota.
Technical report. 1 (1954). (N70-3:3824b).

3181 MINNESOTA geographer. (Minnesota council for geographic education). St. Cloud, Minnesota. 1948-1972. Closed 1972. Mimeographed.

(3182 MISSISSIPPI geographer. (Mississippi council for geographic education [and] University of Southern Mississippi. Department of geography). Hattiesburg. 1- (1973-) Annual. (N75-2:1466c).
[The Mississippi Geographer, Southern Station Box 8468, University of Southern Mississippi, Hattiesburg, Mississippi 39401, U.S.A.]

3183 MISSOURI. University, Kansas City. Department of geology-geography. Kansas City, Missouri.
Geographic publication. 1- (1972-). (B73:128a).
[Department of Geology-Geography, University of Missouri, Kansas City, Missouri 64110, U.S.A.]

3184 MISSOURI council for the social studies. Parkville, Missouri.
Missouri information pamphlets. Series C. Geography. April 1959- . (N70-3:3844c).

(3185 MONADNOCK. (Clark university geographical society). Worcester, Massachusetts. 1- (1927-) Annual. (U-3:2724a).
Mainly newsletter but some papers.

(3186 MONTCLAIR state college. Department of geography and urban studies. Montclair, New Jersey.
Monographs. 1- (1975-). (DP).

(3187 NCGE pacesetters in geography. (National council for geographic education [and] Kendall/Hunt publishing company). Dubuque, Iowa.
v1 (1975-) Irregular. (N77).
v1-2 (1975-1976) issued in Tualatin, Oregon.

3188 NATIONAL council for geographic education. Chicago, Illinois.
"Do it this way" series. no1-8 (1960-1969). (N70-3:4021b).

3189 ______.
Geographic education series. no1-7 (1956-1967). Closed. (N70-2:2280a).

(3190 ______.
Instructional activities series. Irregular.
In four subseries: Elementary section, IA/E-1- (1975-);
Secondary section, IA/S-1- (1975-);
Geographic activities for the elementary classroom, IAS/OE/01- (1977-);
Geographic activities for the secondary classroom, IAS/OS/01- (1977-).
[The National Council for Geographic Education, Department of Geography, Western Illinois University, Macomb, Illinois 61455, U.S.A.]

3191 ______.
Miscellaneous papers. 1-25. Irregular.
Reprints.

3192 ______.
Professional papers. no1-26 (1927-1964) Irregular. Closed. (U-4:2862b).

3193 ______.
Special publication. 1-16 (1951-1970) Irregular. Closed. (N70-3:4021b).
no1 was issued as Professional paper no12.

3194 ______.
Topics in geography. 1-5 (1966-1970) Irregular. Closed.
Reprints of related articles from the Journal of geography [3154].

3195 ______.
Yearbook. 1-2 (1970-1971). Closed 1971. (N70-3:4021b; N75-1:907a; N75-2:2418b; B73:369a).
1- (1970). Geography of population. (N75-1:907a).
Superseded by the Council's NCGE pacesetter series [3187].

(3196 NATIONAL geographic books. (National geographic society. Special publications division). Washington, D.C. (N70-3:4036a).
Also referred to as the society's Special publication.
______. Second series. 1967- Quarterly.

*(3197 NATIONAL geographic (magazine). (National geographic society). Washington, D.C. 1- (1888-) Monthly. (U-4:2877c; B-3:307a; BS:595a; F-3:684b-685a).
Indexes: 1888-1946, 1947-1976.
[National Geographical Society, 17th and M Streets N.W., Washington, D. C. 20036, U.S.A.]

3198 ______.
[Indexes]. Hand key to your National geographics; subject and picture locater. 1- (1954-). (N70-3:4036a).

3199 NATIONAL geographic school bulletins. (National geographic society). Washington, D.C. O 6 1919-My 9 1921; nsv1- (F 6 1922-1975) Weekly October-May, 30 nos. a year. (U-2:1693c; B-2:268b).
v1-20 no12 (1919-1941) as Geographic news bulletin.
Title varies: Geographic school bulletin.
Superseded by World [3298].

(3200 NATIONAL geographic society. Washington, D.C.
Research reports. 1- (1963-) Irregular. (N70-3:4036a).
[National Geographic Society, 17th and M Streets N.W., Washington, D.C. 20036, U.S.A.]

3201 NATIONAL research council. Division of earth sciences. Washington, D.C.
Report. 1919/20- Annual. (U-4:2905c; N70-3:4058a). Mimeographed.
1919/20-1952/53 as council's Division of geology and geography.
Annual report. 1919/20-1924/25 lack title.
[National Research Council, 2101 Constitution Avenue, N.W., Washington, D.C. 20418, U.S.A.]

3202 ______. ______. Foreign field research program.
Report. 1- (1958-) Irregular monographs. Closed. (N70-3:4058a).
A subseries of the council's Publications (not in this list).

3203 NEBRASKA. University, Lincoln. Department of geography. Lincoln, Nebraska.
Occasional papers. 1- (1971-). (B73:240b; DP).

3204 NEBRASKA council of geography teachers.
Bulletin. v1 no1- (1926-). (U-4:2947c).
Mimeographed. Publication suspended during World War II.
v1-12 (1926-1938) published under earlier name: National council of geography teachers. Nebraska chapter.

(3205 NEW ENGLAND-St. Lawrence valley geographical society. Marblehead, Massachusetts.
Proceedings. 1- (1971-) Annual. (N76).
v1 also includes v5 of Association of American geographers. Middle states division.
[Editor, 28 Clifton Avenue, Marblehead, Massachusetts 01945, U.S.A.]

3206 NEW MEXICO. University. Department of Geography. Albuquerque, New Mexico.
University of New Mexico publications in geography. 1 (1969).
Closed. Only one published. (N70-3:4152b).

3207 NEW YORK (State). State university, Binghamton. Department of geography. Binghamton, New York.
Research in contemporary and applied geography: a discussion series.
1 (1977) Irregular. (N79).

(3207a NEW YORK (State). State university, Buffao. Geographic information systems laboratory. Buffalo.
Research report series. 77/1- (1977-) Irregular. Variable number of reports per year.

3208 ______. State university, Oneonta. Department of geography. Oneonta, New York.
Publications in geography. 1- (1969-). (N70-3:4177c; B73:276b).
no1 unpaged reproduction of computer printout.

3209 NEWSMAP magazine. Chicago, Illinois. v1-6 no6 (N 1938-Ap 1944).
Closed 1944. (U-4:3060b).
v1 no1-5 as Monthly newsmap.

*(3210 NORTH CAROLINA. University, Chapel Hill. Department of geography. Chapel Hill, North Carolina.
Studies in geography. 1- (1970-). (N70-4:5593b; DP).
[Department of Geography, University of North Carolina at Chapel Hill, Chapel Hill, North Carolina 27514, U.S.A.]

3211 NORTHWESTERN university, Evanston, Illinois. Department of geography. Evanston, Illinois.
Research report. no1- (1963-) Irregular. (N70-3:4282c; DP).
Mimeographed technical reports.

3212 ______. ______.
Special publication. 1-4 (1967-1970) Irregular. Processed. (DP).

3213 ______. ______.
Technical report. (N70-3:4282c).

*(3214 NORTHWESTERN university studies in geography. (Northwestern university, Evanston, Illinois. Department of geography). Evanston, Illinois.
1- (1952-) Irregular. (N70-3:4283bc; DP).

(3215 OHIO. State university, Columbus. Department of geography. Columbus, Ohio. Discussion paper. 1- (1969-) Irregular but frequent. Partly undated. Processed. (DP).

(3216 OHIO geographers: recent research themes. (Akron. University. Department of geography). 1- (1973-) Annual. (N75-2:1645a).
Also as Ohio. Academy of science. Geography section.
[Department of Geography, The University of Akron, Akron, Ohio 44325, U.S.A.]

3217 OREGON geographer. (Oregon geography council). Eugene, Oregon. 1- (1966-). (N70-3:4423a).

3218 PACIFIC geographic magazine. (Pacific geographic society). Beverly Hills, California. v1 no1-3 (Ap-Je 1936). Closed 1936. (U-4:3232b).
Superseded by Pacific horizons [3221].

3219 PACIFIC geographic news letter. Los Angeles, California. v1 no1-5 (D 1935-Ap 1936). Closed 1936. (U-4:3232b).

3220 PACIFIC geographic society, Los Angeles. Los Angeles, California. Bulletin. no1 (1931). Closed 1931? (U-4:3232b).
See also Pacific geographic magazine [3218]; Pacific horizons [3221].

3221 PACIFIC horizons. (Pacific geographic society). Los Angeles, California. Ja-O 1939. Closed 1939. (U-4:3232c).
No issues for Ap, Jl-S.
Supersedes Pacific geographic magazine [3218].

3222 PACIFIC Northwest geographer. (Central Washington state college, Ellensburg. Department of geography). Ellensburg, Washington. 1- (1962?-) Semiannual. (N70-3:4493b).

3223 PAN AMERICAN institute of geography and history. United States national section. Washington, D.C.
Special publications. 1-6 (1970-1971). Closed. (N75-2:1693a; B73:324b).

3224 PARACAS papers. (Florida Atlantic university. Department of geography). Boca Raton, Florida. v1 no1-2 (M 1968-Ag 1968). Closed? (N70-3:4538c).
v1 no2 issued also as the department's Occasional publications [3089].
Each number has a distinctive title.

3225 PENDULUM. (De Paul geographical society). Chicago, Illinois. 1974- .

3226 PENINSULAR. (Michigan council for geographic education). East Lansing. 1970.

3227 PENNSYLVANIA. State university. Department of Geography. University Park, Pennsylvania.
PSUDOG: authentic newsletter. 1- (1973-).

(3228 ______. ______. ______.
University Park, Pennsylvania.
Papers in geography. 1- (1969-) Irregular. (N75-2:1696b; B73:251a; DP).

*(3229 PENNSYLVANIA geographer. (Pennsylvania council for geography education. Journal). Indiana, Pennsylvania. 1- (1963-) Quarterly. (N70-3:4584a).
Index: v1-14 (1963-1976) in v14 no4 (D1976), p. 23-34.
[Department of Geography, Indiana University of Pennsylvania, Indiana, Pennsylvania 15701, U.S.A.]

(3230 PERSPECTIVE. (National council for geographic education. Newsletter). Macomb, Illinois.
1- (1969-) 5 to 6 nos. per year. (N70-3:4599c).
[James W. Vining, Editor, Perspective, Department of Geography, Western Illinois University, Macomb, Illinois 61455, U.S.A.]

3231 PERSPECTIVES in geography. (Northern Illinois university. Earth science department). DeKalb, Illinois. 1-2 (1971-1972) Irregular. (N75-2:1720c; B73:254b).
[Northern Illinois University Press, DeKalb, Illinois 60115.]

*(3231a PHYSICAL geography. Washington, D.C. 1- (1980-) Semiannual.
[V. H. Winson and Sons, 7961 Eastern Avenue, Silver Springs, Maryland 20910, U.S.A.]

(3231b PIONEER America: the journal of historic American material culture. Falls Church, Virginia; Baton Rouge, Louisiana. 1- (1968-) Semiannual.
[Pioneer America, P.O. Box 22230, Baton Rouge, Louisiana 70893, U.S.A.]

3232 PLACES: a literary journal of geography and travel. Indiana, Pennsylvania. v1-3 (no1-10) (Mr 1974-N 1976). Closed 1976. (N75-2:1739c).

*(3233 POLAR geography and geology. (Scripta publishing). Washington, D.C. 1- (1977-) Quarterly. (N77).
Published in cooperation with the American geographical society.
1-3 (1977-1979) as Polar geography.
[Scripta Publishing Co., 7961 Eastern Avenue, Silver Spring Maryland 20910, U.S.A.]

3234 PORTLAND state university. Department of geography, Portland, Oregon. Occasional publications in geography. 1-2 (1971-1973). (N75-2:1758b; DP).

*(3235 PROFESSIONAL geographer. (Association of American geographers. Journal). Washington, D.C. 1-8 (1943-1948); nsv1- (1949-) Quarterly. (U-4:3449b; B-1:129b).
v1 no1-3 (1943) as American society for geographical research. Bulletin; v2 no1-5 (1944), v3 no1-6 (1945) as American society for professional geographers. Bulletin: v4 (1946), v5-6 (1947) and v7-8 (1948) as Professional geographer. Bulletin of the American society for professional geographers. nsv1- as continuation also of Association of American geographers and American society for professional geographers. Joint newsletter [3050].
Index: 1946-1966 in v19 no2 (Mr 1967).
[Association of American Geographers, 1710 Sixteenth Street, N.W., Washington, D.C. 20009, U.S.A.]

3236 ______.
Supplement: Directory of the members and associates of the Association of American geographers. 1949, 1952.
Supersedes Association of American geographers. List of members [3034] and American society for professional geographers. Directory [3023].
Superseded by the association's Directory [3032].

*(3237 QUATERNARY research: an interdisciplinary journal. (Academic Press). New York. 1- (S 1970) Bimonthly. (N75-2:1812a; B73:281b).
[Academic Press, 111 Fifth Avenue, New York, New York 10003, U.S.A.]

3237a RSEMS: Remote sensing of the electromagnetic specturm. (Association of American geographers. Remote sensing committee). East Lansing, Michigan; Omaha, Nebraska.
v1-5 (1974-1978) Quarterly. Closed 1978.

Additional subtitle varies: Newsletter of the Remote sensing committee of the Association of American geographers; Forum of the Remote sensing committee of the Association of American geographers; Forum of remote sensing in geography.
Superseded by Remote sensing quarterly [3244a].

3238 R S R I Discussion paper series. (Regional science research institute). Philadelphia, Pennsylvania. 1- (1963-). (N70-3:4897c; DP).
[Regional Science Research Institute, G.P.O. Box 8776, Philadelphia, Pennsylvania 19101, U.S.A.]

*3239 REGIONAL science association. Philadelphia, Pennsylvania.
Papers. 1- (1954-) 2 per annum. (N70-3:4956b).
Title varies: v1-7 (1954-1961) as Papers and proceedings of the annual meeting.
Index: 1-20 (1955-1968).
[Regional Science Association, Wharton School, University of Pennsylvania, Philadelphia, Pennsylvania 19104, U.S.A.]

3240 ______. Western section.
Papers [presented at the] annual meeting. 2- (1962-). (N70-3:4956c).
Papers of the first meeting, 1962, not published.

3241 REGIONAL science research institute. Philadelphia, Pennsylvania.
Bibliographical series. no1- (1961-) Irregular. (N70-3:4956c; B68:77b).
[Regional Science Research Institute, G.P.O. Box 8776, Philadelphia, Pennsylvania 19101, U.S.A.]

3242 ______.
Monograph series. 1- (1965-). (N70-3:4956c).
no3 issued in 1965, no2 in 1966, no1 in 1967, no4 in 1970, no5 in 1973, no6 in 1975.

3243 ______.
Occasional papers series. no2 (1965). (N70-3:4956c).
No more published?

*3244 REMOTE sensing of the environment: (an interdisciplinary journal. New York. 1- (1969-) Quarterly. (N70-3:4966b; B73:287b).
[Elsevier North-Holland, Inc., New York, 52 Vanderbilt Avenue, New York, New York 10017, U.S.A.]

*3244a REMOTE sensing quarterly: forum of remote sensing in geography. (Association of American geographers. Remote sensing committee; National council for geographic education. Remote sensing committee; University of Nebraska at Omaha. Remote sensing laboratory and Department of geography). Omaha, Nebraska. 1- (1979-) Quarterly.
Supersedes RSEMS: Remote sensing of the electromagnetic specturm [3237a].
[Ronald Rundquist, Department of Geography-Geology, The University of Nebraska at Omaha, Omaha, Nebraska 68182, U.S.A.]

3245 RHODE ISLAND. University. Department of geography. Kingston, Rhode Island.
Occasional papers. no1 (1976). (N78; DP).
[Department of Geography, University of Rhode Island, Kingston, Rhode Island 02881, U.S.A.]

3246 RHODE Island college of education.
Geography-science bulletin. v1-2 (1943-1945). Closed 1945.

3247 ______.
Papers in geography. no1 (1944). Only one issued?
Mimeographed.

(3248 RUTGERS university, New Brunswick, New Jersey. Department of geography. Discussion paper series. 1- (1970-). (B73:96a).
[Department of Geography, Rutgers University, New Brunswick, New Jersey 08903, U.S.A.]

3249 SEARCHLIGHT books. (D. Van Nostrand Company). Princeton, New Jersey. 1- (1962-) Irregular monographs.
[D. Van Nostrand Company, 120 Alexander Street, Princeton, New Jersey 08540, U.S.A.]

3250 SOCIETY of woman geographers. Washington, D.C.
Bulletin. 1926- Usually annual.

*(3251 SOUTHEASTERN geographer. (Association of American geographers. Southeastern division. Journal). Place of publication varies. 1- (1961-) Semiannual. (N70-4:5445b; B68:519b).
Supersedes Association of American geographers. Southeastern division. Memorandum folio [3049].
[Department of Geography, University of Tennessee, Knoxville, Tennessee 37916, U.S.A.]

*(3252 SOVIET geography: review and translation. Washington, D.C.
1- (1960-) Monthly except July and August. (N70-4:5463ab; BS:825b; B68:520b).
Published in co-operation with the American Geographical Society.
[V. H. Winston and Sons, 7961 Eastern Avenue, Silver Spring, Maryland 20910, U.S.A.]

*(3253 SPECIAL libraries association. Geography and map division. New York, New York.
Bulletin. 1- (N 1947-) Quarterly. (U-5:4042c; BS:829b). Mimeographed.
N 1947-Je 1950 under earlier name: Geography and map group.
Indexes: 1-70 (1947-1967), 71-102 (1968-1975).
[Mrs. Kathleen I. Hickey, Business Manager, Geography and Map Division SLA, 9927 Edward Avenue, Bethesda, Maryland 20014, U.S.A.]

3254 STUDIES in medical geography. New York, New York. 1-14 (1958-1974) Irregular monographs. Closed. (N70-4:5595c).
1-4 as American geographical society. Studies in medical geography.
[Hafner Press, Macmillan Publishing Co., 866 Third Avenue, New York, New York 10022, U.S.A.]

3255 STUDIES in the diffusion of innovation. (Ohio state university, Columbus. Department of geography). Columbus, Ohio.
1-60 (1974-1979). Closed 1979. (DP).

3256 SYRACUSE geographical series. (Syracuse university. Department of geography). Syracuse, New York. 1- (1965-) Irregular monographs. (N70-4:5662b).
[Syracuse University Press, Box 8, University Station, Syracuse, New York 13210, U.S.A.]

(3257 SYRACUSE university. Department of geography. Syracuse, New York.
Discussion paper series. 1- (Mr 1975-). (DP).

3258 TEXAS. University. Department of geography. Austin, Texas.
Technical report. (N70-4:5755a).

3259 TEXAS geographic magazine. (Texas geographic society). Dallas, Texas.
v1-13 no2 (1937-1949). Closed 1949. (U-5:4192b; B-4:304b).

3260 TRANSITION. (Socially and ecologically responsible geographers. Journal). Cincinnati, Ohio. 1- (1971-) Quarterly.
v1-4 (1971-1974) as Newsletter of SERGE, issued irregularly. Present format as quarterly from v5 no2 (Spring, 1975).
[Laurence G. Wolf, 610 Foulke Street, Cincinnati, Ohio 45220, U.S.A.]

3261 TRANSPORTATION geography newsletter. (Temple university. Department of geography). Philadelphia, Pennsylvania.
61 (1977-). (N78).

3262 U.S. Board on geographic names. Washington, D.C.
Bulletin. 1-3 (1890-1891). Closed 1891. (B-4:420a).

3263 ______.
Decision list. s1 1-41 (1918-1934); s2 Decisions rendered (1934/35-1941/43); s3 (1943-) Irregular. (BS:921a).
Through 1968 the first numbers indicate the year; the last two, the report number in that year. Number each year varies; 1969/70- designated by years.
Beginning 1944, cumulations for specific geographical areas issued irregularly with title: Cumulative decision list.

3264 ______.
Decisions on geographic names in the United States. 1959- Irregular.

3265 ______.
Gazetteer. 1-116 (1955-1970) Irregular. (N70-4:5979a). Each no. a separate country.
no105-116 (1968-1970) issued by U. S. Army topographic command.
Superseded by unnumbered gazetteers issued by the U. S. Army topographic command.

3266 ______.
Reports. no1-6 (1890-1932). Closed 1932.
1 (1890-91); 2 (1890-99); 3 (1890-1906); 4 (1890-1916); 5 (1890-1920); 6 (1890-1932). Each number is cumulative.

3267 ______.
Special decision list. 1- (1950-) Irregular. (N70-4:5979b; B-4:420a).

3268 ______.
Special publications. 1- (F 1943-). (B-4:420a).
Numbers published: 1-12, 14-15, 18, 21, 24-25, 30, 40, 42, 44-49, 51, 53, 58, 62-75, 78-87, 89, 91, 140. 86 supplements 1 and 2 (1950-1951).

3269 U. S. Bureau of the census. Washington, D.C.
Geographic reports. 1- (F 1963-). (N70-4:5991b).
[Superintendent of Documents, U.S. Government Printing Office, Washington, D.C. 20402, U.S.A.]

3270 ______.
Geographic reports (Series GEO). 1-6 (1951-1953). (N70-4:5991b).

3271 U.S. Department of State. Bureau of intelligence and research. Washington, D.C.
Geographic bulletin. 1- (1967-) Irregular, infrequent.

3272 ______.
Office of the geographer. Washington, D.C.
Geographic bulletin. 1-8 (1963-1968) Irregular. (N70-4:6015a; B73:128a).
Subseries of U.S. Department of state. Publication (not in this list).
[Superintendent of Documents, U.S. Government Printing Office, Washington, D.C. 20402, U.S.A.]

3273 ______. ______.
Geographic notes. 1- (1964-).

3274 ______. ______.
Geographic report. 1-16 (1961-1971) Irregular. (N70-4:6015a).
Each issue also has distinctive title.

*(3275 ______. ______.
International boundary study. 1- (1961-) Irregular. (N70-4:6015a).

3276 ______. ______.
Limits in the sea. no1- (1970-) Irregular. (N75-2:1352c).

(3277 ______. ______.
Status of the world's nations. v1 1963, 1964, 1965, 1967, 1970, 1976, Ja 1978. Revised irregularly. (N70-4:6015a).
Issued as geographic bulletin no2.

3278 U.S. Geographical and geological survey of the Rocky Mountains regions. Washington, D.C.
Final reports. 1877-1880. Closed 1880. (B-1:650a; F-2:258b).
The Powell reports.

3279 U.S. Geographical surveys west of the 100th meridian. Washington, D.C.
Annual reports.
v1-12 covering years 1869-1884 published 1872-1884. The Wheeler reports. From Appendices to U.S. Chief of engineers. Annual reports. 1869 and 1871 as Preliminary reports of U.S. Engineer department; 1872 not issued separately.

3280 U.S. Geological and geographical survey of the territories. Washington, D.C.
Annual report. v1-3, 5-13 (1867-1883). Closed 1883). (B-4:418a; F-1:327ab).
v4 never published. The Hayden reports.

3281 ______.
Bulletin. v1-6 (1875-1882). Closed 1882. (B-4:418a; F-1:993ab).

3282 ______.
Miscellaneous publications. v1-12 (1873-1880). Closed 1880. (B-4:418a).

3283 U.S. Library of congress. Washington, D.C.
Library of congress catalog. A cumulative list of works represented by library of congress printed cards. Maps and atlases. Washington, D.C. 1953-1955. Closed 1955. Semiannual with annual cumulation. (N70-4:6047ab).
Formerly included in the Library's author catalog and its subject catalog. Beginning with 1956 included in the National union catalog. A cumulative author list representing Library of congress printed cards and titles reported by the American libraries and in the library's catalog. Books. Subjects.

(3284 URBAN concerns. (Urban concerns, Inc.) Washington, D.C. v1 no1 (My/Je 1979; v1 no2- (Ja/F 1980-). Bimonthly.
[Editor, Urban Concerns, P.O. Box 8645, Washington, D.C. 20011, U.S.A.]

*(3284a URBAN geography. Washington, D.C. 1- (1980-) Quarterly.
[V. H. Winston and Sons, 7961 Eastern Avenue, Silver Spring, Maryland 20910, U.S.A.]

3285 UTAH. University. Department of geography. Salt Lake City, Utah.
Research papers. RP76/1- (1976-) Irregular. (DP). Each year numbered separately.

3286 VERMONT geographer. (Vermont. University. Department of geography), Burlington, Vermont. 1-2 (1972-1975). (N78).
[Department of Geography, University of Vermont, 112 Old Mill Building, Burlington, Vermont 05401, U.S.A.]

(3287 VIRGINIA geographer. (Virginia geographical society). Fredericksburg, Virginia. v1-9- (1966-1974-) 2 nos a year. Publication delayed. (N70-4:6202c).
Supersedes Virginia geographical society. Bulletin [3288].
[The Virginia Geographer, Box 3486, College Station, Fredericksburg, Virginia 22401, U.S.A.]

3288 VIRGINIA geographical society. Farmville, Virginia.
[The] Bulletin. 1-18 (1948-1966). Closed 1966. (U-5:4405c).
Superseded by Virginia geographer [3287].

(3288a VIRGINIA polytechnic institute and state university. Department of geography. Blacksburg, Virginia.
Serial publications in geography. 1975- Annual.

3289 WASHINGTON (state). University. Department of geography. Seattle, Washington.
Discussion paper. 1-42 (1958-1961). (N70-4:6277a). Mimeographed working papers.

3290 WEST Georgia college. Department of geography. Carrollton, Georgia.
West Georgia college studies in the social sciences. 1- (My 1962-) (DP).

(3291 WESTERN association of map libraries. Santa Cruz, California, etc.
Information bulletin. 1- (S 1969-) 3 per annum. (N75-2:2281bc).
Titles varies, S1969-Mr 1970 as Newsletter. Issue for S 1969 lacks volume numbering but constitutes v1 no1.
[Western Association of Map Libraries, c/o University Library, University of California, Santa Cruz, Santa Cruz, California 95064, U.S.A.]

(3292 ______. Santa Cruz, California.
Occasional paper. 1- (1973-) Irregular. (N75-2:2281c).
[Western Association of Map Libraries, c/o University Library, University of California, Santa Cruz, Santa Cruz, California 95064, U.S.A.]

3293 WISCONSIN. University, Madison. Cartographic laboratory. Madison, Wisconsin.
Papers. 1- (1975-) Irregular.
Also as Irregular paper series.
[Cartographic Laboratory, Science Hall, University of Wisconsin, Madison, Wisconsin 53706, U.S.A.]

3294 ______. University, Oshkosh. Department of geography. Oshkosh, Wisconsin.
Occasional papers. no1- (1974-). (N77; DP).
[Secretary, Department of Geography, University of Wisconsin-Oshkosh, Oshkosh, Wisconsin 54901, U.S.A.]

3295 WISCONSIN council for geographic education. Milwaukee?
Newsletter. 1970?-. (N77).

(3296 WISCONSIN geographer. (Wisconsin. University, Madison. Department of geography). Madison, Wisconsin. (N70-4:6364c).

(3297 WORLD. (Virginia geographical society). 1- (1966-) Irregular. (N70-4:6374c).
Supersedes Geogram [3096].

(3298 WORLD. (National geographical society). Washington, D.C.
nol (S 1975-) Monthly. (N76; N77).
Supersedes National geographic school bulletin [3199].
Cover title: National geographic world.
[National Geographic Society, 17th and M Streets, N.W., Washington, D.C. 20036, U.S.A.]

URUGUAY (Spanish)

República Oriental del Uruguay

3299 ASOCIACIÓN nacional de profesores de geografía. Montevideo.
Congreso nacional de geografía. 1-3 (1967-1971) Biennial.

3300 ______.
Cuadernos de geografía. 1-6 (n.d.) (N70-2:2235c).
[25 de Mayo 537, Montevideo, Uruguay.]

3301 ______.
Cuadernos de geografía regional. 1-7 (1972-1973).

3302 CUADERNOS estuario. (Centro de estudios geográficos del Uruguay. C.E.G.U. [Ateneo del Uruguay]).
Montevideo. 1-2 (1967-1969).

3303 ESTUARIO. Revista de geografía e historia. Montevideo. v1-2, nos. 1-5 (My 1958-Ag 1959) Quarterly. Publication suspended. (N70-2:1957c; BS:294b).

3304 GEOPOLÍTICA. (Instituto uruguayo de estudios geopolíticos, I.U.D.E.G.). Montevideo. 1- (1976-) 4 nos. a year.
[Casilla correos 5006, Montevideo, Uruguay.]

3305 INSTITUTO histórico y geográfico del Uruguay. Montevideo.
Revista. 1-21 (1920-1954) Annual. (U-3:2026c; B-2:506b; F-4:136b).
Index: v1-20 (1920-1953).

3306 MONTEVIDEO. Instituto de estudios superiores. Sección de investigaciones geográficas. Montevideo.
Boletín. v1-2 (no1-8) (1938-1942). Closed 1942. (B-3:247a; F-1663a).

3307 ______. Universidad. Instituto nacional de investigaciones geográficas.
Boletín. no1-30/31 (N 1942-N/D 1946); no32-50 (1950-1957). Publication suspended. (U-3:2743b).
Title varies.
[Avenida Cataluña 3180, Montevideo, Uruguay.]

3308 ______. ______. ______.
Revista. no1-6 (1943-1944); 1945 no1-4; 1947, 2 special numbers; 1955 no1; 1959 no1-2; 1962 no3. (U-3:2743b).
Numbering irregular.
[Avenida Cataluña 3180, Montevideo, Uruguay.]

3309 NUESTRA tierra. Montevideo. 1-48 (1969-1970) Weekly.
s2. Los departamentos, 1-18 (1970).
s3. Montevideo, 1-10 (1971) . (N70-3:4318b).
[Editorial Nuestra Tierra, Soriano 875, esc. 6, Montevideo, Uruguay.]

3310 REVISTA uruguaya de geografia. (Asociación de geógrafos del Uruguay). Montevideo. no1-9 (1950-1957); s2, no1-3 (1971-1975). (N70-3:5023c; BS:735a).

3311 URUGUAY. Servicio geográfico militar. Montevideo.
Boletín. 1-2 (1919-1924); 3-4 (1944-1945). (ULG-570c).

3312 URUGUAY hoy. (Centro de estudios geográficos del Uruguay. [Ateneo del Uruguay. Departamento de publicaciones Estuario]). Montevideo.
1-5 (1965).

VENEZUELA (Spanish)

República de Venezuela

3313 ACTA venezolana. (Sociedad interamericana de antropologia y geografía. Grupo local de Caracas). Caracas. 1-3 (1945-1948). Closed 1948? (U-1:60c; B-1:47a).

3314 ASOCIACIÓN de geógrafos latinoamericanos. (Mérida. Universidad de los Andes. Escuela de geografía). Mérida. Noticiero AGELA. Ap 1978.

3315 BIBLIOTECA de geografía e historia. Serie Rufino Blanco Fombona. Caracas. 1- (1950-). (N70-1:741b).

3316 GEA; revista venezolana de geografía. Caracas. 1-5 (1961-1965). (N70-2:2260b; B68:222b).

(3317 MÉRIDA (city). Universidad de los Andes. Instituto de geografía y conservación de recursos naturales. Mérida. Cuadernos geográficos. 1- (1961-). Irregular monographs. (N70-3:3733c).

*(3318 REVISTA geográfica. (Mérida. Universidad de los Andes. Facultad de ciencias forestales. Instituto de geografía y conservación de recursos naturales). Mérida. 1- (1959-) 2 per annum. (N70-3:5016c-5017a).
Index: v1-12 (1959-1971).
In Spanish. Abstracts in English and French.

*(3319 SÍNTESIS geográfica. (Venezuela. Universidad central, Caracas. Escuela de geografía). 1- (1977-).

3320 SOCIEDAD bolivariana de geografía. Caracas. Boletín. 1- (1971-).
[Apartado de Correos No. 2446, Caracas, D.F., Venezuela.]

3321 SOCIEDAD interamericana de antropología y geografía. Grupo local de Caracas. Inter-American society of anthropology and geography, Washington. Caracas chapter. Caracas.
Publicaciones. v1 no1-4 (1944-1947). Closed 1947. (U-3:2039b).

3322 TERRA. (Venezuela. Universidad central, Caracas. Facultad de humanidades y educación. Escuela de geografía). Caracas. 1- (1977-) Irregular.

3323 TIERRA y hombre. Revista geográfica. (Colegio de geógrafos de Venezuela). Caracas. v1 no1 (N-D 1969 and Ja 1970).

3324 VENEZUELA. Ministerio de la defensa nacional. Estado mayor conjunto. División de informaciones. Sección geografía. Caracas. Publicación. (N70-4:6151a).

3325 ______. Universidad central, Caracas. Escuela de biblioteconomía y archivos. Caracas.
Colección geografía. 1- (1964-). (N70-4:6153a).

VIET NAM (Vietnamese)

Cộng Hòa Xã Hội Chủ Nghĩa Việt Nam

(3326 ACTA scientiarum vietnamicarum. Sectio scientiarum geologicarum et geographicarum. Hanoi. (BL).
In Western languages.

3327 SOCIÉTÉ de géographie commerciale. Section indochinoise. Hanoi.
Annales. no4 (1909).
In French.

3328 SOCIÉTÉ de géographie de Hanoi. Hanoi.
Cahier. 1-34 (1922-1938). Closed 1938. (U-5:3945b; B-4:135b).
In French.

(3329 TẠP chí Sinh vật - Địa học. (Revue de biologie et de géographie. Journal of biology and geography). Hanoi. Quarterly. (BL).
In Vietnamese. Biology and geography now separate series.

(3330 VIETNAM. Direction du service géographique national. Saigon.
Annual report on activities of the National geographic service of South Vietnam. 1963- . Closed. (N70-4:6187b).
Issued also in French.

3331 ______. ______.
Rapport annuel.
1956- . Closed. (N70-4:6187b).
Title and text also in Vietnamese.

YUGOSLAVIA (Serbo-Croatian, Slovenian, Macedonian)

Socijalistička Federativna Republika Jugoslavija

Социјалистичка Федеративна Република Југославија

3333 БЕОГРАД. Институт екологију и биогеографију. Београд.
Зборник Радова.

3334 ______. Универзитет. Географски Институт. Београд.
Посебна издања.

*(3335 ______. ______. ______. [Географски завод].
Зборник радова

3336 БИБЛИОГРАФИЈА географских радова о Југославији (Савет географских друштава). Београд.

(3340 ГЕОГРАФСКИ годишњак. Крагујевац.

(3342 ГЕОГРАФСКИ лист. (Географско дгуштво Босне и Херцегових). Сарајево.

*(3344 ГЕОГРАФСКИ преглед . (Географско друштво Босне и Херцеговине). Сарајево.

*(3345 ГЕОГРАФСКИ разгледи. (Географско друштво на С. Р. Македонија). Скопје.

(3346a ГЕОГРАФСКИ видик. (Географско друштво на С. Р. Македонија). Скопје.

3348 ГЕОГРАФСКО друштво Босне и Херцеговине. Сарајево. Библиотека географије у школи.

(3349 ______ . Посебна издања. Књига.

3349a Географско друштво Црне Горе. Титоград. Годишнак.

3350 ГЕОГРАФСКО друштво на С. Р. Македонија. Скопје. Билтен.

(3352a ГЛОБУС. Часопис за педагошко-методска питања... (Српско географско друштво). Београд.

*(3354 КОНГРЕС географа Југославије. Зборник. (Савез географских друштва ФНРЈ).

(3354б НАСТАВА географије: часопис за питања дидактике и методике географије. (Географско друштво Босне и Херцеговине). Сарајево.

3355 НИШ. Виша педагошка школа. Катедра за географију. Ниш. Радови.

3356 НОВИ САД. Универзитет. Природно-математички факултет. Нови Сад. Зборник радова. Серије за географију.

(3357 СКОПЈЕ. Универзитет. Географски факултет. Скопје. Годишен зборник.

3358 ______. ______. Природно-математички факултет. Скопје. Годишен зборник. Географија и Геологија.

*(3361 СРПСКА академија наука и уметности. Географски институт Јован Цвијић. Београд. Посебна издања.

*(3362 ______. ______. Зборник радова.

3363 СРПСКО географско друштво. Београд. Атласи.

*(3364 ______ . Гласник.

*(3366 ______ . Посебна издања.

3368 ЗЕМЉА и ЉУДИ: популарно научни зборник. (Српско географско друштво). Београд.

3332 ARGO; Zeitschrift für krainische Landeskunde. (Laibach) Ljubljana. v1-10 no6 (1892-1903). Closed 1903. (U-1:475a; B-1:214a).
Suspended 1896-1902.
Index: 1-10.
In German.

3333 BEOGRAD. Institut za ekologiju i biogeografiju. Beograd. Zbornik radova. Recueil des travaux. 1-8? (1950-1956?) Irregular. (N70-1:684a; BS:269a; R-1:37a). Alternate listing: Srpska akademija nauka i umetnosti, Beograd. Institut za ekologiju i biogeografiju. Zbornik radova. (N70-4:5514c-5515a).
Subseries of Academy's Zbornik radova.
Superseded by Academy's Bioloski institut. Zbornik radova.
Text in Serbian with summaries in English or French.
[ul. 29 Novembra br. 100, Beograd, Yugoslavia.]

3334 ______. Univerzitet. Geografski institut. Beograd. Posebna izdanja. 1- (1967-) Irregular. (N70-1:688b).
In Serbian.

*(3335 ______. ______. ______. [Geografski zavod]. Zbornik radova. Recueil de travaux. 1- (1954-) Annual. (N70-1:688b; R-1:37a).
In Serbian. Table of contents and abstracts in French or occasionally English.
[Geografski Institut Univerziteta, Studentski trg 3/III, Beograd, Yugoslavia.]

3336 BIBLIOGRAFIJA geografskih radova o Jugoslaviji. (Savet geografskih društva). Beograd. 1956? Irregular. (N70-1:725c).

3337 GEOGRAFIA revuo. (Internacia geografi asocio). Beograd. 1-5 (1956-1963). (N70-2:2278a; BS:341a; R-3:267b).
In Esperanto.

(3338 GEOGRAFSKA bibliografija Slovenije. Geographic bibliography of Slovenia. (Ljubljana. Univerza. Inštitut za geografijo). Ljubljana. (1966-). (N70-2:2279c).
Supersedes Slovenska geografska bibliografija [3360].

*(3339 GEOGRAFSKI glasnik; Bulletin de géographie. (Geografsko društvo Hrvatske). Zagreb. 1-10 (1929-1939); 11- (1949-) Annual. (U-2:1693b; B-2:424b; BS:393b; F-2:833b; R-3:268b).
v1-10 (1929-1939) as Hrvatski geografski glasnik.
In Croatian. Supplementary table of contents and abstracts in English or other international languages.
[Geografsko Društvo Hrvatske, Marulićev trg 19/II, Zabreb, Yugoslavia.]

(3340 GEOGRAFSKI godišnjak. (Srpsko geografsko društvo. Podružnica Kragujevac). Kragujevac. 1- (1965-). (N70-2:2279c).
In Serbian.

(3341 GEOGRAFSKI horizont. (Geografsko društvo Hrvatske). Zagreb. 1- (1955-) Quarterly. (N70-2:2279c; R-3:268b).
In Croatian.
[Geografsko Društvo Hrvatske, Marulićev trg 19/II, Zagreb, Yugoslavia.]

(3342 GEOGRAFSKI list. (Geografsko društvo Bosne i Hercegovine). Sarajevo. 1- (1975-) Bimonthly during school year. In Serbo-Croatian.
[Geografsko društvo Bosne i Hercegovine, Vojvode Putnika 43, Sarajevo, Yugoslavia.]

(3343 GEOGRAFSKI obzornik. (Geografsko društvo Slovenije). Ljubljana. 1- (1954-) Quarterly. (N70-2:2279c; R-3:268b).
In Slovenian.
[Zemljepisni Muzej Slovenije, trg Francoske Revolucije 7, Ljubljana, Yugoslavia.]

*(3344 GEOGRAFSKI pregled; Revue de géographie. (Geografsko društvo Bosne i Hercegovine). Sarajevo. 1- (1957-) Annual. (N70-2:2279c; BS:341b; R-3:268b).
Index: 1-20 (1957-1976) in v20 (1976), p. 234-246.
In Serbo-Croatian. Supplementary titles in table of contents and abstracts in English, French, German, or Russian.
[Geografsko Društvo Bosne i Hercegovine, Prirodno-matematički Fakultet, Sarajevo, Yugoslavia.]

*(3345 GEOGRAFSKI razgledi. Revue géographique. (Geografsko društvo na S.R. Makedonija). Skopje. 1- (1962-) Annual. (N70-2:2279c; B73:128a; R-1:26a)
In Macedonian. Abstracts in English, French, German, or Russian.
[Geografsko Društvo na SR Makedonija, Geografski Fakultet, Skopje, Makedonija, Yugoslavia.]

*(3346 GEOGRAFSKI vestnik. Časopis za geografijo in sorodne vede. (Geografsko društvo Slovenije). Ljubljana. 1- (1925-) Annual. (U-2:1693b; B-2:268b; BS:341b; F-2:727b; R-3:268b-269a).
In Slovenian. Supplementary titles in table of contents and abstracts in English.

(3346a GEOGRAFSKI vidik. (Geografsko društvo na SR Makedonija). Skopje. 1- (1970-) Annual.
In Macedonian.
[Geografsko Društvo na Makedonija, Skopje, Yugoslavia.]

*(3347 GEOGRAFSKI zbornik; Acta geographica. (Slovenska akademija znanosti in umetnosti. Inštitut za geografijo Antona Melika, or Geografski institut). Ljubljana. 1- (1952-) Irregular. (N70-4:5355b; B-2:268b; BS:341b; R-3:269a).
In Slovenian. Supplementary titles in table of contents, abstracts, and summaries in English.

3348 GEOGRAFSKO društvo Bosne i Hercegovine. Sarajevo. Biblioteka geografija u školi. 1-2 (1959-1961). All issued.

(3349 ______.
Posebna izdanja. Knjiga. 1- (1959-) Irregular monographs. (N70-2:2280a)
In Serbo-Croatian. Some abstracts in English.
[Geografsko Društvo BiH, Maršala Tita 114, Sarajevo, Yugoslavia.]

3349a GEOGRAFSKO društvo Crne Gore. Titograd.
Godišnak. 1- (1964-).
In Serbo-Croatian. Abstracts in English or French.
[Geografsko Društvo Crne Gore, Titograd, Yugoslavia.]

3350 GEOGRAFSKO društvo na S.R. Makedonija. Skopje.
Bilten. 1- (1969?-). 2 nos. a year.
In Macedonian.
[Prirodno-matematički fakultet, Skopje, Yugoslavia.]

*(3351 GEOGRAPHICA slovenica. (Ljubljana. Univerza. Inštitut za geografijo). Ljubljana. 1- (1971-) Irregular. (N75-1:905c; GZS).
In Slovenian and other languages. Summaries in English and other languages. Abstracts in English and Slovenian.
[Inštitut za Geografijo, Aškerčeva 12, 6100 Ljubljana, Yugoslavia.]

*(3352 GEOGRAPHICAL papers. (Zagreb. Sveučilišta [Univerzitet]. Institut za geografiju). Zagreb. 1- (1970-). (N75-1:906b; N75-2:2327a).
English name of issuing agency: University of Zagreb. Institute of geography.
In English, German, or Croatian.
[Marulićev trg 19/II, Zagreb, Yugoslavia.]

(3352a GLOBUS. Časopis za pedagoško-metodska pitanja: unapredjenje nastave geografije. (Srpsko geografsko društvo). Beograd. 1- (1969-)
Annual.
In Serbo-Croatian.

3353 JUGOSLAVENSKA akademija znanosti i umjetnosti. Academia scientiarum et artium slavorum meridionalium. Zagreb.
Krš Jugoslavije. Carsus iugoslaviae. 1- (1957-) Irregular. (R-4:210a).
In Croatian. Abstracts in English or German.

*(3354 KONGRES geografa Jugoslavije. Zbornik. (Savez geografskih društva FNRJ).
1- (1949-). 1 (1949) Zagreb; 2 (1951) Skopje; 3 (1954) Sarajevo; 4 (1955) Beograd; 5 (1958) Crna Gora; 6 (1961) Ljubljana; 7 (1964) Zagreb; 8 (1968) Skopje; 9 (1972) Sarajevo; 10 (1976) Beograd. (N70-2:3336c).
Place of publication varies.
In Croatian, Macedonian, Serbian, or Slovenian.
Abstracts in English, French, Russian, or German.

(3354a MIGRACIJA: mjesečnik Centra za istraživanje migracija. (Centar za istraživanje migracija). Zagreb. 1- (1972-) Periodical. In Croatian.
[Centar za Istraživanje Migracija, Krčka br. 1, Zagreb, Yugoslavia.]

(3354b NASTAVA geografije: časopis za pitanja didaktike i metodike geografije. (Geografsko društvo Bosne i Hercegovine). Sarajevo. 1- (1977-)
Periodical.
In Serbo-Croatian. Abstracts in English, German, French, or Russian.
[Geografsko društvo Bosne i Hercegovine, Vojvode Putnika 42a, Sarajevo, Yugoslavia.]

3355 NIŠ. Viša pedagoška škola. Katedra za geografiju. Niš.
Radovi. Irregular. (N70-3:4238b).

3356 NOVI SAD. Univerzitet. Prirodno-matematički fakultet. Novi Sad..
Zbornik radova. Serije za geografiju. 1- (1971-). (N75-2:1623b).
Added title page: Review of research.
In Serbo-Croatian or occasionally in English or German.
Summaries in English, French, German, or Serbo-Croatian.

(3356a RASPRAVE o migracijama. (Centar za istraživanje migracija).
Zagreb. 1- (1974-) Periodical.
In Croatian.
[Centar za Istraživanje migracija, Krčka br. 1, Zagreb, Yugoslavia.]

*(3357 SKOPJE. Univerzitet. Geografski fakultet.
Godišen Zbornik. 4[16]-10[22] (1968-1976); 23- (1977-) Annual.
French title: Annuaire. Institut de géographie. 4 [16] -9 [21] (1968-1975) as Prirodno-matematički fakultet. Godišen zbornik. Geografski institut, subseries of faculty's Godišen zbornik.
Supersedes in part Godišen zbornik. Geografija i geologija [3358].
In Macedonian. Abstracts in English, French, German, or Russian.
[Geografski Fakultet, P.O. Box 1139, Skopje 91 010, Yugoslavia.]

3358 ______. ______. Prirodno-matematički fakultet. Skopje.
Godišen zbornik. Geografija i geologija. 1-3 [13-15] (1962-1966). Closed 1966. (N70-4:5348a).
French title: Annuaire. Geographie et géologie. Subseries of faculty's Godišen zbornik, the number of which is also carried. Superseded by Godišen zbornik. Geografski institut, later Geografski fakultet. Godišen zbornik [3357].
In Macedonian with summaries in English, French, German, or Russian.

3359 SLOVENSKA akademija znanosti in umetnosti, Ljubljana. Inštitut za geografijo Antona Melika. Ljubljana.
Dela. 1- (1950-) Irregular. (N70-4:5355b; BS:34b).
Subseries of Slovenska akademija znanosti in umetnosti, Ljubljana. Razred za prirodoslovne in medicinske vede. Dela (not in this list).

(3359a ______. Inštitut za raziskovanje krasa.
Acta carsologica. Krasoslovni zbornik. Pročila. 1- (1955-). (N70-4:5355c; R-5:223a).
In Slovenian. Abstracts in English, French, or German.
[Inštitut za Raziskovanje Krasa pri Slovenski Akademiji Znanosti in Umetnosti, Novi trg 3, Ljubljana, Yugoslavia.]

3360 SLOVENSKA geografska bibliografija. (Ljubljana. Univerza. Inštitut za geografijo). Ljubljana. 1960-1965. Closed 1965. (N70-4:5356c).
Superseded by Geografska bibliografija slovenije [3338].

*(3361 SRPSKA akademija nauka i umetnosti. Geografski institut "Jovan Cvijić." Beograd.
Posebna izdanja. Monographies. 1- (1949-) Irregular. (N70-4:6681b; BS:268b).
17-24 (1962-1972) issued by Institute as an independent institution not part of the Academy.
In Serbian with French summaries.

*(3362 ______. ______.
Zbornik radova. Recueil de travaux. 1- (1951-) Irregular. (N70-1:683b; BS:268b; R-1:37a).
18-24 (1962-1972) issued by Institute as an independent institution, not part of the Academy.
In Serbian. French or English titles in table of contents and abstracts.
[Knez Mihailova 35/III, Beograd, Yugoslavia.]

3363 SRPSKO geografsko društvo, Beograd. (Société Serbe de géographie). Beograd.
Atlasi. 1-13 (1929-1935). Closed 1935. (U-2:1693b; B-4:214a; FC-1:41).
1910-1916 as Geografsko društvo, Beograd.
In Serbian with headings in French or German.

*(3364 ______.
Glasnik; Bulletin. 1- (1912-) 2 nos. a year. (U-2:1693b; B-4:214a; BS:833a; FC-1:163; R-1:27b).
Suspended 1915-1919; 1941-1946.
Index: 1961-1971 in v51 no2 (1971), p. 143-153.
In Serbian. French, English, German, or Russian abstracts listed in table of contents.
[Srpsko Geografsko Društvo, Studentski trg 3, Beograd, Yugoslavia.]

(3365 ______.
Mémoires [de la Société serbe de géographie]. 1-5 (1933-1936); 6- (1950-) Irregular. (U-2:1694b; B-4:214a; F-3:455b; R-4:344a).
Monographs in English, French, or German.
[Srpsko Geografsko Društvo, Studentski trg 3, Beograd, Yugoslavia.]

*(3366 ______.
Posebna izdanja...éditions spéciales. 1- (1927-) Irregular. Monographs. (U-2:1694b; B-4:214a; F-2:518; FC-2:154; R-1:74a).
In Serbian with supplementary titles and abstracts usually in French, occasionally in English or German.

*(3367 ZAGREB. Sveučilista. Institut za geografiju. Zagreb.
Radovi. 1- (1958-) Irregular. (N70-4:6449c; N75-2:2327a; R-5:386a).
1-8 (1958-1969) issued by Geografski institut.
Includes a subseries: Migracije radnika, 1- (1970-).
In Croation. Supplementary titles in table of contents and abstracts in English or another international language.

3368 ZEMLJA i ljudi: popularno naučni zbornik. (Sprsko geografsko društvo, Beograd). Beograd.
1- (1951-) Annual. (N70-4:6469c).
In Serbian.

ZAÏRE (French)

République du Zaïre

3369 INSTITUT géographique du Congo belge. Léopoldville. (now Kinshasa).
Rapport annuel. 1956- Closed. (N70-2:2807c).
Superseded by Institut géographique du Zaïre. Rapport annuel [3370].

3370 INSTITUT géographique du Zaïre. Kinshasa.
Rapport annuel. 1974- (N76).
Continues Institut géographique du Congo belge. Rapport annuel [3369].

ZAMBIA (English)

Republic of Zambia

(3371 UNIVERSITY of Zambia. Department of geography. Lusaka.
Report of the department of geography. 1975/76- (N78).

(3372 ZAMBIA geographical association (ZGA). Lusaka.
Handbook series. 1- (1972-) Annual. (N75-2:2330a; N78).
1-2 (1972-1973) as Conference handbook.
[Zambia Geographical Association, P.O.B. RW 287, Lusaka, Zambia.]

(3373 ______. Zambia geographical association. Lusaka.
Occasional newsletter. no1- (1977-) Irregular.

(3374 ______.
Occasional studies. 1- (1967-) Irregular.

(3375 ______.
Schools supplement. 1- (1973-) Annual. (N78).
Cover title: ZGA schools supplement.
[Zambia Geographical Association, P.O.B. RW 287, Lusaka, Zambia.]

3375a ______.
Special publications. 1-4 (1967-1970). Closed 1970.

(3376 ______.
ZGA bibliographical series. 1- (1974-) Irregular. (N75-2:2330a).

*(3377 ZAMBIAN geographical journal. (Zambia geographical association). Lusaka.
1- (1967-) 2 nos. a year. (N70-4:6455b; N75-2:2330a; N79).
1-9 (O 1967- Ja 1970) as ZGA newsletter; 10-28 (Ap 1970- O 1974) as ZGA magazine; 29/30 (1975-) as Zambian geographical journal.
[Zambia Geographical Association, P.O. RW 287, Lusaka, Zambia.]

ZIMBABWE (RHODESIA) (English)

*(3378 GEOGRAPHICAL association of Zimbabwe Rhodesia.
Proceedings. Salisbury. 1- (1968-) Annual. (N75-1:906a; B76).
1 (1967/68) issued by Geography association of Rhodesia.
2-10 (1968/69-1977) issued by Geographical association of Rhodesia.
In English.
[c/o Department of Geography, University of Zimbabwe, P.O. Box MP 167, Salisbury, Zimbabwe-Rhodesia.]

NOTE ON NUMBER OF ENTRIES

Total number of entries 3,445

Numbered entries 1-3378

13 numbers have been suppressed: 340, 711, 883, 1294, 1352, 1356, 1366, 1820, 1821, 2099, 2199, 3071, and 3120.

80 entries were added after initial numbering of entries: 62a, 65a, 138a, 187a, 351a, 496a, 496b, 698a, 715a, 812a, 1002a, 1003a, 1004a, 1050a, 1106a, 1167a, 1168a, 1176a, 1227a, 1230a, 1231a, 1232a, 1254a, 1328a, 1335a, 1398a, 1398b, 1444a, 1458a, 1506a, 1583a, 1585a, 1591a, 1616a, 1766a, 1800a, 1874a, 1994a, 2000a, 2004a, 2080a, 2090a, 2095a, 2096a, 2209a, 2283a, 2360a, 2361a, 2443a, 2551a, 2683a, 2736a, 2879a, 2879b, 2883a, 2893a, 2934a, 2968a, 2996a, 3024a, 3040a, 3126a, 3126b, 3131a, 3173a, 3207a, 3231a, 3231b, 3237a, 3244a, 3284a, 3288a, 3346a, 3349a, 3352a, 3354a, 3354b, 3356a, 3359a, 3375a.

Of the 3,445 entries, 1,089 are confirmed and recorded as being current, as indicated by underlining of the entry number.

Of the 1,089 geographical serials recorded as current 443 marked by an asterisk, have been chosen for inclusion in the companion volume, Annotated World List of Selected Current Geographical Serials, 4th edition, 1980.

INDEX AND CROSS-REFERENCE

The form of entries in this alphabetical index is identical to that followed in the body of the International List of Geographical Serials. All numbers refer to entry number of the titles, not to page numbers. Numbers in parentheses are cross-references to entries under different titles.

Diacritics have been ignored. Entries are ordered in accordance with the letter sequence of the Latin alphabet without regard to the multi-letter implications of diacritics.

A

F

K

N

S

T

U

V

W

Y

Z

THE UNIVERSITY OF CHICAGO

DEPARTMENT OF GEOGRAPHY

RESEARCH PAPERS (Lithographed, 6×9 inches)

Titles in print are available from Department of Geography, The University of Chicago, 5828 S. University Avenue, Chicago, Illinois 60637, U.S.A. Price: $8.00 each; by series subscription, $6.00 each.

COMPLETE LIST TO JUNE 1980

1. GROSS, HERBERT HENRY. *Educational Land Use in the River Forest-Oak Park Community (Illinois).* 1948. 123 p.
2. EISEN, EDNA E. *Educational Land Use in Lake County, Ohio.* 1948. 161 p.
3. WEIGEND, GUIDO GUSTAV. *The Cultural Pattern of South Tyrol (Italy).* 1949. 198 p.
4. NELSON, HOWARD JOSEPH. *The Livelihood Structure of Des Moines, Iowa.* 1949. 140 p.
5. MATTHEWS, JAMES SWINTON. *Expressions of Urbanism in the Sequent Occupance of Northeastern Ohio.* 1949. 179 p.
6. GINSBURG, NORTON SYDNEY. *Japanese Prewar Trade and Shipping in the Oriental Triangle.* 1949. 308 p.
7. KEMLER, JOHN H. *The Struggle for Wolfram in the Iberian Peninsula, June, 1942—June, 1944: A Study in Political and Economic Geography in Wartime.* 1949. 151 p.
8. PHILBRICK, ALLEN K. *The Geography of Education in the Winnetka and Bridgeport Communities of Metropolitan Chicago.* 1949. 165 p.
9. BRADLEY, VIRGINIA. *Functional Patterns in the Guadalupe Counties of the Edwards Plateau.* 1949. 153 p.
10. HARRIS, CHAUNCY D., and FELLMANN, JEROME D. *A Union List of Geographical Serials.* 1950. 124 p.
11. DE MEIRLEIR, MARCEL J. *Manufactural Occupance in the West Central Area of Chicago.* 1950. 251 p.
12. FELLMANN, JEROME DONALD. *Truck Transportation Patterns of Chicago.* 1950. 109 p.
13. HOTCHKISS, WESLEY AKIN. *Areal Pattern of Religious Institutions in Cincinnati.* 1950. 103 p.
14. HARPER, ROBERT ALEXANDER. *Recreational Occupance of the Moraine Lake Region of Northeastern Illinois and Southeastern Wisconsin.* 1950. 176 p.
15. WHEELER, JESSE HARRISON, JR. *Land Use in Greenbrier County, West Virginia.* 1950. 180 p.
16. MCGAUGH, MAURICE EDRON. *The Settlement of the Saginaw Basin.* 1950. 407 p.
17. WATTERSON, ARTHUR WELDON. *Economy and Land Use Patterns of McLean County, Illinois.* 1950. 154 p.
18. HORBALY, WILLIAM. *Agricultural Conditions in Czechoslovakia, 1950.* 1951. 104 p.
19. GUEST, BUDDY ROSS. *Resource Use and Associated Problems in the Upper Cimarron Area.* 1951. 127 p.
20. SORENSEN, CLARENCE WOODROW. *The Internal Structure of the Springfield, Illinois, Urbanized Area.* 1951. 190 p.
21. MUNGER, EDWIN S. *Relational Patterns of Kampala, Uganda.* 1951. 165 p.
22. KHALAF, JASSIM M. *The Water Resources of the Lower Colorado River Basin.* 1951. Volume I, 234 p.; Volume II, 15 maps in pocket.
23. GULICK, LUTHER H., JR. *Rural Occupance in Utuado and Jayuya Municipios, Puerto Rico.* 1952. 254 p.
24. TAAFFE, EDWARD JAMES. *The Air Passenger Hinterland of Chicago.* 1952. 161 p.
25. KRAUSE, ANNEMARIE ELISABETH. *Mennonite Settlement in the Paraguayan Chaco.* 1952. 143 p.
26. HAMMING, EDWARD. *The Port of Milwaukee.* 1952. 162 p.
27. CRAMER, ROBERT ELI. *Manufacturing Structure of the Cicero District, Metropolitan Chicago.* 1952. 176 p.
28. PIERSON, WILLIAM H. *The Geography of the Bellingham Lowland, Washington.* 1953. 159 p.
29. WHITE, GILBERT F. *Human Adjustment to Floods: A Geographical Approach to the Flood Problem in the United States.* 1942. 225 p. Reprinted 1953.
30. OSBORN, DAVID G. *Geographical Features of the Automation of Industry.* 1953. 106 p.
31. THOMAN, RICHARD S. *The Changing Occupance Pattern of the Tri-State Area, Missouri, Kansas, and Oklahoma.* 1953. 139 p.
32. ERICKSEN, SHELDON D. *Occupance in the Upper Deschutes Basin, Oregon.* 1953. 139 p.
33. KENYON, JAMES B. *The Industrialization of the Skokie Area.* 1954. 124 p.
34. PHILLIPS, PAUL GROUNDS. *The Hashemite Kingdom of Jordan: Prolegomena to a Technical Assistance Program.* 1954. 191 p.
35. CARMIN, ROBERT LEIGHTON. *Anápolis, Brazil: Regional Capital of an Agricultural Frontier.* 1953. 172 p.

36. GOLD, ROBERT N. *Manufacturing Structure and Pattern of the South Bend-Mishawaka Area.* 1954. 224 p.
37. SISCO, PAUL HARDEMAN. *The Retail Function of Memphis.* 1954. 160 p.
38. VAN DONGEN, IRENE S. *The British East African Transport Complex.* 1954. 172 p.
39. FRIEDMANN, JOHN R. P. *The Spatial Structure of Economic Development in the Tennessee Valley.* 1955. 187 p.
40. GROTEWOLD, ANDREAS. *Regional Changes in Corn Production in the United States from 1909 to 1949.* 1955. 78 p.
41. BJORKLUND, ELAINE M. *Focus on Adelaide—Functional Organization of the Adelaide Region, Australia.* 1955. 133 p.
42. FORD, ROBERT N. *A Resource Use Analysis and Evaluation of the Everglades Agricultural Area.* 1956. 127 p.
43. CHRISTENSEN, DAVID E. *Rural Occupance in Transition: Sumter and Lee Counties, Georgia.* 1956. 160 p.
44. GUZMÁN, LOUIS E. *Farming and Farmlands in Panama.* 1956. 137 p.
45. ZADROZNY, MITCHELL G. *Water Utilization in the Middle Mississippi Valley.* 1956. 119 p.
46. AHMED, G. MUNIR. *Manufacturing Structure and Pattern of Waukegan-North Chicago.* 1957. 117 p.
47. RANDALL, DARRELL. *Factors of Economic Development and the Okovango Delta.* 1957. 268 p.
48. BOXER, BARUCH. *Israeli Shipping and Foreign Trade.* 1957. 162 p.
49. MAYER, HAROLD M. *The Port of Chicago and the St. Lawrence Seaway.* 1957. 283 p.
50. PATTISON, WILLIAM D. *Beginning of the American Rectangular Land Survey System, 1784-1800.* 1957. 248 p.
51. BROWN, ROBERT HAROLD. *Political Areal-Functional Organization: With Special Reference to St. Cloud, Minnesota.* 1957. 123 p.
52. BEYER, JACQUELYN L. *Integration of Grazing and Crop Agriculture: Resources Management Problems in the Uncompahgre Valley Irrigation Project.* 1957. 125 p.
53. ACKERMAN, EDWARD A. *Geography as a Fundamental Research Discipline.* 1958. 37 p.
54. AL-KHASHAB, WAFIQ HUSSAIN. *The Water Budget of the Tigris and Euphrates Basin.* 1958. 105 p.
55. LARIMORE, ANN EVANS. *The Alien Town: Patterns of Settlement in Busoga, Uganda.* 1958. 208 p.
56. MURPHY, FRANCIS C. *Regulating Flood-Plain Development.* 1958. 204 p.
57. WHITE, GILBERT F., *et al. Changes in Urban Occupance of Flood Plains in the United States.* 1958, 235 p.
58. COLBY, MARY MCRAE. *The Geographic Structure of Southeastern North Carolina.* 1958. 226 p.
59. MEGEE, MARY CATHERINE. *Monterrey, Mexico: Internal Patterns and External Relations.* 1958. 118 p.
60. WEBER, DICKINSON. *A Comparison of Two Oil City Business Centers (Odessa-Midland, Texas).* 1958. 239 p.
61. PLATT, ROBERT S. *Field Study in American Geography.* 1959. 405 p.
62. GINSBURG, NORTON, editor. *Essays on Geography and Economic Development.* 1960. 173 p.
63. HARRIS, CHAUNCY D., and FELLMANN, JEROME D. *International List of Geographical Serials.* 1960. 189 p.
64. TAAFFE, ROBERT N. *Rail Transportation and the Economic Development of Soviet Central Asia.* 1960. 186 p.
65. SHEAFFER, JOHN R. *Flood Proofing: An Element in a Flood Damage Reduction Program.* 1960. 198 p.
66. RODGERS, ALLAN L. *The Industrial Geography of the Port of Genova.* 1960. 144 p.
67. KENYON, JAMES B. *Industrial Localization and Metropolitan Growth: The Paterson-Passaic District.* 1960. 224 p.
68. GINSBURG, NORTON. *Atlas of Economic Development.* 1961. 119 p. 14 × 9½″. Cloth. University of Chicago Press.
69. CHURCH, MARTHA. *Spatial Organization of Electric Power Territories in Massachusetts.* 1960. 187 p.
70. WHITE, GILBERT F., *et al. Papers on Flood Problems.* 1961. 228 p.
71. GILBERT, EDMUND WILLIAM *The University Town in England and West Germany.* 1961. 74 p.
72. BOXER, BARUCH. *Ocean Shipping in the Evolution of Hong Kong.* 1961. 95 p.
73. ROBINSON, IRA M. *New Industrial Towns of Canada's Resource Frontier.* 1962. 190 p.
74. TROTTER, JOHN E. *State Park System in Illinois.* 1962. 152 p.
75. BURTON, IAN. *Types of Agricultural Occupance of Flood Plains in the United States.* 1962. 167 p.
76. PRED, ALLAN R. *The External Relations of Cities during 'Industrial Revolution'.* 1962. 113 p.
77. BARROWS, HARLAN H. *Lectures on the Historical Geography of the United States as Given in 1933.* Edited by WILLIAM A. KOELSCH. 1962. 248 p.

78. KATES, ROBERT WILLIAM. *Hazard and Choice Perception in Flood Plain Management.* 1962. 157 p.
79. HUDSON, JAMES W. *Irrigation Water Use in the Utah Valley, Utah.* 1962. 249 p.
80. ZELINSKY, WILBUR. *A Bibliographic Guide to Population Geography.* 1962. 257 p.
81. DRAINE, EDWIN H. *Import Traffic of Chicago and Its Hinterland.* 1963. 138 p.
82. KOLARS, JOHN F. *Tradition, Season, and Change in a Turkish Village.* 1963. 205 p.
83. WIKKRAMATILEKE, RUDOLPH. *Southeast Ceylon: Trends and Problems in Agricultural Settlement.* 1963. 163 p.
84. KANSKY, KAREL J. *Structure of Transportation Networks: Relationships between Network Geometry and Regional Characteristics.* 1963. 155 p.
85. BERRY, BRIAN J. L. *Commercial Structure and Commercial Blight.* 1963. 235 p.
86. BERRY, BRIAN J. L., and TENNANT, ROBERT J. *Chicago Commercial Reference Handbook.* 1963. 278 p.
87. BERRY, BRIAN J. L., and HANKINS, THOMAS D. *A Bibliographic Guide to the Economic Regions of the United States.* 1963. 101 p.
88. MARCUS, MELVIN G. *Climate-Glacier Studies in the Juneau Ice Field Region, Alaska.* 1964. 128 p.
89. SMOLE, WILLIAM J. *Owner-Cultivatorship in Middle Chile.* 1963. 176 p.
90. HELVIG, MAGNE. *Chicago's External Truck Movements: Spatial Interactions between the Chicago Area and Its Hinterland.* 1964. 132 p.
91. HILL, A. DAVID. *The Changing Landscape of a Mexican Municipio, Villa Las Rosas, Chiapas.* 1964. 121 p.
92. SIMMONS, JAMES W. *The Changing Pattern of Retail Location.* 1964. 200 p.
93. WHITE, GILBERT F. *Choice of Adjustment to Floods.* 1964. 150 p.
94. MCMANIS, DOUGLAS. R. *The Initial Evaluation and Utilization of the Illinois Prairies, 1815–1840.* 1964. 109 p.
95. PERLE, EUGENE D. *The Demand for Transportation: Regional and Commodity Studies in the United States.* 1964. 130 p.
96. HARRIS, CHAUNCY D. *Annotated World List of Selected Current Geographical Serials in English.* 1964. 32 p.
97. BOWDEN, LEONARD W. *Diffusion of the Decision To Irrigate: Simulation of the Spread of a New Resource Management Practice in the Colorado Northern High Plains.* 1965. 146 p.
98. KATES, ROBERT W. *Industrial Flood Losses: Damage Estimation in the Lehigh Valley.* 1965. 76 p.
99. RODER, WOLF. *The Sabi Valley Irrigation Projects.* 1965. 213 p.
100. SEWELL, W. R. DERRICK. *Water Management and Floods in the Fraser River Basin.* 1965. 163 p.
101. RAY, D. MICHAEL. *Market Potential and Economic Shadow: A Quantitative Analysis of Industrial Location in Southern Ontario.* 1965. 164 p.
102. AHMAD, QAZI. *Indian Cities: Characteristics and Correlates.* 1965. 184 p.
103. BARNUM, H. GARDINER. *Market Centers and Hinterlands in Baden-Württemberg.* 1966. 173 p.
104. SIMMONS, JAMES W. *Toronto's Changing Retail Complex.* 1966. 126 p.
105. SEWELL, W. R. DERRICK, *et al. Human Dimensions of Weather Modification.* 1966. 423 p.
106. SAARINEN, THOMAS FREDERICK. *Perception of the Drought Hazard on the Great Plains.* 1966. 183 p.
107. SOLZMAN, DAVID M. *Waterway Industrial Sites: A Chicago Case Study.* 1966. 138 p.
108. KASPERSON, ROGER E. *The Dodecanese: Diversity and Unity in Island Politics.* 1966. 184 p.
109. LOWENTHAL, DAVID, editor, *Environmental Perception and Behavior.* 1967. 88 p.
110. REED, WALLACE E., *Areal Interaction in India: Commodity Flows of the Bengal-Bihar Industrial Area.* 1967. 209 p.
111. BERRY, BRIAN J. L. *Essays on Commodity Flows and the Spatial Structure of the Indian Economy.* 1966. 334 p.
112. BOURNE, LARRY S. *Private Redevelopment of the Central City, Spatial Processes of Structural Change in the City of Toronto.* 1967. 199 p.
113. BRUSH, JOHN E., and GAUTHIER, HOWARD L., JR., *Service Centers and Consumer Trips: Studies on the Philadelphia Metropolitan Fringe.* 1968. 182 p.
114. CLARKSON, JAMES D., *The Cultural Ecology of a Chinese Village: Cameron Highlands, Malaysia.* 1968. 174 p.
115. BURTON, IAN, KATES, ROBERT W., and SNEAD, RODMAN E. *The Human Ecology of Coastal Flood Hazard in Megalopolis.* 1969. 196 p.
116. MURDIE, ROBERT A., *Factorial Ecology of Metropolitan Toronto, 1951–1961.* 1969. 212 p.
117. WONG, SHUE TUCK, *Perception of Choice and Factors Affecting Industrial Water Supply Decisions in Northeastern Illinois.* 1969. 93 p.
118. JOHNSON, DOUGLAS L.. *The Nature of Nomadism: A Comparative Study of Pastoral Migrations in Southwestern Asia and Northern Africa.* 1969. 200 p.

119. DIENES, LESLIE. *Locational Factors and Locational Developments in the Soviet Chemical Industry.* 1969. 262 p.

120. MIHELIĆ, DUŠAN. *The Political Element in the Port Geography of Trieste.* 1969. 104 p.

121. BAUMANN, DUANE D. *The Recreational Use of Domestic Water Supply Reservoirs: Perception and Choice.* 1969. 125 p.

122. LIND, AULIS O. *Coastal Landforms of Cat Island, Bahamas: A Study of Holocene Accretionary Topography and Sea-Level Change.* 1969. 156 p.

123. WHITNEY, JOSEPH B. R. *China: Area, Administration and Nation Building.* 1970. 198 p.

124. EARICKSON, ROBERT. *The Spatial Behavior of Hospital Patients: A Behavioral Approach to Spatial Interaction in Metropolitan Chicago.* 1970. 138 p.

125. DAY, JOHN C. *Managing the Lower Rio Grande: An Experience in International River Development.* 1970. 274 p.

126. MAC IVER, IAN. *Urban Water Supply Alternatives: Perception and Choice in the Grand Basin Ontario.* 1970. 178 p.

127. GOHEEN, PETER G. *Victorian Toronto, 1850 to 1900: Pattern and Process of Growth.* 1970. 278 p.

128. GOOD, CHARLES M. *Rural Markets and Trade in East Africa.* 1970. 252 p.

129. MEYER, DAVID R. *Spatial Variation of Black Urban Households.* 1970. 127 p.

130. GLADFELTER, BRUCE G. *Meseta and Campiña Landforms in Central Spain: A Geomorphology of the Alto Henares Basin.* 1971. 204 p.

131. NEILS, ELAINE M. *Reservation to City: Indian Migration and Federal Relocation.* 1971. 198 p.

132. MOLINE, NORMAN T. *Mobility and the Small Town, 1900–1930.* 1971. 169 p.

133. SCHWIND, PAUL J. *Migration and Regional Development in the United States.* 1971. 170 p.

134. PYLE, GERALD F. *Heart Disease, Cancer and Stroke in Chicago: A Geographical Analysis with Facilities, Plans for 1980.* 1971. 292 p.

135. JOHNSON, JAMES F. *Renovated Waste Water: An Alternative Source of Municipal Water Supply in the United States.* 1971. 155 p.

136. BUTZER, KARL W. *Recent History of an Ethiopian Delta: The Omo River and the level of Lake Rudolf.* 1971. 184 p.

137. HARRIS, CHAUNCY D. *Annotated World List of Selected Current Geographical Serials in English, French, and German* 3rd edition. 1971. 77 p.

138. HARRIS, CHAUNCY D., and FELLMANN, JEROME D. *International List of Geographical Serials* 2nd edition. 1971. 267 p.

139. MCMANIS, DOUGLAS R. *European Impressions of the New England Coast, 1497–1620.* 1972. 147 p.

140. COHEN, YEHOSHUA S. *Diffusion of an Innovation in an Urban System: The Spread of Planned Regional Shopping Centers in the United States, 1949–1968,* 1972. 136 p.

141. MITCHELL, NORA. *The Indian Hill-Station: Kodaikanal.* 1972. 199 p.

142. PLATT, RUTHERFORD H. *The Open Space Decision Process: Spatial Allocation of Costs and Benefits.* 1972. 189 p.

143. GOLANT, STEPHEN M. *The Residential Location and Spatial Behavior of the Elderly: A Canadian Example.* 1972 226 p.

144. PANNELL, CLIFTON W. *T'ai-chung, T'ai-wan: Structure and Function.* 1973. 200 p.

145. LANKFORD, PHILIP M. *Regional Incomes in the United States, 1929–1967: Level, Distribution, Stability, and Growth.* 1972. 137 p.

146. FREEMAN, DONALD B. *International Trade, Migration, and Capital Flows: A Quantitative Analysis of Spatial Economic Interaction.* 1973. 201 p.

147. MYERS, SARAH K. *Language Shift Among Migrants to Lima, Peru.* 1973. 203 p.

148. JOHNSON, DOUGLAS L. *Jabal al-Akhḍar, Cyrenaica: An Historical Geography of Settlement and Livelihood.* 1973. 240 p.

149. YEUNG, YUE-MAN. *National Development Policy and Urban Transformation in Singapore: A Study of Public Housing and the Marketing System.* 1973. 204 p.

150. HALL, FRED L. *Location Criteria for High Schools: Student Transportation and Racial Integration.* 1973. 156 p.

151. ROSENBERG, TERRY J. *Residence, Employment, and Mobility of Puerto Ricans in New York City.* 1974. 230 p.

152. MIKESELL, MARVIN W., editor. *Geographers Abroad: Essays on the Problems and Prospects of Research in Foreign Areas.* 1973. 296 p.

153. OSBORN, JAMES F. *Area, Development Policy, and the Middle City in Malaysia.* 1974. 291 p.

154. WACHT, WALTER F. *The Domestic Air Transportation Network of the United States.* 1974. 98 p.

155. BERRY, BRIAN J. L., *et al. Land Use, Urban Form and Environmental Quality.* 1974. 440 p.

156. MITCHELL, JAMES K. *Community Response to Coastal Erosion: Individual and Collective Adjustments to Hazard on the Atlantic Shore.* 1974. 209 p.

157. COOK, GILLIAN P. *Spatial Dynamics of Business Growth in the Witwatersrand.* 1975. 144 p.

158. STARR, JOHN T., JR. *The Evolution of Unit Train, 1960–1969.* 1976. 233 p.

159. PYLE, GERALD F. *et al. The Spatial Dynamics of Crime.* 1974. 221 p.

160. MEYER, JUDITH W. *Diffusion of an American Montessori Education.* 1975. 97 p.

161. SCHMID, JAMES A. *Urban Vegetation: A Review and Chicago Case Study.* 1975. 266 p.

162. LAMB, RICHARD F. *Metropolitan Impacts on Rural America.* 1975. 196 p.

163. FEDOR, THOMAS STANLEY. *Patterns of Urban Growth in the Russian Empire during the Nineteenth Century.* 1975. 245 p.

164. HARRIS, CHAUNCY D. *Guide to Geographical Bibliographies and Reference Works in Russian or on the Soviet Union.* 1975. 478 p.

165. JONES, DONALD W. *Migration and Urban Unemployment in Dualistic Economic Development.* 1975. 174 p.

166. BEDNARZ, ROBERT S. *The Effect of Air Pollution on Property Value in Chicago.* 1975. 111 p.

167. HANNEMANN, MANFRED. *The Diffusion of the Reformation in Southwestern Germany, 1518–1534.* 1975. 235 p.

168. SUBLETT, MICHAEL D. *Farmers on the Road. Interfarm Migration and the Farming of Noncontiguous Lands in Three Midwestern Townships, 1939–1969.* 1975. 214 p.

169. STETZER, DONALD FOSTER. *Special Districts in Cook County: Toward a Geography of Local Government.* 1975. 177 p.

170. EARLE, CARVILLE V. *The Evolution of a Tidewater Settlement System: All Hallow's Parish, Maryland, 1650–1783,* 1975. 239 p.

171. SPODEK, HOWARD. *Urban-Rural Integration in Regional Development: A Case Study of Saurashtra, India, 1800–1960.* 1976. 144 p.

172. COHEN, YEHOSHUA S. and BERRY, BRIAN J. L. *Spatial Components of Manufacturing Change,* 1950–1960. 1975. 262 p.

173. HAYES, CHARLES R. *The Dispersed City: The Case of Piedmont, North Carolina.* 1976. 157 p.

174. CARGO, DOUGLAS B. *Solid Wastes: Factors Influencing Generation Rates.* 1978. 100 p.

175. GILLARD, QUENTIN. *Incomes and Accessibility. Metropolitan Labor Force Participation, Commuting, and Income Differentials in the United States, 1960–1970.* 1977. 106 p.

176. MORGAN, DAVID J. *Patterns of Population Distribution: A Residential Preference Model and Its Dynamic.* 1978. 200 p.

177. STOKES, HOUSTON H.; JONES, DONALD W. and NEUBURGER, HUGH M. *Unemployment and Adjustment in the Labor Market: A Comparison between the Regional and National Responses.* 1975. 125 p.

179. HARRIS, CHAUNCY D. *Bibliography of Geography. Part I. Introduction to General Aids.* 1976. 276 p.

180. CARR, CLAUDIA J. *Pastoralism in Crisis. The Dasanetch and their Ethiopian Lands.* 1977. 319 p.

181. GOODWIN, GARY C. *Cherokees in Transition: A Study of Changing Culture and Environment Prior to 1775.* 1977. 207 p.

182. KNIGHT, DAVID B. *A Capital for Canada: Conflict and Compromise in the Nineteenth Century.* 1977. 341 p.

183. HAIGH, MARTIN J. *The Evolution of Slopes on Artificial Landforms—Blaenavon, U.K.* 1978. 293 p.

184. FINK, L. DEE. *Listening to the Learner. An Exploratory Study of Personal Meaning in College Geography Courses.* 1977. 186 p.

185. HELGREN, DAVID M. *Rivers of Diamonds: An Alluvial History of the Lower Vaal Basin, South Africa.* 1979. 389 p.

186. BUTZER, KARL W., *editor. Dimensions of Human Geography: Essays on Some Familiar and Neglected Themes.* 1978. 190 p.

187. MITSUHASHI, SETSUKO. *Japanese Commodity Flows,* 1978. 172 p.

188. CARIS, SUSAN L. *Community Attitudes toward Pollution.* 1978. 211 p.

189. REES, PHILIP H. *Residential Patterns in American Cities, 1960.* 1979. 405 p.

190. KANNE, EDWARD A. *Fresh Food for Nicosia.* 1979. 106 p.

191. WIXMAN, RONALD. *Language Aspects of Ethnic Patterns and Processes in the North Caucasus.* 1980. 243 p.

192. KIRCHNER, JOHN A. *Sugar and Seasonal Labor Migration: The Case of Tucumán, Argentina.* 1980. 174 p.

193. HARRIS, CHAUNCY D. and FELLMANN, JEROME D. *International List of Geographical Serials, Third Edition, 1980.* 1980. 457 p.

194. HARRIS, CHAUNCY D. *Annotated World List of Selected Current Geographical Serials, Fourth, Edition, 1980.* 1980. 165 p.

195. LEUNG, CHI-KEUNG. *China: Railway Patterns and National Goals.* 1980. 233 p.